Jörg Friedrich

The Plasma Chemistry of Polymer Surfaces

Related Titles

Parvulescu, V. I., Magureanu, M., Lukes, P. (eds.)

Plasma Chemistry and Catalysis in Gases and Liquids

2012

ISBN: 978-3-527-33006-5

Rauscher, H., Perucca, M., Buyle, G. (eds.)

Plasma Technology for Hyperfunctional Surfaces

Food, Biomedical, and Textile Applications

2010

ISBN: 978-3-527-32654-9

Schlüter, D. A., Hawker, C., Sakamoto, J. (eds.)

Synthesis of Polymers

New Structures and Methods

Series: Materials Science and Technology

2 Volume Set

2012

ISBN: 978-3-527-32757-7

Hippler, R., Kersten, H., Schmidt, M., Schoenbach, K. H. (eds.)

Low Temperature Plasmas

Fundamentals, Technologies and Techniques

2nd, revised and enlarged edition

2008

ISBN: 978-3-527-40673-9

Knoll, W. Advincula, R. C. (eds.)

Functional Polymer Films

2 Volume Set

2011

ISBN: 978-3-527-32190-2

Ostrikov, K.

Plasma Nanoscience

Basic Concepts and Applications of Deterministic Nanofabrication

2008

ISBN: 978-3-527-40740-8

Harry, J. E.

Introduction to Plasma Technology

Science, Engineering and Applications

2010

ISBN: 978-3-527-32763-8

Coqueret, X., Defoort, B. (eds.)

High Energy Crosslinking Polymerization

Applications of Ionizing Radiation

2006

ISBN: 978-3-527-31838-4

Kawai, Y., Ikegami, H., Sato, N., Matsuda, A., Uchino, K., Kuzuya, M., Mizuno, A. (eds.)

Industrial Plasma Technology

Applications from Environmental to Energy Technologies

2010

ISBN: 978-3-527-32544-3

Lazzari, M., Liu, G., Lecommandoux, S. (eds.)

Block Copolymers in Nanoscience

2006

ISBN: 978-3-527-31309-9

Jörg Friedrich

The Plasma Chemistry of Polymer Surfaces

Advanced Techniques for Surface Design

WILEY-
VCH

WILEY-VCH Verlag GmbH & Co. KGaA

The Author

Prof. Dr. Jörg Friedrich
BAM – Bundesanstalt für
Material forschung u. -prüfung
Unter den Eichen 87
12205 Berlin

Library of Congress Card No.: applied for

British Library Cataloguing-in-Publication Data
A catalogue record for this book is available from the British Library.

Bibliographic information published by the Deutsche Nationalbibliothek
The Deutsche Nationalbibliothek lists this publication in the Deutsche Nationalbibliografie; detailed bibliographic data are available on the Internet at <http://dnb.d-nb.de>.

Typesetting Toppan Best-set Premedia Limited, Hong Kong
Printing and Binding Markono Print Media Pte Ltd, Singapore
Cover Design Grafik-Design Schulz, Fußgönheim

Printed in Singapore
Printed on acid-free paper

Print ISBN: 978-3-527-31853-7
ePDF ISBN: 978-3-527-64803-0
oBook ISBN: 978-3-527-64800-9
ePub ISBN: 978-3-527-64802-3
mobi ISBN: 978-3-527-64801-6

Contents

Preface

Some 40 years experience with plasmas applied to polymers and the special view of a polymer chemist are the motivation for writing this book. The rapid growth of applications of plasma processes on an industrial scale is connected with the pioneering work of engineers. Basic research into plasmas and their properties is associated with plasma and astrophysics. Pure plasmas of noble gases under well-defined conditions in exactly determined geometries are traditional objects of plasma physics. Thus, chemical processes are out of view. However, in such simple systems the chemistry of irradiation and release of degradation products also play an important role, as do the polymer surface, near-surface layers, plasma boundary layer, and plasma bulk. Organic and polymer chemistry often dominate the use of molecular plasmas for polymer surface treatment and modification. A much more complicated and complex situation is found for plasma polymerization processes, which can often be described only by formal kinetics as the elementary and chemical processes are not known in exact detail. Electrical low- and atmospheric-pressure plasmas are characterized by a surplus in energy and enthalpy needed for simple chemical processes. The chemistry of excess energy allows endothermic reactions to be performed because the dose rate exceeds all necessary enthalpies of reaction pathways known in chemistry or even in radiation chemistry. Thus, random, statistic, and exotic processes dominate and, therefore, the reaction products are most often chemically irregular in terms of structure and composition. Additionally, the polymer products are unstable because of plasma-produced metastable radicals that are trapped in the polymer bulk and which subsequently remain capable of undergoing oxidation on exposure to oxygen from air. Therefore, the plasma product is unstable and changes continuously during storage. A nice example may illustrate such a "terrible" plasma. At the beginning of my work, in the early 1970s, I had scraped plasma polymers from the wall of the plasma reactor for infrared analysis. The plasma polymer flakes were collected, cooled with liquid nitrogen, and then ground for production of polymer powder. This powder was disseminated in KBr powder, which is necessary for KBr disk preparation. After evaporation of nitrogen the sample begun to smolder and became black. The technical assistant was stunned and did not want to continue his work with other samples. The behavior of the sample was, in fact, due to the fast reaction of radicals that came into contact with oxygen from the air after the plasma polymer layer was disintegrated. Peroxide formation and undefined autooxidation were initiated.

Organic chemists or polymer chemists turn away from such “black box chemistry,” labeling it as impure chemistry, far from regular chemistry, that does not follow a defined chemical mechanism. Thus, the pure chemist is shocked and all his knowledge is superfluous. If a polymer chemist must accept that chemically inert gases, such as methane or benzene, can be polymerized or polymers exposed to plasmas are destroyed, degraded, etched, and so on, any previous thinking, any knowledge, is of no help.

The task of this book is to bring together physicists, engineers, chemists, and polymer researchers, looking preferentially from the chemical and especially from the polymer chemical point of view into plasma processes and the reactions in the polymer body. Here, a new type of plasma chemist, who treats and produces polymers, is created or, better, a plasma polymer chemist is born.

Forty years of experience with plasma and polymer chemistry, analysis, and polymer degradation have been concentrated in this book. It discusses important findings in this field from all parts of the world.

Berlin, 20th September 2011 *Jörg Friedrich*

1
Introduction

The interaction of polymers with different materials such as metals, ceramics, other polymers, coatings, or inorganics is crucial for the adhesion at interfaces in polymer composite structural elements. The absence or weakness of interactions as well as any lack of durability are responsible for the collapse of load-bearing composite components. In 2005 the ice rink in Bad Reichenhall (Germany) collapsed, burying several people, because of adhesion failure (fatigue of the interface bonds).

Many polymers, in particular polyolefins, such as polyethylene and polypropylene are chemically inert and cannot strongly interact with other materials. The reason for this is the absence of polar and reactive functional groups in their structure. Thus, interactions with other materials are poor and so too is adhesion. Weak physical interactions only occur. J. D. van der Waals found their existence in 1879 [1]. These forces are electrostatic, induced and permanent dipoles, dispersion interactions, and hydrogen bonds. They are very weak and operate over a short range [2]. Polyolefins show only dispersion interactions among their own molecules and, thus, they are often difficult to wet or bond because of the absence of polar groups, which are able to promote interactions to the other material. Dipole or induced-dipole interactions or even chemical bonds between polymer and coating at the interface require the existence of functional groups.

Polar groups are often introduced by flaming [3] or plasma exposure [4]. Such oxidations form various oxidized polar species at the polyolefin surface, which can undergo the desired interactions to other materials. The introduction of chemical bonds at the interface is more efficient because of the much higher binding energies [5]. To install such covalent bonds between polymers and coatings, most often the production of monotype functional groups at the polyolefin surface is a necessary precondition. Such monosort functionalization is extraordinarily difficult. New processes have been developed for its realization, that is, exposure of the polyolefin surface to brominating plasma [6]. The C–Br groups could be converted into amino, carboxyl, or hydroxyl groups or consumed by amines, alcohols, and glycols [7]. The additional introduction of flexible, water-repellent, and metal-binding spacer molecules by grafting onto C–Br groups produced highly adhered and durable polyolefin composites [8].

The Plasma Chemistry of Polymer Surfaces: Advanced Techniques for Surface Design, First Edition.
Jörg Friedrich.

In highly stressed polymer components for structural assemblies all forces are applied to the interface and distributed to the interfacial bonds. Either a large number of weak physical interactions or a smaller number of strong chemical bonds is needed to withstand the disruption under mechanical load along the interface. However, in general, chemically and structurally, completely different materials need to be joined together. Polymers, in particular polyolefins, show very low surface energy and metals or inorganics a much higher one. The difference amounts to two orders of magnitude (original value for polymers 30–40 $mN\,m^{-1}$ and for metals 1000–3000 $mN\,m^{-1}$), which is nearly the same difference in surface energy as before polymer treatment (40–50 and 1000–3000 $mN\,m^{-1}$) [9]. At the molecular level, interactions are absent due to the chemical inertness of polyolefins.

Post-polymerization introduction of functional groups onto polyolefin surfaces has a principal problem. The (radical) substitution of H by any functional group is accompanied by C–C bond scissions of the polymer backbone because of equivalent (or lower) binding energies [10]. Thus, degradation occurs simultaneously, although C–C bonds were partially shielded from attack. Nevertheless, such a disruption of the polymer surface produces anchoring points for physical and chemical interactions but also a weak boundary layer, which is mechanically, chemically, and thermally unstable (low molecular weight oxidized material, LMWOM) [11]. Moreover, polymers, metals, or inorganics have thermal expansion coefficients that differ by two orders of magnitude. Therefore, the thus produced mechanical stress is focused onto the monolayer of interactions along the interface. As mentioned before, spacer introduction can balance this mechanical stress along the interface.

The surface modification of polyolefins must be also considered within the framework of 100 Mio tons production of polyethylene and polypropylene per year worldwide. Several technical applications demand a solution to the adhesion problem. Mechanical interlocking, chemical roughening by etching, ion and electron beam modification, UV irradiation, UV-induced graft copolymerization, laser beam or excimer lamp irradiation, ^{60}Co irradiation, flaming, corona treatment, use of adhesion promoters, glues, adhesives, etc. were successfully tested to modify polyolefin surfaces for adhesion [2]. However, all these pretreatments produce a broad variety of different functional groups.

As mentioned before, the formation of monotype functional groups followed by spacer grafting can solve the problem of moderate adhesive bond strength and durability. However, the great energy and enthalpy excess present in a plasma is most often responsible for non-selective reactions and the formation of a broad variety of products [12].

The dream of all plasma chemists is to achieve monosort functionalized polyolefin surfaces. The excess energy present in the plasma state [13] and the equivalency of C–C and C–H dissociation energies make it difficult to realize this dream [10]. However, a few chemical reactions produce end-products that are also stable towards plasma. Examples of such stable end-products are (i) in the case of bromination the electronic state of the neighboring noble gas (krypton) and (ii) silica-like SiO_x layers formed in the oxidation of Si compounds in an oxygen plasma [14].

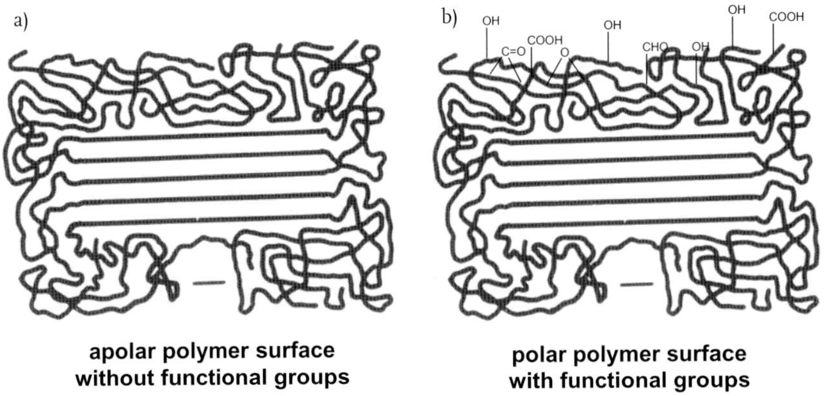

Figure 1.1 Assumed structure of polyethylene at the surface without functional groups (a), and after surface oxidation and introduction of oxygen-containing functional groups groups (b) and the behavior after wetting with a drop of water.

This book presents several variants of such surface techniques with monotype functional groups, such as chemical post-plasma reduction, pulse-pressure plasma polymerization, underwater plasma and glow discharge electrolysis, and deposition of functionalized prepolymers and oligomers by aerosol plasma and electrospray [15].

Polyolefins have a semi-crystalline structure, which can be represented by the model of "Fransenmicelle" as shown in Figure 1.1.

Amorphous regions are characterized by random localization of macromolecular chains, whereas crystalline regions show the parallel and close orientation of the all-trans configuration of the chain with folded loops, thus forming the lamellae as present in polyethylene [16].

The concept of polymer functionalization by plasma exposure is to attach atoms or fragments of the dissociated plasma gas as functional group by H substitution at the polymer chain. Since there are there many different fragments and atoms present in the plasma a broad variety of related functional groups is produced. The formation of at least 12 oxygen-containing groups at the surface of poly(ethylene terephthalate) has been shown after oxygen plasma exposure [17].

There is also an interrelation between plasma, polymer, surface charging, surface cleaning, surface functionalization, etching, and emission of degradation products as well as changing of plasma by the appearance of oxygen-containing groups in the gas phase and so on (Figure 1.2).

The substrate, here the polymer, gives a specific response to plasma exposure. Polymers react very sensitively to any exposure to plasmas. This is due to their complex and supermolecular structure. Polymers have some common features with living matter and therefore they are very sensitive, in almost the same manner, towards particle or radiation exposure. Thus, special knowledge of polymer chemistry, physics, and technology is necessary to understand the specific

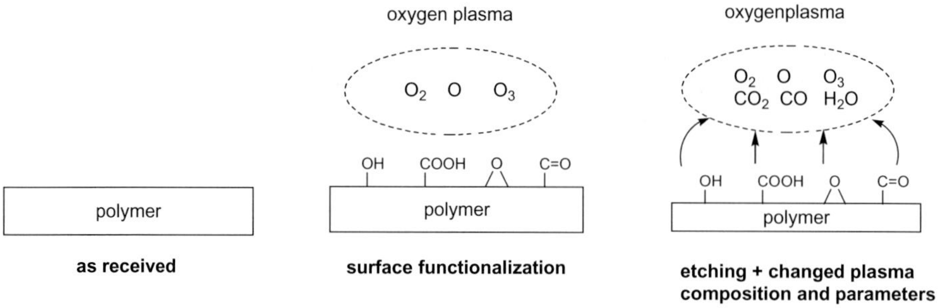

Figure 1.2 Changes in plasma phase upon polymer etching.

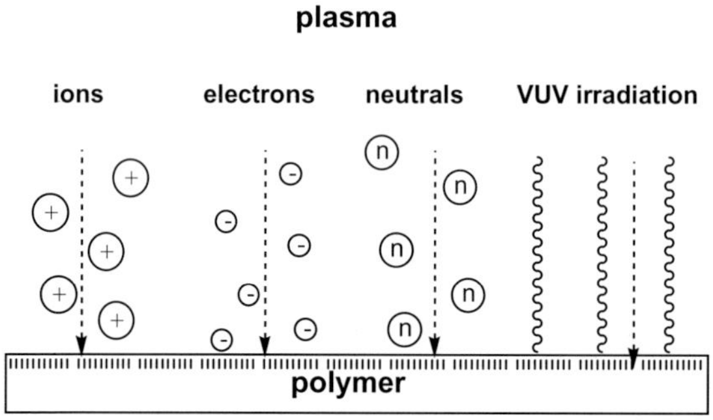

Figure 1.3 Plasma particle shower and vacuum UV (VUV) irradiation of polymer surfaces during plasma exposure.

and complex behavior of polymer surfaces on plasma exposure. Starting from plasma physics and taking simple atomic (noble) or molecule gas plasmas, which are well-defined and well-characterized but, nevertheless, are associated with high power consumption and high average electron energy the contradictoriness of flow from plasma to polymer, thus the confrontation is perfect. A shower of high-energy particles and photons bombards the polymer surface. A result of this bombardment is the formation of degraded or crosslinked products with the complete loss of original structure (Figure 1.3) [18].

As a matter of course, as a precondition, the plasma gas temperatures should be near room temperature or, in the case of energy-rich hot plasmas, a very short residence time in the plasma zone is mandatory. Low gas temperature is characteristic for low-pressure glow discharges, also known as non-isothermal plasmas or colloquially as "cold" plasmas [19]. Figure 1.4 shows schematically the prototype of such a plasma, namely, the low-pressure DC (direct current) glow discharge. The volume between the two electrodes is filled with the uniform plasma of the "positive column," which is the most suitable place for polymer treatment.

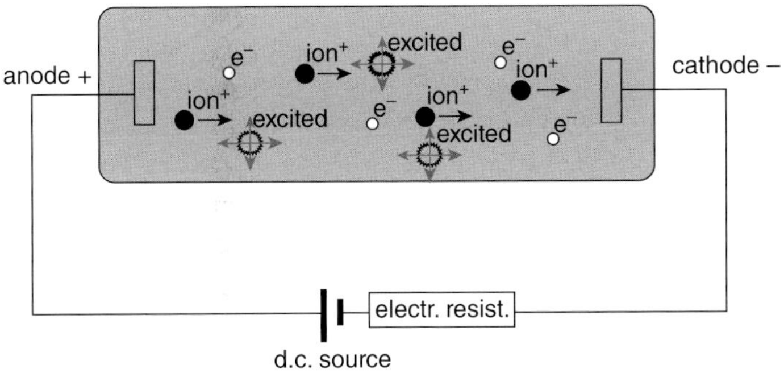

Figure 1.4 Principle of plasma formation: direct current (DC) discharge tube with electrons, ions, and energy-rich neutrals as excited states.

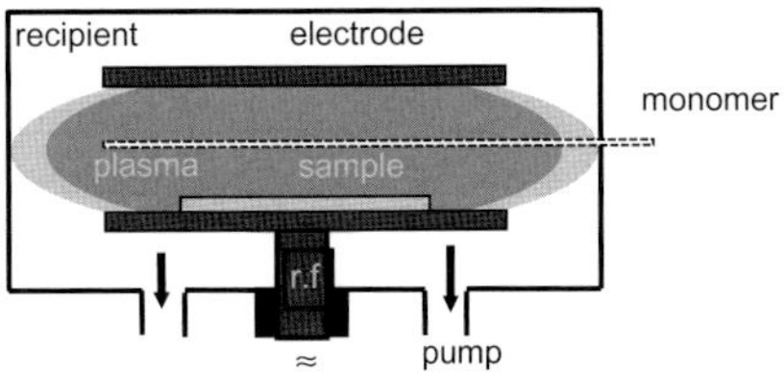

Figure 1.5 Example of a plasma reactor: a diode-like plasmatron that produces a low-pressure plasma suitable for polymer surface modification.

This type of discharge is very seldom used today because of the permanent danger of electrode contamination and coating, which influence the plasma characteristics. Capacitively or inductively coupled radio-frequency (rf) or microwave (mw) generated plasmas are used more often. Figure 1.5 shows the often used diode-like reactor type, which is the most important among the broad variety of plasmatron constructions. It is an example for the production of low-pressure plasmas, called diode-like or parallel-plate reactor. Normally, one electrode is mass (asymmetric coupling).

In contrast, the plasma bombardment of polymer surfaces (cf. Figure 1.3) is an efficient, easy, clean, comfortable, and fast way to create reactive centers at which plasma gas fragments or atoms can stick as new functional groups. The processes starts, again, with desorption and functionalization and is continued by modification of near-surface layers, etching at the surface, and photo-modification of far-from surface layers considering an overlap of all processes. The surface functionalization is limited to an O/C ratio of about 0.28; a steady-state process between continuation of introduction of functional groups and polymer etching is then established. The maximum density of functional groups at the outermost polymer surface is completed after a few seconds, most often after only 2 s [20]. Limiting the plasma exposure to such short treatment protects the polymer surface against undesired advanced degradation and formation of defects. Moreover, applying minimal energy also preserves the original polymer structure. *Preservation of original polymer* structure is always the *best guarantee for maximum mechanical properties* and chemical integrity.

Unfortunately, the average and, especially, the maximal energy level of plasma particles and radiation, particularly that of electrons, is about an order of magnitude higher than that of the binding energies in polymers. Thus, plasma chemistry in a low-pressure glow discharge is equivalent to chemistry with a high excess of energy. This discrepancy is the most important hindrance to polymer modification by the use of plasma. Moreover, the plasma energy is continuously delivered as electrical current from a power plant and introduced into the plasma from the electrical power supply. Therefore, the electrically produced plasma is effectively an inexhaustible source of energy/enthalpy.

The list of binding energies for different chemical bonds (in $kJ\,mol^{-1}$) at 100 kPa and 298 K reflects the situation in polymers: HC≡CH (963), N≡N (950), $H_2C{=}CH_2$ (720), N=N (418), O=O (498), N–N (163), $H_3C{-}CH_3$ (368), $H{-}C_2H_5$ (410), $H{-}CH_3$ (435), $H{-}CCl_3$ (402), $H{-}C_6H_5$ (460), $H{-}CH_2OH$ (402), H–Cl (431), $H{-}OCH_3$ (440), $H{-}N(CH_3)_2$ (398), and C–O (358) (3–6 eV binding energy). These binding energies match exactly the average energies in electrical glow discharges working at low pressures in the range 10^0–10^3 Pa. Considering the high energy tail of the electron energy distribution function in a gas plasma (E_{kin} = 3–>20 eV) these electrons have enough energy to break all chemical bonds present in polymers (Section 4.1). It must be also considered that all electrons slowed or stopped by inelastic collisions with atoms or ions in the plasma are re-accelerated to high energies due to the applied electrical field. Thus, as noted, the plasma appears as a source of (nearly) unlimited enthalpy/energy flow.

On the other hand, plasma is a very easy tool with which to form reactive sites as radicals or functional groups independent of the inertness of polymer. In each case the plasma forms anchoring sites for further chemical reactions.

Atoms at the topmost layer of solid surfaces have generally unsaturated valences towards the gas phase. These unsaturated binding forces are responsible for the surface energy of solids. This surface energy determines wetting and gluing properties. In alkyl chains any significant dipole moment is absent; thus, polyolefins, silicones, or perfluorinated polymers have very low surface energies, composed of the dispersion component without an appreciable polar component. The introduction of functional groups forms dipoles at the surface and, thus, the polar component of the surface energy is increased strongly, as presented schematically in Figure 1.6 on an atomic level.

Drost, McTaggart, and Venugopalan, illustrated several traditional atmospheric and low-pressure glow discharge processes in industry [19, 21, 22]. Polyolefin or polymer modifications by surface functionalization were performed in atmospheric-pressure plasmas (corona, barrier, and glow discharges) [23] or low-pressure glow discharges [4]. This is because of the non-isothermal (non-equilibrium, nt) character of such plasmas, which show low gas temperatures ("cold" plasma) [24]. Thus, both the atmospheric and low-pressure plasmas were generally suited for any polymer pretreatment. Very fast and continuous corona or barrier discharge

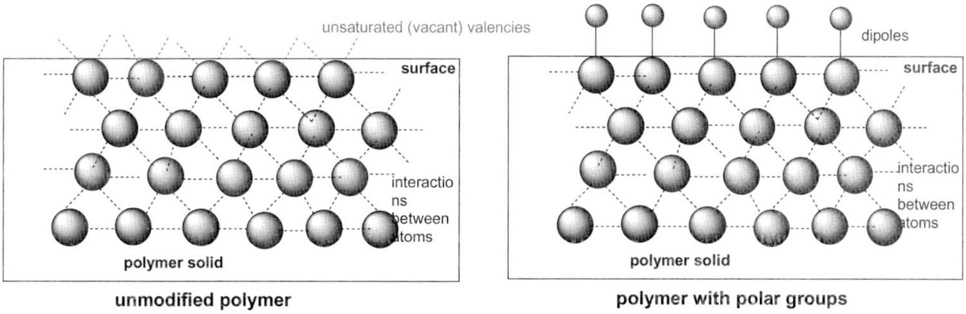

Figure 1.6 Schematics of polymer surfaces.

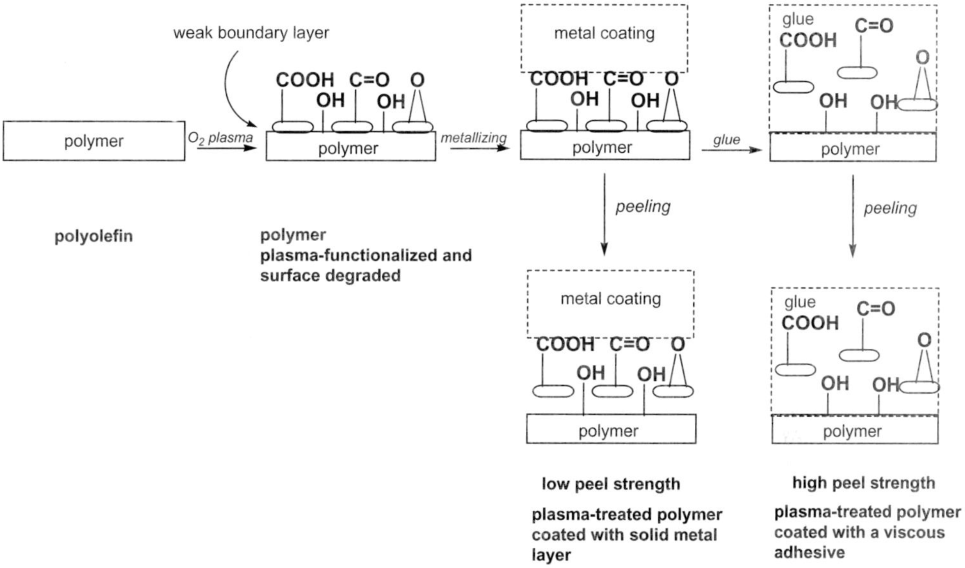

Figure 1.7 Schematic representation of a polyethylene surface before and after plasma treatment as well as after coating with a viscous adhesive.

treatments of polymer foils at atmospheric pressure are traditional plasma applications to polymers on an industrial scale [25, 26]. However, the field of applications of this atmospheric plasmas enlarged slowly because the surface modification could not be well-controlled, is afflicted with a few inhomogeneities, and degraded material (LMWOM) is formed. More recently, the atmospheric-pressure glow discharge (APGD) has compensated these disadvantages [27, 28]. The non-selective surface functionalization accompanied by uncontrolled degradation need not be accepted, as demonstrated by use of the newly developed atmospheric-pressure

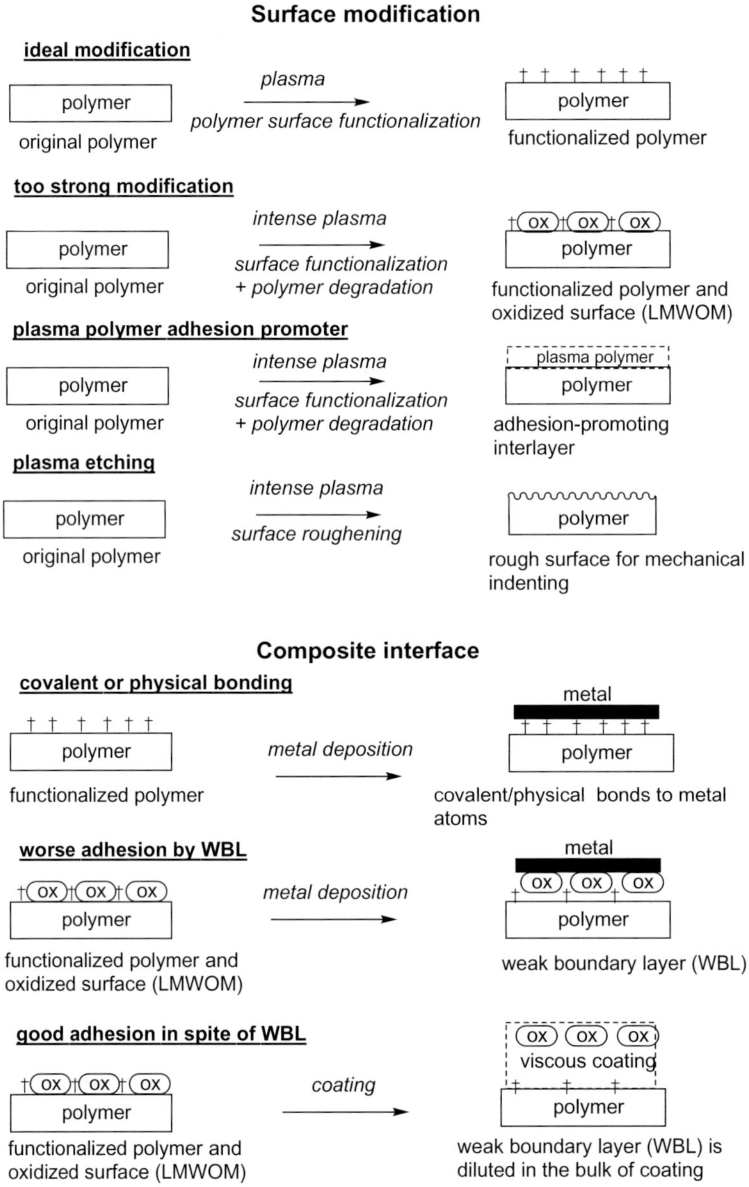

Figure 1.8 Schematics of adhesion promotion in polymer composites using the plasma technique.

plasma techniques, such as aerosol barrier discharge and electrospray ionization (ESI) film deposition [15].

The improved new plasma pretreatments aim to favor the introduction of (most advantageously: monosort) functional groups onto the surface to enhance the interactions to coatings or biomolecules or to form anchoring points for further graft synthesis. In addition, surface roughening, crosslinking of surface-near polymer layers, polymer etching, polymer ashing, and so on are also being examined by the plastic and electronic industries and in medical techniques.

Considering the use of plasma for adhesion promotion in more detail, the more ambitious improvement of metal adhesion to polymers demands the application of low-pressure glow discharge plasmas. The polyfunctional O-containing groups are rapidly produced on exposure to the oxygen low-pressure plasma. However, polymer degradation products at the interface of a polymer–metal composite hinder the adhesion by forming a "weak boundary layer" (WBL) [29] that originates from the LMWOM. This "debris" hinders adhesion to solid-phase metal layers. In contrast, liquid paints or glues can assimilate these loosely bonded fragments, dissolve and distribute them in the liquid phase of the adhesive. Thus it does not strongly hinder the adhesion between the coating and the polymer. Now, the coating can interact directly with the non-degraded polymer surface (Figure 1.7). Roughening and crosslinking as well as plasma-chemical deposition of a thin adhesion-promoting interlayer are alternatives (Figure 1.8).

References

1 van der Waals, J.D. and Kohnstamm, P. (1908) *Lehrbuch der Thermodynamik: Teil 1*, Johann-Ambrosius Barth Verlag.

2 Wu, S. (1982) *Polymer Interface and Adhesion*, Marcel Dekker, New York, Basel.

3 Kreidl, W.H. and Hartmann, F. (1955) *Plast. Technol.*, **1**, 31.

4 Rossmann, K. (1956) *J. Polym. Sci.*, **19**, 141.

5 Endlich, W. (1998) *Kleb- und Dichtstoffe*, Vulkan-Verlag, Essen.

6 Friedrich, J., Wettmarshausen, S., and Hennecke, M. (2009) *Surf. Coat. Technol.*, **203**, 3647–3655.

7 Friedrich, J., Mix, R., and Wettmarshausen, S. (2008) *J. Adhesion Sci. Technol.*, **22**, 1123.

8 Huajie, Y., Mix, R., and Friedrich, J. (2011) *J. Adhesion Sci. Technol.*, **25**, 799–818.

9 Enzyklopädie, K. (1970) *Atom-Struktur der Materie*, VEB Bibliographisches Institut, Leipzig.

10 Fanghänel, E. (2004) *Organikum*, 22nd edn, Wiley-VCH Verlag GmbH, Weinheim.

11 Strobel, M., Corn, S., Lyons, C.S., Korba, G.A., and Polym, J. (1987) *J. Polym. Sci., A.*, **25**, 1295.

12 Friedrich, J., Kühn, G., Mix, R., Fritz, A., and Schönhals, A. (2003) *J. Adhesion Sci. Technol.*, **17**, 1591–1618.

13 Friedrich, J., Kühn, G., Mix, R., and Unger, W. (2004) *Plasma Process. Polym.*, **1**, 28–50.

14 (a) Wettmarshausen, S., Mittmann, H.-U., Kühn, G., Hidde, G., and Friedrich, J.F. (2007) *Plasma Process. Polym.*, **4**, 832–839; (b) Vasile, M.J. and Smolinsky G. (1972) *J. Electrochem. Soc.* **119**, 451–455.

15 Friedrich, J.F., Mix, R., Schulze, R.-D., Meyer-Plath, A., Joshi, R., and Wettmarshausen, S. (2008) *Plasma Process. Polym.*, **5**, 407–423.

16 Wunderlich, B. (1976) *Macromolecular Physics*, vol. 2, Academic Press, New York.

17 Friedrich, J., Loeschcke, I., Frommelt, H., Reiner, H.-D., Zimmermann, H., and Lutgen, P. (1991) *J. Polym. Degrad. Stabil.*, **31**, 97–119.

18 Friedrich, J., Gähde, J., Frommelt, H., and Wittrich, H. (1976) *Faserforsch. Textiltechn./Z. Polymerenforsch.*, **27**, 604–608.

19 Drost, H. (1972) *Plasma Chemistry*, Akademie-Verlag, Berlin.

20 Friedrich, J.F., Unger, W.E.S., Lippitz, A., Koprinarov, I., Weidner, S., Kühn, G.,

and Vogel, L. (1998) *Metallized Plastics 5 & 6: Fundamental and Applied Aspects* (ed. K.L. Mittal), VSP, Utrecht, pp. 271–293.
21 McTaggart, F.K. (1967) *Plasma Chemistry in Electrical Discharges*, Elsevier, Amsterdam.
22 Venugopalan, M. (1971) *Reactions under Plasma Conditions*, John Wiley & Sons, Inc., New York.
23 Kim, C.Y., Evans, U., and Goring, D.A.I. (1971) *J. Appl. Polym. Sci.*, **15**, 1357.
24 Hertz, G. and Rompe, R. (1973) *Plasmaphysik*, Akademie-Verlag, Berlin.
25 Owens, D.K. (1975) *J. Appl. Polym. Sci.*, **19**, 265.
26 Owens, D.K. (1975) *J. Appl. Polym. Sci.*, **19**, 3315.
27 Guimond, S., Radu, I., Czeremuszkin, G., Carlsson, D.J., and Wertheimer, M.R. (2002) *Plasmas Polym.*, **7**, 71.
28 Klages, C.-P. and Grishin, A. (2008) *Plasma Process. Polym.*, **5**, 368–376.
29 Bikerman, J.J. (1972) *Angew. Makromol. Chem.*, **26**, 177.

2
Interaction between Plasma and Polymers

2.1
Special Features of Polymers

The class of organic polymers is closely related to natural macromolecular products. Numerous natural polymers are produced by animals or plants. Thus, polymers are positioned in nature at the borderline between organic materials and living species. This special position determines the properties and the sensitive response of polymers to chemical and physical interactions. In contrast to ionic (inorganic, ceramic, glass) or metallic materials polymers are not able to distribute reaction energy over the bulk, especially if introduced precisely by physical or chemical processes. Consequently, there are local energy-rich spots, characterized by increased movement of polymer chains, side-chains, or groups in the solid and by supply of excess energy in bond scissions followed by extensive chemical reactions.

Natural polymers also exhibit in many cases very simple structures, such as waxes, which are also extensively produced industrially by synthesis, such as the polyolefin polymers. The performance of polymers is strongly determined by several important parameters such as chemical composition, structure of monomer unit, polymerization degree or chain length or molar mass, and its distribution, tacticity, branching, crosslinking, supermolecular structure, crystallinity, stretch orientation, and so on.

Often, beginning at the surface, an *oxidative aging* of polymers is observed as part of natural weathering. It is characterized by the attachment of molecular oxygen as oxygen-containing polar group onto radicals, which are formed mechanically, by irradiation or by plasma exposure. On a low energy level, chromophore groups (defects) in the polymer are starting points for oxidation. Radiative processes are also often associated with the formation of unsaturation and crosslinking in deeper layers of the polymer. Unsaturations are also chromophoric centers and can also undergo slow oxidation on exposure to air. The radical-promoted oxidation is often self-accelerated by the auto-oxidation process (see below).

There is also a physical aging, which is important for surface modification. Functional groups at the surface are not permanently fixed. They can slowly move and diffuse from the topmost layer to the bulk. Moreover, complete polymer

The Plasma Chemistry of Polymer Surfaces: Advanced Techniques for Surface Design, First Edition.
Jörg Friedrich.

segments equipped with functional groups diffuse into the bulk. This mobility of functional groups and macromolecule segments is called "surface dynamics" or "hydrophobic recovery" [1–3]. The driving force is thermodynamics. The high concentration of functional groups becomes adjusted to the zero concentration in the bulk by diffusion [4]. Figure 2.1 shows schematically the slow diving that occurs upon rotation around the C–C bond and the segmental movement from the surface to the bulk.

While diffusion and balancing of concentration are one driving force, the interaction of polar groups at polymer surface with water or other polar media is the counterpart. If air at the surface is replaced by water the polar group returns to the surface and forms hydrogen bonds with water molecules (Figure 2.2) [2].

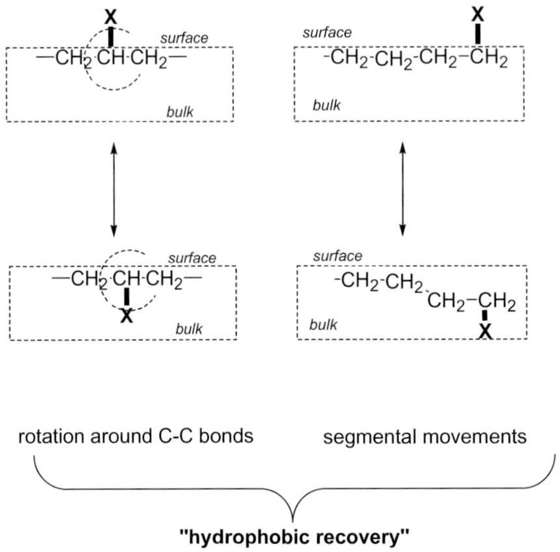

Figure 2.1 Schematic of "hydrophobic recovery" and surface dynamics.

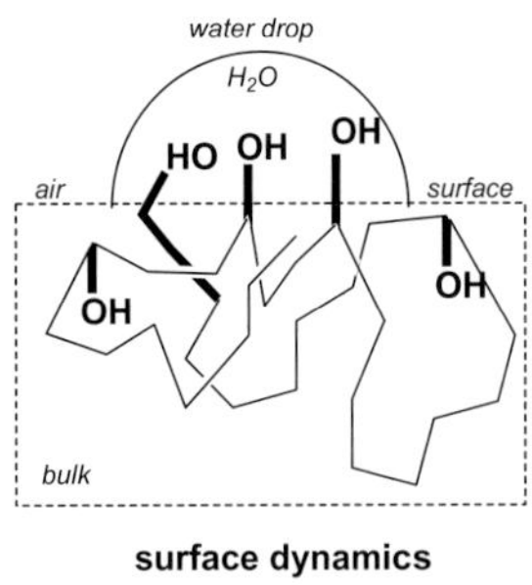

Figure 2.2 Surface dynamics caused by different environments.

The formation of interactions between polar groups and water and the respective energetic profit overcompensates the dilution tendency [5]. More complicated is the situation when functional groups are protonated, find counter ions, are rearranged, or even grafted – then their mobility is strongly changed [6]. The existence or absence of functional groups at a polymer surface determines its "communication" with the environment and, thus, the adhesion properties.

Another important factor for the mechanical behavior of polymers is the *molar mass* and its *distribution*. Changes in molar mass may provoke dramatic alterations in mechanical properties of polymers. Polymers exposed to plasmas often show a change of molar mass [7].

A general problem of any polyolefin surface modification is the *equivalence* of C–C and C–H *bond strengths*. The binding energies (dissociation energies) of the C–H- and C–C-bonds, which dominate the structure of polyolefins, are close together as 435 kJ mol^{-1} for H–CH_3, 411 kJ mol^{-1} for H–CH_2–CH_3, 396 kJ mol^{-1} for $(CH_3)_2$CH–H, and 385 kJ mol^{-1} for $(CH_3)_3$C–H. The corresponding C–C bond has similar (standard) dissociation energy, for example, 370 kJ mol^{-1} for CH_3–CH_3 (ca. 3 eV). In addition, these C–H- and C–C-binding energies vary strongly with the type of substitution or residues and, thus, the ranges of binding energies overlap widely. Therefore, specific chemical attack of the C–H bond in alkyl chains (polyethylene) is not possible and selective reactions are impossible. Consequently, only the simultaneous attack of all C–H and C–C bonds occurs. This undesired behavior is supported by exothermic substitution reactions, which produce additional thermally-initiated chain scissions. Nevertheless, surprisingly, in practice the shielding of hydrogen atoms hinders slightly the preferred C–C bond scission and, thus, degradation does not dominate too strongly [8].

The (standard) reaction enthalpy (R) of the radical H substitution reaction (S_R) by a plasma-formed functional group producing an atom or radical is given as the difference of (standard) heats of formation (B) of the end product (E) and starting material (S): $\Delta_R H^0 = \Delta_B H^0{}_E - \Delta_B H^0{}_S$. Considering Hess rule $\Delta_R H^0$ is the difference of dissociation enthalpies of the scissioned (S) and formed (F) bonds: $\Delta_R H^0 = \Sigma^S \Delta_D H^0 - \Sigma^F \Delta_D H^0$. Therefore, for a raw estimation, the heat of formations can be replaced by dissociation enthalpies [8]. Because of the gas–solid type of reaction the entropy must be considered (Gibbs–Helmholtz): $\Delta_R G^0 = \Delta_R H^0 - T\Delta_R S^0$. The entropy term $\Delta_R S^0$ often shows ambivalent behavior [9]. This term is not important if the number of products and educts are equal, phase transitions not possible, and so on.

Oxidation and fluorination show strong exothermic character during the substitution reaction and, therefore, the reaction proceeds as an auto-oxidative process, as demonstrated for fluorination:

$$F^\bullet + H{-}CH_2{-}CH_3 \rightarrow F{-}H + {}^\bullet CH_2{-}CH_3$$

$$^\bullet CH_2{-}CH_3 + F_2 \rightarrow F{-}CH_2{-}CH_3 + F^\bullet$$

$$\Delta_R H^0 = \Delta_D H^0{}_{CH3-CH2-H} - \Delta_B H^0{}_{H-F} = 411 - 566 \text{ kJ mol}^{-1} = -155 \text{ kJ mol}^{-1}$$

The negative reaction enthalpy signals the spontaneous and auto-accelerating proceeding of the reaction but also indicates the probable appearance of a broad product spectrum.

On the other hand, the introduction of a new substituent X for H atoms positioned in alkyl chains ($CH_2 \rightarrow CHX$) leads to activation of the remaining hydrogen atom (*activation of C–H bonds by functionalization*). The new polar C–X bond may produce also a shift of C–H bond electrons. Thus, the C–H bond is weakened and may be substituted by a second X more easily and selectively. Simultaneously, the C–C chain scissions are minimized. Such a more selective process without significant polymer degradation is the chlorination of poly(vinyl chloride) to higher chlorinated species [10].

If the plasma delivers the enthalpy and dissociates the halogen molecules and the C–H bonds in the polymer molecule the reaction enthalpy becomes much more negative. Therefore, the selectivity of the substitution reaction decreases and the polymer degradation increases strongly if plasma is used. *Plasma processes produce generally low selectivity and polymer degradation.* Chemical-oxidative attack, excimer lamp or laser irradiation, and flame or mechanochemical exposure of polymer surfaces work in an analogous manner. Thus, the formation of functional groups within strongly exothermic reactions is accompanied by random polymer degradation and etching.

2.2
Processes on Polymer Surfaces during Plasma Exposure

Summarizing the above-discussed extraordinary sensitive behavior of polymers on exposure to a high-energy particle shower from plasma and short-wavelength plasma radiation, the energy introduction must be minimized as much possible to produce selectivity and to hinder polymer degradation. However, the energy of particles and radiation in the plasma cannot be decreased arbitrarily. The discharge needs energy to start and a minimum energy to sustain the plasma state, a necessary energy level that is much higher than the C–H and C–C binding energies. This is the dilemma.

Moreover, the C–H substitution in alkyl chains is principally accompanied by C–C bond scissions, but with increasing number of new substituents the substitution may become more selective automatically. Nevertheless, chain degradation always occurs and loosely bonded polymer fragments are formed at the surface. Ideally, only the topmost (one) atom layer, or one molecular layer, must be equipped with functional groups. These functional groups are sufficient to form physical or, preferentially, chemical links to other materials.

Scheme 2.1 shows the idealized replacement of H atoms from CH_2 groups of alkyl chains with atoms or radicals (from plasma) along with the associated polymer chain scissions occurring in particular for exothermic reactions.

Such scissions of polymer backbones decrease the cohesive (mechanical) strength and durability (oxidation and migration) of polymer surface layers.

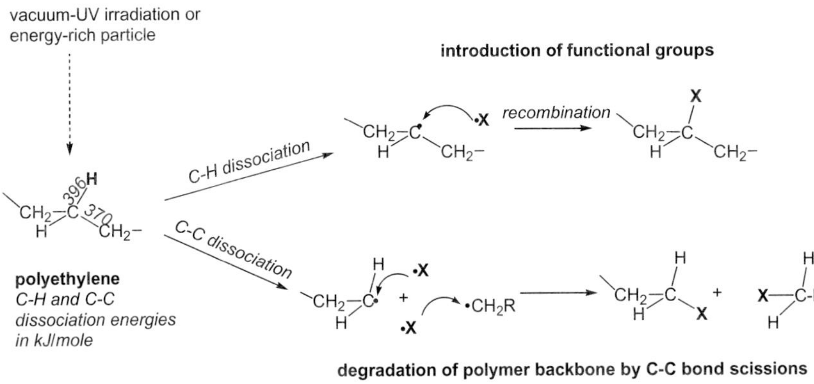

Scheme 2.1 Response of polymer to energy introduction (X = introduced functional group).

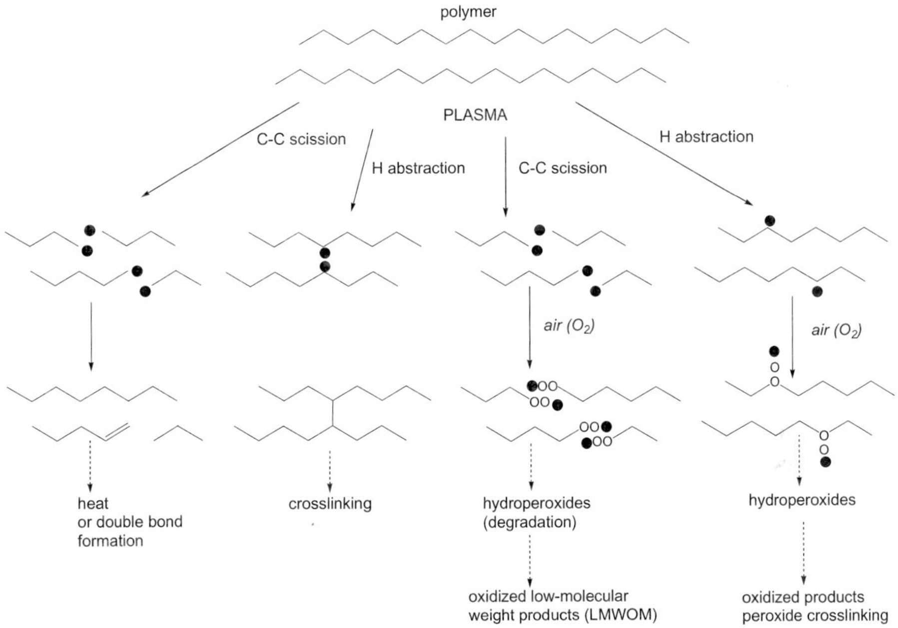

Scheme 2.2 Chain scissions in polyethylene upon exposure to a plasma.

However, as seen in Scheme 2.1 these reactions are radical processes. Thus, degradation occurs only if the two C radicals formed after C–C bond scission cannot immediately recombine. Such recombination produces heat within the polymer. The C radicals formed by H abstraction can also recombine but they form crosslinking, while neighboring C radicals produce C double bonds. The C radicals also react rapidly with traces of oxygen. If the radicals are shielded by substituents and therefore trapped they can react with oxygen from air after hours, days, or weeks (Scheme 2.2) [11–13]:

$CH_2 \rightarrow CH^\bullet + {}^\bullet H$	H-abstraction
$CH_2{-}CH_2 \rightarrow CH_2^\bullet + {}^\bullet CH_2$	C–C scission
$CH^\bullet + {}^\bullet X \rightarrow CH\text{-}X$	attachment of functional groups
$CH_2{-}CH_2 \rightarrow CH{=}CH + H_2$	formation of double bonds
$2CH_2{-}CH_2^\bullet \rightarrow {-}CH{=}CH_2 + \text{-}CH_3{-}CH_2$	disproportionation
$CH^\bullet + {}^\bullet CH \rightarrow CH{-}CH$	recombination
$CH^\bullet + {}^\bullet O{-}O^\bullet \rightarrow CH{-}O{-}O^\bullet$	peroxy formation

Taken together with the (always) produced broad variety of plasma-formed functional groups the plasma functionalization process must be characterized as completely unselective (Scheme 2.1). Nevertheless, if one accepts these insufficiencies it is possible to influence the type of produced functional group roughly by appropriate choice of the "monomer" X. The term *monomer* may be better replaced by *precursor* because monomer refers to a molecule that can easily undergo a chemical chain-growth polymerization reaction. However, a precursor molecule becomes more or less dissociated in plasma, often independently, if it is saturated or unsaturated and thus polymerizable or not. The dissociated atoms or fragments form the functional groups (or plasma polymer).

This characteristic behavior of plasma, which produces a broad variety of functional groups (unspecific, polysort), may be presented schematically for the oxygen plasma and a polyolefin polymer:

$$O_2 + \textit{plasma} \rightarrow O_2^+, O^+, O^-, O_2^-, O^\bullet, O^*, O_2^*, O_3, \text{and so on} + \text{polymer} \rightarrow C{-}OH,$$
$$C{-}O{-}C, C{-}O{-}OH, \text{epoxy}, {>}C{=}O, CHO, \text{aryl-}OH, COOH,$$
$$COOR, C(O){-}O{-}OH, C(O){-}O\text{-}OR, O{-}CO{-}O, \text{and so on } [14, 15].$$

This fragmentation–recombination mechanism is open for all random processes and selectivity is most often completely absent.

Nevertheless, the plasma gas determines roughly the type, or more exactly the family, of different but related functional groups (Figure 2.3, cf. "polysort functionalization"). Prominent examples for the production of families of similar products are oxygen and ammonia plasma treatments of polymers. With oxygen plasma exposure nearly all known oxygen-containing groups are produced (family of plasma gas-specific *polysort* oxygen-containing *groups*), sometimes also as secondary products of auto-oxidation. Examples are hydroxyl and carboxylic groups, which need hydrogen, which was abstracted before from the polymer chains, for their formation. The spectrum of functional groups produced at polyolefin surfaces using ammonia plasma exposure involves different families of functional groups (polysort oxygen-containing and polysort nitrogen containing groups) because of secondary reactions with oxygen from air to form N-, NO_x- and O-containing functional groups.

Besides the desired polymer surface functionalization several undesired side-reactions must be accepted. This was shown for ammonia plasma exposure. The

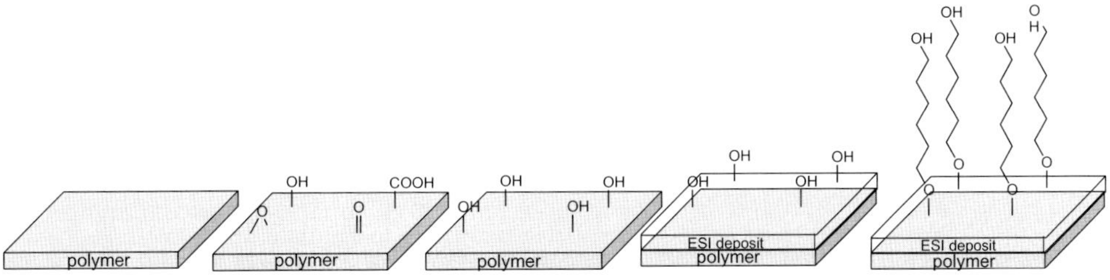

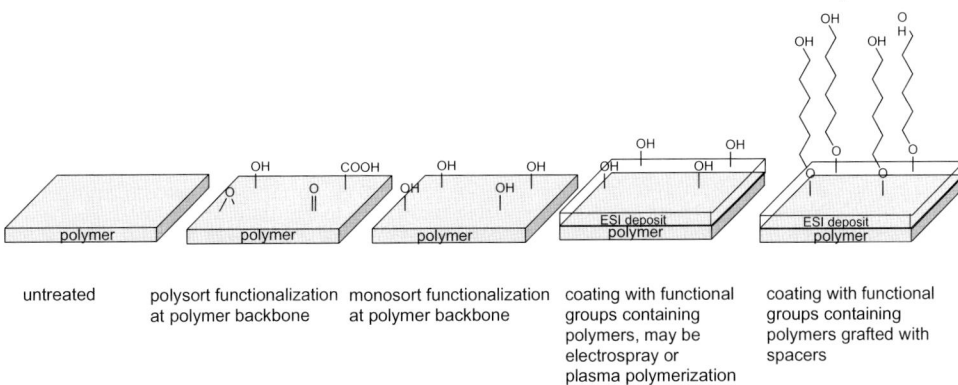

Figure 2.3 Possibilities for the introduction of functional groups onto a polymer surface.

most prominent side-reactions, especially when exposed to oxygen plasma, are the above-mentioned *degradation* of molar masses and *crosslinking*. Because of the existence of trapped C radical sites in the polymer, bulk *oxidation* and sometimes *auto-oxidations* occur, as shown before.

Another route to surface functionalization is the chemical substitution of C–H-bonds available for exothermic reactions without any assistance by a plasma, as shown above for polyolefin fluorination.

The most ambitious intention of plasma pretreatment is to introduce *monosort (monotype) functional groups* onto the polymer surface, such as OH, COOH, CHO, NH_2, epoxy, SH, CN, Br, Cl, F, and so on, which may serve as chemical anchoring points for (chemical) interactions (polar groups) or graft reactions (Scheme 2.3).

All introduced functional groups onto a polymer surface influence the wetting behavior, which is reflected in polymer *surface energy*. The surface energy can easily be measured by using several test liquids, measuring the contact angles, and calculation using one of the theoretical models [16–20]. Rough information on surface energy, but much easier to measure, is obtained using test inks [21].

Thus, we have seen that molar mass degradation is inevitable. This degradation is accompanied by a loss in mechanical strength. The reason for this is that the cohesive strength of a polymer is strongly dependent on the molar mass and its distribution. The macromolecules located in the topmost surface layer are predominantly affected by this degradation. Low-molecular weight oxidized material (LMWOM) is formed ("molecular debris") that has no bonding or (weak) adhesion to the unmodified polymer material in deeper layers (bulk) [22]. Usually, these degradation products form a dust-like layer at the polymer surface, which is often observed when the surface is exposed to a corona discharge or to dielectric barrier discharge in air. These degraded macromolecule fragments at polymer surfaces are of technical importance because they interfere with adhesion to coatings, inks, or metals. They form a *weak boundary layer* at the interface between the polymer and coating [23, 24].

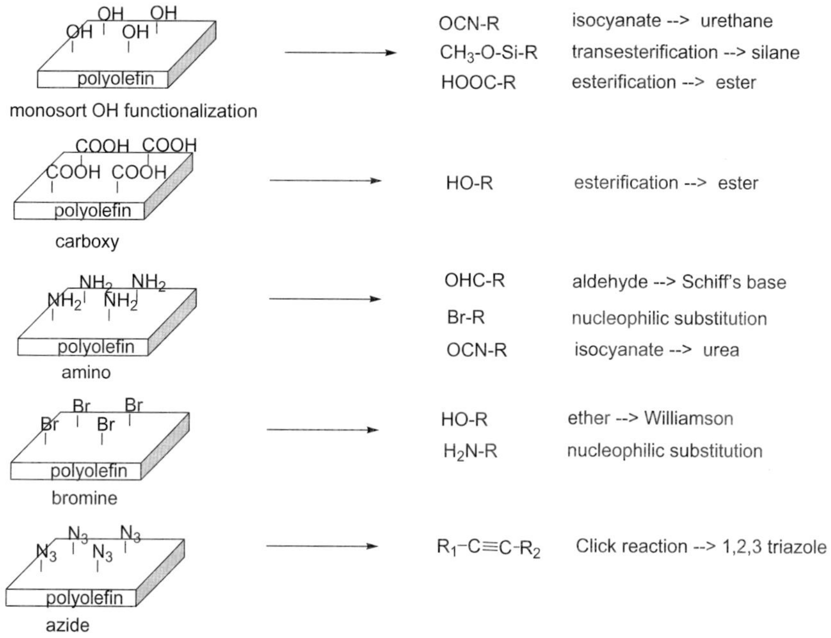

Scheme 2.3 Examples of graft reactions at monosort functional groups on polyethylene surfaces.

If plasma exposure to polymers is continued the process passes over to an *etch process* (Figure 2.4). Using as an example the oxygen plasma, the pre-degraded fragments at the surface are gasified during the etching process to CO_2, CO, H_2O, formaldehyde, and so on.

If the polymer has a homogenous structure, it is anisotropic, that is, it does not possess a supermolecular structure with highly ordered (crystalline) zones, then a steady-state etching process is established with constant etching rate. The etching rate in the plasma often follows the ease of formation of gaseous degradation products in the polymer structure, for example, polymers with structures that contain pre-formed degradation products, which are formed by scission of only one or two bonds, are preferably etched. For example, Norrish I and II rearrangements were observed for poly(ethylene terephthalate) on exposure to oxygen plasma [25]. It was a dream of polymer researchers to produce a dry etching process that would enable excavation of supermolecular structures in crystalline polymers; such structures could then be pictured by electron microscopy. Many attempts were made to find these supermolecular structures by plasma etching; however, the plasma is to insensitive and changes the original structure by forming irregular plasma-created structures [26].

A supermolecular structure or layer-like inhomogeneities or the ability to crosslink influence the etching rate. Thus, amorphous components are preferably etched while crosslinked structures resist the etching attack longer (Figure 2.5)

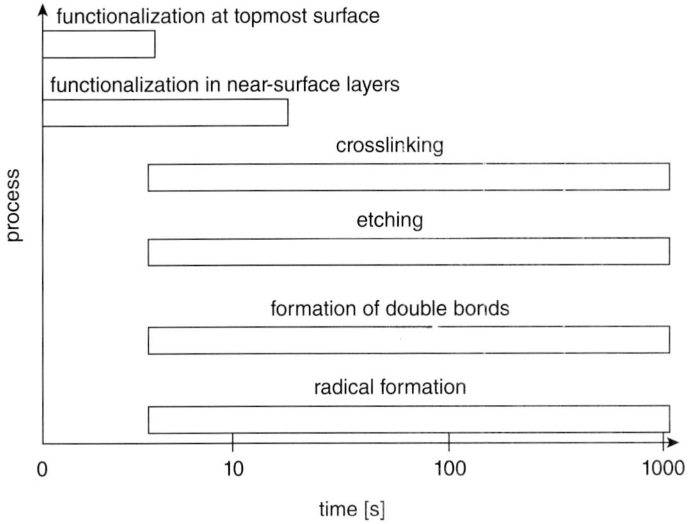

Figure 2.4 Time scale of plasma-induced processes at polymer surfaces.

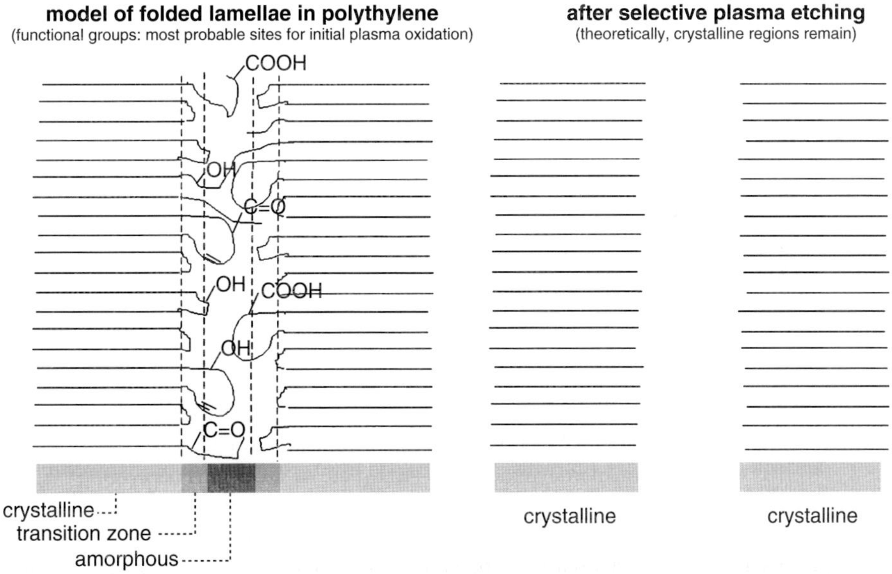

Figure 2.5 Idealized selective (oxygen) plasma etching of amorphous regions in polymers.

[27]. With regard to the adhesion properties of polymers such a roughening helps the mechanical anchoring of metal layers, top coatings, or paints. The roughening also increases the specific surface of the polymer target. In terms of adhesion, this means a higher specific surface, more interactions, and better adhesive bond strength.

Plasma exposure of polymers is always associated with *crosslinking* of the polymer surface layer, as was discovered at the end of 1960s and the beginning of the 1970s [28, 29]. As argued more recently, it is not the plasma particle bombardment that is responsible for crosslinking but the always-emitted plasma-vacuum UV irradiation of extreme short wavelength (20–200 nm) [30]. It must be considered that the energy content of this plasma radiation correlates to the electron energy distribution function in the low-pressure discharge. All radiative processes are coupled with the kinetic energy of electrons. Owing to inelastic collisions of atoms or molecules with electrons, excitations to different electronic and vibration states occur:

$$A + e^- \rightarrow A^* + e^{-\prime}$$

$$A^* \rightarrow A + h\nu$$

The shortest wavelength of this radiation (ca. 20–100 nm), occurring from the deactivation of activated states of singly and, much more, doubly charged atoms:

$$A^{(2+)*} \rightarrow A^{(2)+} + h\nu$$

penetrates polymers unhindered because of the absence of absorbing structures within polymers. Such radiation up to 20 nm wavelength is emitted by low-pressure plasmas [30]. Moreover, radiation in the range $\lambda = 100$–180 nm has much greater intensity. Note that it can also penetrate a few micrometres into the polymer bulk and it can also produce σ_{C-C}- and σ_{C-H}-transitions. These excitations produce dehydrogenation, crosslinking, and chain scissions (degradation) of the macromolecule backbone [31, 32]. Highly absorbing structures (>C=O, C=C, and so on) or the formation of polyene structures by dehalogenation or dehydrogenation and so on limit the penetration depth of radiation at wavelengths greater than 180 nm.

Generally, the polymer responds to this irradiation by etching in the presence of oxygen, radical formation (isolated or "trapped" radical sites) in near-surface layers, and degradation, but also with partial crosslinking (radical recombination). The crosslinking produces high densities, as discussed later. The trapped radicals also reach high concentrations, of a few percent of all carbon atoms [33]. These isolated and trapped radicals, which are not able to immediately recombine under room temperature conditions, react preferably with oxygen dissolved in the polymer or diffusing into the bulk when exposed to ambient air [34]. This extensive reaction causes a pronounced *aging sensitivity* of all polymers that have been exposed to plasma. The reaction of oxygen, which slowly diffuses into the polymer, with the radicals produces *post-plasma oxidative reactions* and may be also initiate a few *auto-oxidative reactions* of polymers during their storage in ambient air. The auto-oxidation is a self-accelerating process because of its chain reaction character (Scheme 2.4).

Thus, the original composition of the plasma-modified polymer becomes permanently changed in the sense of a progressive propagation of oxidation. In summary, any plasma treatment, also in the absence of oxygen or oxygen-containing plasma gases, produces automatically metastable trapped radicals that

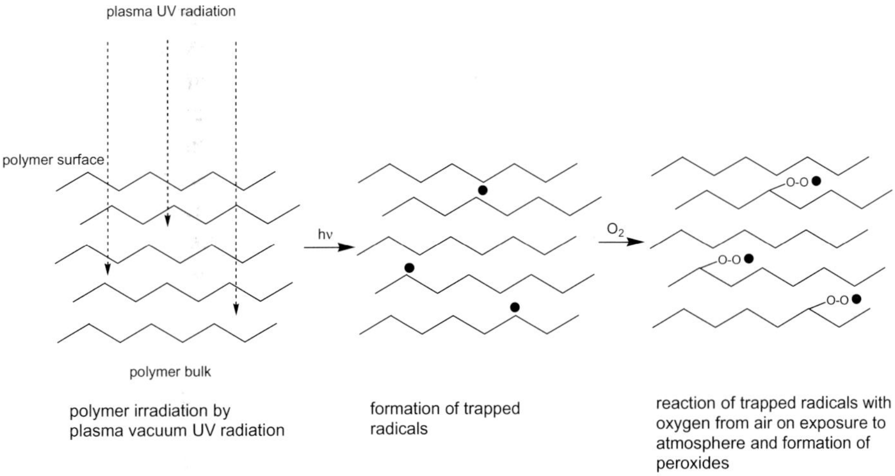

Scheme 2.4 Formation of trapped radicals in polymers during plasma exposure.

always react with oxygen on exposure to air. The radicals always react in the same manner, thus giving a similar product variety independent of the type of plasma used [11]. Each contact with air results in the introduction of oxygen into the near-surface layers of polymers, a process known as *post-plasma oxidation.* First, C radical sites are produced, which react with molecular oxygen in the ground state to give peroxy radicals that are chemically transferred into hydroperoxides within the chain-propagation reaction [35–37]. As a result of this peroxide formation and the further reaction of hydroperoxides a large number of different O-functional groups are formed.

This auto-oxidation is superposed on plasma-oxidation in the oxygen plasma and may explain the similar qualitative and quantitative distribution of oxidation products in polymers found with different plasmas. An exponential increase of oxygen introduction can be observed during exposure to an oxygen plasma or during subsequent storage under ambient air and also during weathering under climatic conditions (Figure 2.6) [38].

The exposure of polypropylene to a radio-frequency discharge in nitrogen for 10 min is characterized by the undesired co-introduction of oxygen (Figure 2.7). The oxygen concentration amounts to more than 200% of nitrogen incorporation. Approximately, oxygen was bonded to each third carbon atom [9].

Similar observations were made using carbon fibers exposed to ammonia plasma [39].

Comparing the time-scale of plasma oxidation (Figures 2.4 and 2.6) with photochemical oxidation or in particular with weathering, the former is clearly compressed by several orders of magnitude. Moreover, the oxidation was caused by both the oxygen particle shower from the plasma accompanied by oxygen introduction to the polyolefin surface and the irradiation with energetic radiation produced

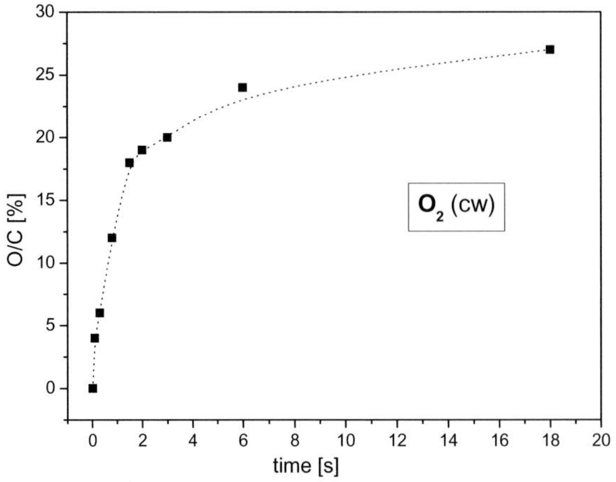

Figure 2.6 XPS-measured oxygen introduction in polypropylene upon exposure to oxygen continuous-wave (cw rf) plasma.

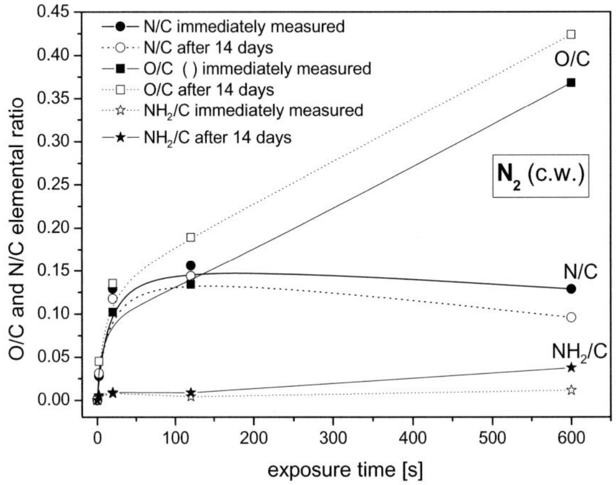

Figure 2.7 XPS-measured nitrogen and oxygen introduction in polypropylene upon exposure to a nitrogen continuous-wave (cw rf) plasma for 10 min and after 14 days storage in ambient air.

in the plasma [40–42]. This was motivation for using oxygen plasma exposure and the post-plasma auto-oxidation to simulate and accelerate the natural weathering of polymers (*"accelerated plasma aging of polymers"*). In particular, the plasma-UV irradiation and the associated radical formation in the oxygen plasma was used to simulate polymer aging by weathering characterized by oxygen introduction and, more specifically, by increasing the carbonyl index as well as by loss in mechanical strength and elasticity. For aromatic polyester materials such an artificial plasma aging has shortened the aging time to 8–30 h, which corresponds 1–2 years weathering in a Florida climate [43–45].

In addition to radical formation, degradation and/or crosslinking and auto-oxidation are further and general responses of polymers to plasma exposure. The polymer becomes *dehydrogenated,* which is caused by crosslinking and also by the formation of *olefinic unsaturations* as well as *cracking of aromatic rings* [46].

2.3 Influence of Polymer Type

The type of polymer and, therefore, structure, composition, functional groups, supermolecular structure, and so on determines significantly the chemical processes on exposure to the plasma. In similar manner polymers are categorized in two groups when exposed to ionizing irradiation [47]. Thus, the general response of polymers on plasma exposure can be summarized as follows. Strongly different etching rates for polymers were measured [48]. Concerning the functional groups of polymers a few of them contain a preformed structure similar to that of (stable) gases such as CO, CO_2, H_2O, N_2, and so on. Therefore, these groups can be easily split off as gaseous degradation products by scission of one or two bonds. The (negative) heat of formation assists separation from the polymer, such as is obvious for carbonate, ester, acid, or acetal groups. Aromatic and conjugated unsaturated systems stabilize towards photon absorption or energy transfer by collisions with excited or charged particles. Aliphatic chains tend to H-abstraction and formation of C radical sites, auto-oxidation, formation of olefinic double bonds, polymer degradation, and/or crosslinking. Other types of polymer depolymerize through recovery of the stable monomer structure. Monomer unit by monomer unit is split off successively. This process can proceed only at active chain ends (P_i*), for example, at living anionic polymers as:

$$P_i^* \rightarrow P_{i-1}^* + M$$

Dead polymers must first undergo a homolytic chain scission:

$$P_{i+j} \rightarrow P_i^\bullet + P_j^\bullet$$

The radical chain ends make the chain active for depolymerization:

$$P_i^\bullet \rightarrow P_{i-1}^\bullet + M, \text{and so on}$$

The zip length (Ξ) presents the kinetic chain length of depolymerization in analogy to the chain length of chain propagation during the polymerization. The zip length reflects the number of monomer units removed from the polymer chain before the radical is annihilated by chain transfer or termination. PMMA [poly(methyl methacrylate)] and PAMS [poly(α-methylstyrene)] are very easily depolymerized ($\Xi > 200$) and PS (polystyrene) and PIB [poly(isobutylene)] ($\Xi \approx 3$) with more moderate ease. For example, PE shows a zip length of only $\Xi = 0.01$.

Thus, both similarities and differences to the behavior of polymers under ionizing irradiation and plasma exposure were obvious. The response to ionizing radiation is classified in two general categories, depolymerizing or crosslinking behavior. Plasma exposure is much more complex. First, the energy dose consumed in the uppermost surface layer is at least six orders of magnitudes greater with the plasma than with γ-irradiation even though the energy of the primary species is much higher in the case of γ-irradiation (MeV) in relation to average plasma energies (in or below the 10 eV region). Nevertheless, the much higher dose in all secondary processes in the plasma compared to the same processes along the spur of γ-quanta in polymers dominates and produces more side and exotic products as well as the more complex reactions that are observed.

The main degradation mechanisms have been identified:

1) random degradation,
2) (random) degradation after preceding crosslinking (and formation of macrocycles),
3) depolymerization,
4) photo-oxidative degradation [49].

Mechanisms 1 and 2 are random processes characteristic for plasma-dominated reactions. In contrast, mechanisms 3 and 4 are well known from chemistry [50] and, therefore, are pure chemical processes. These photo-oxidation routes are well known and were characterized by carbonate decay, Norrish I and Norrish II-type chain scissions, or photo-Fries rearrangements, for example, leading to phenolic (aryl-OH) or benzoic acid (aryl-COOH) structures in PET or PC [51].

The dominance of pure chemical processes during the degradation of, for example, PET or PMMA, is caused by the higher rate of chemical degradation process compared to the plasma-specific (random degradation) processes. Plasma processes are based on atom and small-fragment formation and rearrangement as well as random recombination of them. Additionally, a purely physical phenomenon must be considered, namely, sputter etching. Here, only the transfer of charges as well as that of mechanical or thermal energy of the plasma particles impinging on the surface atoms produces the release of atoms, fragments, molecules, or molecule clusters from the polymer surface. The etching rate of PE with an oxygen plasma is about tenfold greater than with an argon plasma [52]. Self-bias and applied bias voltage support the sputter effect of ions.

Summarizing the above-mentioned polymer characteristics under plasma exposure, a few *degradation-sensitive building units* or functional groups exist that trigger the rapid etching of polymers as acid or ester [48], pyranose ring and acetal [53], and carbonate groups [54].

2.4
Methods, Systematic, and Definitions

Two different procedures can be used to modify polymer surfaces by plasma treatment: surface modification or surface functionalization and coating of surface (cf. Scheme 2.3).

2.4.1
Surface Modification (Functionalization)

The aim of surface modification of polymers is to introduce functional groups to the surface (*surface functionalization*). As described above, H-abstraction and attachment of plasma fragments to the C radical sites at the polymer surface produce the functionalization. By proper selection of the plasma gas its main element or one of its characteristic groups can be attached (recombined) to the polymer surface. Therefore, this process is called as *plasma-gas specific* functionalization of polymer surfaces. Table 2.1 lists a few examples of plasma gases/vapors and the functionalities they predominantly produce.

This type of polymer surface functionalization has the disadvantage that in nearly all cases a broad spectrum of different functional groups are formed containing the mother elements, which are provided by the plasma gas, most often connected with post-plasma introduced oxygen. A previously mentioned example is the oxygen-plasma treatment of polymers, [55] which produces an ensemble of O-functional groups from the oxygen plasma treatment itself and from auto-oxidation (Scheme 2.5) [15]. *Chemical transformation into monotype OH groups* was possible by post-plasma wet-chemical reduction with diborane or $LiAlH_4$ to about 60% (Scheme 2.5) [56].

Table 2.1 Gases or vapors used to produce plasma-gas-specific functional groups.

Plasma gas	Functional group
O_2, H_2O, H_2O_2	–OH
NH_3, N_2H_4, N_2	$-NH_2$
CO_2	–COOH
CS_2, H_2S	–SH
SO_2, SO_3	$-SO_3H$
CF_4, SF_6, XeF_2, NF_3, BF_3, SiF_4	–F
CCl_4	–Cl
Br_2, $HCBr_3$	–Br

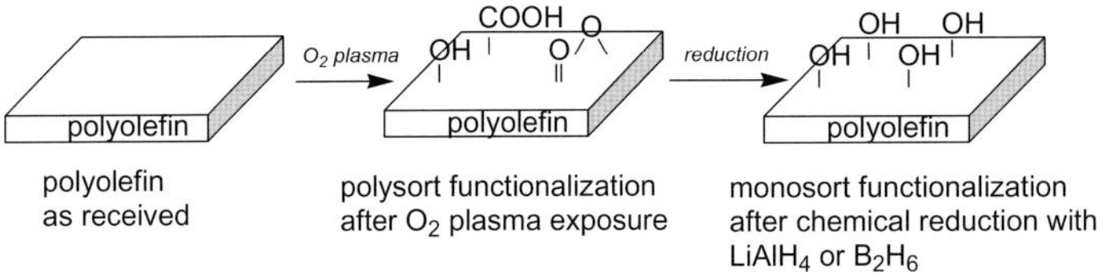

Scheme 2.5 Specific and unspecific functionalization.

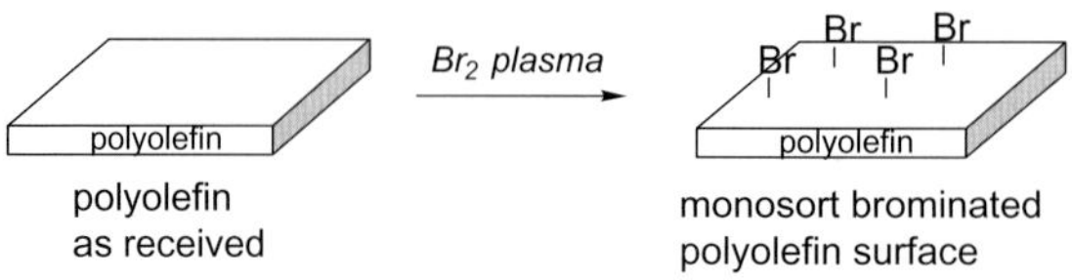

Scheme 2.6 Brominated polyolefin.

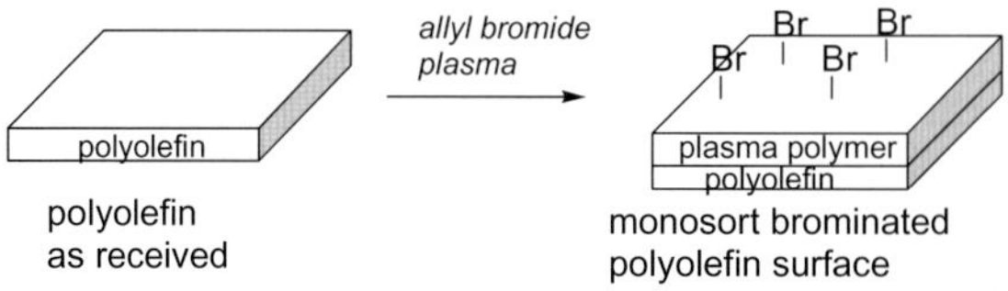

Scheme 2.7 Plasma-polymer coated polymer substrate with monofunctional groups (specific functionalization).

The special case of *monosort functional groups* was achieved with bromoform and bromine plasmas [57–60]. Here, nearly quantitatively, C–Br groups are formed (specific functionalization or monotype functionalization) (Scheme 2.6) [61].

2.4.2 Coating of Polymer Surfaces with Functional Group-Bearing Plasma Polymers

The deposition of very thin layers of *plasma polymers* (ca. 20–100 nm), which carry functional groups in their structure, is also used to functionalize solids (Scheme 2.7). A precondition is sufficient adhesion between the polymer, metal, or inorganic substrate and the deposited plasma polymer. Most often one finds very good adhesive bond strength between plasma polymers and polymer substrates caused by plasma exposure of the polymer substrate and the monomer molecules, which form the plasma polymer.

Two types of plasma polymerization have to be distinguished: plasma-chemical and pulsed-plasma.

2.4.2.1 Plasma-Chemical Polymerization

Plasma-chemical polymerization (of any organic evaporable molecule or a few inorganic molecules) occurs by partial or, under rigorous plasma conditions, total fragmentation of the organic molecules in the plasma and the subsequent deposition of a polymer-like layer by the more or less complete recombination of all these fragments (*fragmentation and random polyrecombination*):

$$R_1R_2R_3R_4R_5 + \text{plasma} \rightarrow R_1^\bullet + R_2^\bullet + R_3^\bullet + R_4^\bullet + R_5^\bullet \quad \text{fragmentation process}$$

$$R_1^\bullet + R_2^\bullet + R_3^\bullet + R_4^\bullet + R_5^\bullet \rightarrow R_5R_2R_3R_1R_4 \quad \text{random recombination.}$$

2.4.2.2 **Pulsed-Plasma Polymerization**

In contrast to a random process leading to plasma polymers with irregular structure and composition, a hybrid process exists that is positioned between such random plasma polymerization and classic radical polymerization. This pulsed plasma process produces *plasma-initiated chemical gas phase polymerization*. This process starts with plasma-chemical activation of classic (chemically polymerizable) monomers such as vinyl or acrylic monomers during the plasma-on periods. The radical chain propagation occurs mainly during the plasma-off periods, that is, the main contribution to polymer formation should be dominated by chemical chain propagation. Only a small fraction, produced during the plasma-on pulses, is produced by the fragmentation–polyrecombination process. This plasma-initiated radical gas-phase polymerization can only be applied to classic monomers. Notably, this process was applied very successfully to the allyl monomer polymerization, which is known to be strongly hindered by radical transfer processes [50]. If only one sort of classic monomer is introduced into the pulsed plasma the denotation is "plasma-initiated polymerization." This process produces *homopolymers*. When two or more (co)monomers are present in the pulsed plasma the resulting process is called *"plasma-chemically-initiated copolymerization"* (Figure 2.8).

A precondition for chemically dominated copolymerization is the use of classic co-monomers. Moreover, they have to possess adequate copolymerization tendencies, which are expressed in the "copolymerization parameters" [50, 62]. In contrast to this chemically-dominated copolymerization, the process of

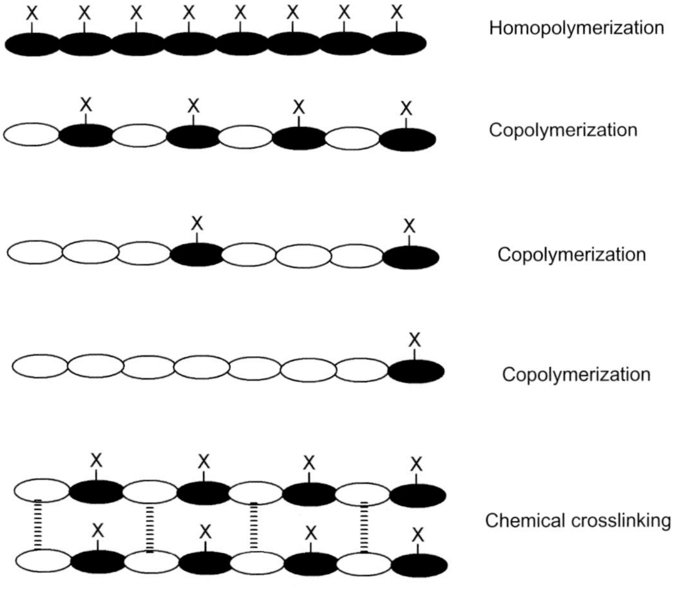

Figure 2.8 Homo- and (alternating) copolymerization.

plasma-chemical fragmentation of "co-monomers" in the plasma allows their complete dissociation and fragmentation to atoms and fragments, which can randomly recombine subsequently (fragmentation–polyrecombination process) independently of the original structure. An undefined randomly composed and structured "copolymer" is formed by this plasma-dominated process. Thus, chemically inert substances, such as saturated alkanes, benzene, halides, and so on, can be forced to "polymerize" or "copolymerize" in the plasma. In the past, this process was generally indicated as "plasma copolymerization," and is far from any chemical processes.

2.4.3 Other Polymer Process

2.4.3.1 Polymer Etching

Another important process with polymers is polymer etching, for example, caused by removal of material in chemical and/or physical ways. At the beginning of this process it is associated with surface functionalization, for example, characterized by O-containing groups when using the oxygen plasma. The unspecific surface functionalization is the preliminary step followed by materials ablation to form gaseous degradation products (Figure 2.9).

The etch front consists of the oxygen-functionalized polymer surface. Short-wavelength vacuum ultraviolet irradiation modifies additionally the surface-near polymer structure (pre-aged layer). Degradation products, formation of double bonds, and radicals characterize this modified layer.

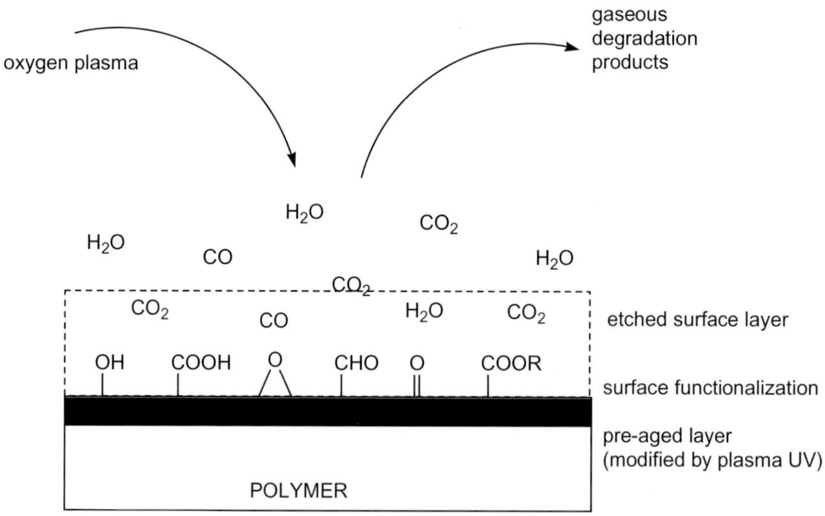

Figure 2.9 Scheme of plasma etching as a steady-state process consisting of permanent polymer surface functionalization and further oxidation to gaseous etching products.

2.4.3.2 Crosslinking

All plasma processing with polymers is additionally associated with crosslinking, which takes place simultaneously with other processes. These processes are located in a sub- or μm thick surface layer. The layer thickness varies strongly with the absorption properties of the polymer and the wavelength of plasma-emitted UV radiation (Figure 2.10). The dominance of crosslinking in degradation using chemically inert plasma gases has been used as a technical process to harden polymers at the surface. This process was entitled as CASING (crosslinking by activated species of inert gases) in 1967 and found a strong echo in the literature at that time [28].

However, the explanation of crosslinking by particle bombardment could not be verified. Later, Hudis showed that this crosslinking was caused by the plasma UV radiation [29]. As also shown later, crosslinking via peroxy (-O–O-) links is possible [15].

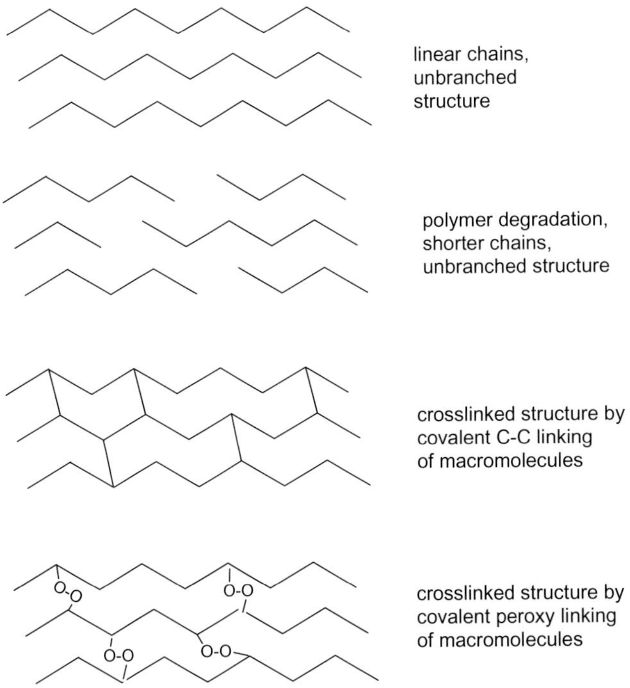

Figure 2.10 Schematic structures of original, degraded, and crosslinked polymers.

2.5 Functional Groups and Their Interaction with Other Solids

Polymer surfaces are poor in functional groups. The number of (chemically reactive) end-groups is limited because of the macroscopic length of a polymer chain.

The polymer chain and its building blocks are most often chemically inactive, as is obvious for polyolefins, siloxane, or perfluorinated polymers. In contrast, a few building blocks of macromolecules possess functional groups, such as poly(acrylic acid) and poly(vinyl alcohol).

The aim of any polymer surface activation by exposure to plasmas is to introduce (polar) functional groups by attachment of atoms or fragments of the plasma gas. As mentioned before the formation of bromine groups and subsequent hydrolysis to monotype OH groups has the advantage that all functional groups have the same reactivity because only OH groups are produced (Scheme 2.8) [63].

Another route was oxygen-plasma treatment followed by wet-chemical reduction to OH groups. This monosort functionalization is important if these groups are to be grafted chemically (cf. Scheme 2.5).

Moreover, using this surface functionalization the surface energy can be increased if polar and unshielded groups are introduced. Thus, surface dipoles are formed or hydrogen bonds become possible. In total, the interaction between a polar polymer surface and coatings such as metal layers can be enhanced, resulting in stronger adhesive bond strength (Scheme 2.9).

Dipole–dipole or acid–base interactions as well as hydrogen bonds are obviously not as efficient as regular chemical bonds (Table 2.2).

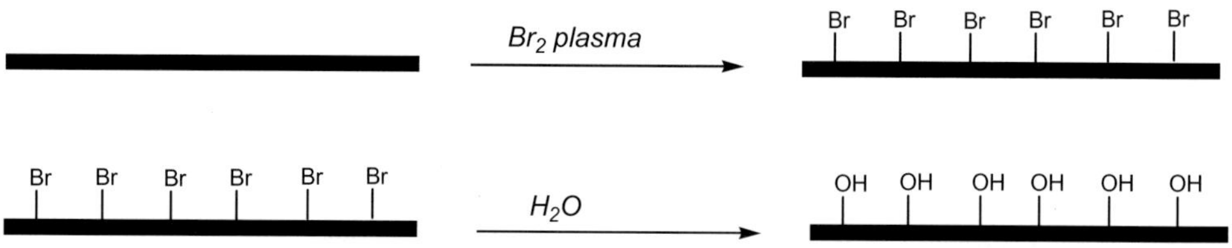

Scheme 2.8 Bromine plasma treatment of polyolefins and subsequent wet-chemical hydrolysis of Br groups.

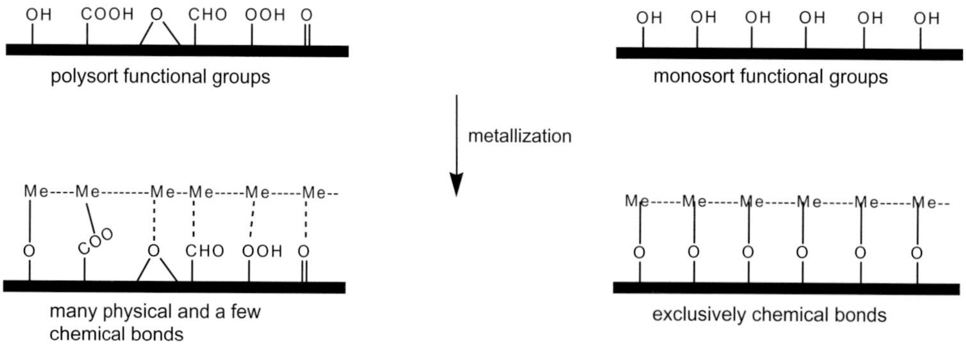

Scheme 2.9 Possible interactions of metals with oxygen functional groups produced by an oxygen plasma.

Table 2.2 Binding energies and bond length or operation range for physical forces and for chemical bonds.

	Binding energy (kJ mol^{-1})	Bond length (Å)
Physical forces		
van der Waals forces		
Dipole–dipole (Keesom)	20	3–5
Dipole-induced dipole (Debye)	0.08–40	3–5
Dispersion forces (Heitler–London)	2	3–5
Hydrogen bonds	50	3–5
Chemical bonds		
Ionic	560–1000	1–2
Covalent	100–680	1–2
Metallic	110–260	1–2

References

1 Yasuda, T., Okuno, T., Yoshida, K., and Yasuda, H. (1988) *J. Polym. Sci., B: Polym. Phys.*, **26**, 1781.

2 Garbassi, F., Morra, M., and Occhiello, E. (1998) *Polymer Surfaces–From Physics to Technology*, John Wiley & Sons, Ltd, Chichester.

3 Behnisch, J., Holländer, A., and Zimmermann, H. (1993) *Surf. Coat. Technol.*, **59**, 356.

4 Truica-Marasescu, F., Guimond, S., Jedrzejowski, P., and Wertheimer, M.R. (2005) *Nucl. Instrum. Methods Phys. Res. B*, **236**, 117–122.

5 Chilkoti, A. and Ratner, B.D. (1993) *Surface Characterization of Advanced Polymers* (eds L. Sabbatini and P.G. Zambonin), VCH, Weinheim, p. 221.

6 Everhart, D.S. and Reilley, C.H. (1981) *Surf. Interface Anal.*, **3**, 258.

7 Friedrich, J., Kühn, G., Mix, R., Retzko, I., Gerstung, V., Weidner, S., Schulze, R.-D., and Unger, W. (2003) *Polyimides and Other High Temperature Polymers: Synthesis, Characterization and Applications* (ed. K.L. Mittal), VSP, Utrecht, pp. 359–388.

8 Beckert, R., Fanghänel, E., Habicher, W.D., Metz, P., Pavel, D., and Schwetlick, K. (2004) *Organikum*, 22nd edn, Wiley-VCH Verlag GmbH, Weinheim.

9 Friedrich, J.F., Wettmarshausen, S., Hanelt, S., Mach, R., Mix, R., Zeynalov, E., and Meyer-Plath, A. (2010) *Carbon*, **48**, 3884–3894.

10 Schönburg, C. and Wick, G. (1934) Internal Report of IG Farben.

11 Jesch, K., Bloor, J.E., and Kronick, P.L. (1966) *J. Polym. Sci. A*, **1**, 1487.

12 Denaro, A.R., Owens, P.A., and Crawshaw, A. (1968) *Eur. Polym. J.*, **4**, 93.

13 Denaro, A.R., Owens, P.A., and Crawshaw, A. (1969) *Eur. Polym. J.*, **5**, 471.

14 Sabadil, H. (1966) *Beitr. Plasmaphys.*, **6**, 305.

15 Friedrich, J., Kühn, G., and Gähde, J. (1979) *Acta Polym.*, **30**, 470–477.

16 Young, T. (1805) *Philos. Trans. R. Soc. London*, January, 65.

17 Fowkes, F.M. (1964) *Ind. Eng. Chem.*, **56**, 40–52.

18 Owens, D.K. and Wendt, R.C. (1969) *J. Appl. Polym. Sci.*, **13**, 1741.

19 Rabel, W. (1971) *Farbe Lack*, **77**, 997–1009.

20 Kaelble, D.H. (1970) *Physical Chemistry of Adhesion*, Wiley-Interscience, New York.

21 Strobel, J.M., Strobel, M., Lyons, C.S., Dunatov, C., and Perron, S. (1991) *J. Adhesion Sci. Technol.*, **5**, 119–130.

22 Strobel, M., Corn, S., Lyons, C.S., and Korba, G.A. (1987) *J. Polym. Sci., Part A: Polym. Chem.*, **25**, 1.
23 Bikerman, J.J. (1968) *The Science of Adhesive Joints*, 2nd edn, Academic Press, New York.
24 Wu, S. (1982) *Polymer Interface and Adhesion*, Marcel Dekker, New York.
25 Friedrich, J., Loeschcke, I., Frommelt, H., Reiner, H.-D., Zimmermann, H., and Lutgen, P. (1991) *J. Polym. Degrad. Stabil.*, **31**, 97–119.
26 Friedrich, J., Gähde, J., and Pohl, M. (1980) *Acta Polym.*, **31**, 312–315.
27 Lipatow, J.B., Bezruk, L.I., Lebedew, E.V., and Gomza, J.P. (1974) *Vysokomol. Soedin.*, **5**, 328.
28 Hansen, R.H. and Schonhorn, H. (1966) *J. Polym. Sci. B*, **4**, 203.
29 Hudis, M.V. (1972) *J. Appl. Polym. Sci.*, **16**, 2397.
30 Clark, D.T. and Dilks, A. (1980) *J. Polym. Sci.: Polym. Chem. Ed.*, **18**, 1233.
31 Weidner, S., Kühn, G., Friedrich, J., Unger, W., and Lippitz, A. (1996) *Rapid Commun. Mass Spectrom.*, **10**, 727–737.
32 Friedrich, J., Unger, W., Lippitz, A., Koprinarov, I., Ghode, A., Geng, S., and Kühn, G. (2003) *Composite Interface*, **10**, 139–172.
33 Inagaki, N. (1997) *Plasma Surface Modification and Plasma Polymerization*, Technomic, Basel.
34 Krüger, S., Schulze, R.-D., Brademann-Jock, K., and Friedrich, J. (2006) *Surf. Coat. Technol.*, **201**, 543–552.
35 Kuzuya, M., Kondo, S., Sugito, M., and Yamashiro, T. (1998) *Macromolecules*, **31**, 3230–3234.
36 Berezin, I.V. and Denisov, E.T. (1996) *The Oxidation of Cyclohexane*, Pergamon Press, New York.
37 Kang, E.T. and Zhang, Y. (2000) *Adv. Mater.*, **12**, 1481.
38 Friedrich, J., Unger, W., Lippitz, A., Gross, T., Rohrer, P., Saur, W., Erdmann, J., and Gorsler, H.-V. (1996) *Polymer Surface Modification: Relevance to Adhesion* (ed. K.L. Mittal), VSP, NL-Utrecht, pp. 49–72.
39 Friedrich, J., Ivanova-Mumeva, V.G., Gähde, J., Andreevskaya, G.D., and Loeschcke, I. (1983) *Acta Polym.*, **34**, 171–177.
40 Friedrich, J. (1980) *Chem. Technol.*, **32**, 393–403.
41 Friedrich, J., Gähde, J., and Nather, A. (1981) *Plaste Kautsch.*, **28**, 420–425.
42 Stiller, W. and Friedrich, J. (1981) *Z. Chem.*, **21**, 91–118.
43 Friedrich, J. and Gähde, J. (1980) DD-WP 224 824. Oct. 30, 1980.
44 Friedrich, J. (1986) *Wissenschaft Fortschritt*, **36**, 311–319.
45 Friedrich, J., Loeschcke, I., Reiner, H.-D., Frommelt, H., Raubach, H., Zimmermann, H., Elsner, T., Thiele, L., Hammer, L., and Merker, E. (1990) *Int. J. Mater. Sci.*, **13**, 43–56.
46 Stille, J.K., Sung, R.L., and van der Kooi, J. (1965) *J. Org. Chem.*, **30**, 3116.
47 Schnabel, W. (1981) *Polymer Degradation*, Hanser International, Munich.
48 Yasuda, H., Lamaze, C.E., and Sakaoku, K. (1973) *J. Appl. Polym. Sci.*, **17**, 137.
49 Friedrich, J.F., Unger, W.E.S., Lippitz, A., Koprinarov, I., Weidner, S., Kühn, G., and Vogel, L. (1998) *Metallized Plastics 5 & 6: Fundamental and Applied Aspects* (ed. K.L. Mittal), VSP, Utrecht, p. 271.
50 Elias, H.-G. (1997) *An Introduction to Polymer Science*, Wiley-VCH Verlag GmbH, Weinheim.
51 Rabek, J. (1995) *Polymer Photodegradation*, Chapman & Hall, London.
52 Friedrich, J., Gähde, J., Frommelt, H., and Wittrich, H. (1976) *Faserforsch. Textiltechn./Z. Polymerenforsch.*, **27**, 604–608.
53 Friedrich, J., Throl, U., Gähde, J., and Schierhorn, E. (1982) *Acta Polym.*, **33**, 405–410.
54 Lippitz, A., Koprinarov, I., Friedrich, J.F., Unger, W.E.S., Weiss, K., and Wöll, C. (1996) *Polymer*, **37**, 3157.
55 Rossmann, K. (1956) *J. Polym. Sci.*, **19**, 141.
56 Kühn, G., Weidner, S., Decker, R., Ghode, A., and Friedrich, J. (1999) *Surf. Coat. Technol.*, **116–119**, 796–801.
57 Friedrich, J., Loeschcke, I., and Lutgen, P. (1990) *Proceeding in Adhesion and Surface Analysis, Loughborough April 1990* (ed. D.M. Brewis), The Adhesion Society, pp. 125–127.
58 Friedrich, J. (1991) Plasma modification of polymers, in *Polymer-Solid Interfaces*

(eds J.J. Pireaux, P. Bertrand, and J.L. Bredas), Institute of Physics Publishing, Bristol, pp. 443–454.
59 Wettmarshausen, S., Mittmann, H.-U., Kühn, G., Hidde, G., and Friedrich, J.F. (2007) *Plasma Process. Polym.*, **4**, 832.
60 Kiss, E., Samu, J., Toth, A., and Bertoti, I. (1996) *Langmuir*, **12**, 1651.
61 Friedrich, J., Wettmarshausen, S., and Hennecke, M. (2009) *Surf. Coat. Technol.*, **203**, 3647–3655.
62 Braun, D. and Hu, F. (2006) *Progr. Polym. Sci.*, **31**, 239.
63 Friedrich, J.F., Mix, R., Schulze, R.-D., Meyer-Plath, A., Joshi, R., and Wettmarshausen, S. (2008) *Plasma Process. Polym.*, **5**, 407–423.

3
Plasma

3.1
Plasma State

Plasma ("figure," "structure") is defined in physics as a gas in the (partial) ionized state. The term plasma was introduced by Langmuir in 1928 [1, 2]. A plasma consists of electrons, ions, energy-rich neutrals, molecules, fragments, atoms, and photons. It exists under low, atmospheric, and high pressure. The significant deviations of the properties of plasma from those of gases led to the postulation that plasma is the "fourth state of matter" [3]. The existence of charge carriers as electrons and ions produces conductivity. Collisions of charge carriers are the reason for the luminous glow. Most plasma types are produced with an electrical current. Consequently, they possess a nearly infinite source of continuously provided energy (enthalpy) because of the continuous supply of electrical energy. The most important properties of plasmas are:

- increased content of charge carriers (electrons and ions);
- high content of high energetic neutrals (atoms, radicals, complexes);
- electrically conducting;
- glowing;
- form a magnetic field;
- "quasi-neutrality" in macroscopic dimensions (electrons = ions);
- microscopic dimension space charging;
- stationary plasma state possesses high energy with respect to enthalpy (energy is continuously supplied by electrical current, high dose rates);
- stationarity between formation and quenching of charge carriers (ion avalanche ↔ recombination);
- velocity and energy distributions;
- emission of high-energetic vacuum-UV irradiation (2–200 nm);
- the differences between isothermal ("hot," high-pressure) and non-isothermal (glow) plasmas (at low pressures).

The interpretation and discussion of plasma physics is not the focus of this book. Therefore, only important correlations between plasma physical properties and chemical or polymer chemical aspects are briefly discussed. In this context, all

The Plasma Chemistry of Polymer Surfaces: Advanced Techniques for Surface Design, First Edition.
Jörg Friedrich.

discussed plasma-physical facts are highlighted under the aspect of polymer modification and plasma polymer deposition. Thus, this book is focused on plasmas that are well-suited to process polymers. Most work in the context of polymers is reported on the low-pressure glow discharge plasma, which is considered in this book. Atmospheric-pressure plasmas have become of increasing interest in recent years because of their high technical importance and compatibility; therefore, a special section is dedicated to atmospheric plasmas (Section 3.5).

Generally, low-pressure plasmas are associated with a low density of atoms in the gas phase and therefore with low collision rates among the particles. They are characterized by non-isothermal (sometimes also: non-thermal or "cold") behavior of the diverse subsystems in the plasma: electrons, ions, energy-rich neutrals, gas molecules or atoms, and photons. This non-isothermal behavior is characterized by a mismatch between the kinetic energies of electrons and photons on the one hand and those of the heavy particles such as ions, molecules, and atoms on the other hand. The difference is caused by the strongly differing masses of electrons compared with those of ions and atoms. Electrons are not able to transfer efficiently their kinetic energy to heavy particles by elastic collisions caused by the energy and impulse conservation law. In contrast, heavy particles in the plasma can transfer their kinetic energies to another heavy particle or to the wall of the plasma recipient with high efficiency. Consequently, heavy particles easily distribute uniformly their energy among themselves and then transfer this energy to the walls of the plasma recipient. Thus, the subsystems of ions, atoms, and molecules are cooled to temperatures not far from room temperature. However, the electrons distribute their kinetic energy among themselves following a Maxwellian or related distribution. They are not able to transfer energy to the heavy particle subsystems. The reason is, again, the difference in mass because energy cannot be efficiently transferred from electrons to heavy particles. The energy remains in the subsystem of electrons; new energy is absorbed from the applied electrical field. Thus, the electrons become "heated" to kinetic energies of several ten thousands Kelvin [or several electron volts (eV)]. Following kinetic gas theory the kinetic energy (E_{kin}) can be expressed by their base equation $E_{kin} = m/2v^2 = 1.5kT$. Thus, the kinetic energy can be transformed into an equivalent "temperature:" $kT = 1\,\text{eV} = 11\,600\,\text{K}$ considering the most probable velocity (v_p) [4] or referencing it to the average particle energy $1\,\text{eV} \equiv 7733\,\text{K}$ [5]. Therefore, the plasma temperature is commonly measured in Kelvin (K) or electronvolt (eV) and is (roughly speaking) a measure of the thermal kinetic energy per particle. In most cases the electrons are close enough to thermal equilibrium and, therefore, their temperature is relatively well-defined, even when it significantly deviates from a Maxwell energy distribution function, for example, due to UV radiation, energetic particles, or strong electric fields. Because of the large difference in mass, the electrons come to thermodynamic equilibrium among themselves much faster than they come into equilibrium with the ions or neutral atoms. For this reason the ion temperature may be very different from (usually lower than) the electron temperature. This is especially common in weakly ionized technological plasmas, where the ions are often at near ambient temperature. For such non-isothermal low-pressure glow discharge plasmas the following succession of temperatures (or kinetic energies) is representative: $kT_e >> kT_{ion} \approx kT_{gas}$ [6]. Typical temperatures are $T_e = 10^4$–10^5 K for electrons and $T_{gas} \approx T_{ions} = 300$–$400$ K.

The Maxwell distribution function $F(v) = dn/dv$ can be expressed as:

$$F(v) = 4nv^2/\pi^{-1/2}\,(m/2kT)^{-3/2}\exp[-mv^2/2kT]$$

The average of velocity (v_{aver}) is given by $v_{aver} = (8kT/\pi m)^{-1/2}$; the most likely velocity (v_{ml}) is $v_{ml} = (2kT/m)^{-1/2}$. The occupation of excited states follows the Boltzmann distribution $n_{ij}/n_i = g_{ij}/g_i \exp(-E_{ij}/kT)$ [7]. Figure 3.1 shows the calculated Maxwell–Boltzmann energy distribution function for electrons with a density of $10^{16}\,m^{-3}$ and temperature of 3 eV as an example.

The degree of ionization of a plasma is the proportion of atoms that have lost (or gained) electrons, and is controlled mostly by the temperature. Even a partially ionized gas in which as little as 1% of the particles are ionized can have the characteristics of a plasma (i.e., respond to magnetic fields and be highly electrically conductive). The degree of ionization (α) is defined as $\alpha = n_i/(n_i + n_a)$, where n_i and n_a are the number density of ions and neutral atoms, respectively. The electron density is related to this by the average charge state $\langle Z \rangle$ of the ions through $n_e = \langle Z \rangle n_i$, where n_e is the number density of electrons.

The common diffusion of charge carriers, for example, electrons and ions, known as ambipolar diffusion, is characterized by the higher mobility of electrons in comparison to the ions due to the lower mass of electrons. However, the removal of electrons from the plasma cloud causes a net positive charge, which will act to hold the electrons back. The more electrons are moved away the larger the remaining positive charges and the stronger the thus formed positive electrical field. Consequently, the electrons and ions tend to stay together and to move or diffuse together. The conclusion is that the plasma does not allow a charge separation in larger volumes and therefore the plasma is quasi-neutral. This is valid for

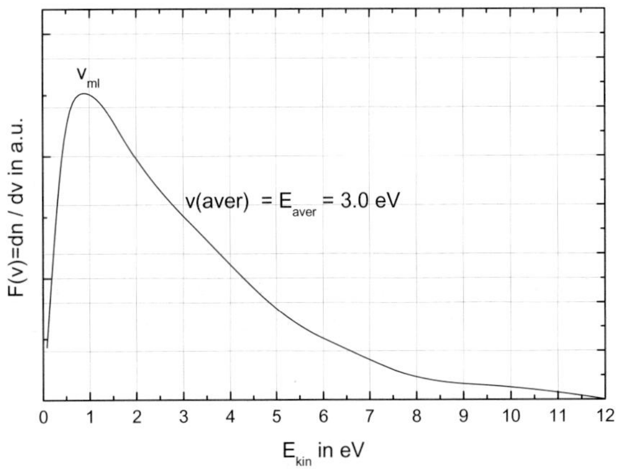

Figure 3.1 Calculated velocity distribution of electrons in a plasma based on the Maxwell–Boltzmann distribution.

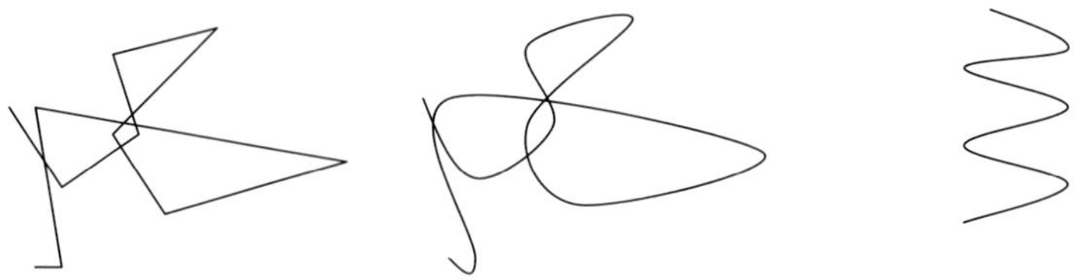

Figure 3.2 Curved orbits of charge carriers in a plasma as a consequence of Coulomb interactions.

plasmas far from electrodes and walls of the reactor such as the positive column in direct-current (DC) glow discharges or in radio-frequency (rf) plasmas.

One characteristic feature of any plasma is the occurrence of the Maxwell velocity distribution for all subsystems in the plasma. The energy distribution in each system arises by elastic collisions. For charge carriers distinct differences exist. Energy transfer via elastic collisions goes into the background and the energy transfer via electrostatic fields (Coulomb fields) comes to the fore. Since Coulomb forces have a wider range than van der Waals forces, the path of charge carriers changes from linear to curved (Figure 3.2).

Helical orbits of charge carriers are found in an additional magnetic field, such as that used in electron-cyclotron resonance (ECR) plasma sources.

An electrical micro-field is established around such a charge carrier [8]. Statistical calculation shows the correlation $E = sen^{3/2}$, where s is a constant depending on the charge configuration, and ranges between 10 and 20, and n is the density of charge carriers.

Because of the much higher mass the ions can be assumed to be static while the electrons can easily move in the electrical field, with a drift velocity 100-times higher than that of ions [9], around the ions and therefore they produce a potential Φ. The electron's response to the potential will depend on their kinetic energy. The Coulomb potential (Φ_C) is superposed by the space charge potential (Φ_{sc}) so the effective potential is $\Phi = \Phi_C + \Phi_{sc}$. For a plasma exhibiting the Boltzmann equilibrium the electron temperature can be introduced $e_0 V << kT_e$. If the Boltzmann equilibrium is valid [8], the electron energy (n_e), expressed in terms of electron energy (T_e), at the perturbation is given by $n_e = n_\infty \exp\,(e\Phi/kT_e)$, where n_∞ is the undisturbed density far from the negative charge perturbation. The plasma shields the potential formed by the perturbation so that at a distance x away the potential is:

$$\Phi = -\Phi_0 \exp(-x/\lambda_d)$$

where λ_d is the Debye length and is given by $\lambda_d = (\varepsilon_0 kT_e)^{-1/2}\ (e^2 n_e)$, in which ε_0 is the permittivity of free space. A more practical approach is to use $\lambda_d = 7400\ (kT_e/n_e)^{-1/2}$.

After any perturbation of the plasma system a relaxation length or relaxation time is needed to return to the state of equilibrium. The relaxation times for electron–electron, ion–ion, electron–neutral, and electron–ion interactions range from 10^{-8} to 10^{-6} s.

Temporary fluctuations of the charge carrier density will set up electrical fields in the plasma. The electrons will move first in response to the field but they will then be returned to the original condition by the newly formed opposite field. Thus, oscillations are set up in the plasma with a typical plasma frequency that is given by the Langmuir equation $\omega^2 = e^2 n_e / \varepsilon_0 m_e$, where ω is the plasma frequency, m_e is the electron mass, and n_e is the electron density.

A plasma is ignited by applying an electrical field and the existing charge carriers are accelerated. In the process of the ion avalanche the number of charge carriers increases exponentially on considering the first Townsend coefficient (α_T), that is, the number of ions and electrons formed during 1 cm electron drift in the electrical field. At ignition of the discharge the number of charger carriers (I) depends exponentially on the Townsend coefficient: $I = I_0 \exp(\alpha_T d)$, where I_0 is the starting concentration and d is the electrode distance [8]. This chain reaction is limited by the neutralization of charge carriers at the electrodes or at the walls of plasma. Plasma structure in contact with the wall is determined by the appearance of the plasma edge sheath. If the wall has a floating potential and consists of an isolating material such as polymers, which are the focus of this work, the plasma edge sheath plays an important role. The electrons and ions diffuse ambipolar to the walls. However, electron diffusion is much greater than that of ions. The plasma self-balances these separations. The electron depleted and therefore positively charged sheath is concentrated near the solid. The isolating solid surface is negatively charged by the flux of electrons. This sheath is in the range of millimeters, is not quasi-neutral, and shows a potential gradient. It reflects electrons to the bulk of plasma and accelerates ions to the surface up to the establishment of a steady-state. This plasma edge sheath is often visible as a glow. The positively charged sheath in contact with the solid induces the plasma potential. Electrically isolating walls such as polymers exposed to the plasma will attain a negative potential relative to the plasma, thus balancing the ion and electron flux. This potential is known as the floating potential. Figure 3.3 depicts the course of electron and positively charged ion density in contact with the plasma and formation of the plasma edge sheath. This sheath also determines the transport phenomena of charges, mass, and energies in the plasma. Otherwise, the energetic neutrals, such as excited atoms or molecules, radicals, complexes, and so on are not influenced by this electrically charged sheath.

The plasma loses charge carriers permanently by homogeneous recombination ($A^+ + e^- \rightarrow A^*$) or by diffusion and neutralization at the walls. Further losses are heat conduction to the walls and convective and radiative deficits. The charge carriers migrate in the direction of the applied electrical field, thus producing anisotropic plasma. A steady-state between loss and production of charge carriers is established. The charge carriers move in the field with an average drift velocity, which is caused by the electrical field. Moreover, this drift is superimposed on the ordinary diffusion in the gas phase (and charging of walls, bias voltage). Thus, the diffusion depends on the type of gas, its diffusion coefficient, and the density

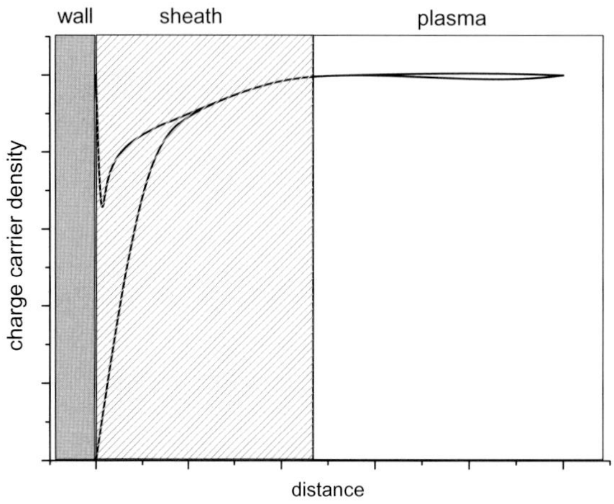

Figure 3.3 Schematic view of the plasma edge sheath during contact with isolating polymer surfaces.

gradient. However, the diffusion in the plasma is also influenced by the existence of charge carriers and its quasi-neutrality. The agile electrons hurry ahead of the ions and produce electrical fields and so on as described for the ambipolar diffusion.

The velocity of electrons and therefore their kinetic energy in the electrical field under low-pressure conditions is so high that electron impacts with atoms or molecules produce ionization and the release of a secondary electron by inelastic collisions: $A + e \rightarrow A^+ e' + e''$. This is the basic process for the initiation of the ion avalanche at the ignition of plasma and for sustaining the plasma. Such collisions, which lead to transfer of the kinetic energy of electrons by changing the inner energy of the impinged atoms ($E_{kin} \rightarrow E_{pot}$), are called inelastic collisions. The ionization of atoms or molecules in the plasma phase is caused by such inelastic collisions. The inelastic electron impacts begin with kinetic energies of a few electron volts. The main fraction of kinetic energy is transferred to the inner energy of the heavy particle. Such energy absorption occurs in discrete levels. As explained before, the transfer of additional kinetic energy is not possible. This process is characterized by an energy balance that follows the equation $m/2\,v^2 = eV_i$, where V_i is the ionization potential; dn/n electrons produce along their path x subsequent ionizations: $dn/n = \alpha dx$, where α is the ionization coefficient. Electron impacts followed by twofold or multiple ionization, ionization by fast heavy particles, or photo-ionization in the gas phase (limited by selection criteria) seldom occur in the plasma. Often, the ionization of molecules is associated with their dissociation (dissociative ionization):

$$AB + e \rightarrow A^+ + B + e' + e''$$

Ionization can be also follow by "superelastic" impacts with excited atoms (collisions of second order):

$$A^* + B \rightarrow A + B^+ + e$$

This process is also known as Penning ionization and is carried out using rare gas plasmas and an admixture of organic substances [10]. Superelastic impacts can also increase the kinetic energy of the impinged electron:

$$A^* + e \rightarrow A + e'$$

Other processes caused by inelastic collisions are dissociations and excitations:

$$AB + e \rightarrow A + B + e'$$

The electron energy necessary for excitation is lower than for ionization:

$$A + e \rightarrow A^* + e'$$

In addition, ions can also be excited by electron impact:

$$A^+ + e \rightarrow A^{+*} + e'$$

Another inelastic electron impact is the ion–molecule reaction:

$$A^+ + A \rightarrow AA^+$$

or the re-charging:

$$A^+ + B \rightarrow A + B^+ + \Delta E$$

The excited state A* can deactivate radiatively:

$$A^* \rightarrow A + h\nu$$

Such emission is not possible in every case. There are forbidden transitions to the ground state:

$$A^* \rightarrow A^{met} + \Delta E$$

Here, intersystem crossing (ISC) occurs and transition to the triplet system under the formation of a metastable species (Jablonski scheme) [11]:

$$A^{met} \neq A^0 + h\nu$$

However, a time-delayed transition, known as phosphorescence, is observed. Deactivation of metastable excited species demands further collisions with atoms or molecules. Thus, metastable excited species can also undergo Penning ionization or triplet–triplet annihilation:

$$A^{met} + A^{met} \rightarrow A^* + A \rightarrow 2A + h\nu$$

The radiative processes, starting from excited states, follow for all organic substances (and polymers) a general principle that can be summarized in a Jablonski scheme. In such a scheme IC denotes the internal conversion and signifies thermal equilibration while ISC stands for intersystem crossing and signifies electron spin decoupling and the selection rule-forbidden transition to the triplet system in one of the T_x levels. The transition from T_1 to the ground state S_0 is also forbidden but occurs nevertheless as time-delayed phosphorescence.

The types of plasma-emitted radiation and their relevance for polymers are discussed in Section 3.2. Here, some basic facts are mentioned. There are discrete bond–bond related radiative transitions that occur as line and/or resonance radiation or free-bond and free-free transition related transitions (bremsstrahlung and recombination radiation) as continuums radiation.

The self-absorption of the plasma is a further fact that must be mentioned. However, it is only important for optically thicker plasmas. Self-absorption means that the emitted lines are absorbed and, subsequently, excitation occurs with emission of the same line. Consequently, the light diffuses to the outer sphere of the plasma glow.

Recombination processes occur as charge recombination of electrons with positively charged ions:

$$A^+ + e \rightarrow A + h\nu$$

However, direct recombination processes must be able to distribute the high energy involved. Consequently, only large molecules with a high number of degrees of freedom can recombine directly:

$$ABC^+ + e \rightarrow ABC^*$$

Usually, a three-body process is necessary to distribute the energy:

$$A^+ + e + M \rightarrow A + M^*$$

The easiest case is energy transfer to the walls ("M") of the plasmatron.

Another related process is dissociative recombination:

$$AB^+ + e \rightarrow (AB)^* \rightarrow A^* + B$$

Because the same quantities of negatively and positively charged particles recombine the recombination rate can be formulated as:

$$-dn^+/dt = -dn^-/dt = \beta n^+ n^- = \beta n^2$$

where β is the recombination coefficient.

It should be added that in plasmas of electronegative atoms such as oxygen or fluorine high quantities of negatively charged ions exist. Those ions can also recombine with positively charged ions:

$$A^+ + B^- \rightarrow A^* + B$$

or:

$$A^+ + B^- + M \rightarrow A + B + M^*$$

As mentioned above, the formation of negative ions is focused on halogens, oxygen, hydrogen, and so on [12, 13]:

$$A + e \rightarrow A^- + h\nu$$

Negative ions are also formed by dissociative electron capture, but this is associated with dissociation of the molecule:

$$AB + e \rightarrow A + B^-$$

An alternative route is the three-body impact:

$$A + e + M \rightarrow A^- + M^*$$

Only at high electron energies (>>10 eV) is ion-pair formation observed:

$$AB + e \rightarrow A^+ + B^-$$

Most of electronegative ions are unstable (life time ≈ 10^{-10} s). Benzene also forms negative ions, which decay within 10^{-12} s, but a fraction of energy remains in the molecule as resonance energy and thus a highly vibration-excited, and therefore reactive, benzene molecule is formed [14, 15].

Thus, elastic collisions are responsible for the distribution of energy within the subsystems of electrons and heavy particles, whereas inelastic collisions provoke chemical reactions. In addition, an enormous fraction of introduced electrical energy in the low-pressure glow discharge is consumed by radiation processes [16]. Indeed, roughly 70% of the energy is consumed in the emission of vacuum UV irradiation (resonance radiation), 10% in visible light, 15% is consumed at the walls, and 5% in the volume for mercury discharge at 10 Pa [17–19]. By using low current and low pressure in the glow discharge, secondary processes do not play an important role and the quantum emission is proportional to the number of excitation processes [20, 21]. Thus, the emission intensity of spectral lines I_v is: $I_v = \alpha n h \nu$, where n the concentration of atoms and α is a coefficient for the number of excitations depending on excitation potential, cross-section, and the electron energy distribution function [19].

The electron temperature and density are dependent on the plasma parameters. Among these parameters, the pressure plays an important role. The collision rate becomes smaller with falling pressure (p) and increasing mean-free path (λ_{mfp}):

$$\lambda_{mfp} = kT/2^{-1/2}\,\pi^{-1}\,d^{-2}\,p^{-1}$$

The collision rate (υ) depends on the type of gas and also on the cross-section (σ). Using τ as the time between two collisions gives $\upsilon = 1/\tau = \bar{u}/\lambda$. After introduction of the collision cross-section $\sigma = \pi d^2$ the collision rate is $\upsilon = \sigma n \bar{u} = \pi d^2 n \bar{u}$, where n is the number of molecules per cm^3. The cross-section of most plasma gases for ionization is maximal between 100 and 200 eV [22] and excitation ranges from 10 to 20 eV [23]. For the mean-free path of gas molecules in the plasma, the introduction of a factor $2^{-1/2}$ allows us to write the equation $\lambda = 2^{-1/2}\pi^{-1}d^{-2}n$. Table 3.1 shows the mean-free path of different gases [24].

The electron density (n_e) (and the current density i in the positive column of a DC low-pressure glow discharge) increases with pressure (p) (and the current, I). The electron density varies from 10^3 to $10^{12}\,cm^{-3}$. The ionization degree is very low in such a plasma and ranges roughly from 10^{-10} to 10^{-4} [19]. The very low

Table 3.1 Mean-free path of atoms at different pressures.

Gas	Ionizing potential (V)	Constant (c)	Diameter (d) (cm)	Mean-free path (cm) at 271 K	
				Atmospheric pressure	Low pressure (133 Pa)
He	24.6	0.004	2.18×10^{-8}	1.72×10^{-5}	13.7×10^{-3}
Ne	21.5	0.006	2.58×10^{-8}	1.25×10^{-5}	9.56×10^{-3}
Ar	15.8	0.040	3.66×10^{-8}	0.63×10^{-5}	4.82×10^{-3}
Kr	14.0	–	4.14×10^{-8}	0.49×10^{-5}	3.70×10^{-3}
Xe	12.1	–	4.88×10^{-8}	0.35×10^{-5}	2.66×10^{-3}
H_2	13.5 (H), 15.4 (H_2)	0.010	2.72×10^{-8}	1.13×10^{-5}	8.54×10^{-3}
N_2	14.5 (N), 15.8 (N_2)	0.040	3.78×10^{-8}	0.58×10^{-5}	4.56×10^{-3}
O_2	13.6 (O), 12.5 (O_2)	0.029	3.62×10^{-8}	0.64×10^{-5}	4.92×10^{-3}

charge carrier concentration decreases strongly the role of volume recombination processes by inelastic collisions. Therefore, most charge carriers migrate to the walls of the discharge tube by ambipolar diffusion and recombine there. Thus, the smaller the diameter of the discharge tube the higher is the field strength (X) and, therefore, the electron temperature. Within the discharge tube the electron temperature is much higher around the axis than in the vicinity of the sheath in front of the walls.

The theory of the DC low-pressure positive column was established by Schottky, considering the ambipolar diffusion to the wall, the carrier balance, and the electron drift, with the existence of quasi-neutrality and constant electron temperatures. Under these presumptions such a radial electron density profile follows a Bessel function J_0: $n(r) = n_0 J_0(2.405r/R)$, where n_0 the electron density in the axis (Figure 3.4).

There have been attempts to reduce the widely used radio-frequency (rf) E_{rf} discharge to the equivalent DC discharge E_{dc}, postulating Maxwell distribution and considering the collision frequency (ν) and the rf-frequency (ω): $E_{dc}{}^2 = E_{rf}{}^2 \nu^2/2$ $(\nu^2 \omega^2)$. In rf discharges is $\nu^2 >> \omega^2$, thus $E_{dc} = E_{rf}/2$. Thus, the Schottky theory is also valid for rf discharges: $(ev_i/kT_e)^{-1/2} \times \exp(-ev_i/kT_e) = 1.16 \times 10^7 \times c^2 p^2 R^2$, where R is the tube diameter, p the pressure, and c a gas-specific constant considering the atom dimensions and mobility (cf. Table 3.1 and Section 3.4). From this equation the electron temperature as function of pressure can be derived (Figure 3.5).

According to its very high ionization potential and smallest atom diameter among the noble gases, the helium plasma shows the highest energy level of all gas plasmas, followed by the neon plasma. According to the Schottky equation the hydrogen plasma also has a relatively high electron energy level due to the very small diameters of protons, hydrogen atoms, and molecules (cf. Section 3.4). These plasmas also emit energy-rich vacuum-UV radiation in the range 2–200 nm.

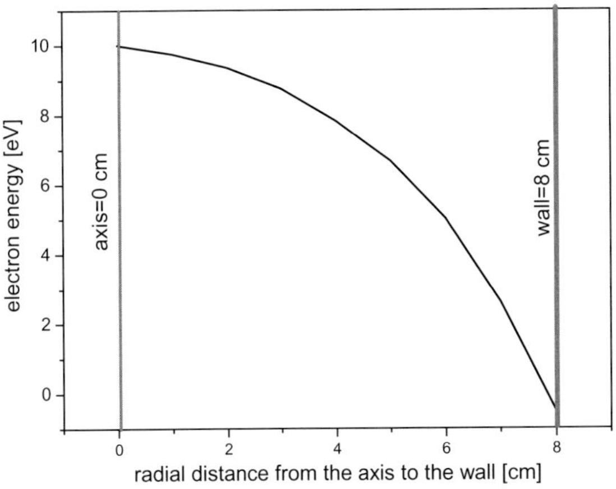

Figure 3.4 Radial dependence of electron temperature in a DC glow discharge tube.

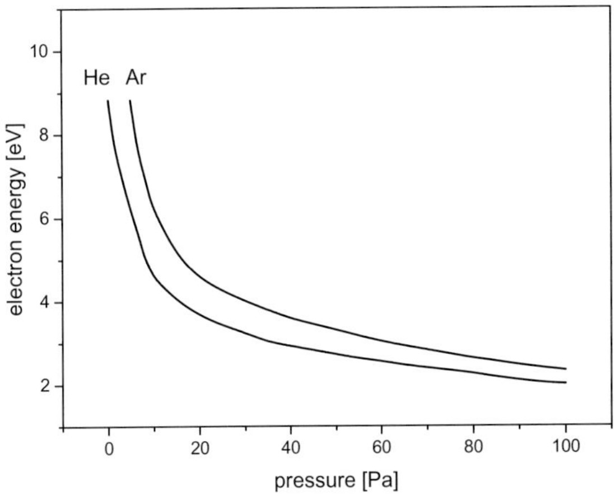

Figure 3.5 Dependence of electron temperature on the pressure of a DC glow discharge.

Table 3.2 summarizes the elementary collision processes in low-pressure plasma.

3.2
Types of Low-Pressure Glow Discharges

Glow discharges can be produced in the pressure range of about 1 to a few 100 Pa by high voltage flow of current with different frequencies. DC-produced discharges

Table 3.2 Inelastic collision processes in the plasma.

	Elementary process	Name given to process
1	$e^- + A \Rightarrow A^+ + e^-_1 + e^-_2$	Ionization
2	$e^- + A \Rightarrow A^* + e^-_1$	Excitation
3	$A^* + B \Rightarrow A + B^+ + e^-_1$	Penning ionization
4	$A^* \Rightarrow A + h\nu$	Radiative deactivation
5	$A^+ + B \Rightarrow A + B^+$	Recharging
6	$A^+ + e^- + M \Rightarrow A + M^*$	Recombination by three-body collision
7	$A^+ + e^- \Rightarrow A + h\nu$	Radiative recombination
8	$A^+ + B \Rightarrow AB^+$	Ion–molecule reaction
9	$A^* + B \Rightarrow AB^+ + e^-$	Hornbeck–Molnar process
10	$ABCD^+ + e^- \Rightarrow ABCD^*$	Recombination with internal excitation of vibrations
11	$AB^+ + e^- \Rightarrow A^* + B$	Dissociative recombination
12	$A^+ + B^- + M \Rightarrow AB + M^*$ or $\Rightarrow A + B + M^*$	Recombination of ions
13	$A + e^-$ (+M) $\Rightarrow A^-$ (+ $h\nu$, M^*)	Formation of negative ions
14	$AB + e^- \Rightarrow A^+ + B^- + e^-_1$	Ion-pair formation (at high kinetic energies)

show a complex structure. Different sheaths around the electrodes are formed. Commonly, the positive column is the major part of the discharge and consists of the stationary, quasi-neutral plasma of low ion energy. The electron energy distribution is Maxwellian or shows Druyvestein behavior and, thus, this plasma region is well suited for polymer surface modifications.

Plasma production requires the existence of internal electrodes (cathode and anode). Film forming plasmas also deposit layers onto the surface of electrodes; the transition resistance through these layers is permanently increased with growing layer thickness. Therefore, the voltage must be increased to overcome the increasing resistance. If the isolating deposition too thick the current comes to a standstill or a breakdown with an electrical arc occurs. Thus, the use of DC plasmas for polymer modifications is practicable if exclusively using non-depositing gas plasmas.

Kilo-hertz plasmas are sometimes used but more often radio-frequency produced plasmas are brought into action. Most often 13.6 or 27.2 MHz produced radio-frequency (rf) plasmas are applied. Several variations are possible for coupling the rf power into the plasma and for reactor design. The rf power production and transfer is possible by coaxial cables or metallic foils of $\lambda/2$ or $\lambda/4$ width. The two power supply cables may transfer the rf power or one cable supplies the "hot electrode" (or "powered electrode") and the other one is mass. Generally, the power can be supplied capacitively by putting the discharge tube between the plates of a capacitor (capacitive coupling) or within a coil (inductive coupling). In the case of capacitive coupling the electrodes are placed as a "hot electrode" within the plasma reactor and the counter-electrode is opposite or the walls of the metallic reactor form the mass counter-electrode. The special properties of rf power also make it possible to place the electrode outside the reactor when using glass or quartz. In this electrode-less capacitive or inductive rf coupling the deposition of isolating films in the plasma is not important. In this way, chemical aggressive plasmas can also be easily applied and contaminations by sputtered or etched metals from the electrodes can be avoided. Microwave plasmas are often used instead of rf plasmas for high-rate depositions of films or etching of layers.

Polymer samples can be placed in different positions in the glow discharges, either onto the electrodes or without connection with any electrode. Polymers are usually non-conducting materials. They preferably adopt negative charges transferred by the electrons. Because of the charges at the isolating polymer surfaces, which can only slowly flow off, a negative "floating" potential (bias) of a few volts or a few ten volts is permanently formed. Thus, ions from the plasma sheath are accelerated to the polymer and neutralize a fraction of negative charges. In this way the negative charging of the polymer surface is slightly reduced and, thus, quicker electrons impinge the surface.

The application of the plasma process determines the construction of the plasma reactor. Glass bell jar, parallel plate (diode type), or barrel reactors, beam-like acting or remote plasma systems, and two chamber equipment with pre-vacuum and main chamber are widely in use (Figure 3.6).

There are also several types of power input:

- with internal electrodes,
- electrode-less,
- capacitive coupling into the plasma (glass and quartz),
- inductive coupling (glass and quartz),
- remote (afterglow or downstream).

Moreover, a few variants of power supply are possible:

- DC (direct current with internal electrodes),
- rf in the continuous wave mode (cw),
- rf in pulsed mode (pp),
- rf in pulsed plasma and pressure mode (ppp).

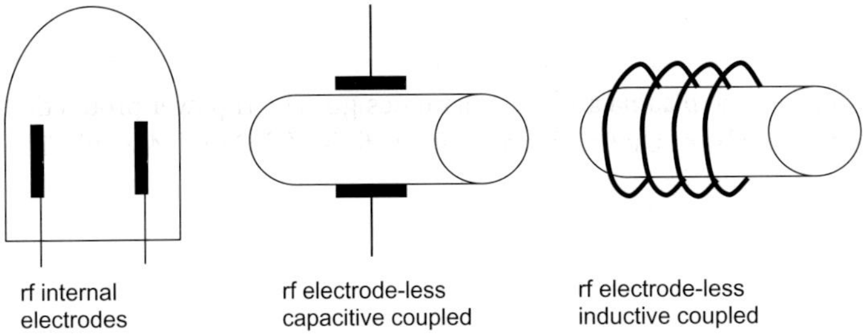

Figure 3.6 A few principal types of plasma devices.

3.3
Advantages and Disadvantages of Plasma Modification of Polymer Surfaces

Polyolefins and polyolefin-related polymers (polyethylene, polypropylene, polyisobutylene, polystyrene), polydiene (polybutadiene, polyisoprene), siloxane polymers (polydimethylsiloxane), and perfluorinated polymers (polytetrafluoroethylene) do not possess polar groups or chemically reactive groups. Therefore, they have a low surface energy, with a minimal polar contribution. Several techniques have been developed to increase the surface energy of such apolar polymers by introduction of polar groups onto their surface. Surface oxidations that introduce O-functional groups are the most common approach, with flame treatment, irradiation with ionizing, ultraviolet (excimer) or laser light, treatment with oxidizing reagents such as chromic acid, ozone, fluorine, and so on as the methods of choice. However, all treatments produce a broad variety of O-functional groups. Moreover, oxidative treatments with oxygen are generally not successful in modifying fluorine polymers.

Plasma-assisted oxidations produced in oxygen-low pressure glow discharges may be also possible in air, NO_2, H_2O, H_2O_2, CO_2, and so on. The result of such oxidation is similar to that of flame treatment. The same is true for treatments in air with atmospheric-pressure plasmas using atmospheric-pressure glow discharges (APGD), dielectric barrier discharges (DBD), corona, or all types of oxidative plasma jets.

Inert gas plasma treatment, which may lead to CASING (crosslinking by activated species of inert gases) [25, 26], produces a high concentration of radicals at surface and surface-near layers of polymers. These radicals react slowly with oxygen on exposure to air in a post-plasma reaction. Thus, a secondary oxidation of polyolefin surfaces occurs after exposure to air for some days or weeks, very similar to that of direct exposure to oxygen plasma. However, direct exposure to the oxygen plasma produces a (nearly) stable and unalterable oxidation while the inert gas plasma exposure achieves an equilibrium or holdup only after several days, weeks, or months.

All the mentioned irradiative, chemical, and plasma treatments produce a broad variety of O-containing functional groups, which all increase the polar contribution to the surface energy. These treatments are sufficient to improve the adhesion property temporarily. However, only weak interactions are formed in composites, which are sensitive to traces of humidity. These weak interactions are electrostatic, dipole–dipole, or induced dipole interactions as well as hydrogen bonds. Moreover, often, post-plasma reactions proceed and change the permanence of adhesive bonds.

The main advantages of using the plasma technique for oxidation:

- all inert polymer surfaces can be oxidized easily, with the exception of fluorinated polymers,
- well-controllable oxidation,

- (most often) no thermal exposure of polymers,
- polymers can also be etched easily, that is, roughened.

Disadvantages are:

- no selectivity in oxidation reactions;
- broad variety of O-functional groups and, therefore, no selective chemical consumption of O-functional groups by grafting;
- damaging of surface-near polymer layers by plasma irradiation (radical formation, polymer degradation, crosslinking);
- post-plasma reactivity, often aging.

Selective plasma reactions, which produce only one sort of functional groups, are rare. Such reactions are necessary to graft complex molecules onto polyolefin surfaces, such as is needed for medical materials, sensors, membranes, and so on.

3.4
Energetic Situation in Low-Pressure Plasmas

As mentioned in different sections of this book there is a strong general discrepancy between the energetic level of the high-energy tail of the electron distribution function, the potential energy of ions and electrons, and the plasma-emitted vacuum ultraviolet radiation, on the one hand, and the common C–C and C–H binding energies in polymers as well as the exothermal polymerization reactions with low activation energy for polymerization on the other hand. In principle, plasma and polymers do not match. Plasma is a source of permanent energy flow, chemical reactions most often work batch-wise with controlled energy addition as chemical reaction potential or heat.

Polyolefins, perfluorinated polymers, and siloxane resins are chemically inert and need strong chemical power for activation. However, oxidations proceed easily. Plasma exposure cleaved all chemical bonds immediately, easily, and completely. Only a few waste gases are formed, such as carbon dioxide or water. This may be the main reasons for preferred application of gas plasmas for modification of polymer surfaces.

The span of standard dissociation energies amounts to 3.6–4.3 eV for C–C and C–H σ-bonds. Often, C=C double bonds react exothermally, with the loss of the π-bond and retention of one σ-bond. Complete dissociation of olefinic C=C (6.4 eV) and (acetylenic) C≡C (8.6 eV) bonds is unlikely. More probable is the cracking of the aromatic C=C bonds in benzene (5.4 eV), as shown by its easy plasma polymerization [27]. Incidentally, electrophilic aromatic substitution is also possible with retention of the aromatic system, as is the radical addition of chlorine to lindane with loss of the aromatic system. Cracking of the benzene ring requires about 5.4 eV and benzene ionization needs about 9.2 eV [28].

3.5
Atmospheric and Thermal Plasmas for Polymer Processing

Based on the relative temperatures of the electrons, ions, and neutrals, plasmas are classified as thermal or non-thermal. Non-thermal plasmas are particularly suited for polymer treatment because of their low gas temperatures. High temperature plasmas are not completely inappropriate, if the residence time in the plasma amounts to a few milliseconds so as to avoid thermal cracking. Often, these short treatment times are satisfactory in achieving an unspecific surface functionalization. The parallel process consists of flame treatment [29, 30]. Here, chemical plasma is produced by the combustion of a fuel gas. Nevertheless, low-pressure plasmas are most often used to modify polymers. Non-thermal plasmas are non-equilibrium plasmas, which are often designated as non-isothermal plasmas. Here, the temperatures (or energies) of all types of plasma species differ strongly, as mentioned before: $T_e >> T_{ion} \approx T_{gas}$. This difference in energy is dependent on the collision rate between electrons among themselves and with heavy particles. Kinetic energy can only be transferred efficiently from electrons to atoms by inelastic collisions. The number of elastic and inelastic collisions and therefore the energy distribution between electrons, ions, and neutrals depends, generally, on the pressure (Figure 3.7).

Several distinct differences differentiate the plasma state from that of the gas phase. The most important point is the excess of energy in the plasma. To overcome and to sustain the plasma state, energy-rich species are needed. These species have high potential and kinetic energies that exceed by far the dissociation energies of any chemical bond in polymers, monomers, and organic substances.

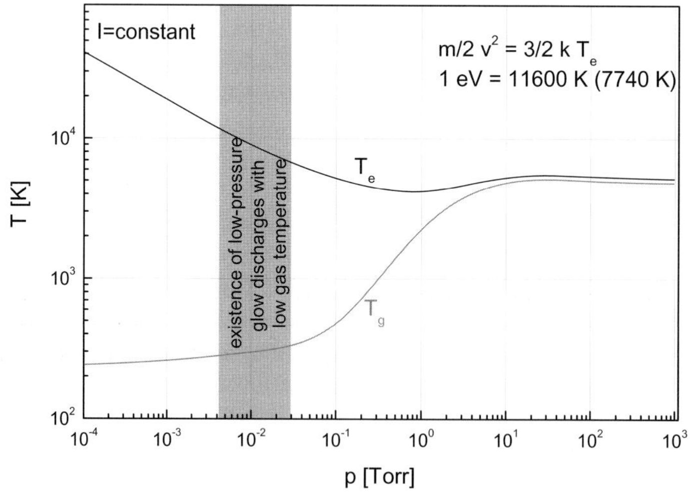

Figure 3.7 General behavior of electron and gas temperature on pressure in glow discharges.

Thus, the plasma is the ideal tool to activate molecules or to form reactive sites at the surface of polymers but it does not work selectively (Figure 3.8).

3.6 Polymer Characteristics

Polymers are the multiple of one or more "monomers" (Table 3.3). A synonym of polymer is the term macromolecule, which have a molecular weight hundreds of times that of the monomer. They do not participate in plasma as a whole because macromolecules cannot be transferred into the gas phase without degradation, fragmentation, or depolymerization. Although their molar mass is normally too high to deal with, exceptions have been detected recently for atmospheric processes by electrospray ionization (ESI) and matrix-assisted laser desorption ionization (MALDI) processes.

However, all interactions between plasma and polymer surface produce degradation and the release of low-molecular weight gas molecules and solid fragments, as discussed elsewhere in this book.

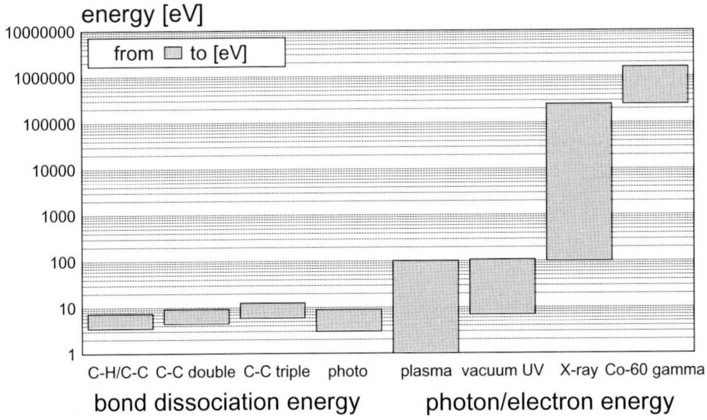

Figure 3.8 Schematic comparison of different treatment media for modifying polymers with different process energies.

Table 3.3 Overview of molar masses and polymerization degrees of oligomers and polymers.

Substance	Molecular weight (g mol^{-1} or Da)	Polymerization degree
Monomer	≈100	1
Oligomer	100–10 000	1–100
Polymer	>10 000–10 000 000	1000–100 000
Thermoset	Indefinite	Crosslinked

Polymers consist of recurrent (monomer) units corresponding to the polymerization degree. In contrast to such homopolymers, copolymers consist of two or more different polymer sequences that are linked by covalent bonds and appear as one molecule.

Polymer molecules are near relatives of biomolecules. Therefore, they have a much stronger response to any exposure to energy-rich radiation or plasma than inorganic or metallic materials. Polymers cannot easily distribute excess energy. Aromatic ring-containing polymers and conjugated polymers can distribute energy slightly better. However, any significant energy excess leads to unintentional degradation, oxidation, or crosslinking reactions. Such reactions are dehydrogenation, depolymerization, decarbonylization, decarboxylation, condensation reactions, random fragmentation, cyclization, crosslinking, formation of trapped radical sites, or auto-oxidation; polymers with high oxygen content, such as carbonate, ester, OH groups, and aliphatic segments may also be formed. In addition, aromatic rings are very sensitive to plasma exposure if the striking particles have energies >8 eV. Another of the most significant features of most polymers is their sensitivity towards heat impact, such as occurs in thermochemistry or in thermal plasmas.

Another sensitivity of polymers is caused by the supermolecular structure of polymers that often exist. Such structures are associated with orientation phenomena such as crystallinity, spherulites, and so on. Such ordered crystalline regions coexist with unordered amorphous fractions. Therefore, one has, strictly speaking, partial crystallinity in polymers. Such crystalline regions, which are found in discrete crystals, also show transition zones to the amorphous phase, for example, that formed by the loops of the folded macromolecules at the margins of lamellae, part of the crystal. There are different opinions and variations concerning ordered regions. Such supermolecular aggregates are micelles, spherulites, dendrites, shish kebab structures, and so on. Moreover, stretching of the amorphous phase leads to a parallelization of macromolecules and, therefore, to their orientation. Nearly all polymer foils are stretched uni-axially and most often two-dimensionally. Biaxially oriented (stretched) polypropylene (BOPP) or Mylar foils [poly(ethylene terephthalate)] are such examples and are well known and often applied.

The orientation after stretching or that in crystalline regions of polymers reduces the distance between two neighboring macromolecule chains, lying side by side, and thus make it easier to form strong intermolecular interactions. Therefore, all suited-functional groups in the polymer chain interact and, thus, maximal interactions are observed. In contrast, such interactions are scarcer and random in the amorphous zones. Plasma-introduced polar groups in polymers can also produce “orientation” because of newly introduced dipole–dipole or electrostatic interactions.

Polymers “live,” in the sense that numerous diffusion processes of polymer segments, side-chains, groups, additives, and contaminations change permanently the structure and composition of the polymer surface.

3.7 Chemically Active Species and Radiation

Here, only a short overview of the most important species is given. Details appear in the other sections of this book. Electrons and ions determine the reactivity in plasmas towards polymers. However, as mentioned before, radiation emitted by plasma consumes the majority of energy in plasmas and plays an important role in plasma–polymer interactions. Plasmas are an efficient source of vacuum UV light with wavelengths of a few nanometres to 200 nm [26]. This radiation causes bond scission in polymers ($\sigma \rightarrow \sigma^*$ transitions) within a few μm of the polymer surface. Metastables, such as noble gas metastable excited states, are of chemical importance, as shown by the CASING process [25]. Last but not least, the UV-radiation produced C-radical sites react quantitatively with oxygen from air in a post-plasma reaction [31].

References

1 Langmuir, I. (1928) Oscillations in ionized gases. *Proc. Natl. Acad. Sci. USA*, **14**, 628.
2 Langmuir, I. (1961) *The Collected Works of Irving Langmuir*, vol. 5 (ed. C.G. Suits), Pergamon, New York, pp. 111–120.
3 Crookes, W. (1879) "On radiant matter". A lecture delivered to the British Association for the Advancement of Science, at Sheffield, Friday, August 22, 1879.
4 Graham, W.G. (2007) *Plasma Technologies for Polymers* (ed. R. Shishoo), Woodhead Publishing, Boca Raton, p. 3.
5 Rutscher, A. and Deutsch, H. (1983) *Wissensspeicher Plasmatechnik*, VEB Fachbuchverlag, Leipzig.
6 Hertz, G. and Rompe, R. (1968) *Plasmaphysik*, Akademie-Verlag, Berlin.
7 Weissmantel, C., Lenk, R., Forker, W., Ludloff, R., and Hoppe, J. (1970) *Atom-Struktur Der Materie*, VEB Bibliographisches Institut, Leipzig.
8 Drost, H. (1978) *Plasma Chemistry*, Akademie-Verlag, Berlin.
9 Bär, R. (1927) The glow discharge, in *Handbuch der Physik*, Springer, Berlin, p. 14.
10 Francis, G. (1956) The glow discharge at low pressure, in *Handbuch der Physik* (ed. S. Flügge), Springer, Berlin, p. 22.
11 Wöhrle, D., Tausch, M.W., and Stohrer, W.-D. (1998) *Photochemie–Konzepte, Methoden, Experimente*, Wiley-VCH Verlag GmbH, Weinheim (in German).
12 Liepins, R. and Yasuda, H. (1971) *J. Appl. Polym. Sci.*, **15**, 2957.
13 McCallum, C.T. and Rankin, J.R. (1971) *J. Polym. Sci.*, **139**, 751.
14 Boones, M.J., Larkin, I.W., Hasted, J.B., and Moore, I. (1967) *Chem. Phys. Lett.*, **24**, 292.
15 Compton, R.N., Christophorou, L.G., and Huebner, R.R. (1966) *Chem. Phys. Lett.*, **23**, 656.
16 Agobian, R., Otto, J.L., Eckhard, R., and Gagnard, R. (1963) *Compt. Rend. Acad. Sci.*, **257**, 3844.
17 Klarsfeld, B.N. (1940) *Electron and Ion Devices (Russ.)*, Moscow-Leningrad, Akademisdat.
18 von Engel, A.V. and Fowler, R.G. (1956) *Handbuch der Physik* (ed. S. Flügge), Springer, p. 22.
19 Hantsche, E., Hertz, G., and Rompe, R. (1968) *Plasmaphysik*, Akademie-Verlag, Berlin.
20 Tüxen, F. (1936) *Z. f. Physik*, **103**, 436.
21 Bayet, M. and Guerineau, F. (1954) *Compt. Rend. Acad. Sci.*, **239**, 1029.
22 Kiefer, L.J. and Dunn, G.H. (1966) *Rev. Mod. Phys.*, **38**, 1.

23 Scholz, O. (1972) *Atomphysik*, VEB Fachbuchverlag, Leipzig.
24 Alekhshkhin, I.S., Zapesochnyi, I.P., and Shpenik, O.P. (1967) *Proceedings of V International Conference on the Physics of Electronic and Atomic Collisions*, Nauka, Leningrad, p. 499.
25 Hansen, R.H. and Schonhorn, H. (1966) *J. Polym. Sci. B*, **4**, 203.
26 Hudis, M.V. (1972) *J. Appl. Polym. Sci.*, **16**, 2397.
27 Deutsch, H., Sabadil, H., and Görss, E.-W. (1975) *Contr. Plasma Phys.*, **15**, 183–190.
28 Lechner, M.D., Lax, E., and D'Ans, J. (1992) *Taschenbuch Taschenbuch Für Chemiker und Physiker: Physikalisch-Chemische Daten*, Springer, Berlin.
29 Kreidl, W.H. (1953) US-Patent 2632921.
30 Kreidl, W.H. (1959) *Kunststoffe*, **49**, 71.
31 Friedrich, J., Kühn, G., and Gähde, J. (1979) *Acta Polym.*, **30**, 470–477.

4
Chemistry and Energetics in Classic and Plasma Processes

4.1
Introduction of Plasma Species onto Polymer Surfaces

To attach fragments or atoms produced in a plasma, as functional groups on aliphatic (olefin) polymer surfaces by radical recombination process, hydrogen atoms must be removed from polymer backbone or side-chains (Scheme 4.1). This removal of a H-atom and replacement by a fragment or heteroelement is called substitution. Such a substitution reaction takes place as a radical recombination process (S_R). The concurrent process is chain-scission of the polymer backbone (Scheme 4.1).

Non-aliphatic polymers also allow chemical S_N1 or S_E1, and S_N2 reactions at activated H atoms as well as electrophilic or nucleophilic additions (A_E and A_N) onto double bonds (cf. Scheme 4.2).

Non-equilibrium and non-isothermal low-pressure glow discharge plasmas allow us to modify the surface composition and the introduction of functional groups onto the surface of natural and synthetic polymer materials by their high-energetic and chemically active species in the gas (or liquid) phase. The very high chemical activity of such plasmas (continuous flow of energy and enthalpy) is a very important tool for altering the surface energy and chemistry of polymer surfaces (Schemes 4.1 and 4.2).

The binding energies in polymers, for example, aliphatic polyolefins, are 375 kJ mol^{-1} for the CH_2–CH_2 bond and 396 kJ mol^{-1} for the CH–H bond [1]. Hydrogen abstraction at saturated aliphatic chains proceeds easily, while C–C bond scissions of the chain are rare. This is in contrast to the bond dissociation energies. It is assumed that the hydrogen atoms shield the C–C bonds from attack [1].

Oxidation is one of the most important processes of introducing polar O-functional groups onto polymer molecules. The oxidation tendency depends on the dissociation energy ($\Delta_D H^0$). The reaction $X^\bullet + R{-}H \rightarrow X{-}H + R^\bullet$ must be exothermic and, thus, $\Delta_D H^0_{X-H} > \Delta_D H^0_{R-H}$ [1].

The formation of O-functional groups at polyolefin surfaces requires strong oxidants. However, such oxidants have no selectivity and their application is inevitably accompanied by molar mass degradation of the polymer substrate caused by C–C bond scission and etching. The oxidation progresses as follows:

The Plasma Chemistry of Polymer Surfaces: Advanced Techniques for Surface Design, First Edition.
Jörg Friedrich.

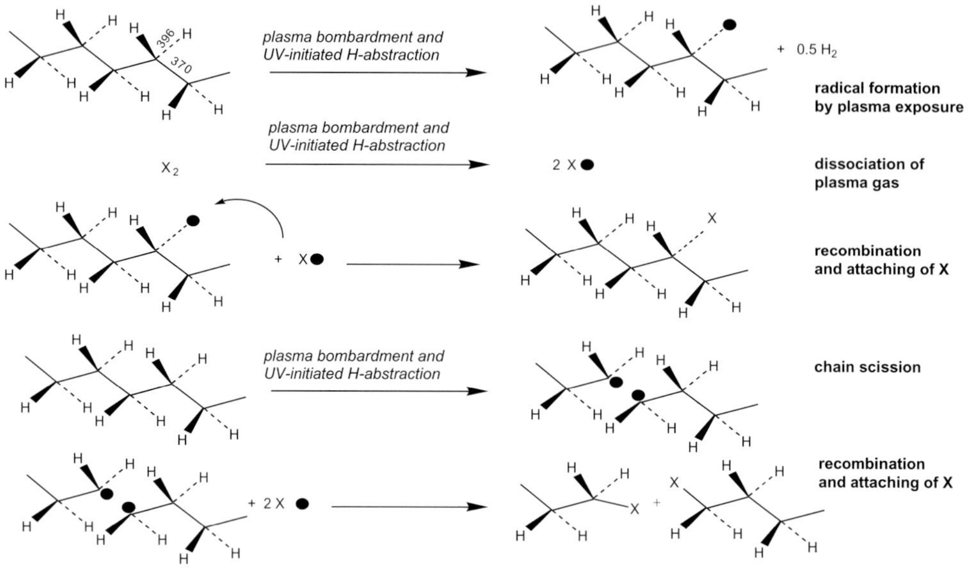

Scheme 4.1 Introduction of a plasma gas-specific fragment as functional group onto an aliphatic polymer chain by substitution of hydrogen.

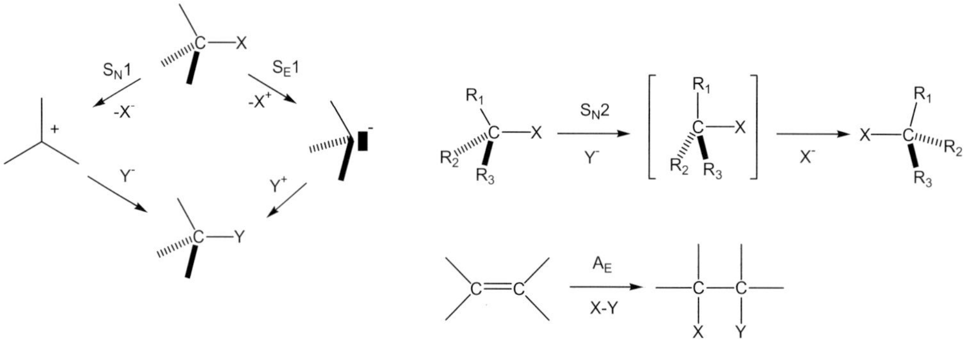

Scheme 4.2 Non-radical substitution and addition mechanism.

$$CH_3 \rightarrow CH_2OH \rightarrow CHO \rightarrow COOH \rightarrow (CO_2 + H_2O)$$

$$>CH_2 \rightarrow >CH{-}OH \rightarrow >C{=}O \rightarrow (COOH \text{ and } C{-}C \text{ scission}) \rightarrow (CO_2 + H_2O)$$

Generally, the permanent energy excess in low-pressure glow-discharge plasmas also hinders any selectivity of oxidation. Therefore, both C–H and C–C bond dissociation occur simultaneously (cf. Scheme 4.1):

$$C{-}H + \textit{plasma} \rightarrow C^{\bullet} + H^{\bullet}$$

$$C{-}C + \textit{plasma} \rightarrow C^{\bullet} + C^{\bullet}$$

The lack of selective plasma processes for the introduction of monotype functional groups onto the polymer surfaces of aliphatic polymer chains strongly handicaps applications of plasma processes in fields such as biochemistry and medical techniques. Such monotype functional groups are the anchoring points for further chemical processing such as covalent grafting of any spacers, prepolymers or functional biomolecules onto the polymer surface. These selective monotype functionalizations introduce NH_2, OH, COOH, epoxy, carboxylic acid, sulfonic acid, double bonds, aldehyde, or other groups (Figure 4.1).

Such selective plasma processes should lead to the dominant production of one type of surface functionalization accompanied by very low concentrations of other types of functional groups also derived from the plasma gas or by post-plasma introduced oxygen reacting with plasma-produced trapped radical sites over a period of weeks or months [2, 3]. There are two types of selectivity for plasma polymerization (Figure 4.2):

1) selective formation of functional groups (retention of all functional groups of the monomer and their appearance in the plasma-polymer) independent of the layer structure;

2) total regular formation of known polymer structure with retention of all functional groups ("overall selectivity").

Selective plasma processes using the plasma technique are rare due to the permanent excess of energy delivered from the plasma, the high degree of freedom for the attachment of plasma species from the gas phase onto the surface, and the unavoidable, permanent, and energy-rich plasma UV irradiation. Therefore, during plasma polymerization a broad number of different functional groups is often formed. Nearly all plasma-activated processes are radical reactions and the

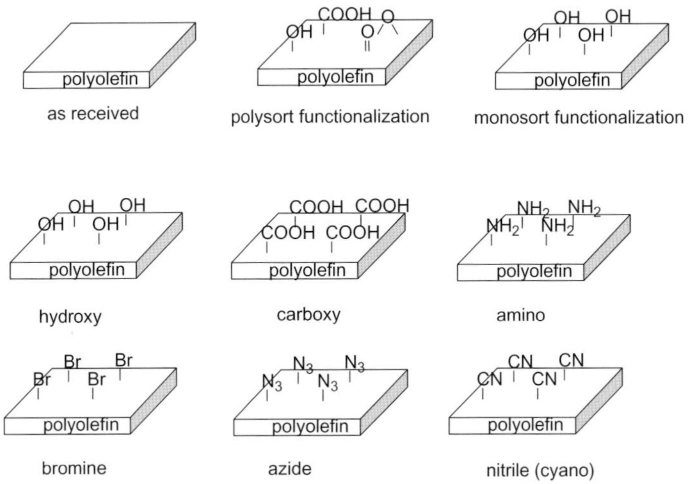

Figure 4.1 Schematic view of functionalized polyolefin surfaces.

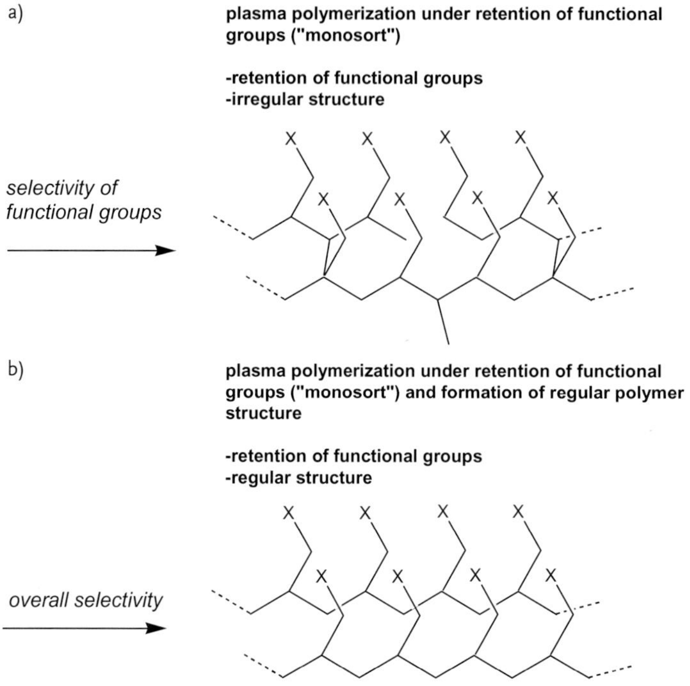

Figure 4.2 Two types of selectivity: (a) related to functional groups and (b) related to the whole polymer structure.

additional plasma UV irradiation further produces radicals at the surface, in surface-near layers, or even in the bulk of the polymers [4]. The remaining unsaturated and metastable trapped radicals are responsible for post-plasma oxidation. This auto-oxidation is strongly non-selective and leads additionally to the appearance of a broad spectrum of O-functional groups. Nevertheless, a few plasma polymerization processes are more or less selective, such as the plasma polymerization of allyl alcohol with retention of >90% of all OH-groups in the deposited plasma polymer layer. However, total retention of the monomer structure in a deposited polymer film has never been found. An exception, though, may be the plasma polymerization of hexamethyldisiloxane in the presence of oxygen to SiO_2 layers.

The surface functionalization of polymers also shows different selectivity (cf. Figure 4.1). Unspecific (non-selective) surface functionalization is the normal situation. Optimization of plasma parameters makes little sense because the selectivity can only be marginally improved. This stems from the general conflict between the too-high energy excess introduced from the plasma, on the one hand, and the low energies necessary to dissociate chemical bonds or make them reactive on the other hand. In particular, the high-energy tail of the electron energy distribution function exceeds significantly the binding energies in polymers (Figure 4.3).

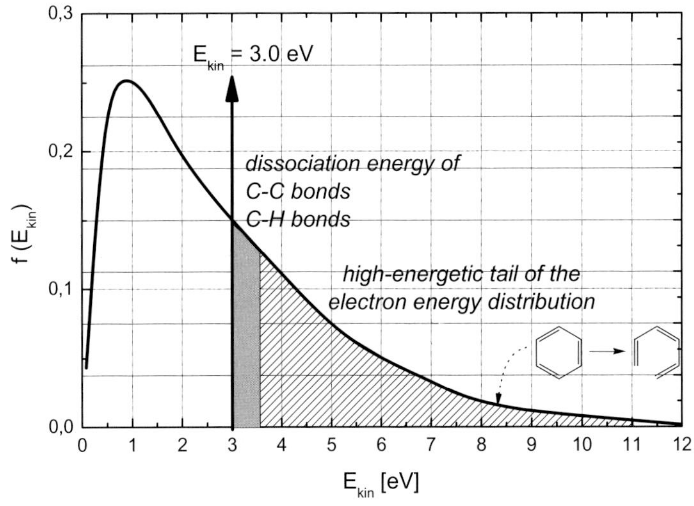

Figure 4.3 Schematic plot of the electron energy distribution function in a plasma in relation to C–C and C–H dissociation energies.

The ionization energies and also the corresponding energies of other plasma-relevant features such as metastables, the short wavelength of radiation, and the permanent power input make it difficult to adjust the two energy levels, that is, that of plasma and that of polymer binding energies (cf. Figure 4.3). The use of pulsed plasmas and remote plasmas as well as lowering the wattage may improve the situation, but only marginally.

Thus, the process of polymer surface functionalization can be divided into three cases:

1) selective monosort functionalization of polymer (polyolefin) surfaces without any changes in polymer structure;
2) monosort surface functionalization with changes in the structure of polymer substrate;
3) formation of a broad spectrum of different functional groups at the polymer surface along with changes in the structure of polymer substrate.

In contrast, a few plasma-initiated functionalization reactions are relatively selective if the energy level in the plasma is low and, therefore, chemical processes can dominate. If the end-product is thermodynamically stable there are no possibilities for side-reactions and selective functionalization occurs. Such cases are rare. One example is plasma bromination using bromoform ($HCBr_3$) or bromine (Br_2) as precursor for the production of free bromine atoms. This process produces bromine monotype functionalization with covalent bonded Br. The reason for this is that halogens seek to fill their outer electron shell completely as the next noble gas, which is realized by C–Br bonds. Therefore, all halogen introductions are selective because the chemistry of halogens only allows the formation of C–X or

X^-. The halogenide anions X^- can be removed from the surface by any wash process. Other side reactions, the trapping of radicals and their slow auto-oxidation, are strongly present in the related plasma fluorination and chlorination but are minimal for bromination [5]. Bromine atoms quench all radical sites in the polymer: $C^{\cdot} + {}^{\cdot}Br \rightarrow C{-}Br$.

The extraordinary selectivity of the Br_2 or $CHBr_3$ bromination process of polymer surfaces compared to similar plasma functionalization processes such as amination by ammonia plasma or hydroxylation by oxygen plasma exposure is also attributed to the plasma-physical component. The lower energy level of bromine-containing plasmas is due to the low ionization energy of Br-containing organic precursors compared to the noble gases (Table 4.1).

The kinetic energy of electrons ("electron temperature" $\rightarrow m/2v^2 = 1.5kT$) also determines the type of chemical reactions, which are initiated by plasma bombardment and plasma-UV irradiation [6].

Following the Schottky theory of the positive column (DC plasma) the location of the electron energy distribution on the energy scale depends on the ionization potential (E_{ion}) (cf. Table 4.1 and Figure 4.4) of the plasma gas as given by Shahin (Maxwellian energy distribution, ambipolar diffusion) [7, 8]:

Table 4.1 Ionization potentials of halogen-containing plasma gases and noble gases and also the factor *c*, which is presented for noble gases.

Gas	E_{ion} (eV)	Gas	E_{ion} (eV)	*c* (movability)
CF_4	17.8	He	24.5	0.004
CCl_4	11.5	Ne	21.5	0.006
$CHCl_3$	11.4	Ar	15.8	0.040
$CHBr_3$	10.5	H_2	15.4	0.010
CH_2I_2	9.3	Hg	10.4	0.070

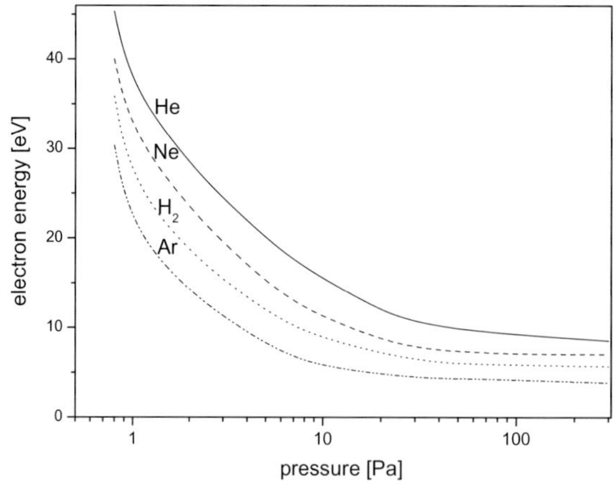

Figure 4.4 Dependence of electron energy on pressure in noble gas and hydrogen glow discharge plasmas.

$$(eE_{ion}/kT_e)\exp(-1/2)\times\exp(eE_{ion}/kT_e)=1.16\times10^7\ c^2p^2r^2$$

where

e = elementary charge,
T_e = electron temperature,
c = gas constant related to atom mobility,
p = pressure,
r = tube diameter.

Therefore, the average electron energy in the bromoform plasma should be lower than that of the CF_4 plasma (or noble gas plasmas), as was confirmed by measurement of electron temperatures using a Langmuir probe [9]. This is expected to lead to:

- more selective reactions because of the low energy level in the plasma;
- emission of low-energy UV radiation with less defects;
- low concentration of (trapped) radicals and limited post-plasma auto-oxidation.

From Table 4.1 the ionization energy of the bromoform molecule is evidently much lower than that of comparable F- and Cl-containing molecules. The same is true for noble gases. With increasing mass number the ionization potential decreases (Table 4.1). By introducing these mass numbers in the above equation, the electron energy (temperature) can be calculated for different pressure as shown in Figure 4.4. The average electron temperature decreases with decreasing ionization potential. Thus, the energetic level in the $CHBr_3$ plasma should be much lower than in the CF_4 plasma. Additionally, chemically, fluorine atoms are much more reactive than bromine atoms [1].

It was proposed that such softer plasma conditions (low level of electron temperature) lower the extent of monomer fragmentation in the plasma or the destruction of polymer substrate surface under the plasma particle bombardment. Generally, energetically softer plasmas promise more retention of structure and composition of monomer and substrate than energetically harder plasmas. Moreover, with the bromoform plasma, bromine atoms quench all radicals by formation of stable covalent C–Br bonds. In the polymer matrix trapped free radicals are the reason for the auto-oxidation of polymers on exposure to the ambient air (cf. Chapter 2, Section 2.2).

The chemical contribution to functionalization or plasma-polymerization starts after plasma-induced radical production and is responsible for all regular and defined reaction products, unless the chemical reaction itself is not selective and does not produce a strong excess in reaction enthalpy (Figure 4.5).

This simple idealized scheme of plasma processes was the background for the introduction of pulsed plasma polymerization, that is, plasma generates radicals from monomers and these radicals initiate chain growth propagation to high molecular weight macromolecules. As explained later, the low pressure hinders such a chain growth polymerization because of monomer deficiency at the active chain center. Moreover, all ideal polymer structures are continuously bombarded

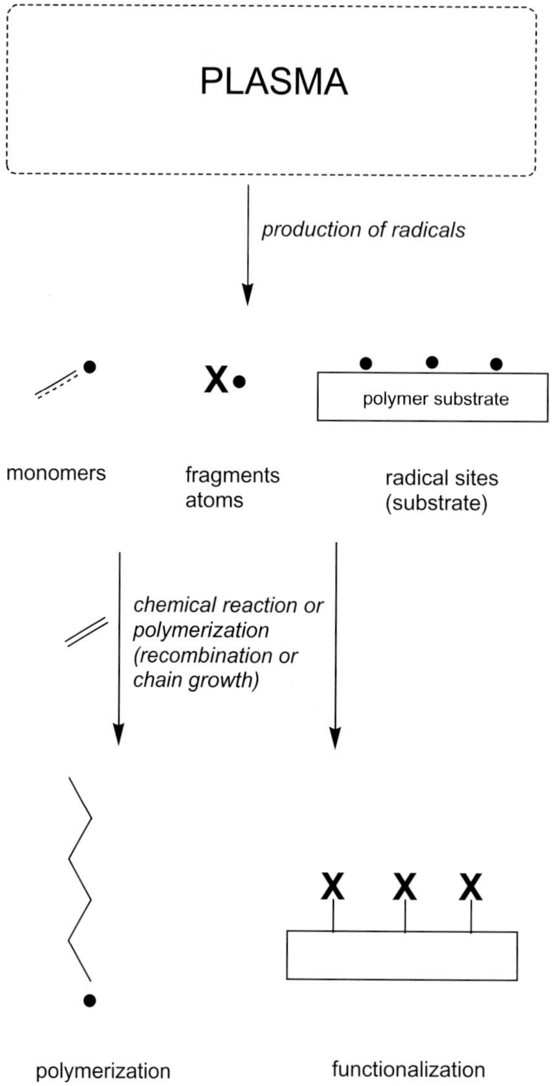

Figure 4.5 Idealized schematic of plasma-induced radical generation and subsequent chemical reaction of radicals.

with high-energy particles and irradiated with high-energy vacuum UV radiation. Thus, the defined product of any chemical reaction is secondarily re-arranged.

The action of a high excess in reaction enthalpy of chemical reactions or polymerizations lies in its chemical consequences for the polymer surfaces, which are comparable to the physical effects (particle bombardment) of the plasma (broad variety of products, degradation, fragmentation, crosslinking, post-plasma oxidation, etc.). It must be considered that under vacuum conditions thermal energy cannot be removed easily. Local thermally-induced reactions are expected.

Among the chemical routes to products chain reactions play the most important role. Chain reactions are present in plasma polymerization as well as in fluorination, chlorination, bromination, or oxidation. The selective character of such chemical chain reactions depends on the reactivity of radicals, which are, therefore, responsible for the uniformity of products. For polymer surface halogenation the chemical reactivity of radicals follows the order:

$$F^{\bullet} > HO^{\bullet} > Cl^{\bullet} > CH_3^{\bullet} > Br^{\bullet} > R\text{-}O\text{-}O^{\bullet} > I^{\bullet}$$

Evidently, the use of bromine or iodine is needed for high selectivity. The weak reactivity of bromine atoms is due to the low standard reaction enthalpy of the reaction:

$$X^{\bullet} + {-CH_2{-}CH_2{-}} \rightarrow X{-}H + {-CH^{\bullet}{-}CH_2{-}}$$

The standard enthalpy of this reaction is given by:

$$\Delta_R H^{\ominus}{}_{298} = \Delta_R H^{\ominus}{}_{R-H} - \Delta_R H^{\ominus}{}_{X-H}$$

Bromination is slightly endothermic, while fluorination shows a strong exothermic standard reaction enthalpy:

$$\Delta_R H^{\ominus}{}_{298} = \Delta_R H^{\ominus}{}_{R-H} - \Delta_R H^{\ominus}{}_{H-F} = 396{-}566\ \text{kJ mol}^{-1}$$

$$\text{fluorination} = -170\ \text{kJ mol}^{-1}$$

$$\text{chlorination} = -32\ \text{kJ mol}^{-1}$$

$$\text{bromination} = 33\ \text{kJ mol}^{-1}$$

$$\text{iodination} = 101\ \text{kJ mol}^{-1}$$

Strongly activated C–H bonds react with bromine and have weak exothermic character. However, the kinetic chain length is low. The selectivity is high. Non-activated H is difficult to substitute and the introduction of multiple Br is also difficult or hindered. For example, the bromination of cyclohexane under UV irradiation shows a quantum yield of about 2 [1].

Because of the positive standard reaction enthalpy the assistance of UV (or plasma) is desired:

$$Br_2 + h\nu \rightarrow 2\,{}^{\bullet}Br$$

$${}^{\bullet}Br + RH \rightarrow R^{\bullet} + HBr$$

$$R^{\bullet} + Br_2 \rightarrow R\text{-}Br + {}^{\bullet}Br\ \text{(chain growing)}$$

or:

$$R^{\bullet} + {}^{\bullet}Br \rightarrow R\text{-}Br\ \text{(chain termination)}$$

Iodine is non-reactive and cannot attack C–H bonds; therefore, the opposite reaction occurs: $RI + HI \rightarrow RH + I_2$.

4.2
Oxidation by Plasma Fluorination and by Chemical Fluorination

On considering the reaction enthalpies it becomes evident that the strong exothermic fluorination of polymer surfaces produces non-selectivity and, therefore, side-reactions such as auto-oxidation occur. The mechanism of chemical fluorination can be ascribed as follows: $F_2 \rightarrow 2\ F^{\cdot}$; $F^{\cdot} + RH \rightarrow R^{\cdot} + HF$; $R^{\cdot} + F_2 \rightarrow RF + F^{\cdot}$. This reaction proceeds as a chain reaction because the reaction enthalpy is strongly negative. This negative reaction enthalpy (exothermal reaction), $\Delta_R H^{\ominus}{}_{298} = -155\,\text{kJ}\,\text{mol}^{-1}$, is also the reason why selectivity is largely absent during chemical polymer surface fluorination. As C–C-chain scission occurs, unreacted C-radical sites are generated that can subsequently react with oxygen from ambient air.

Plasma assistance of fluorination with F_2 is not necessary. It only increases the number of side-reactions. Using chemically unreactive fluorine precursor molecules, such as CF_4, SF_6, NF_3, BF_3, and so on plasma-assistance is necessary for the generation of fluorine atoms. The precursor dissociates in the plasma to release of fluorine atoms: $CF_4 + \textit{plasma} \rightarrow {}^{\cdot}F + {}^{\cdot}CF_3$ or at high plasma power: $CF_4 + \textit{plasma} \rightarrow 4{}^{\cdot}F + C$. Because of the strong chemical reactivity of fluorine atoms (stronger oxidant than oxygen itself) an overlap of chemical and plasma-chemical reactions takes place as described before.

In contrast, defluorination of perfluoro alkanes (e.g., PTFE) is difficult to realize chemically [10]. Only the use of sodium or potassium allows us to remove fluorine from aliphatic chains by chemical reduction: $-CF_2- + 2Na \rightarrow 2C + 2NaF$ [11]. The same process can be also performed under plasma-assistance [12]. It must be considered again that fluorine is the strongest oxidation agent and therefore the C–F bond has the highest oxidation ratio. Using plasma assistance, hydrogen (or ammonia) can also reduce C–F bonds: $H_2 + \textit{plasma} \rightarrow 2{}^{\cdot}H$; $2{}^{\cdot}H + -CF_2- \rightarrow 2HF + 2C$.

Evidently, substitution or addition reactions of polymers by plasma gas species have a significant contribution from pure chemical reactions. Plasma can facilitate chemical reactions by dissociation of (neutral) precursor molecules; however, the selectivity becomes much worse. Therefore, plasma assistance is very similar to photochemical assistance of reactions, but the energy excess is much higher.

Plasma polymerizations do not need continuous plasma-assistance – they work chemically as chain reactions. The chemical polymerization is exothermic and needs only a few deliveries of activation energy by initiator decay or by short-time plasma initiation. Continuous plasma presence is counterproductive. It increases the number of defects and irregularities in the deposited polymer layers very strongly. On the other hand, plasma-assistance is essential for "polymerization" of chemically non-polymerizable "monomers," such as benzene. However, the putative advantage comes with strong disadvantages because of the irregular polymer structure and numerous defects in the surface layer. As will be shown later, plasma assistance is also required for repeated new initiation of chain-growth polymerization because the supply of monomers to the growing radical is too low under low-pressure conditions.

4.3 Comparison of Plasma Exposure, Ionizing Irradiation, and Photo-oxidation of Polymers

Photo-oxidation of polymers occurs under a flux of ultraviolet photons of wavelengths between ca. 200 and 400 nm. The energy of this radiation is roughly the same as that of binding energies in polymers. Thus, more or less specific reactions can be initiated by applying a specific wavelength of the radiation [13]. The efficiency of such a photo-oxidation depends on, among other factors, the flux of photons and the quantum yield. The flux of photons is maximal using LASER. The quantum yield (Φ_j) of a chemical process j is defined as the number of molecules A (n_A) that undergo this process divided by the number of absorbed quanta (n_Q), $\Phi_j = n_A/n_Q$. Table 4.2 gives examples for the backbone scission of several polymers [14, 15].

Photosensitizers absorb radiation and transfer it to the polymer for the activation of chemical reactions at the polymer surface, possibly for the introduction of functional groups [16]. Moreover, excimer or exciplex radiation of laser-active fluorine (F_2) (157 nm), argon fluoride (ArF) (193 nm), krypton fluoride (KrF) (248 nm), xenon chloride (XeCl) (308 nm), or xenon fluoride (XeF) (351 nm) is also often used [17–23]. Vacuum UV irradiation via LiF or MgF_2 windows is also often applied to irradiate polyolefins [24–35].

Ionizing radiation is based on the extraordinary excess of energy accumulated in one particle or quantum. The reactions of such a high-energetic species after passing through the spur and colliding with atoms in the polymer solid are photo-ionization, Compton dissociation, and ion pair formation. Corresponding to the high energy of species a large number of collisions with atoms are possible along the spur of such a primary species. In this way energy degradation occurs up to the level of that of ordinary chemical reactions [36–45]. Such secondary effects are dissociation, fragmentation, recombination, disproportionation, chain reactions, luminescence emission, ion–molecule reactions, energy transfer, and so on. The energy available in irradiation largely exceeds the amount required to break any C–H or C–C bond (4.3 and 3.7 eV, respectively, to about 6 eV for double/triple C–C bonds). In contrast to photochemical quantum yields, *G*-values are formulated for chemical chain-reactions under irradiation with ionizing radiation. The *G*-value is the quotient of the number of chemical events per 100 eV. Chain reactions have

Table 4.2 Quantum yield for scission of the main-chain of various polymers.

Polymer	Quantum yield
Poly(methyl methacrylate) (PMMA)	3×10^{-2}
Polystyrene (PS)	5×10^{-4}
Polyacrylonitrile (PAN)	5×10^{-4}
Poly(ether sulfone) (PES)	7×10^{-4}
Poly(ethylene terephthalate) (PET)	5×10^{-3}

high *G*-values, stoichiometric reactions low values. Under low or moderate irradiation doses chemically specific reactions occur, leading to well-defined products, as also shown for photochemistry. The selectivity rules, however, do not only follow bond energy considerations [39]. Only high irradiation doses lead to crosslinking reactions in polymers or to radiation pyrolysis.

Plasma chemistry is the most aggressive chemistry, working with the highest energy flow to the polymer in thin layers or to the topmost surface. Westwood had pointed out that the energy doses needed for plasma polymerization are a few order of magnitude higher than that for radiation polymerization [46]. It was shown that 10–100 eV per molecule were needed for complete polymerization of vinyl monomers and about 1000 eV per molecule for polymerization of aliphatic monomers [47]. These values are valid for continuous plasma polymerization. Using pulsed plasma technology the *G*-values could be improved up to 2500 for styrene homopolymerization, clearly indicating the existence of a radical chain propagation (growth) polymerization. Thus, different efforts have been made to make the plasma processes more selective. Nevertheless, also the minimizing of energy uptake by low-wattage, use of plasma-pulse, and pressure pulse technique are not suited to producing plasma polymers of chemically defined composition and structure. Here, the second important component, apart from the plasma particle bombardment, is the vacuum UV irradiation, which occurs in the plasma with very short wavelengths and in high intensity. Thus, the defined products were also crosslinked and degraded by this irradiation.

Alongside these disadvantages, in comparison to photo- and radiation chemistry, plasma chemistry focuses the most energy to the surface and near-surface layer, thus allowing activation of this layer for chemical reactions. The plasma energies are high enough to activate very inert surfaces of solids; because of the energy focusing to a very thin topmost layer all reactions are very fast in comparison to photo- and radiation chemistry.

There are many common irradiative, photochemical, and plasma-chemical effects on polymers (Figure 4.6).

In summary, high energy is dissipated and at each level of energy (MeV to 1 eV) the process of polymer surface modification can be started. After the energy has been largely dissipated, all treatments discharge into the basic processes of excitation, ionization, and dissociation of molecules and macromolecules. Therefore, only the number/concentration of such elementary processes (excitation, ionization, dissociation), that is, the dose or energy flow, is of importance. The starting energy of particles or photons is not very important. In contrast, there is polymer weathering. The energetic level of photons that initiate polymer degradation and aging is not of importance but concomitant factors such as humidity, mechanical loading, oxygen presence, acidic rain, dust, coldness–hotness changes, and so on are of great importance.

Thus, an important chemical indication of the difference between high-energy processes and natural weathering is the dominance of:

1) $\sigma \rightarrow \sigma^*$ electron transitions in high-energy processes under vacuum conditions;

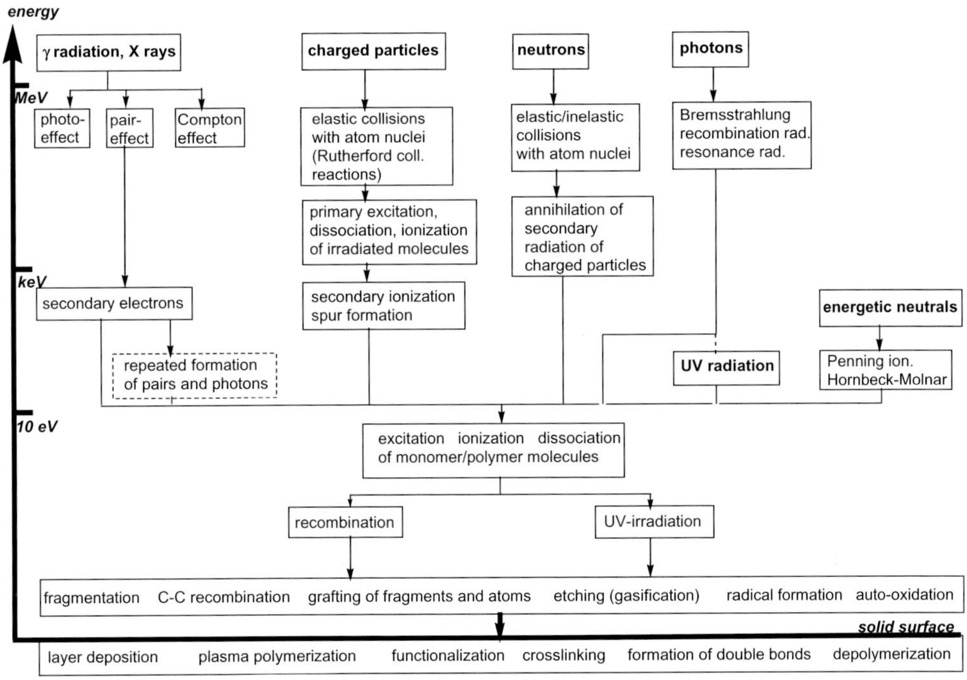

Figure 4.6 Schematic of energy level, flow, and basic processes during exposure of polymers or organic molecules to ionizing radiation, UV irradiation, and plasma [42].

2) $\pi \rightarrow \pi^*$ electron transitions, which are intrinsically reversible but in weathering irreversible by reactions of chromophores with oxygen, water, and so on.

References

1 Beckert, R., Fanghänel, E., Habicher, W.D., Metz, P., Pavel, D., and Schwetlick, K. (2004) *Organikum*, 22nd edn, Wiley-VCH Verlag GmbH, Weinheim.

2 Kühn, G., Ghode, A., Weidner, S., Retzko, I., Unger, W.E.S., and Friedrich, J.F. (2000) *Polymer Surface Modification: Relevance to Adhesion*, vol. *2* (ed. K.L. Mittal), VSP, Utrecht, pp. 45–64.

3 Wann, J.-H., Chen, X., Chen, J.-J., Calderon, J.G., and Timmons, R.B. (1997) *Plasmas Polym.*, **24**, 245.

4 Friedrich, J.F., Kühn, G., and Gähde, J. (1979) *Acta Polym.*, **30**, 47.

5 Wettmarshausen, S., Mittmann, H.-U., Kühn, G., Hidde, G., and Friedrich, J.F. (2007) *Plasma Process. Polym.*, **4**, 832–839.

6 Hertz, G. and Rompe, R. (1968) *Plasmaphysik*, Akademie-Verlag, Berlin.

7 von Engel, A. (1965) *Ionized Gases*, Clarendon Press, Oxford.

8 Shahin, M.M. (1971) *Reactions under Plasma Conditions* (ed. M. Venugopalan), Wiley-Interscience, New York.

9 Friedrich, J., Wettmarshausen, S., and Hennecke, M. (2009) *Surf. Coat. Technol.*, **203**, 3647–3655.

10 Friedrich, J.F., Koprinarov, I., Giebler, R., Lippitz, A., and Unger, W.E.S. (1999) *J. Adhesion*, **71**, 297–310.

11 Friedrich, J.F., Unger, W.E.S., Lippitz, A., Giebler, R., Koprinarov, I., Weidner, S., and Kühn, G. (2000) *Polymer Surface Modification: Relevance to Adhesion*, vol. *2*

(ed. K.L. Mittal), VSP, Utrecht, pp. 137–172.
12 Friedrich, J., Unger, W., Lippitz, A., Koprinarov, I., Kühn, G., Weidner, S., and Vogel, L. (1999) *Surf. Coat. Technol.*, **116–119**, 772–782.
13 Owens, D.K. (1975) *J. Appl. Polym. Sci.*, **19**, 3315.
14 Rabek, J.F. (1996) *Polymer Photodegradation*, Chapman & Hall, Chichester.
15 Blais, P., Day, M., and Wiles, D.M. (1973) *J. Appl. Polym. Sci.*, **17**, 1985.
16 Wöhrle, D., Tausch, M.W., and Stohrer, W.-D. (1998) *Photochemie*, Wiley-VCH Verlag GmbH, Weinheim.
17 Basov, N.G., Danilychev, V.A., Popov, Y.M., and Khodkevich, D.D. (1970) *JETP Lett.*, **12**, 329.
18 Esrom, H. and Kogelschatz, U. (1992) *Thin Solid Films*, **218**, 231–246.
19 Mroszewski, K., Schliefer, K., Heidemann, G., and Schollmeyer, E. (1981) *Melliand Textilber.*, **62**, 742.
20 Bellobono, I.R., Tolusso, F., Selli, E., Calgari, S., and Berlin, A. (1981) *J. Appl. Polym. Sci.*, **26**, 619.
21 Breuers, W., Klee, D., Plein, P., Richter, H.A., Menges, G., Mittermayer, C., and Höcker, H. (1987) *Kunststoffe*, **72**, 1273.
22 Bahners, T. and Schollmeyer, E. (1987) *Angew. Makromol. Chem.*, **151**, 19.
23 Bahners, T., Knittel, D., Hillenkamp, F., Bahr, U., Benndorf, C., and Schollmeyer, E. (1990) *J. Appl. Phys.*, **68**, 1854.
24 Hudis, M. and Prescott, L.E. (1972) *Polym. Lett.*, **10**, 179.
25 Hudis, M. (1972) *J. Appl. Polym. Sci.*, **16**, 2397.
26 Hudis, M. and Prescott, L.E. (1972) *J. Appl. Polym. Sci.*, **19**, 451.
27 Clark, D.T. and Dilks, A. (1977) *J. Polym. Sci. Polym. Chem. Ed.*, **15**, 2321.
28 Matienzo, L.J., Zimmerman, J.A., and Egitto, J.F.D. (2009) *J. Vac. Sci. Technol. A*, **12**, 2662–2671.
29 Chen, J.X., Tracy, D., Zheng, S., Xiaolu, L., Brown, S., VanDerveer, W., Entenberg, A., Vukanovic, V., Takacs, G.A., Egitto, F.D., Matienzo, L.J., and Emmi, F. (2003) *Polym. Degrad. Stabil.* **79**, 399–404.
30 Fozza, A.C., Klemberg-Sapieha, J.E., and Wertheimer, M.R. (1999) *Plasma Polym.*, **4**, 183–206.
31 Fozza, A.C., Roth, J., Klemberg-Sapieha, J.E., Kruse, A., Holländer, A., and Wertheimer, M.R. (1997) *Nucl. Instrum. Methods Phys. Res., Sect. B*, **131**, 205–210.
32 Takacs, G.A., Vukanovic, V., Tracy, D., Chen, J.X., Egitto, F.D., Matienzo, L.J., and Emmi, F. (1993) *Polym. Degrad. Stabil.*, **40**, 73–81.
33 Baydarovtsev, Y.P., Vasilets, V.N., Ponomarev, A.N., Dorofeev, Y.N., and Skurat, V.E. (1984) *Khim. Fiz.*, **3** (**N10**), 1405.
34 Vasilets, V.N., Hirata, I., Iwata, H., and Ikada, Y. (1998) *J. Polym. Sci. Part A*, **36**, 2215–2222.
35 Liston, E.M., Martinu, L., and Wertheimer, M.R. (1993) *J. Adhesion Sci. Technol.*, **7**, 1091–1127.
36 Aleksandrow, A.P. (1958) *Atomnaya Energ.*, **25**, 356.
37 Charlesby, A. (1960) *Atom Radiation and Polymers*, Pergamon Press, London.
38 Charlesby, A. (1977) *Radiat. Phys. Chem.*, **9**, 17.
39 Chapiro, A. (1962) *Radiation Chemistry of Polymeric Systems*, Interscience, New York.
40 Chapiro, A. (1988) *Nucl. Instrum. Methods Phys. Res., Sect. B*, **32**, 111–114.
41 Henglein, A., Schnabel, W., and Wendenburg, J. (1969) *Einführung in Die Strahlenchemie*, Verlag Chemie, Weinheim.
42 Stiller, W. and Friedrich, J. (1981) *Z. Chem.*, **21**, 91–100.
43 Makhlis, F.A. (1975) *Radiation Physics and Chemistry of Polymers*, John Wiley & Sons, Inc., New York.
44 Schnabel, W. (1981) *Polymer Degradation*, Carl Hanser, Munich.
45 Seguchi, T. (2000) *Radiat. Phys. Chem.*, **57**, 367–371.
46 Westwood, A.R. (1971) *Eur. Polym. J.*, **7**, 363.
47 Friedrich, J., Gähde, J., Frommelt, H., and Wittrich, H. (1976) *Faserforsch. Textiltechn./Z. Polymerenforsch.*, **27**, 517–522.

5
Kinetics of Polymer Surface Modification

5.1
Polymer Surface Functionalization

Functionalization denotes the formation of special, most polar, and chemically reactive groups at polymer surfaces. There is a broad variety of O-, N-, S-, or halogen-containing groups. Among them the functional groups shown in Figure 5.1 have been examined for polymer surface modification.

5.1.1
Kinetics of Surface Functionalization

On using ultra-clean polyolefin surfaces (polypropylene or polyethylene) and exposing them to a cw-rf (continuous-wave radio-frequency) discharge at 13.56 MHz with 100 W power-input the formation of O-functional groups at the polymer surface is nearly complete after 2 s exposure time [1] (Figure 5.2).

After 2 s, the XPS-measured concentration of O-functional groups increases only marginally. This is assumed to be due to the required penetration of reactive oxygen species into deeper layers of the polymer. As explained later, the post-plasma process of auto-oxidation of these plasma-irradiated surface-near layers may also be responsible. Associated with this effect, polymer degradation and etching begin. The plateau in oxygen introduction to the polymer surface after about 10 s is assigned to the transition to the etching. Then, a steady-state of surface functionalization, penetration of the oxidized front into the bulk, etching, and the gasification of the polymer material (etching) to CO_2, CO, and H_2O are established. Thus, the course of XPS-measured oxygen concentration over time is determined by:

1) achieving the maximal O-concentration at the topmost surface layer of about 24–28 O/100 C before the degradation starts;
2) migrating of oxidation front into the bulk;
3) passing of oxidation front over the XPS sampling depth;
4) continuous etching of the polymer at the surface.

The Plasma Chemistry of Polymer Surfaces: Advanced Techniques for Surface Design, First Edition.
Jörg Friedrich.

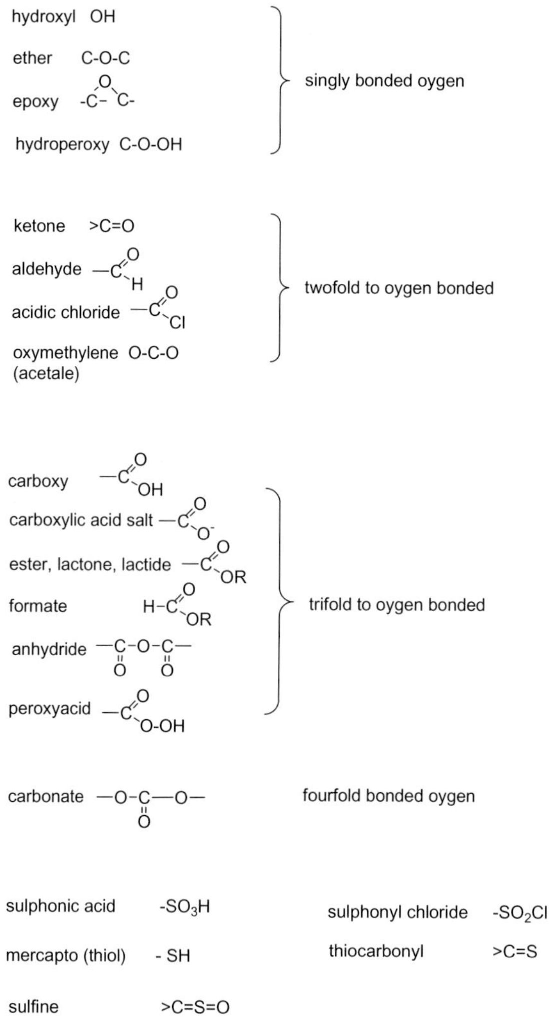

Figure 5.1 Prominent functional groups that occur in polymers or may be introduced by plasma treatment.

These processes are shown schematically in Figure 5.3. In the first step the topmost molecular layer of polymer surface is functionalized following the kinetics presented in Figure 5.2. Then, the next layers are functionalized successively (2–10 s) with, however, a decreasing rate. Meanwhile the topmost layer has achieved the maximum oxidation ratio (O/C > 0.24–0.28). Further oxidation is accompanied by the formation of CO_2, H_2O, and CO occurs (etching).

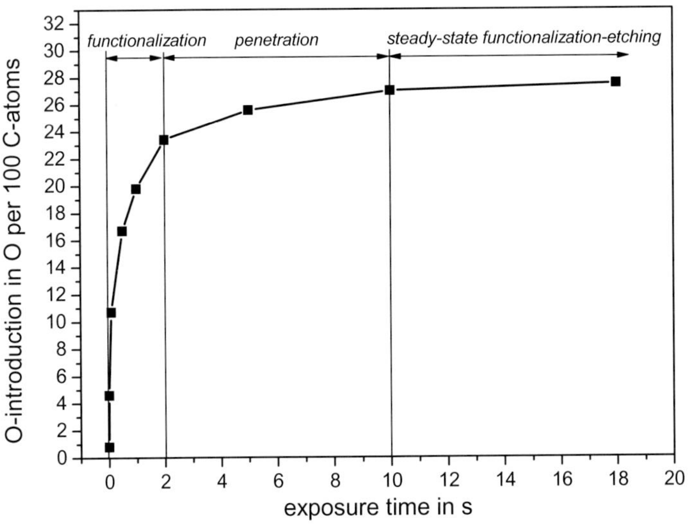

Figure 5.2 Exponential increase of introduced oxygen onto polyethylene surfaces exposed to an O_2 plasma (cw-rf, 6 Pa, 100 W).

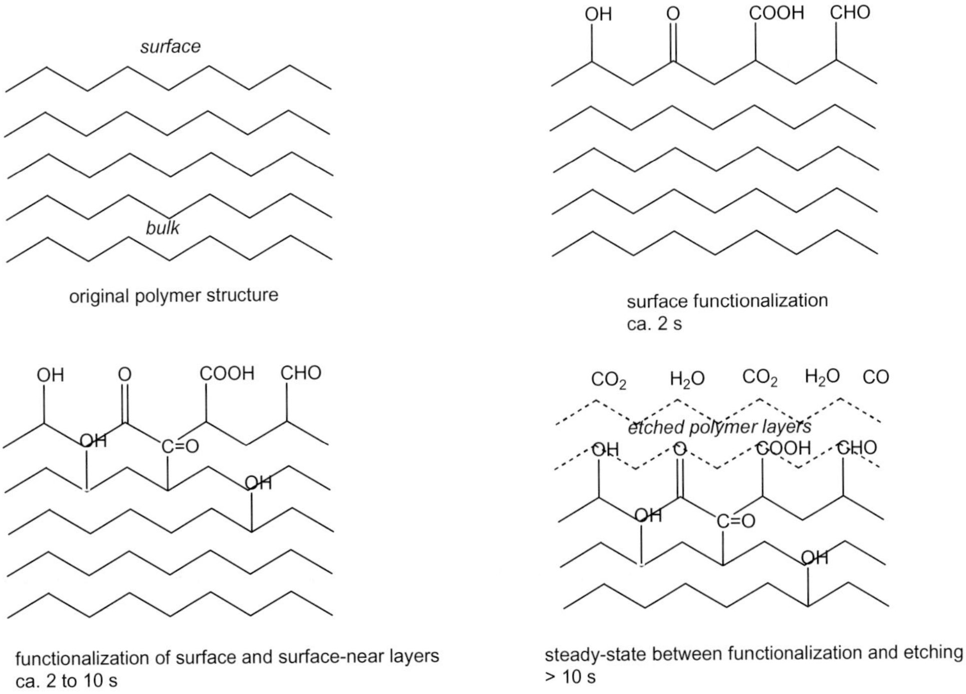

Figure 5.3 Proposed model of surface oxidation of polymers during exposure to oxygen plasma [1].

5.1.2
Unspecific Functionalizations by Gaseous Plasmas

As described below (Section 5.2), different fragments and atoms of the plasma gas are attached to surface vacancies of the macromolecular substrate. Because of the complex composition of such a gas plasma a broad variety of functional groups is produced. The ammonia plasma may serve as an example:

1) different fragments of the gas exist in the plasma (e.g., $NH_3 \rightarrow {}^{\bullet}NH_2$, NH, N, $^{\bullet}H$, N_2H_2, etc.);
2) build up of functional groups due to the synthesis of functional groups from single atoms of the plasma gas ($N_2 + 2H_2 \rightarrow 2{}^{\bullet}NH_2$);
3) formation of reactive (radical) sites at the polymer surface by the plasma particle bombardment, by irradiation with the plasma UV radiation, by chemical reactions as rearrangements, and so on ($\sim\sim\sim\sim + h\nu \rightarrow \sim\sim\sim\sim\bullet + {}^{\bullet}H$);
4) their sticking on reactive sites at the polymer surface ($\sim\sim\sim\sim\bullet + {}^{\bullet}NH_2 \rightarrow \sim\sim\sim\sim NH_2$);
5) post-plasma conversion of non-reacted (trapped) C-radical or peroxy-radical sites (cf. auto-oxidation $\sim\sim\sim\sim\bullet + {}^{\bullet}O\text{-}O^{\bullet} \rightarrow \sim\sim\sim\sim O\text{-}O^{\bullet}$).

Thus, a broad variety of different functional groups at the polyethylene surface is established: NH_3^+, NH_2, NH, N, C≡N, $CONH_2$, C=N–OH, NO, NO_2, COC, OH, >C=O, CHO, COOH, and C–O–OH [2].

5.2
Polymer Surface Oxidation

5.2.1
Polyolefins

Like all paraffin hydrocarbons, polyethylene plastic is chemically inert. Probably as a result of this inertness printing ink will not adhere to the surface of an ordinary polyethylene film. However, if the plastic surface is subjected to certain treatments, printing on the surface becomes possible. Since this is of great importance to the packaging industry, an investigation of the effect of these treatments on the molecular structure of polyethylene was of interest.

An area of considerable importance is the functionalization of polymer surface by the oxygen plasma to improve the surface free energy. As result of the oxygen low-pressure (may be also atmospheric-pressure plasma in air) plasma treatment a broad variety of different O-containing functional groups appears.

Considerable interest is focused on polyolefin surfaces, which do not possess any functional groups. The simplest way to introduce polar groups onto polyethylene and polypropylene surfaces, to enhancing wettability and adhesion properties, is by their oxidation. Flaming of polyolefins is well known as the Kreidl process [3–5]. Alternatively, dielectric barrier discharge treatment (DBD) in air was introduced (may be also called "corona") [6–9], as was low-pressure-plasma treatment ("glow discharge treatment") [10]. The first plasma treatment of polymers and paper was developed in Germany, in 1930, using such treated materials as dielectrics in condensers (private communication). The wettability with isolating oils was improved.

5.2.2 Aliphatic Self-Assembled Monolayers

Basic research was performed using Langmuir–Blodgett (LB) and self-assembled-monolayer (SAMs) as model surfaces for polyolefins exposed to low-pressure plasma. Aliphatic self-assembled monolayers of octadecyltrichlorosilane (OTS) with a methyl end group serve as model for polyethylene surfaces [11]. Such well-defined aliphatic C_{18} aliphatic chains with CH_3 end-groups deposited onto Si-wafers evaporated with ellipsoid silver clusters were exposed to oxygen low-pressure plasma and the oxidation process was observed by XPS and surface-enhanced infrared absorption (SEIRA). The formation of oxygen-containing functional groups was more or less complete within 2 s as shown by XPS (Figures 5.4 and 5.5).

Polyethylene surface oxidation shows the same behavior as depicted in Figure 5.4 (cf. Figure 5.9) [12].

Subsequent transition to the etching process after 2 s exposure to the oxygen plasma can be deduced from SEIRA (Figure 5.6) and XPS spectra.

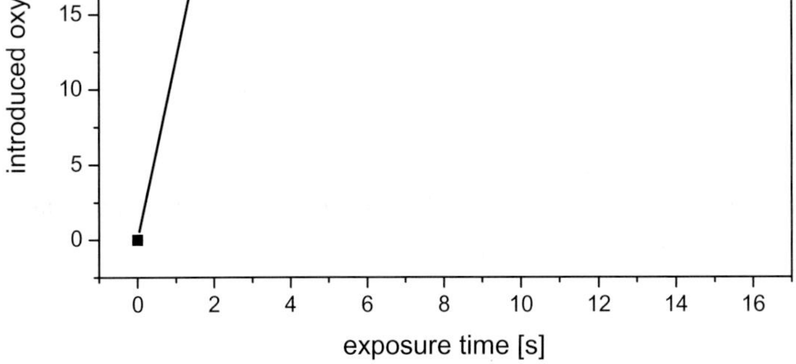

Figure 5.4 Oxygen introduction into octadecyltrichlorosilane self-assembled monolayers during exposure to oxygen plasma (DC, 10 Pa).

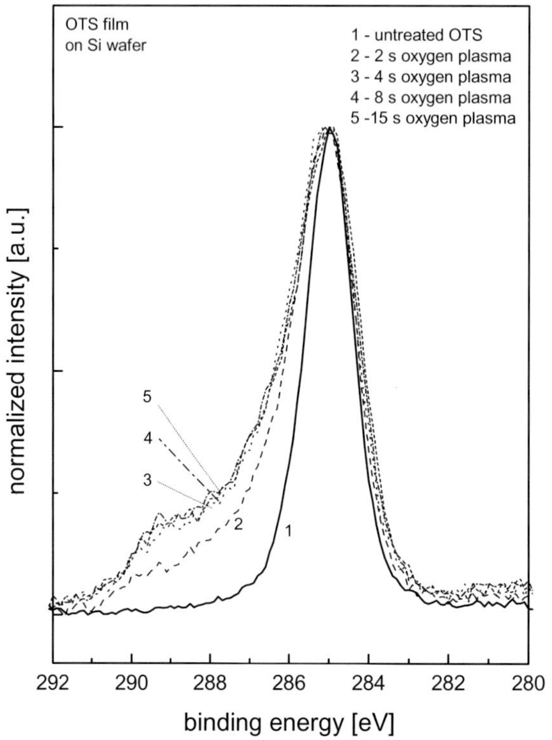

Figure 5.5 C1s signals of self-assembled monolayers of octadecyltrichlorosilane (OTS) exposed to low-pressure oxygen rf plasma (p = 10 Pa).

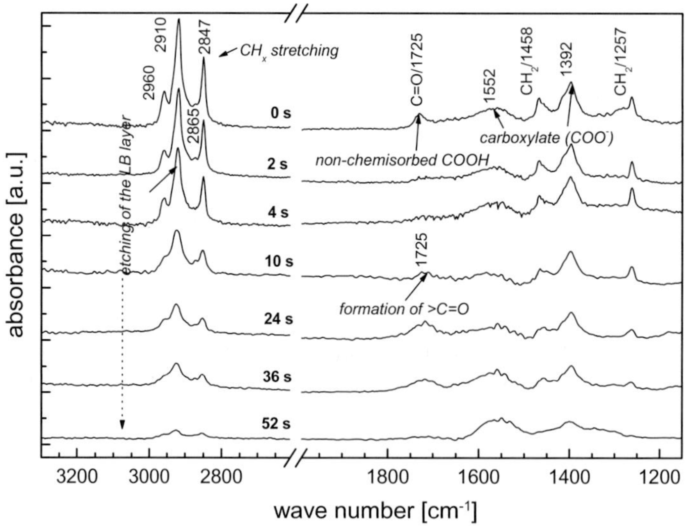

Figure 5.6 Surface-enhanced infrared absorption (SEIRA) spectra of a Langmuir–Blodgett layer of stearic acid deposited onto Ag cluster coated Si-wafer and exposed to oxygen plasma for different lengths of time (DC, 10 Pa).

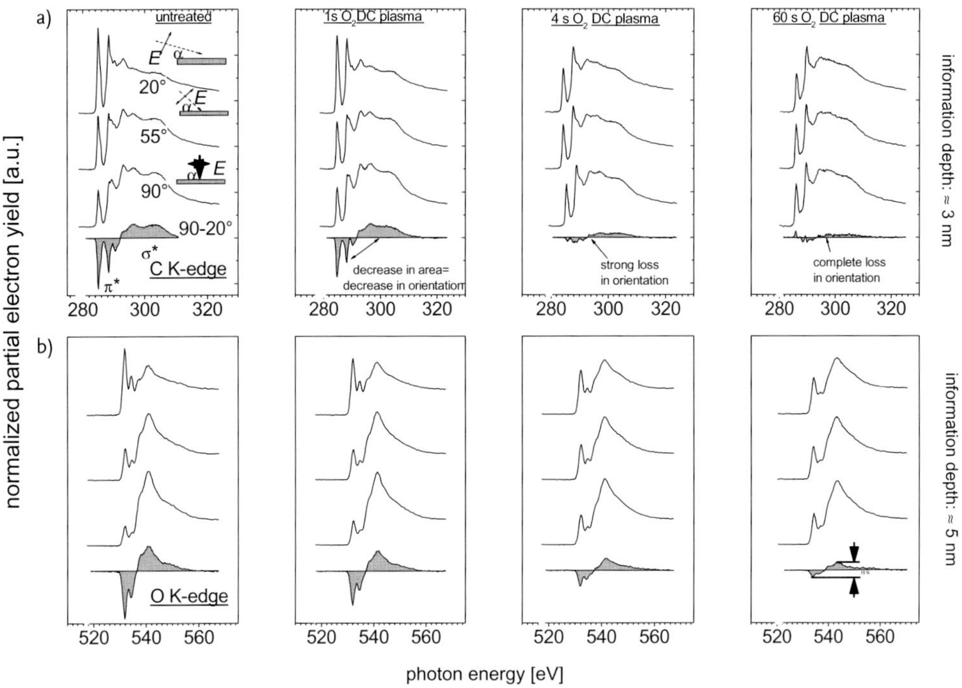

Figure 5.7 C_K-edge (a) and O_K-edge (b) of self-assembled octadecyltrichlorosilane self-assembled monolayers before and after exposure to low-pressure oxygen DC plasma.

The IR spectra show, by the CH_2 stretching vibrations near $2910\,cm^{-1}$, the successive etching of CH_2 units during exposure to the oxygen plasma [13]. Moreover, free carboxylic end-groups were preferentially etched. Instead of the carbonyl unit bond in carboxylic groups a new sort of carbonyl groups was formed, as indicated by a new band in the region of $1700\,cm^{-1}$ (Figure 5.6). Notably, the $\nu_{as}CH_3$ vibration ($2960\,cm^{-1}$) increases in relation to $\nu_{as}CH_2$ ($2910\,cm^{-1}$), evidencing chain scissions and branching.

The same is reflected in the near-edge X-ray absorption fine structure (NEXAFS) spectra of OTS monolayers. Within 2 s of oxygen plasma exposure the orientation of the self-assembled molecules is completely destroyed (Figures 5.7 and 5.8). The comparison of C_K- and O_K-edge spectra shows that in the O_K-spectrum residual occurrence of orientation is present. This is due to the greater sampling depth (≈5 nm) and, therefore, the complete OTS molecule is detected. It can be concluded that the disorientation proceeds from the surface of the SAM layer to the anchor groups on the substrate [14–15].

5.2.3
Polyethylene

Polyethylene is composed of an aliphatic backbone made of methylene groups (-[CH_2–CH_2]$_n$-). Clark had investigated the structural and compositional changes

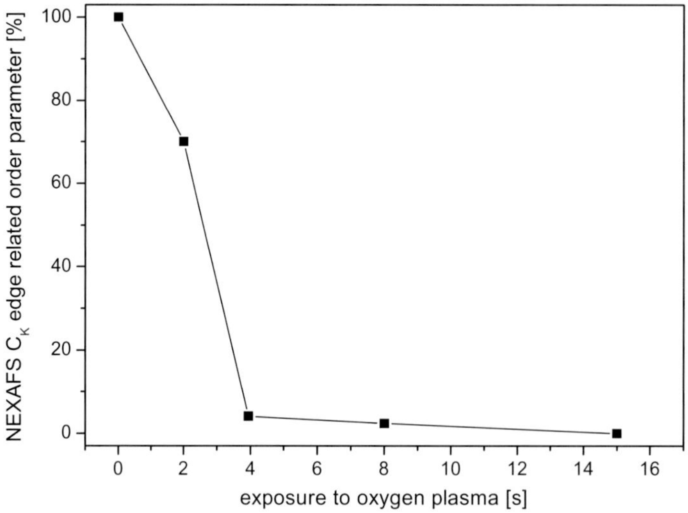

Figure 5.8 Dependence of the C_K-edge related order parameter of octadecyltrichlorosilane self-assembled monolayers on exposure to oxygen plasma.

of polyethylene on exposure to oxygen plasma [16]. C1s signal broadening was observed with increasing plasma exposure time. The fitting procedure of the C1s signal and the assignment of fitted components were adapted from Beamson and Briggs [17]. Figure 5.9 shows the fitting to four oxygen-containing groups.

Nowadays, the fitting procedure conforms to the peak positions in commercial polymers. The standard fitting of singly, doubly, and triply C-to-O bonded species shows all three components: C–O (~286.2 eV, C–O–C, C–OH, epoxides, C–O–OH), C=O [~288.0 eV, $R_1R_2C{=}O$, CHO, and possibly CONH– = (amide)] and O–C=O (~289 eV, COOH, COOR) (Figure 5.9). Moreover, peaks for hydrocarbons and carbons bonded in aromatic rings are positioned at 285.0 eV. Additionally, phenols (287.5 eV), benzoic acids (289.8 eV), and carbonates (290.5 eV) may also be fitted. Notably, most of these groups are also present at the surface after DBD treatment [18]. The number of peaks that could be fitted and their relative areas depend on the energy density (wattage × time) of DBD treatment. At higher energy levels an additional peak at higher binding energy (-O–CO–O) can also be fitted (Figure 5.10) [19].

Figure 5.11 presents the dependences of the fitted C1s components on time of exposure to atmospheric air plasmas and to low-pressure oxygen plasma. Characteristically, atmospheric-pressure plasmas incorporate oxygen functional groups into the polyethylene surface faster than the low-pressure oxygen plasma. Moreover, air plasmas also introduce nitrogen from air onto the polymer surface. Jet plasmas sometimes also deposit traces of metal oxides sputtered from the metal electrodes.

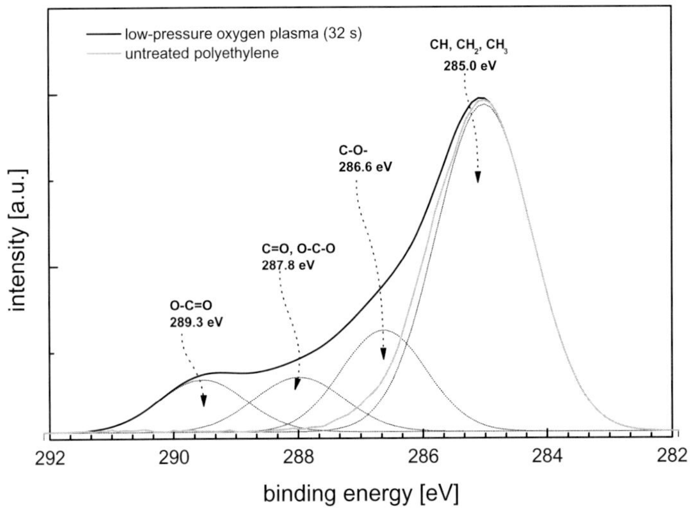

Figure 5.9 C1s peak of polyethylene after O_2 plasma treatment (15 s).

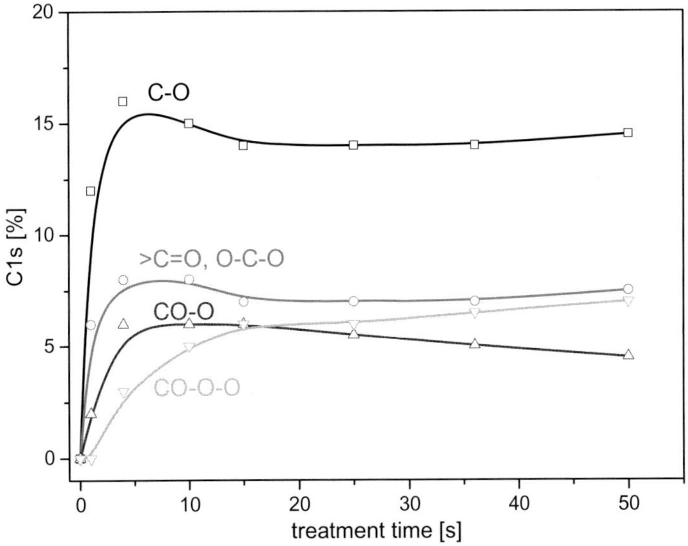

Figure 5.10 Oxygen introduction onto polyethylene surfaces and fitted C1s peak-features according to exposure to the cw-rf plasma (0.1 W, 30 Pa) [16].

Low-pressure oxygen plasma also produces olefinic double bonds and crosslinking. Using hexatriacontane (HTC, $C_{36}H_{74}$) as model substance for polyethylene, the formation of olefinic double bonds and the appearance of dimers and trimers, indications of crosslinking, were shown by matrix-assisted laser desorption ionization time-of-flight (MALDI-ToF) mass spectrometry (Figure 5.12).

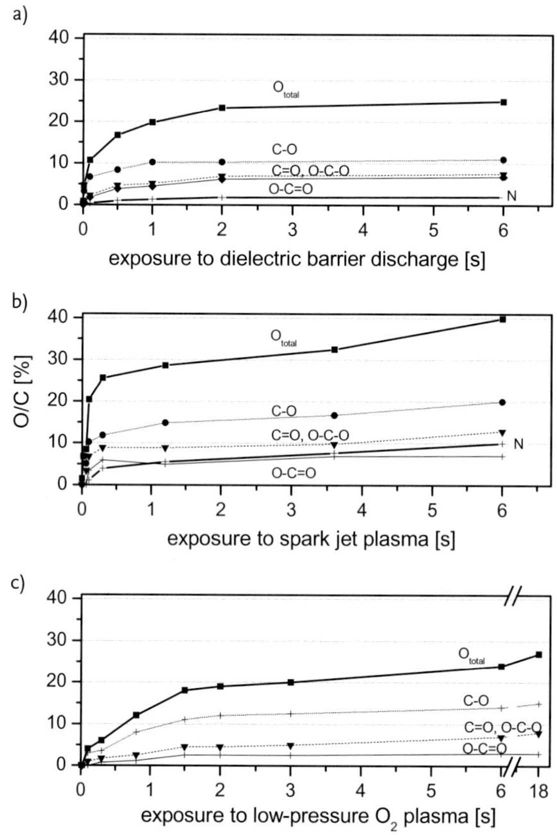

Figure 5.11 Oxygen introduction onto polyethylene surfaces and fitted C1s components according to exposure to (a) atmospheric dielectric barrier discharge, (b) atmospheric spark jet plasma, and (c) cw-rf low-pressure oxygen plasma (30 W, 6 Pa) [20].

The MALDI-ToF mass spectrum of HTC was obtained by drizzling a solution of dissolved matrix (UV absorber) with an additional silver salt onto the HTC. The thus matrix-impregnated HTC could be now analyzed by irradiation with a UV laser (cf. Scheme 6.1). Two double bonds were formed in HTC within the 180 s exposure to oxygen plasma. Such assumed crosslinking in strongly etching oxygen plasma was surprising but was ascribed to the action of plasma vacuum ultra-violet (VUV) radiation [21].

5.2.4
Polypropylene

Polypropylene is composed of an aliphatic backbone with methyl side-groups (-$[CH(CH_3)–CH_2]_n$-). Its C1s signal consists of four oxidized carbon species, which can be attributed to hydrocarbon and singly, doubly, and triple bonded C–O

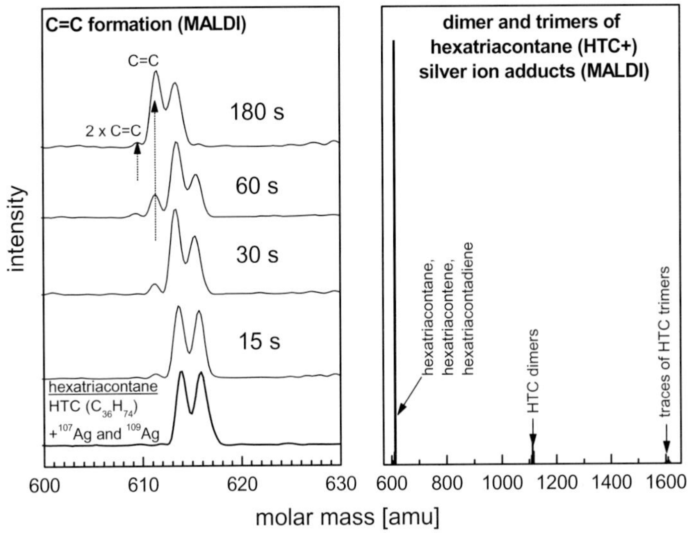

Figure 5.12 MALDI spectra of hexatriacontane (HTC) (additions of silver ions) after exposure to cw-rf low-pressure oxygen plasma.

groups. More generally, these species can be also assigned to the different oxidation ratios of carbon. The following product distribution was very often found when PP (polypropylene) or PE (polyethylene) was exposed to oxygen plasma: C–H, C–C ≡ 100% (hydrocarbon), C–O ~ 15%, C=O/O–C–O ~9%, -O–C=O ~ 6% (normalized to C–H) [22].

The proportion of singly, doubly, or triply bonded oxygen to carbon, as shown in Figure 5.13, is nearly constant for polyethylene and polypropylene using the low-pressure glow discharge oxygen plasma treatment [22]. It amounts roughly to a ratio of singly : doubly : triply bonded C–O species of 10 : 6 : 4.

However, also in the initial period of plasma-gas specific functionalization (<60 s) a weight loss, for example, etching of the polymer, is measured [23]. Therefore, a steady-state of plasma gas attachment and gasification of the polymer to small and stable degradation products is established.

Near-edge X-ray absorption fine structure (NEXAFS) spectra of the C_K-edge show the formation of olefinic double bonds, as seen by the splitting of the newly appeared pre-edge C1s → π* transition (cf. Section 5.6). Moreover, newly formed carbonyl-containing functions are produced during exposure to the low-pressure oxygen rf plasma (30 s) (Figure 5.14).

5.2.5
Polystyrene

Polystyrene (PS) is composed of an aliphatic backbone with attached phenyl side-groups (-[CH(phenyl)-CH_2]$_n$-). The aromatic ring guarantees greater stability

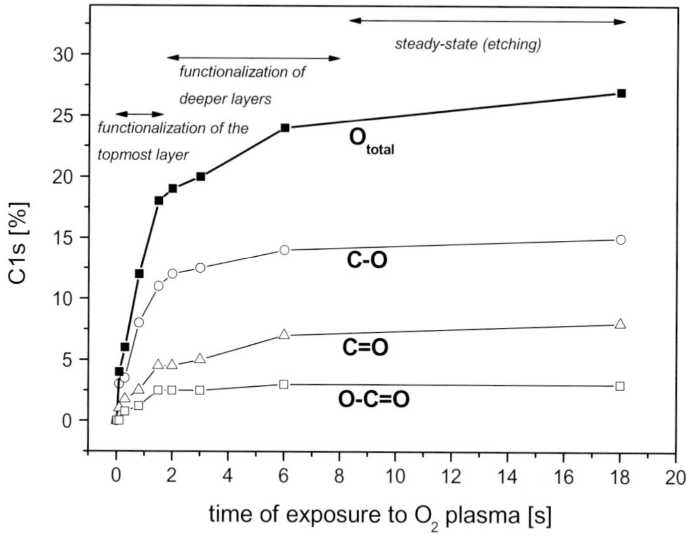

Figure 5.13 Oxygen introduction and fitted C1s on polypropylene surfaces according to exposure time to cw-rf plasma (100 W, 6 Pa).

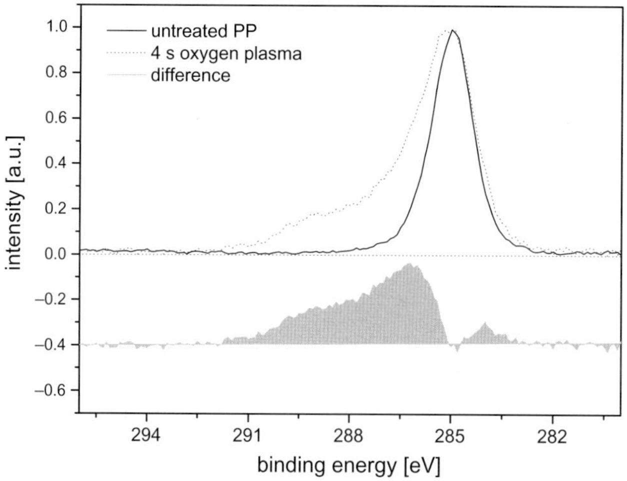

Figure 5.14 XPS-C1s signal of polypropylene before and after exposure to oxygen plasma (4 s, 6 Pa, cw-rf plasma, 30 W).

towards the picking up of charge and radiation from the plasma. The charge or energy can be delocalized over the whole ring and is thus better distributed and stored without bond scission. About 6 eV are needed to crack the aromatic ring system [24]. However, the weakest points of polystyrene molecule are the tertiary C–H bond of the backbone and the neighboring single C–C bond, which

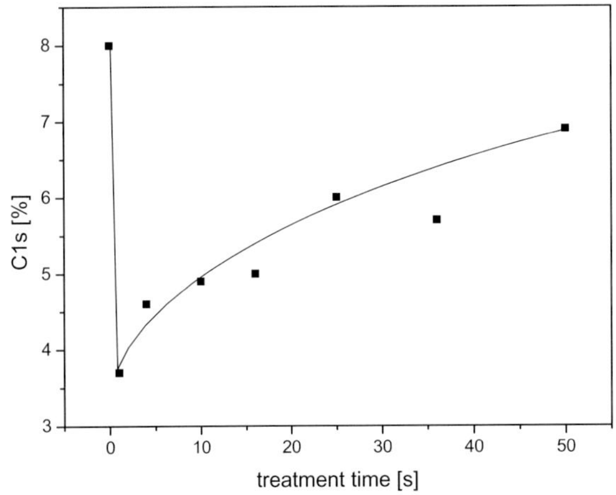

Figure 5.15 Shake-up intensity of polystyrene versus time of exposure to the oxygen plasma (0.1 W, 30 Pa).

links the ring to the polymer backbone. By introducing too much energy to the system the ring can split off or even crack as mentioned before. Another variant of degradation is the release of a complete monomer unit of the polystyrene macromolecule (depolymerization). The susceptibility of polystyrene towards oxidation is evidenced by an intense oxygen signal. Even at low power loading the reaction is rapid and is completed within a few tens of seconds. Clark emphasizes that only the outermost atomic monolayer of PS is strongly modified. This may be caused by phenyl rings at the surface shielding the subsurface and bulk against further irradiation; thus, the plasma effect is concentrated to the topmost monolayer [25].

Information on newly formed olefinic double bonds in polystyrene (originally only aromatic double bonds are present in the phenyl rings) can be obtained by changes in the shake-up satellite of the C1s signal (Figure 5.15).

The shake-up intensity decreased significantly in the first few seconds of oxygen plasma exposure due to the loss of phenyl rings. As the exposure precedes the intensity of the shake-up peak begins to re-increase (5 s). It then increases to nearly the value for unmodified polystyrene (Figure 5.15) [26].

This behavior may be explained by formation of an additional bond of the ring to a neighboring backbone, predominantly in para-position, through a second covalent bond, as demonstrated by IR spectroscopy in the range 1700–2000 cm^{-1}. Cracking of the ring produces conjugated and non-conjugated unsaturations and may contribute to crosslinking [27].

Figure 5.16 shows the dependence of the various carbon–oxygen features of polystyrene on exposure time to oxygen low-pressure plasma.

Oxidation at the polystyrene surface proceeds faster than that of polyethylene. The percentage of singly, doubly and triply bonded carbon–oxygen species after

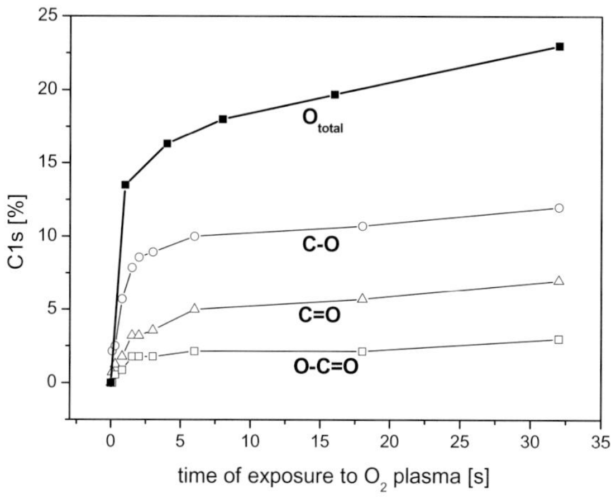

Figure 5.16 Oxygen introduction into polystyrene exposed to oxygen plasma (cw-rf, 50 W, 6 Pa).

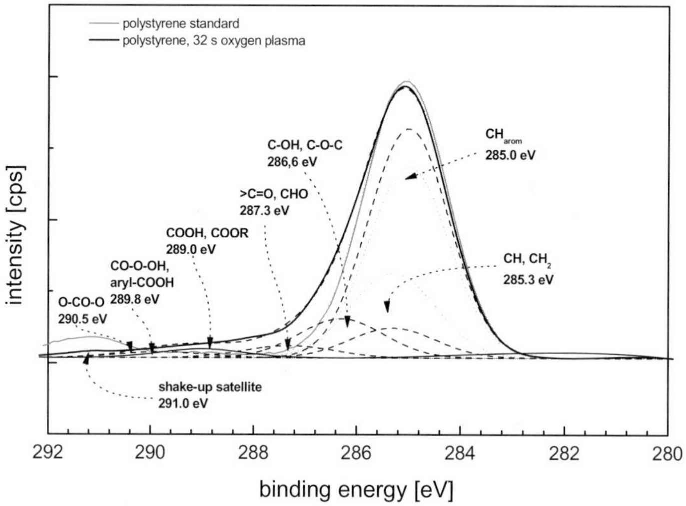

Figure 5.17 C1s peak of polystyrene standard (gray) and after exposure to oxygen plasma (in black, cw-rf plasma, 50 W, 6 Pa, 32 s).

fitting the C1s peak is nearly the same as with polyethylene. Additionally, carbonate-like (-O–CO–O-, fourfold bonded to oxygen, 290.5 eV) and benzoic acid structures (aryl-COOH, 298.8 eV) may be also present (Figure 5.17).

Importantly, polystyrene rapidly loses its aromatic rings. After 16 s exposure to low-pressure oxygen plasma nearly 50% of the aromatic rings have been removed,

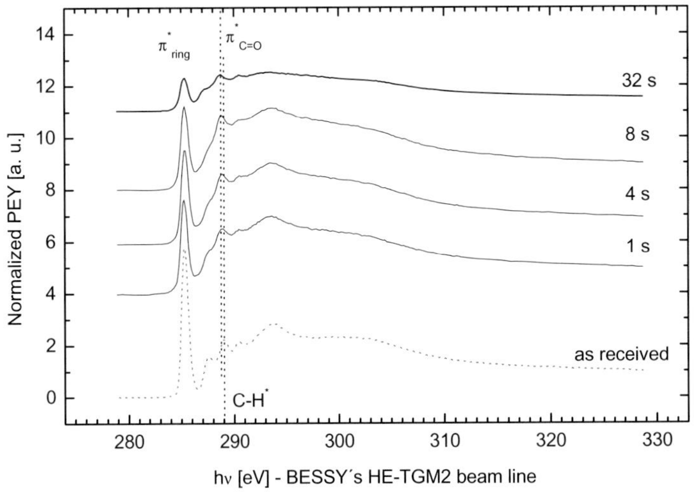

Figure 5.18 C K-edge for polystyrene (PS) on DC oxygen plasma treatment.

as evidenced by the shake-up signal within the C1s peak. It can be assumed that they are split off as benzene or are cracked [28].

NEXAFS spectra of the C1s → π* transition confirm the loss in aromatic rings and the formation of new carbonyl-containing functional groups, as seen by the loss in π^*_{ring} intensity and the appearance of a $\pi^*_{C=O}$ transition (Figure 5.18).

In addition to the XPS-detected features substantial changes in polymer structure are produced by exposure to low-pressure oxygen plasma within a few seconds. Degradation caused by C–C chain scissions is obvious. Degradation products can be detected using size exclusion chromatography (SEC, which is also known as GPC–gel permeation chromatography) and weakly crosslinked fractions by thermal field flow fractionation (ThFFF) [29]. Combining the results of these two methods the schematic diagram in Figure 5.19 was constructed. It demonstrates that both degradation and crosslinking are present simultaneously.

Calibration of the evaporative light scattering device (ELSD) used as detector in ThFFF shows the presence of ultrahigh molar mass species, which form the majority of macromolecules, and a high number of only slightly modified molecules as well as oligomers (and monomers) as typical degradation products. Figure 5.19 demonstrates significant changes in the molar mass distribution after exposure to low-pressure oxygen plasma for 10 s.

ThFFF measurements of the soluble and dispersible fractions of polystyrene show the existence of strongly crosslinked agglomerates, which represent about 90% of the mass of a 280 nm thick polystyrene film (Figure 5.20) [30].

This diagram confirms that the crosslinking proceeds very rapidly and achieves a steady-state already after 10 s exposure to the low-pressure cw-rf oxygen plasma. This behavior is astonishing because of the oxidative action, and therefore

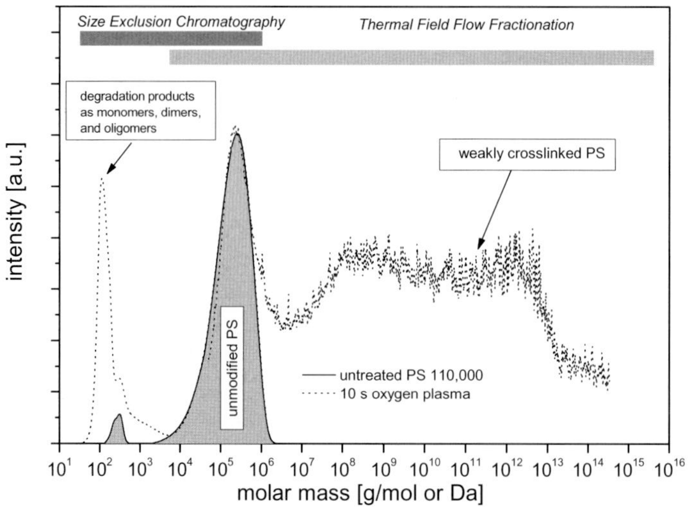

Figure 5.19 Schematic plot of changes in molecular weight of polystyrene (110 000 g mol^{-1}) on exposure to low-pressure cw-rf oxygen plasma, using SEC and ThFFF.

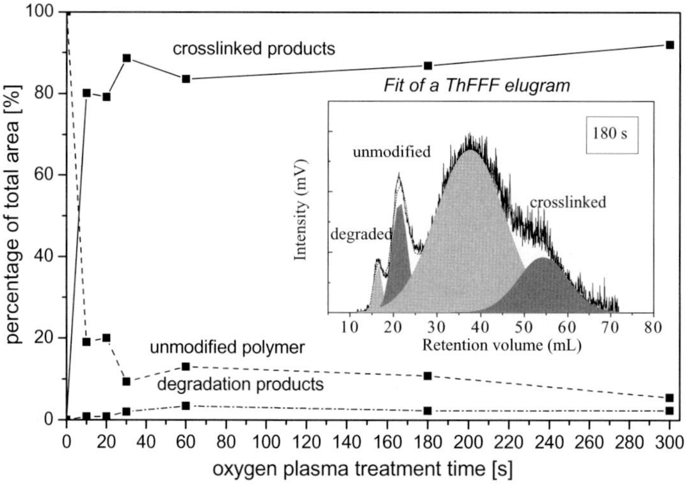

Figure 5.20 Variation of the concentration of the crosslinked fraction in polystyrene with exposure times to low-pressure oxygen plasma, using ThFFF.

degrading action, of oxygen plasma and the weak depolymerization tendency of polystyrene. IR spectroscopy also indicates crosslinking via aromatic rings by two- or threefold ring substitution.

It is also of interest to obtain information on the depth of crosslinking in polystyrene during plasma treatment. For this purpose polystyrene layers of different

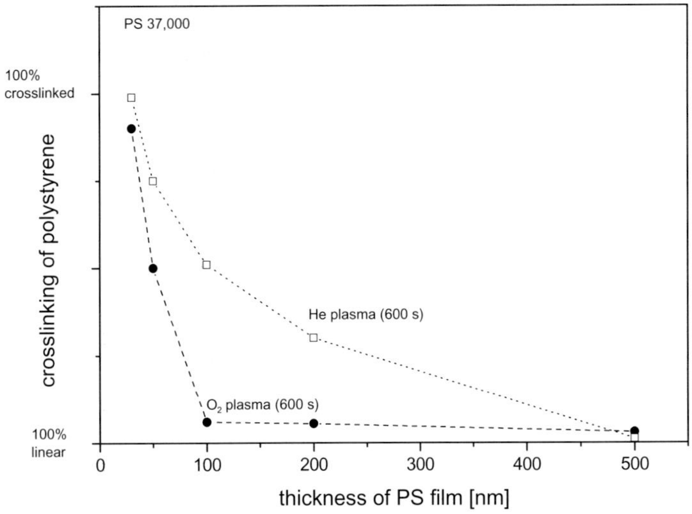

Figure 5.21 Dependence of crosslinked fraction in polystyrene on layer thickness (spin coating) after 600 s plasma exposure, using ThFFF.

thickness (50–500 nm) were deposited on Si-wafers by spin-coating [31]. It could be demonstrated that oxygen and helium plasmas crosslink thick polymer layers within 10 min exposure time (Figure 5.21).

The 50 nm layers are strongly crosslinked by both oxygen and helium plasmas; however, thicker layers (100–200 nm) are only crosslinked by helium plasma.

5.2.6 Polycarbonate

Figure 5.22 presents the C1s signal of polycarbonate before and after low-pressure oxygen plasma exposure [32].

The characteristic double peak in the O1s signal is also strongly changed to a uniform signal. The C1s and O1s signals reveal that oxygen plasma exposure has destroyed significant portions of the carbonate group as well as aromatic structures of the bisphenol A unit. A broad variety of newly formed oxygen-containing groups appears, as shown by the broad shoulder at the high-binding energy side of the C1s peak. Different C–O-containing groups, such as ether, epoxy, peroxy, and hydroxyl, may be produced in addition to carbonyl-containing groups, such as aldehydes and ketones, superposed by phenol groups, and threefold C-to-O bonded features, such as acids, peroxyacids, and esters. The corresponding O1s peak also shows strong changes and appears as a symmetric signal after plasma exposure. Thus, it can be assumed that the carbonate group as well as the aromatic structure is destroyed to a significant degree [33].

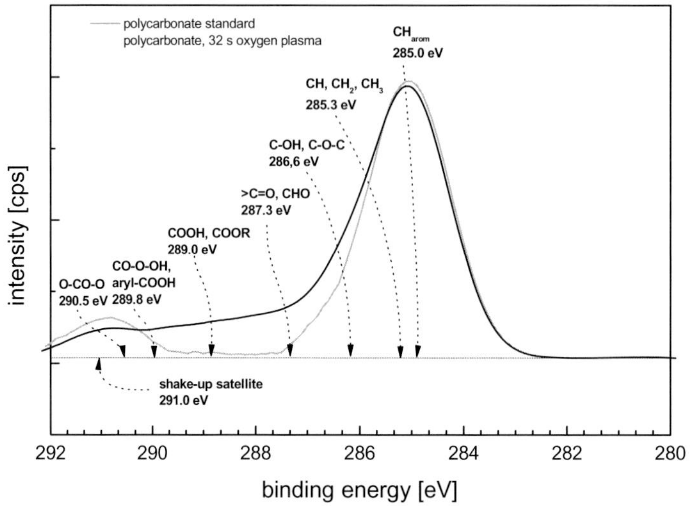

Figure 5.22 C1s peak of polycarbonate standard before (gray) and after exposure to oxygen plasma (black, cw-rf plasma, 50 W, 6 Pa, 32 s).

Polycarbonate shows a similar response to oxygen plasma exposure as poly(ethylene terephthalate) which is discussed below [32]. Polycarbonate has the carbonate group as characteristic group of polymer backbone. Here, the C atom has the highest oxidation state (4+) and, moreover, in the carbonate group the plasma-stable gaseous degradation product CO_2 is preformed. Therefore, also in this case, oxygen introduction during plasma exposure onto the polycarbonate surface is concomitant with the splitting-off of the carbonate group:

$$\text{C–O–CO–O–C} + O_2\ \textit{plasma} \rightarrow \text{-C-O}^\bullet + {}^\bullet\text{O-O–C-} + \text{CO}_2$$

This preferred fragmentation of the carbonate group is reflected in the rapid decrease of the C1s component at 290.5 eV and in those of C1sπ^* and O1s $\rightarrow \pi^*$ signals of NEXAFS (Figure 5.23).

In addition, the aromatic rings, bonded within in the backbone, were cracked rapidly. After 16 s exposure to low-pressure oxygen plasma the corresponding C1s $\rightarrow \pi^*$ resonance signal in the NEXAFS spectra and the corresponding C1s component at 290.5 eV both decrease by about 25% [33].

ThFFF-analysis showed that polycarbonate also crosslinks to a significant extent (Figure 5.24) [21].

5.2.7
Poly(ethylene terephthalate)

After exposure to oxygen plasma, the original structure of the C1s and O1s signals of poly(ethylene terephthalate) (PET) was strongly changed (Figure 5.25).

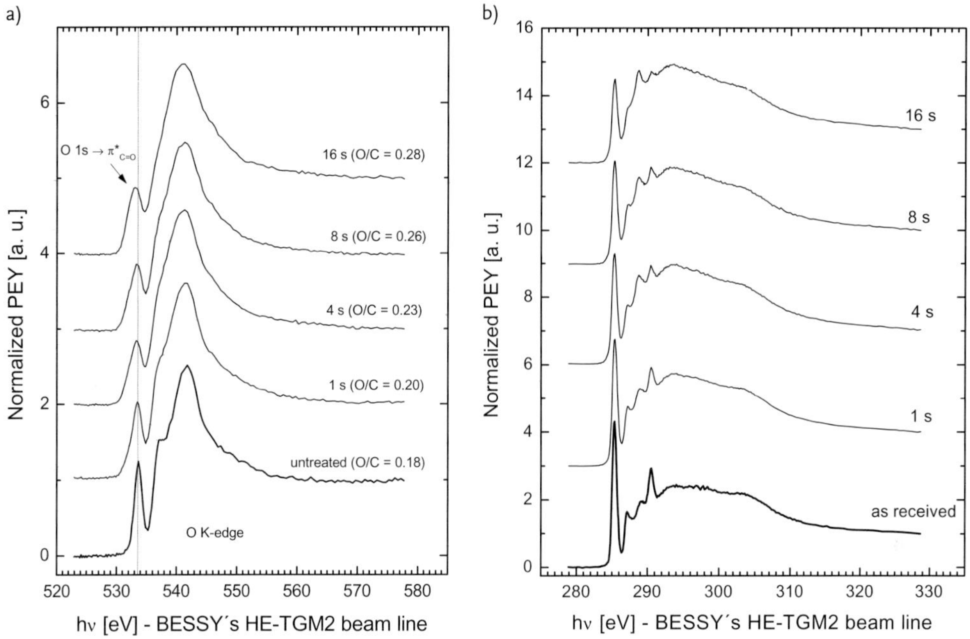

Figure 5.23 NEXAFS O_K edge (a) and C_K edge (b) spectra of low-pressure oxygen DC plasma treated poly(bisphenol-A-carbonate) (PEY = partial electron yield).

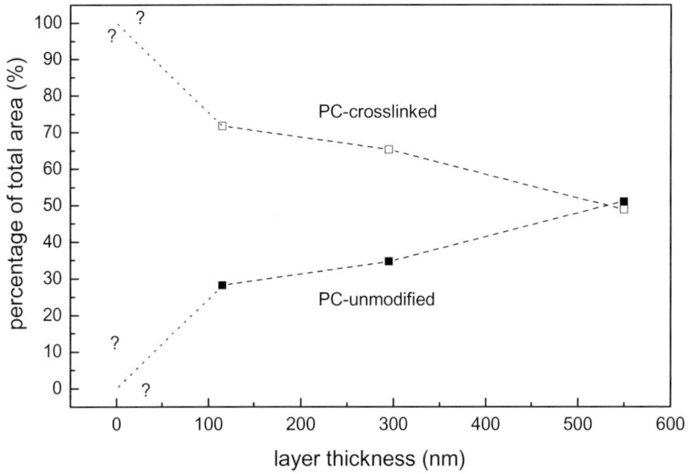

Figure 5.24 Dependence of crosslinked fraction in bisphenol A-polycarbonate on layer thickness (spin coating) after 600 s plasma exposure, obtained using ThFFF.

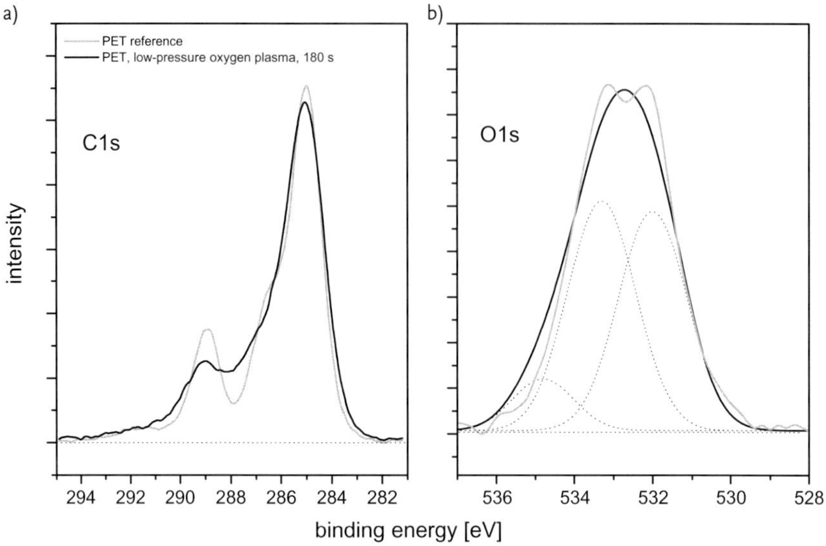

Figure 5.25 Changes in C1s and O1s signals of poly(ethylene terephthalate) after exposure to oxygen plasma (cw-rf plasma, 180 s, 6 Pa, 50 W).

PET shows a rapid decrease in concentration of the original structural elements of the molecule, for example, aromatic rings (shake-up), ester group, and glycol unit (Figure 5.25) during exposure to the oxygen plasma, as shown in the XPS C1s difference spectrum before and after plasma exposure [34].

The characteristic structures of the C1s and O1s peaks are already removed after exposure to the low-pressure rf oxygen plasma for 32 s. In particular, the characteristic shape of the O1s signal (camel-like) is completely diminished after 0.3 s plasma exposure (Figure 5.25) [35]. The loss of original ("old") structural elements and the "new" ones are reflected in the PET-C1s plot as difference spectrum between O_2 plasma modified and original PET (Figure 5.26).

The total amount of oxygen introduced into PET was superimposed by the simultaneous loss of oxygen caused by degradation and the splitting off of oxygen-containing degradation products from the polymer. Therefore, an increase or a decrease in the total oxygen percentage of PET that depended on the type of plasma and its conditions was found (Figure 5.27).

Low-pressure oxygen DC plasma did not significantly change the oxygen percentage of PET (theoretically 40% O : C) but rf plasma decreased the O : C ratio. Spark jet plasma increased significantly the O : C ratio, possibly due to metal oxide deposition originating from electrode material.

The ester group is especially attacked, in a manner comparable to the degradation of the carbonate group in polycarbonate. These two groups are most sensitive to exposure to oxygen plasma. The C1 signal of PET becomes more contour-less and broader, the aromatic $\pi \rightarrow \pi^*$ shake-up satellites diminish with treatment

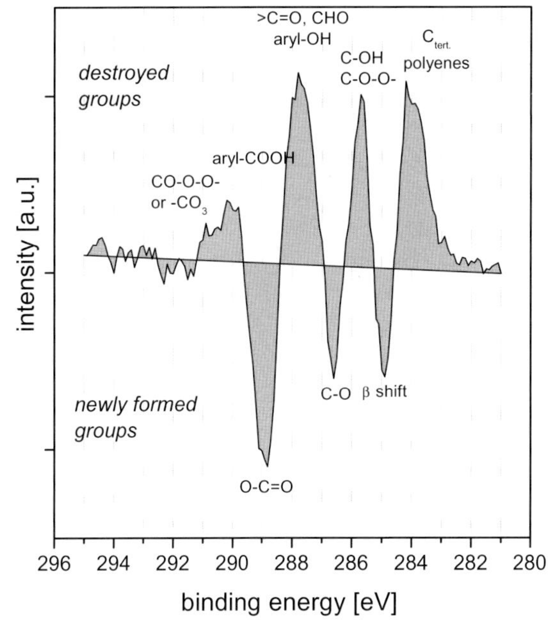

Figure 5.26 Changes in the C1s signals of poly(ethylene terephthalate) before and after exposure to oxygen plasma (cw-rf plasma, 180 s, 6 Pa, 50 W) presented as a difference spectrum.

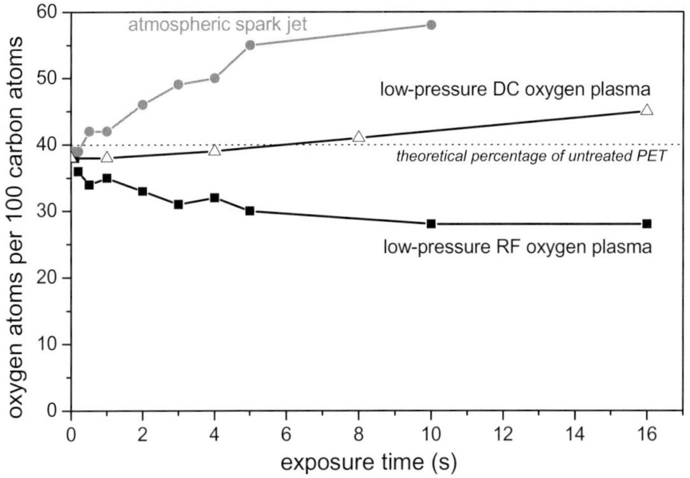

Figure 5.27 XPS measurements of the introduction of oxygen onto PET surfaces using different types of oxygen plasma.

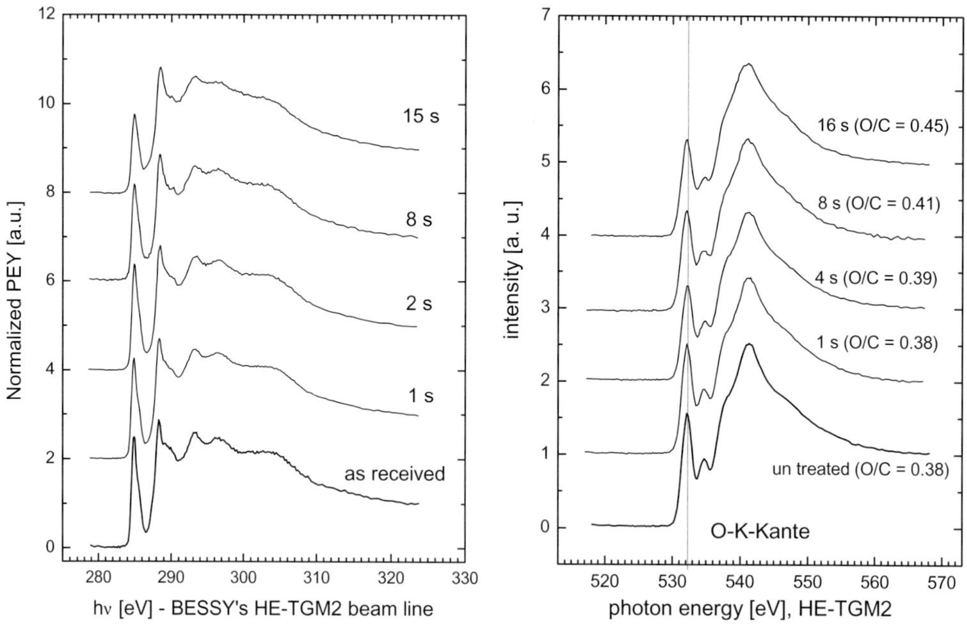

Figure 5.28 C_K-edge NEXAFS spectra of poly(ethylene terephthalate) spin coating films according to exposure to DC low-pressure oxygen plasma.

[33, 36]. On the other hand, new functional groups are formed, such as peroxyacids, carbonyls (aldehyde or ketone), phenols, and polyenes (Figure 5.26) [36–37].

The C_K-edge NEXAFS spectra confirm the loss of aromatic rings and ester groups (Figure 5.28).

Summarizing quantitatively the results of XPS and NEXAFS, the following dependencies on time of exposure to low-pressure oxygen plasma are evident (Figure 5.29).

PET spin coating film and PET biaxially stretched foil (MYLAR) show different supermolecular structures. Using prolonged exposure to the oxygen plasma the oxygen percentage exceeds the original O/C percentage of below 40% and achieves a maximum of more than 50%. Therefore, new carbonyl-containing (and other O-containing) groups are formed in both the spin-coating film and the foil. In both layers aromatic rings were destroyed by about 20% for each 16 s exposure to the oxygen plasma (Figure 5.29).

The following summary gives a picture of the broad variety and complexity of processes during the exposure of PET to oxygen plasma [34]:

Scission of aromatic rings:

phenylene ring → $^{\bullet}$CH=CH(R)-CH=CH-CH(R)=CH$^{\bullet}$

Fragmentation of aromatic rings:

phenylene ring → 3HC≡CH → unsaturated polymers

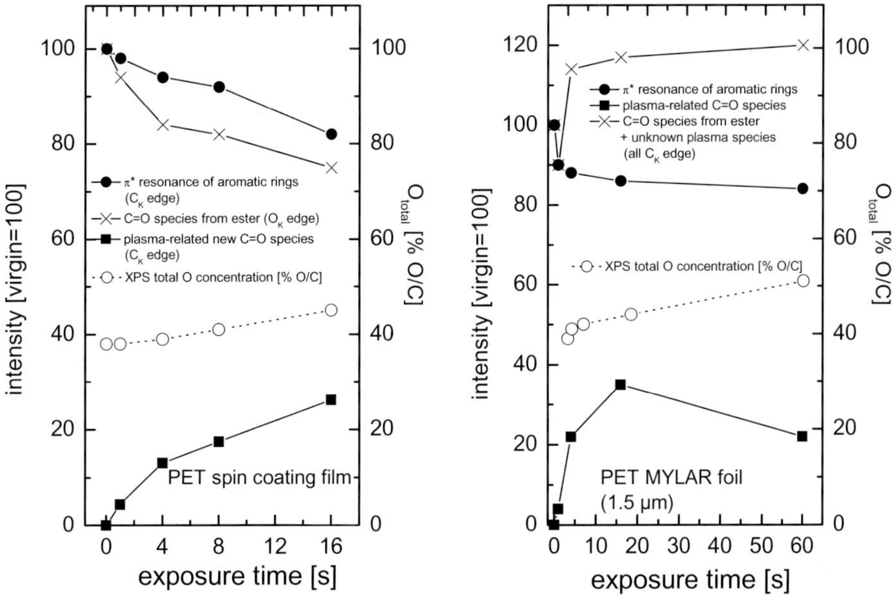

Figure 5.29 Dependence of relative concentration of structural elements in poly(ethylene terephthalate) on time of exposure to low-pressure oxygen DC plasma.

Ester group destruction:

phenylene ring-COO-R $\rightarrow$ phenylene ring-CO$^\bullet$ $\rightarrow$ phenylene ring$^\bullet$ + CO

phenylene ring-COO-R $\rightarrow$ phenylene ring$^\bullet$ + CO_2 + $^\bullet$R

phenylene ring-COO-R $\rightarrow$ phenylene ring$^\bullet$ + CO + $^\bullet$OH

Rearrangements:

phenylene ring-COO-CH_2-CH_2-OOC-phenylene ring
$\rightarrow$ phenylene ring-COO$^\bullet$ + $^\bullet CH_2$-CH_2-OOC-phenylene ring

phenylene ring-COO$^\bullet$ + $^\bullet CH_2$-CH_2-OOC-phenylene ring
$\rightarrow$ phenylene ring-COOH + CH_2=CH-OOC-phenylene ring

phenylene ring-COO-CH_2-CH_2-OH + CH_2=CH-OOC-phenylene ring
$\rightarrow$ products + CH_3CHO

Side reactions:

$n$$CH_3CHO \rightarrow CH_3(CH{=}CH)_{n-1}$-CHO

$n$$CH_2$=CH-OOC-phenylene ring $\rightarrow$ vinyl-ester polymers

vinyl-ester polymers $\rightarrow$ unsaturated linear polymers + terephthalic acid

Auto-oxidation of polymeric ether peroxides:

R-COO-CH_2-CH_2-OOC-R → R-COO-CH_2-•CH-OOC-R →

R-COO-CH_2-CH(O-O•)-OOC-R + RH → R-COO-CH_2-CH(O-OH)-OOC-R →

phenylene ring-COO-R → phenylene ring-CO• + •O-O•
→ phenylene ring-C(O)O-O• → phenylene ring-C(O)O-OH(R)

aliphatic-CHO → aliphatic-CO• → aliphatic-C(O)O-OH(R)

aliphatic-CHO → aliphatic-CO• + RH → aliphatic C-OH + R•

Hydrolysis:

phenylene ring-COO-R + H_2O → phenylene ring-COOH + HO-R

Phenol formation and photo-Fries:

phenylene ring + •OH -•H → phenylene ring-OH

phenylene ring• + •OH → phenylene ring-OH

phenylene ring• + •O-O• → phenylene ring-O-O• + RH
→ phenylene ring-O-OH → phenylene ring-OH

-phenylene-COO- → phenylene(OH)-CO-

Most of the equations have defined chemical origin; other reactions are only initiated by assistance of plasma. Thus, a co-existence of (dominating) chemical reactions occurs as Photo-Fries, Norrish I and II, ester hydrolysis, rearrangements, polymerizations, auto-oxidations, and so on with plasma-assisted (and often exotic) reactions such as the cracking of aromatic rings or the formation of peroxyacids as well as formation of phenols (ca. 287.5 eV) or benzoic acids (ca. 289.8 eV). The dominance of either chemical or plasma-assisted reactions among this broad variety of possible reactions also depends on the plasma parameters [34, 38].

Biaxially-stretched poly(ethylene terephthalate) foil (MYLAR) shows a rapid loss in orientation if exposed to low-pressure-oxygen plasma (Figure 5.30). This was evaluated by calculating the C1s → π^*_{ring} related order parameter (90–20° difference spectra) in the NEXAFS C_K-edge spectrum (Figure 5.30).

After 4 s exposure to the low-pressure oxygen plasma the film orientation has nearly disappeared and after 60 s exposure it is zero. The sampling depth of the C_K-edge spectra is about 3 nm and that of the O_K-edge about 5 nm. The disorientation is not complete within the sampling depth of the O_K-edge. Thus, it is demonstrated that the loss in order is limited to a few nanometres of the topmost surface layer [39].

Molecular weight degradation occurs on exposure to low-pressure oxygen plasma, as detected using MALDI-ToF mass spectrometry [28–29, 39–41]. SEC and ThFFF analysis also confirm the degradation as well crosslinking [30]. Figure 5.31 presents the SEC-detected changes in molecular weight during short-time expo-

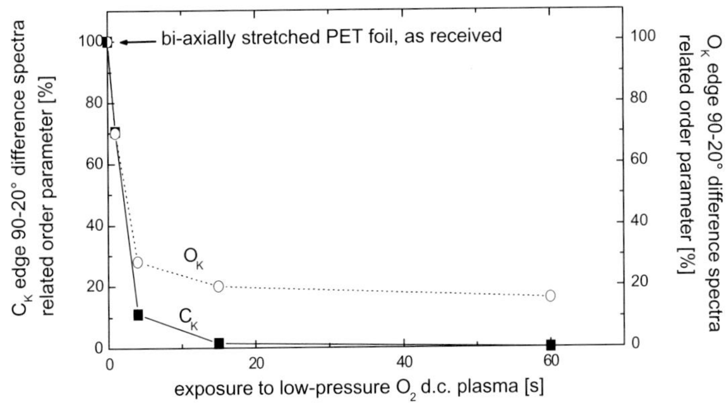

Figure 5.30 Dependence of the C1s → π^*_{ring} related order parameter of poly(ethylene terephthalate) on exposure to low-pressure oxygen DC plasma.

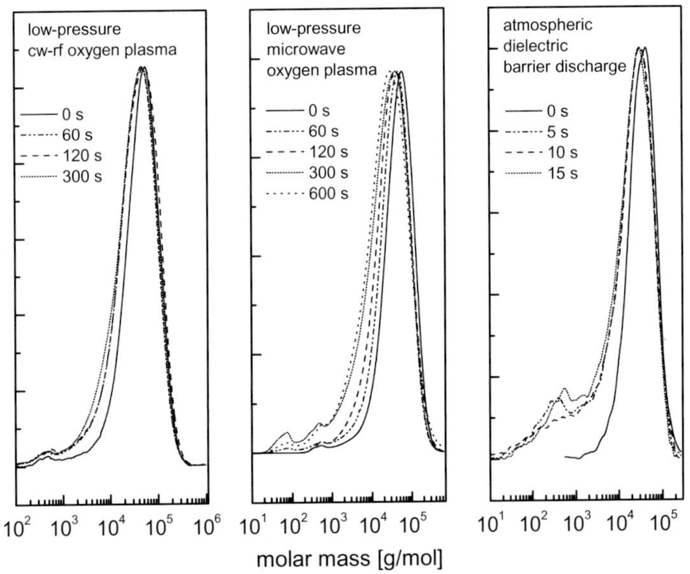

Figure 5.31 Change in molar mass distribution of poly(ethylene terephthalate) exposed to oxidizing gas plasmas as measured by size-exclusion chromatography (SEC).

sure of poly(ethylene terephthalate) spin-coating films (500 nm) to low-pressure oxygen cw-rf plasma [30].

Note that the high-molecular weight tail of the distribution function is cut because the polymer solution was filtered to remove all gel particles and agglomerates so as to avoid blocking of chromatographic columns, that is, only degradation

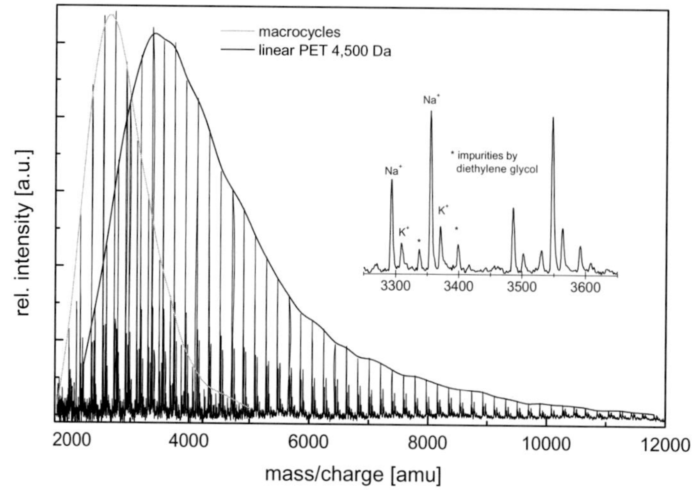

Figure 5.32 On exposure to an oxidizing gas plasma PET shows significant formation of macrocycles in the 2000–4000 Da region.

products are shown. The cw-rf oxygen plasma produces chain scissions appearing as shoulder of reduced molecular weights. It can be added that treatment in atmospheric-pressure plasma also induces significant polymer degradation. Small peaks in the 100–1000 region are assigned to fragments.

A special feature of nearly all polymers is the appearance of macrocycles (cf. Ziegler–Ruggli principle). PET also shows significant formation of macrocycles in the 2000–4000 Da region (Figure 5.32). Macrocycles were produced by the interaction of a thin PET film with oxygen low-pressure plasma. To confirm this, the original PET was chromatographically purified and all macrocycles were removed before plasma exposure was started [38].

5.2.8
Summary of Changes at Polymer Surfaces on Exposure to Oxygen Plasma

It can be generally stated that hydrogen atoms are abstracted from each polymer molecule exposed to oxygen plasma. Therefore, many unsaturations (C=C double bonds) and C-radical sites are formed, causing crosslinking, post-plasma oxidation, and auto-oxidation at ambient air. On the other hand, hydrogen in the plasma phase changes the physical properties of plasma and its chemical reactivity ("quasi-hydrogen plasma"). However, the hydrogen-rich plasma works by hydrogenating and forms CH_3 groups, which are indicators of chain scissions and branching. Analysis of the gas phase during the exposure of polypropylene to low-pressure cw-rf oxygen plasma shows the formation of hydrogen besides other degradation products such as water, carbon dioxide, and carbon monoxide (Figure 5.33).

Degradation of polymer chains is evidenced by a shift of molar mass distribution. Normally, on plasma exposure, the molar mass distribution is strongly broad-

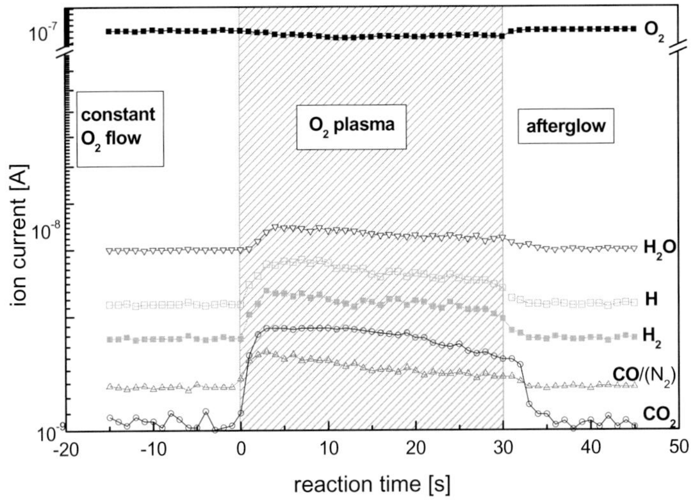

Figure 5.33 Gas-phase analysis of products formed during exposure of polypropylene to a low-pressure cw-rf oxygen plasma.

ened because of simultaneous degradation and crosslinking. In particular, dimers, trimers, oligomers are often produced (Figure 5.34). Macrocycles were formed for thermodynamic reasons. They hinder adhesion and act as an end-group-free separating agent.

Different types of degradation are obvious. Polystyrene forms monomers, dimers, or trimers more than polycarbonate and poly(ethylene terephthalate).

Functional groups were removed from polymer surfaces by exposure to oxygen plasma – in particular those with pre-formed degradation products in the structure, such as carbonate (CO_3) and ester/acid (COO) (Figure 5.35). In addition, other types of oxygen-rich structures were preferably degraded on exposure to any plasma. Dehydrogenation of NH_2 and CH_2 functional groups produces nitrile (see next section) and acetylenic groups.

Here, aromatic rings are stable towards plasma attacks. Otherwise, the plasma has enough energy to crack aromatic rings and form unsaturated structures. Another way of losing aromatic rings is to split-off phenyl groups from the backbone. The single C–C bond between backbone and aromatic ring only must be broken and the phenyl ring can be removed easily as gaseous benzene. Doubly or triply bonded aromatic rings within the polymer backbone are more difficult to remove.

The supermolecular structure, such as partial crystalline spherulitic or biaxially stretching, is rapidly destroyed and transformed into the amorphous state for both crystalline and stretch orientation.

Ether, polyenes, ketones, aldehydes, phenols, benzoic acids, peroxyacids, and hydroperoxides were formed as secondary products by re-arrangement of polymer structure, by post-plasma oxidation, or by abstraction and re-incorporation (Figure 5.35).

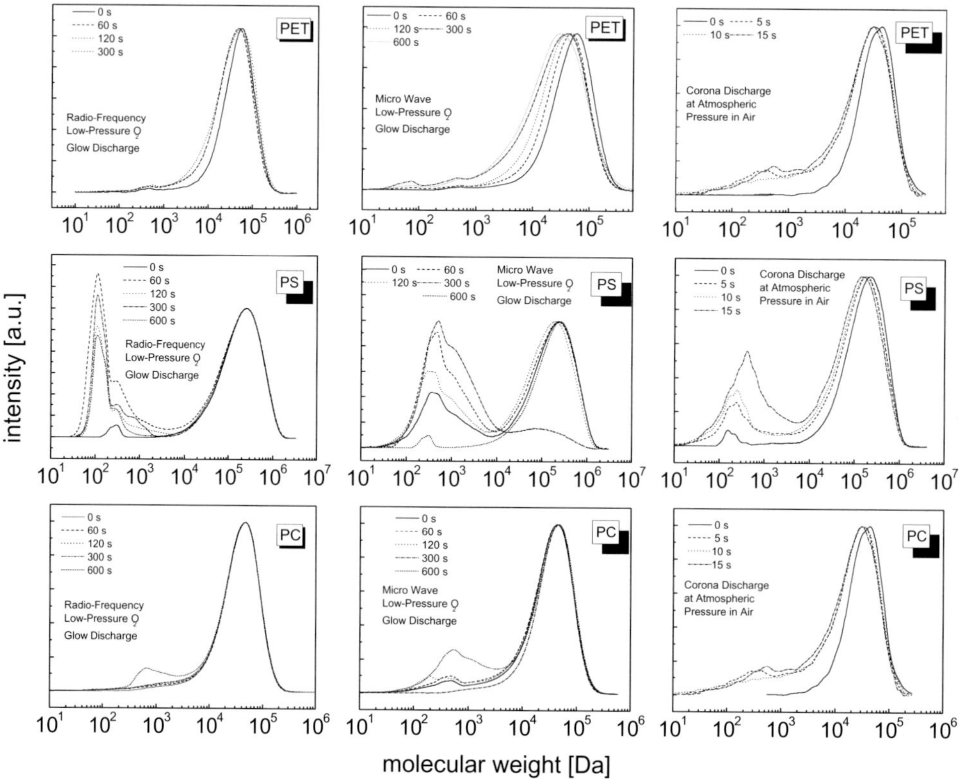

Figure 5.34 Shift of molar mass distribution for various polymers upon plasma exposure.

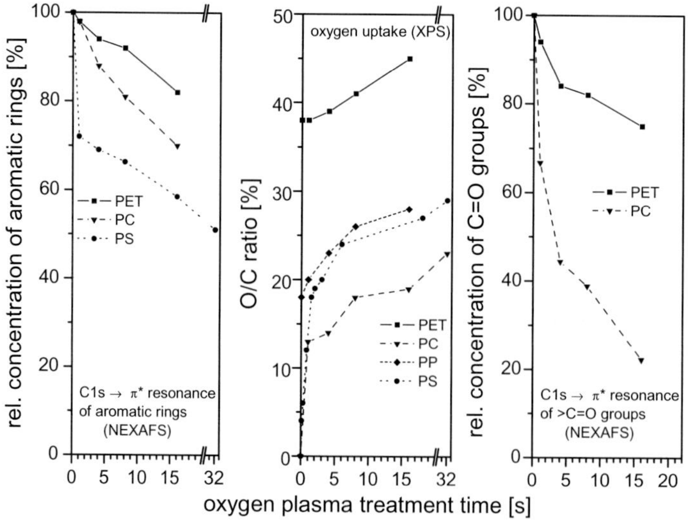

Figure 5.35 Loss of aromatic rings and carbonate/ester groups as well as oxygen incorporation in different polymers on exposure to low-pressure oxygen plasma.

5.2.9
Categories of General Behavior of Polymers on Exposure to Oxygen Plasma

Similar to polymer behavior exposed to ionizing radiation by crosslinking or depolymerization (Figure 5.36), these categories were also found on exposure to oxygen plasma. Two additional categories were photo-oxidative degradation (Figure 5.37) and random degradation [poly(ethylene glycol)] (Figure 5.38).

Poly(ethylene terephthalate) shows degradation behavior to clearly defined products (Figure 5.37), which is assigned to a photo-oxidative degradation mechanism [28].

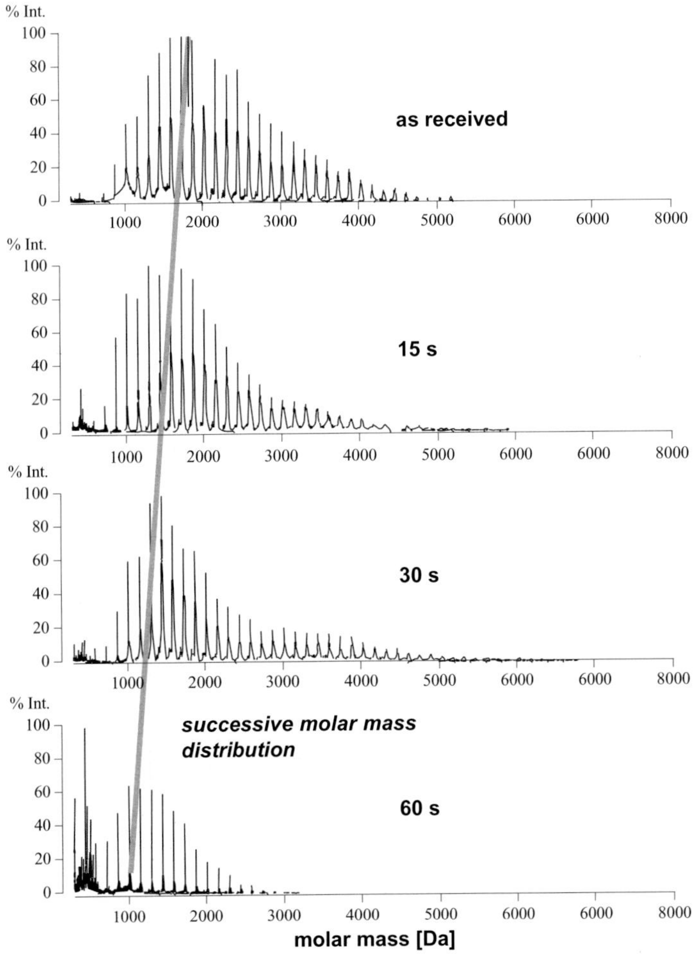

Figure 5.36 Shift of molar mass distribution of poly(methyl methacrylate) on exposure to low-pressure oxygen plasma shown by MALDI mass spectrometry.

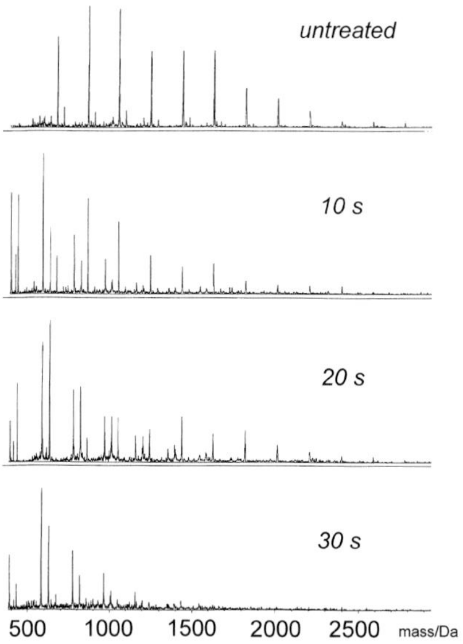

Figure 5.37 Shift of molar mass distribution of oligomeric poly(ethylene terephthalate) on exposure to low-pressure oxygen plasma shown by MALDI mass spectrometry.

The following polymer behavior can now be attributed to exposure to oxygen plasma:

1) preferred crosslinking as observed with polyolefins;
2) depolymerization as observed with poly(methyl methacrylate) or partially with polystyrene;
3) random degradation to undefined products as found with poly(ethylene glycol)s;
4) photo-oxidative dominated degradation processes as detected with poly(ethylene terephthalate).

Another criterion for polymer classification is that of etching rates. Yasuda *et al.* have published a list of polymers arranged according to etching rate [42] because the etching rate is independent of layer thickness [43]. It was found that oxygen-rich polymers were etched most rapidly, such as polyoxymethylene, polyacrylic acid, or poly(methyl methacrylate). Polyolefins, aromatic polyesters, and polyamides lower efficiently the etching process in low-pressure oxygen plasma. Friedrich *et al.* has confirmed this succession of etching rates and found that poly(vinyl acetate) and poly(methyl methacrylate) [23] as well as cellulose [44] had highest etching rates depending on plasma condition and used plasma gas as well as on polymer crystallinity [45–47]. Obviously, the oxygen percentage in the polymer structure, the pre-formation of oxygen-containing degradation products (CO, CO_2, H_2O), and the absence of pronounced crosslinking possibilities play the most important roles [48].

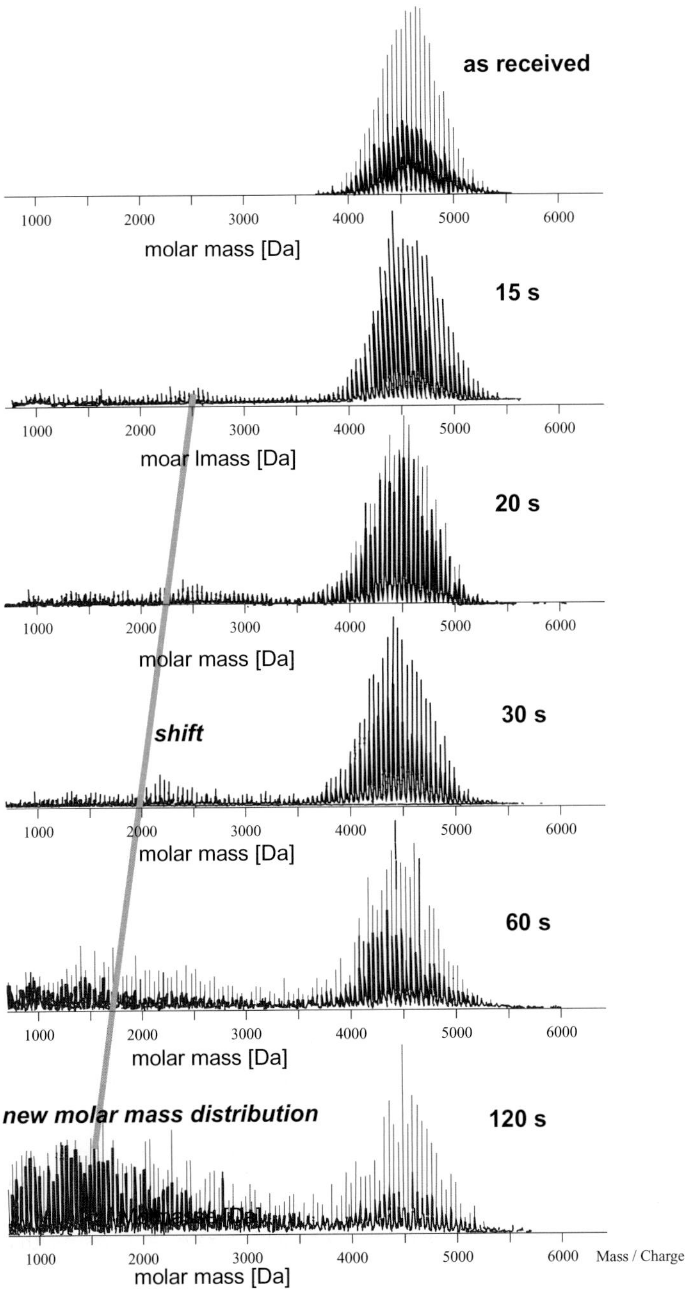

Figure 5.38 Random degradation of poly(ethylene glycol) on exposure to low-pressure oxygen plasma shown by MALDI mass spectrometry.

5.2.10
Role of Contaminations at Polymer Surfaces

Information on the time needed to functionalize the topmost polymer surfaces is strongly dependent on the actual situation at the surface. Its (undesirable) enrichment by inert macrocycles has been discussed with poly(ethylene terephthalate). Most often, other contaminations are also present, such as slip agents, UV stabilizers, catalysts, and contaminations from handling and ambient industrial environment, and are inhomogeneously distributed at the polymer surface. The XPS survey spectra in Figure 5.39 show the contamination of an industrially produced

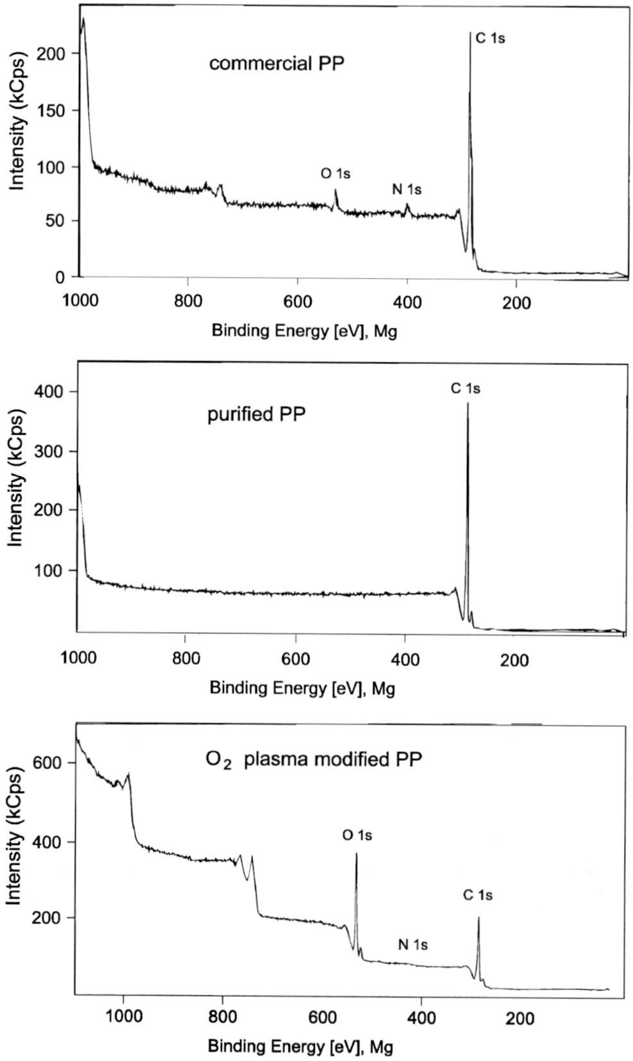

Figure 5.39 Comparison of XP survey spectra of polypropylene (PP) before and after purification and oxygen plasma modification.

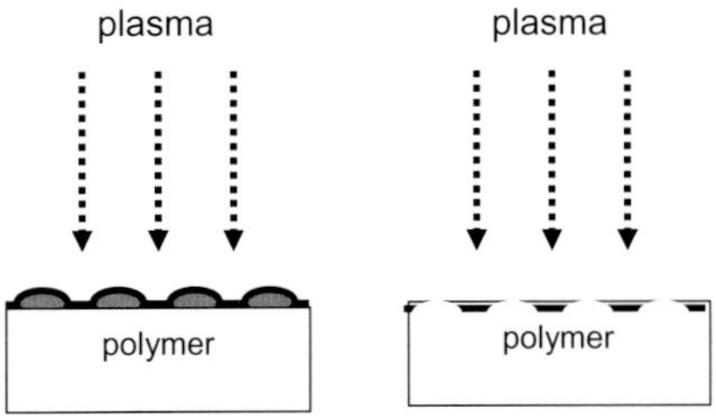

Figure 5.40 Etch effects on inhomogeneously contaminated polymer.

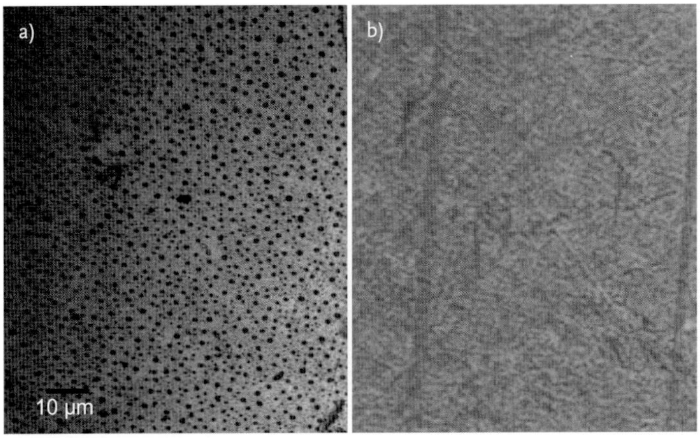

Figure 5.41 A polypropylene surface re-contaminated after exposure for two months to air by diffusion of UV absorber from the polymer bulk to the surface (a) and the topography immediately after exposure to oxygen plasma (b).

PP foil. Such contaminations amount to 3–8 at.% at the surface of polyethylene and polypropylene.

Most often the surface contaminations are located in a bubble- or string-like distribution. Thus, a short and careful plasma oxidation with a O_2 plasma functionalizes the uncovered areas of the polymer film and the bubbles consisting of agglomerated contaminations (Figure 5.40) [49].

However, below the bubbles the polymer substrate is shielded from exposure to plasma. After removal (etching, dissolving or wiping) of contaminations these regions are unmodified. Moreover, uncovered clean area is simultaneously etched, degraded, and/or crosslinked (Figure 5.40). Generally, a too long and too intense plasma etching of polymers is called *over-etching,* meaning that the modification itself has destroyed parts of the polymer structure (cf. Figure 5.40).

Moreover, it must also be considered that additives migrate from the bulk to the polymer surface, re-cover the plasma-activated polymer surface, and make it chemically inactive (Figure 5.41).

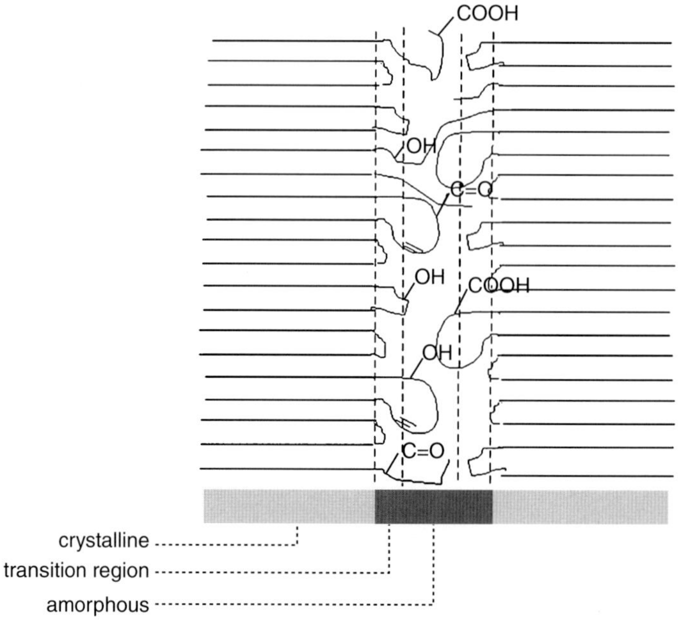

Figure 5.42 Model of a folded lamella of polyethylene and the most probable point of oxidation (in gray: introduced O-functional groups).

Figure 5.41a shows sweated stabilizer bubbles on oxygen plasma exposed polypropylene surface after a two month storage period. The stabilizer is migrated from the bulk to the plasma-activated polymer surface within a few weeks [50].

The polymer supermolecular structure also influences surface functionalization. The primary attack of the oxygen plasma is directed to the amorphous region of the folded parts of the crystalline region of the lamella (Figure 5.42). More details are given in the next section.

5.2.11
Dependence of Surface Energy on Oxygen Introduction

Oxygen introduction by low-pressure plasma treatment is associated with the formation of O-polar groups, which is reflected in increasing of surface energy and its polar contribution (Figure 5.43). Remarkably, the polar contribution decreases after a few 10 s plasma exposure in these examples.

The polar contribution of surface energy shows a characteristic dependence on oxygen concentration at the polymer surface. The dependence is more or less linear up to a maximal value for the polar contribution. After that maximum the polar contribution curves decrease rapidly despite increasing oxygen concentration (Figure 5.44). This behavior is attributed to degradation at polymer surfaces and the formation of low-molecular weight oxidized material. This degraded material

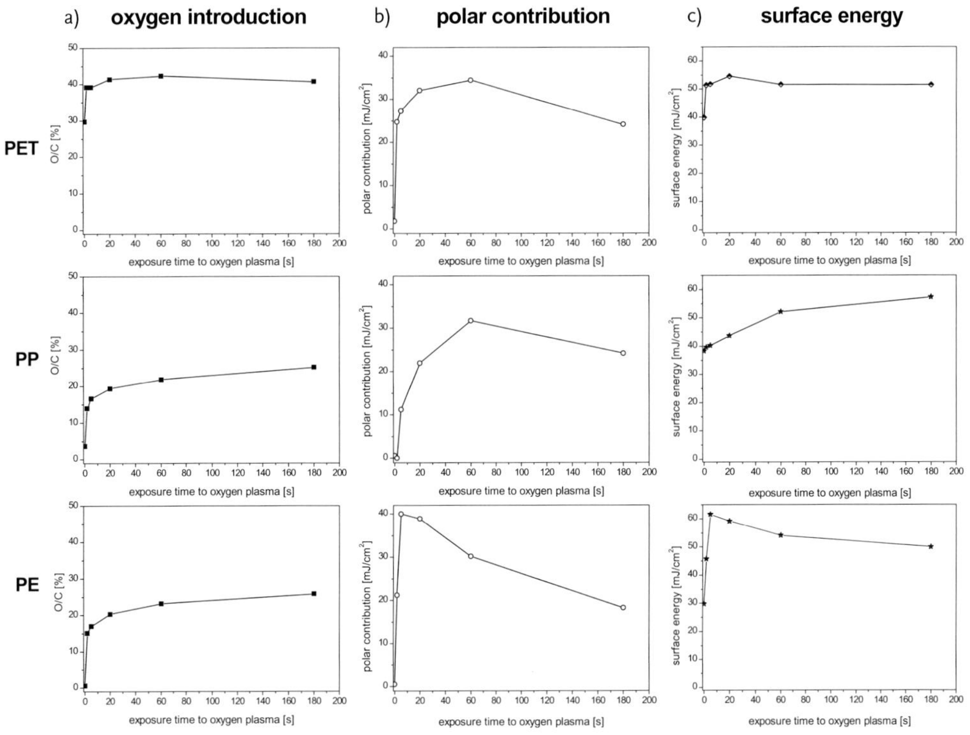

Figure 5.43 Polar contribution and surface energy according to exposure time to low-pressure oxygen plasma for (a) polyethylene terephthalate (PET), (b) polypropylene (PP), and (c) polyethylene (PE).

may be also partially dissolved in a drop of test liquid and, thus, can adulterate the measured contact angle.

5.3
Polymer Surface Functionalization with Amino Groups

5.3.1
Ammonia Plasma Treatment for Introduction of Amino Groups

Amino groups play an important role in nature. They are constituents of amino acids, which are the building blocks of all proteins. Amino groups react with carboxylic groups to give peptide or amide bonds: $R{-}NH_2 + HOOC{-}R^* \rightarrow R{-}NH{-}CO{-}R^*$.

Pure amines are formed from amino acids by decarboxylation within the metabolism. They occur in foods or are formed secondarily in it by microbiological degradation of proteins. In 1953 Miller and Urey explained experimentally the

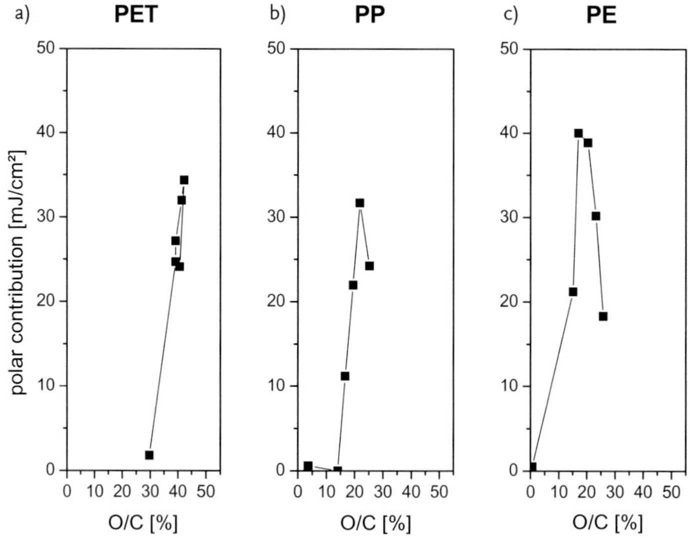

Figure 5.44 Polar contribution of surface energy versus oxygen concentration at the polymer surface introduced by low-pressure oxygen plasma exposure for (a) PET, (b) PP, and (c) PE.

NH_3	$-NH_2$	-NH-	>N-	$>N^+<$
ammonia	primary amine (amino)	secondary amine (imino)	tertiary amine	ammonium

Figure 5.45 Different N-functional groups.

genesis of life. They showed that amino acids are produced on plasma exposure of a liquid-phase mixture of water, hydrogen, ammonia, methane, and carbon monoxide. This mixture simulated the early Earth atmosphere [51–52]. At the beginning of this experiment hydrogen cyanide (HCN) and aldehydes (R-CHO) were formed followed by their reaction to 28 amino acids:

$$\text{R-CHO} + \text{HCN} + \text{H}_2\text{O} \rightarrow \text{H}_2\text{N-CHR-COOH}$$

$$\text{R-CHO} + \text{HCN} + 2\text{H}_2\text{O} \rightarrow \text{HO-CHR-COOH} + \text{NH}_3$$

Because of their chemical reactivity amino groups are of special interest. However, several different N-functional groups exist besides the most important and most reactive primary amino groups (Figure 5.45).

As already noted, the primary amino groups play an important role in biochemistry. First, Hollahan *et al.* used an ammonia plasma to introduce (primary) amino groups onto polymer surfaces such as polypropylene, poly(vinyl chloride), polytetrafluoroethylene, polycarbonate, polyurethane, and poly(methyl methacrylate)

[53]. The primary amino groups made the polymer surface biocompatible by reaction with heparin.

Hollahan had proposed that the ammonia molecule dissociates in the plasma phase by abstraction of a hydrogen atom, thus forming the NH_2 radical. Simultaneously, one hydrogen atom is also removed from the polymer backbone to produce a C-radical site. Thus, NH_2 radical and C-radical can recombine on impinging the polymer surface [53]:

$$NH_3 + e^- \rightarrow {}^{\bullet}NH_2 + H^{\bullet}$$

$$\text{-}[CH_2{-}CH(CH_3)]_{n-} + h\nu \rightarrow \text{-}[CH_2{-}C^{\bullet}(CH3)]_{n-} + H^{\bullet}$$

$$CH_2{-}C^{\bullet}(CH_3)]_{n-} + {}^{\bullet}NH_2 \rightarrow \text{-}[CH_2{-}C(NH_2)(CH_3)]_{n}\text{-}$$

It is shown below that this assumption is too optimistic and that the ammonia molecule tends to form preferably NH and N species. Therefore, it is of note that the yield in primary amino group formation in plasmas of ammonia, nitrogen, or nitrogen-hydrogen mixtures is less than 10% of all inserted nitrogen components. ToF-SSIMS (time-of-flight static secondary ion mass spectrometry) analysis identified $2NH_2/100$ C after exposure of polystyrene to ammonia plasma [54]. NH_3/H_2-mixtures did not improve the yield in NH_2 groups significantly [55–57] and neither did N_2/H_2-mixtures [58].

The reason may be situated in the thermodynamics. The dissociation energies of $N{-}H_x$ bonds favors the formation of secondary and tertiary N products (NH and N), as shown by the standard bond dissociation values ($kJ\,mol^{-1}$) given by Wedenejew *et al.* [24]: N–H = 348, NH–H = 381; NH_2–H = 440, and CH–H = 395. Generally, the formation of NH_2 groups in ammonia plasma is not preferred thermodynamically, as reflected by the low yield of hydrazine and shown by the different dissociation energies [59]:

$$NH_3\ \textit{plasma} \rightarrow {}^{\bullet}NH_2 + {}^{\bullet}H\ 461\ kJ\ mol^{-1}$$

$$NH_3\ \textit{plasma} \rightarrow {}^{\bullet}NH + H_2\ 209\ kJ\ mol^{-1}$$

The formation of desired $^{\bullet}NH_2$ radicals needs more than twice the energy required for NH formation, which explains the high percentage of nitrogen incorporation into polymer surfaces but the low yield of NH_2 group formation [59].

Nevertheless, to inspect this amino group functionalization process in detail polyolefin surfaces have been exposed to ammonia, nitrogen or N_2/H_2 plasmas. NH_2 species may be attached to polyolefin surfaces by radical recombination or by substitution. Such substitution (S_R mechanism) consists of the following processes:

$$NH_3 + e^- \rightarrow {}^{\bullet}NH_2 + H^{\bullet}$$

$$\text{-}[CH_2CH(CH3)]_{n-} + {}^{\bullet}NH_2 \rightarrow \text{-}[CH_2C^{\bullet}(CH_3)]_{n-} + NH_3$$

or:

$$\text{-}[CH_2CH_2CH_2]_{n-} + {}^{\bullet}H \rightarrow \text{-}[CH_2C^{\bullet}(CH_3)]_{n-} + H_2$$

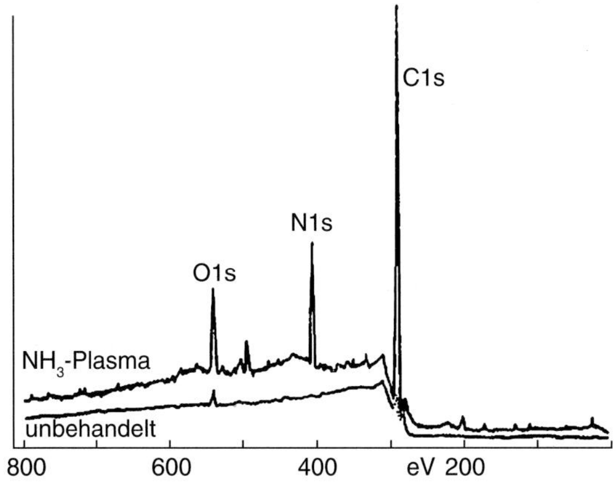

Figure 5.46 XPS-survey scan of a polypropylene surface exposed to ammonia gas plasma (cw-rf, 300 W, 27 Pa, 50 min) [2].

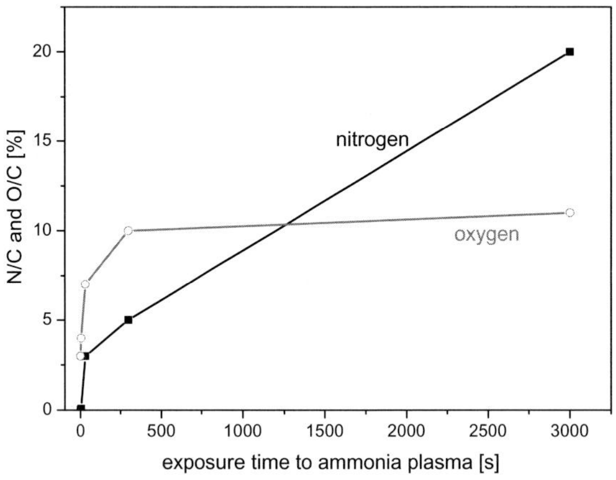

Figure 5.47 Dependence of N and (dominantly post-plasma) O introduction onto a polypropylene surface on exposure time to an ammonia plasma (cw-rf, 300 W, 27 Pa).

$$\text{-[CH}_2\text{-C}^\bullet\text{(CH}_3\text{)]}_{n-} + {}^\bullet\text{NH}_2 \rightarrow \text{-[CH}_2\text{-C(NH}_2\text{)(CH}_3\text{)]}_n\text{-}$$

Compared with the time-dependence of oxygen introduction in the oxygen plasma the ammonia plasma treatment of polypropylene films produces a slower surface functionalization. However, in addition to the introduction of N-functional groups high concentrations of post-plasma attached oxygen were measured (Figure 5.46).

The time-dependences of N- and O-introduction onto polypropylene surfaces show generally the same behavior as that presented in Figure 5.47 using the XPS as detection method. In the following this process is discussed in more detail. Here, only general aspects are mentioned.

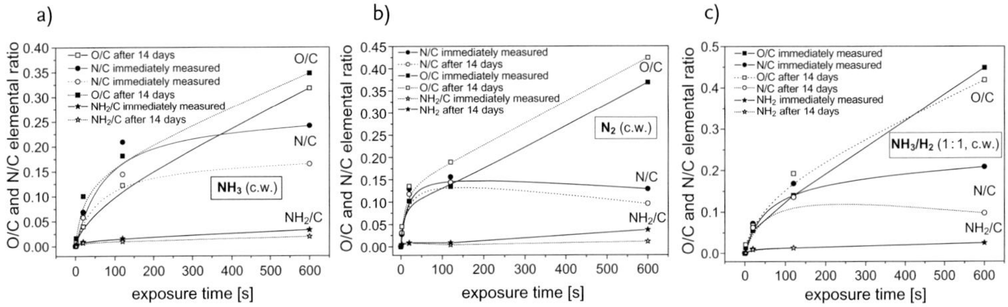

Figure 5.48 XPS-measured N, O, and NH_2 concentrations of polypropylene surfaces exposed to (a) NH_3, (b) N_2, and (c) $N_2 + H_2$ cw-rf plasmas (6 Pa, 30 W).

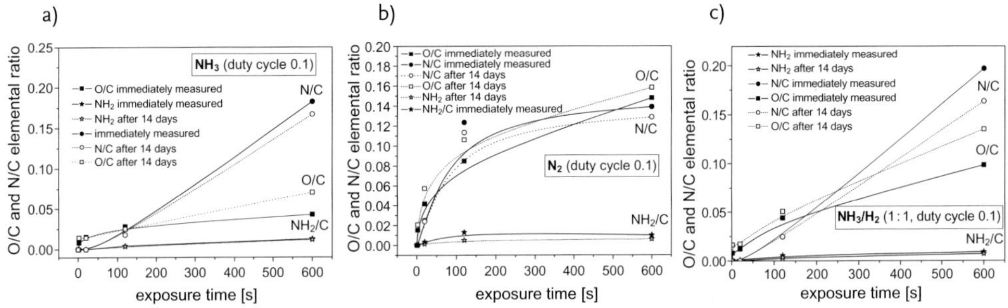

Figure 5.49 XPS-measured N, O, and NH_2 concentrations of polypropylene surfaces exposed to (a) NH_3, (b) N_2, and (c) $N_2 + H_2$ pulsed-rf plasma (6 Pa, 30 W, pulse frequency 1000, duty cycle 0.1).

The N-concentration, introduced by exposure to the ammonia plasma, often increases more slowly with treatment time than that of oxygen upon exposure to oxygen plasma. The undesired co- or post-plasma introduction of oxygen in the post-plasma auto-oxidation on exposure of polyolefin samples to ambient air is very high and exceeds 10 O per 100 C-atoms. Similar behavior is found for nitrogen and nitrogen + hydrogen plasma produced in a continuous-wave plasma (Figure 5.48). Moreover, the yields in NH_2 groups were in the same range for all three gas plasmas.

The addition of hydrogen to the nitrogen plasma did not increase the yield in NH_2 groups, as can be expected for chemical equilibrium processes following the Le Chatelier principle. In contrast, similar NH_2 yields argue for the same NH_2 steady state concentration in the different plasmas. On replacing the cw-rf plasma with a pulsed rf plasma (pulse frequency 10^3, duty cycle 0.1), the nitrogen introduction increases, the undesired oxygen incorporation decreases, but the desired NH_2 group formation remains nearly constant (Figure 5.49).

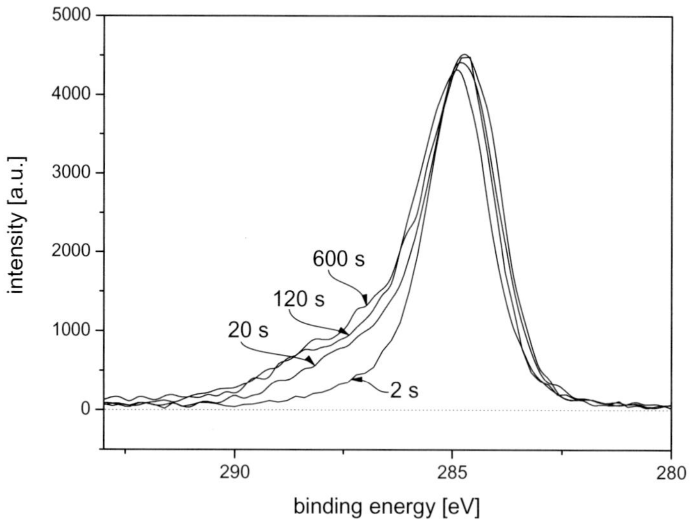

Figure 5.50 Dependence of C1s peaks of ammonia plasma treated PP on treatment time (cw-rf, 100 W, 6 Pa).

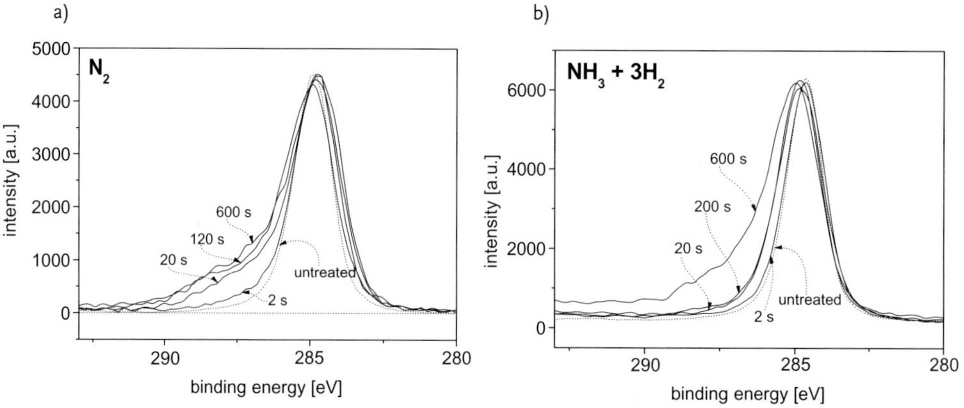

Figure 5.51 C1s peaks of (a) N_2 and (b) $NH_3 + 3H_2$ plasma treated PP (cw-rf, 100 W, 6 Pa).

In the C1s peak of polypropylene after ammonia plasma exposure a fair degree of peak broadening is observed. The appearance of a separate C1s subpeak at 288.9 eV was assigned to carboxylic groups, which are formed within the auto-oxidation process after exposure of the plasma treated samples to air before XPS measurements (Figure 5.50). The peak broadening can be attributed to the attached N-functional groups at slightly increased binding energies as well as to singly C–O bonded functional groups. In principle, N_2 plasma shows similar results to those of $NH_3 + 3H_2$ plasma (Figure 5.51).

The concentrations of primary amino groups among all N-functionalities was estimated to be 1–4 NH_2 groups per 100 C-atoms by derivatization using pentafluorobenzaldehyde and XPS for polypropylene exposed to ammonia, nitrogen, or N_2/H_2 plasmas. These yields were not significantly influenced by either changing the plasma parameters or by replacing the continuous wave plasma with a pulsed or microwave or radio-frequency plasma [60].

5.3.2
Side Reactions

Several newly formed features were identified using IR and XPS when polypropylene was exposed to the ammonia plasma. In the IR-difference spectrum (ammonia-treated – as received) shown in Figure 5.52 the main features can be identified. Hydrogenation products (CH_3 groups as a consequence of polymer chain scissions in polyethylene), C=C double bonds or amide or primary amino groups (1640 cm^{-1}), and hints of oxidation are visible that may indicate amide formation (1711 cm^{-1}) and other carbonyl features. Characteristic absorptions found between 2100 and 2250 cm^{-1} were assigned to nitriles, isonitriles, and acetylenes.

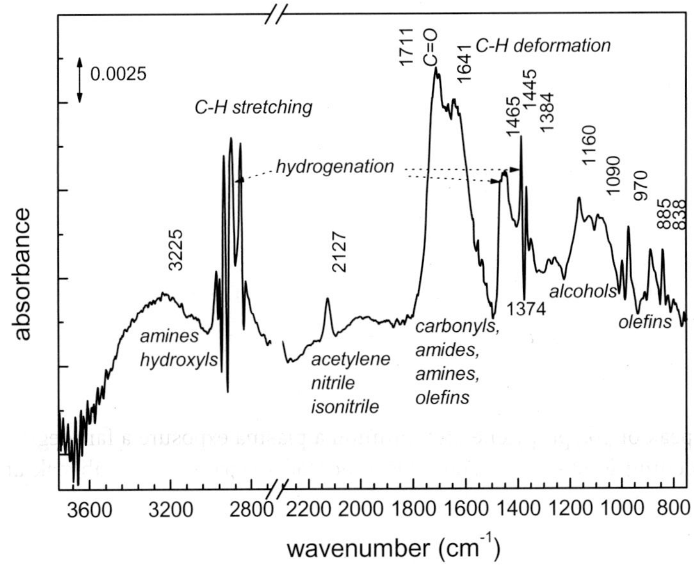

Figure 5.52 Polypropylene exposed to ammonia plasma (cw-rf plasma, 100 W, 10 Pa). The spectrum is the difference between treated and untreated samples. Spectra were recorded using the SEIRA (surface-enhanced infrared absorption) technique [13].

The formation of nitrilo groups demands an excess of energy because dehydrogenation steps are necessary and it results in a chain scission:

$$>CH{-}NH_2 \rightarrow {-}C{\equiv}N + H_2\uparrow$$

Moreover, acetylene bonds are probably formed by a similar process:

$$-CH_2{-}C(CH_3)H{-}CH_2{-}R \rightarrow {-}C{\equiv}C{-}CH_3 + CH_3{-}R + H_2\uparrow$$

Therefore, acetylenes, isonitriles, and nitriles are characteristic products of plasma processes under the conditions of significant excess of enthalpy/energy. Notably, nitrilo groups were also formed during the plasma polymerization of amino-group containing monomers (Section 12.10.2).

5.3.3 Instability Caused by Post-Plasma Oxidation

The general oxidation ability increases in the order [61]: R-H < R-OH < R-NH_2. Therefore, primary amino groups are sensitive towards oxidation. However, the exact mechanism is still unknown. The formation of N-oxide is one reason why amino group containing polymer surfaces oxidize very rapidly. The N-oxide may be a fragile group and the precursor for further oxidation and it is often discussed in the context of pyridine oxide and poly(phenylquinoxaline) [62]:

$$C{-}NH_2 + O_2 \rightarrow C{-}NH \blacktriangleright O$$

Moreover, it is known that the next oxidation step is formation of the hydroxylamine group [11]:

$$C{-}NH_2 \blacktriangleright O \rightarrow C{-}NH{-}OH$$

Often, amide group formation was found after exposure of amino group functionalized polymers to air. In this case, oxime structures may be the starting point of such a Beckmann rearrangement [61]:

$$R_1R_2C{=}N{-}OH \rightarrow R_2CO{-}NH{-}R_2$$

The oxime may originate from a nitroso group:

$$CH{-}NO \leftrightarrow C{=}N{-}OH$$

Under low-pressure plasma conditions it is also possible that the oxime dehydrates to nitrile:

$$-CH{=}N{-}OH \rightarrow {-}C{\equiv}N + H_2O$$

The oxidation sensitivity of primary amino groups can be minimized by their protection through the introduction of reversible blocking groups such as:

$$R{-}NH_2 + (CH_3)C{-}O{-}CO{-}O{-}(CH_3)_3 \rightarrow R{-}NH{-}CO{-}O{-}(CH_3)_3$$

and using pyridine and HCl to reverse this:

$$R{-}NH{-}CO{-}O{-}(CH_3)_3 \rightarrow R{-}NH_2 + (CH_3)C{-}O{-}CO{-}OH$$

5.3.4
Exposure of Self-Assembled (SAM) and Langmuir–Blodgett (LB) Monolayers to Ammonia Plasma

It is also possible to identify the effect of ammonia plasma exposure on aliphatic model surfaces of self-assembled monolayers (SAMs) by means of FTIR (Figures 5.53 and 5.54).

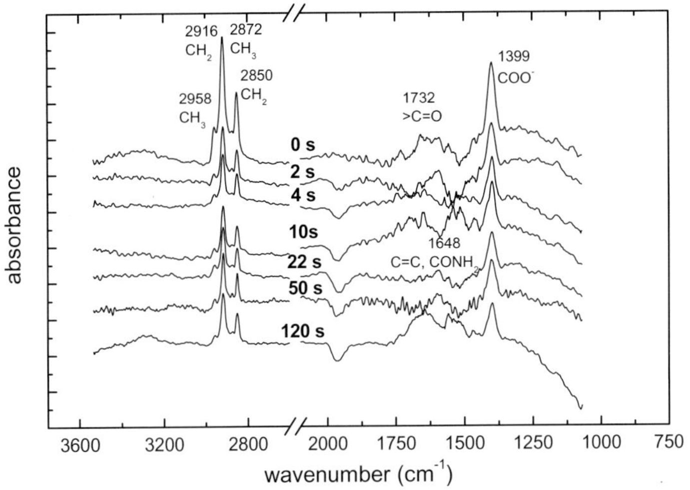

Figure 5.53 Self-assembled monolayer of stearic acid exposed to ammonia plasma (cw-rf plasma, 10 W, 10 Pa).

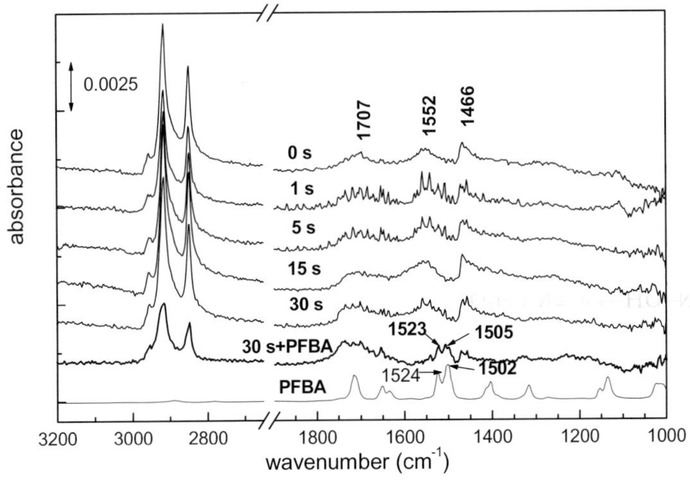

Figure 5.54 Self-assembled monolayer of arachinic acid exposed to ammonia plasma (cw-rf plasma, 10 Pa).

Langmuir–Blodgett monolayers of aliphatic acids were deposited onto silver-cluster modified silicon wafers, which were prepared by thermal evaporation. Then, surface-enhanced IR absorption (SEIRA) spectra were recorded [13]. After their preparation, stearic and arachinic acid monolayers were exposed to ammonia plasma for 120s each. To estimate the number of introduced NH_2 groups these modified layers were exposed to the vapor of pentafluorobenzaldehyde (PFBA) and the thus-introduced fluorine percentage was measured by XPS. Schiff's base (azomethine) bonds were formed, which fix one PFBA at one NH_2 group:

$$\text{R-NH}_2 + \text{OHC-C}_6\text{F}_5 \rightarrow \text{R-N=CH-C}_6\text{F}_5 + \text{H}_2\text{O}$$

A few NH_2 groups (1–4) were among the introduced N- (and O) functionalities (Figures 5.53 and 5.54). The broad $\nu_{C=O}$ stretching vibration around 1700 cm^{-1} allows us to also assume that amide groups are formed during contact of the samples with air.

5.3.5 XPS Measurements of Elemental Compositions

In contrast to other methods such as FTIR-ATR (Fourier-transform infrared–attenuated total reflectance) or NMR, X-ray photo-electron spectroscopy (XPS) using Mg or Al K_α X-ray excitation can detect just a few layers of atoms (about 6 nm) of the substrate. It gives information on the chemical composition. Figure 5.55 shows changes in the elemental composition of hexatriacontane (HTC, $C_{36}H_{74}$)

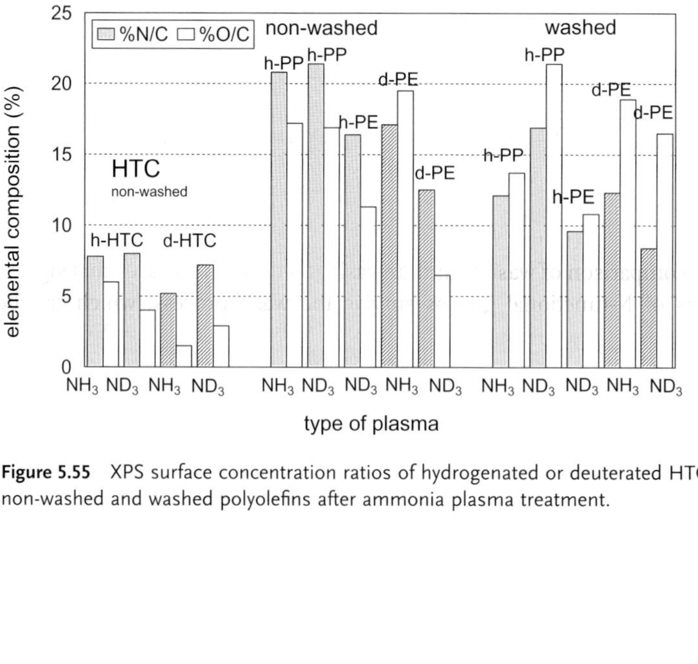

Figure 5.55 XPS surface concentration ratios of hydrogenated or deuterated HTC and of non-washed and washed polyolefins after ammonia plasma treatment.

and polyolefins (PE and PP) treated in ammonia (NH_3) or deuterated ammonia (ND_3) in low-pressure plasma at a pressure of 8 Pa (3 min treatment time, 100 W power) and also for species washed with THF for 20 min after plasma treatment. This investigation aimed to detect N and NH_2 introduction into the topmost surface as well as the depth of H↔D exchange.

Hexatriacontane (H_3C-$(CH_2)_{34}$-CH_3), as low-molecular weight model of polyethylene, shows much lower N-incorporation because it is a powder and, therefore, is not exposed to the plasma at shadowed interfaces. The N-attachment onto polypropylene surfaces is slightly higher than that of polyethylene; d- and h-PE showed no significant differences in N-incorporation.

Noticeably, high undesired O-incorporation occurs of the same order as N-introduction. More recently, plasma polymerization of allylamine has revealed that oxygen introduction is a post-plasma process and did not occur during transfer of samples under vacuum conditions [63, 64]. Therefore, the transfer time of about 1 or 2 h to the spectrometer with exposure to ambient air (gassing the plasma reactor with air–evacuation of sample in the spectrometer) may be the reason for high O/C values.

The corresponding C1s peaks do not give information on primary amino groups or other N-containing groups because of the superposition of many N- and O-containing functional groups.

As is also known, the N1 signal cannot be resolved for most of N-containing groups such as tertiary, secondary, primary amines, Schiff's base, enamines, nitriles, or may be isonitriles. Only oxidized N-species such as amides, nitroso, nitro, oximes, and so on are shifted to higher binding energies (401–408 eV) [65].

Furthermore, this work addresses the low selectivity of such common plasma process, as confirmed by numerous side- and post-plasma reactions and the occurrence of several products at the plasma-treated surfaces. A further side-reaction to be considered is the extensive scission of polymer backbones on ammonia plasma exposure and the formation of low-molecular weight oxidized material (LMWOM) [66, 67]. Nevertheless, in consequence, the introduced N-functional groups may be attached to degraded low-molecular-weight material or to the original backbone of the polyolefin. Therefore, significant amounts of all functional groups could be attached to degraded or partially degraded polymer molecules. This material can be removed by washing in slightly polar solvents. On the other hand, crosslinking may be also occur by plasma-ultraviolet irradiation and hinder the removing of all N-functional groups by the wash process. It could be shown by comparison of washed and unwashed polyolefin surfaces that significant amounts of N-functional groups survive the wash process, which can be interpreted as evidence for strong bonding to polymer chains and/or to the crosslinked network.

Unfortunately, new undesired side-reactions were initiated by the wash process. Solvent-induced swelling of the polymer surface layer makes it easier for oxygen from the air to reach trapped C-radical sites in near-surface layers of polymer. Thus, the oxygen incorporation is further increased (Figure 5.55).

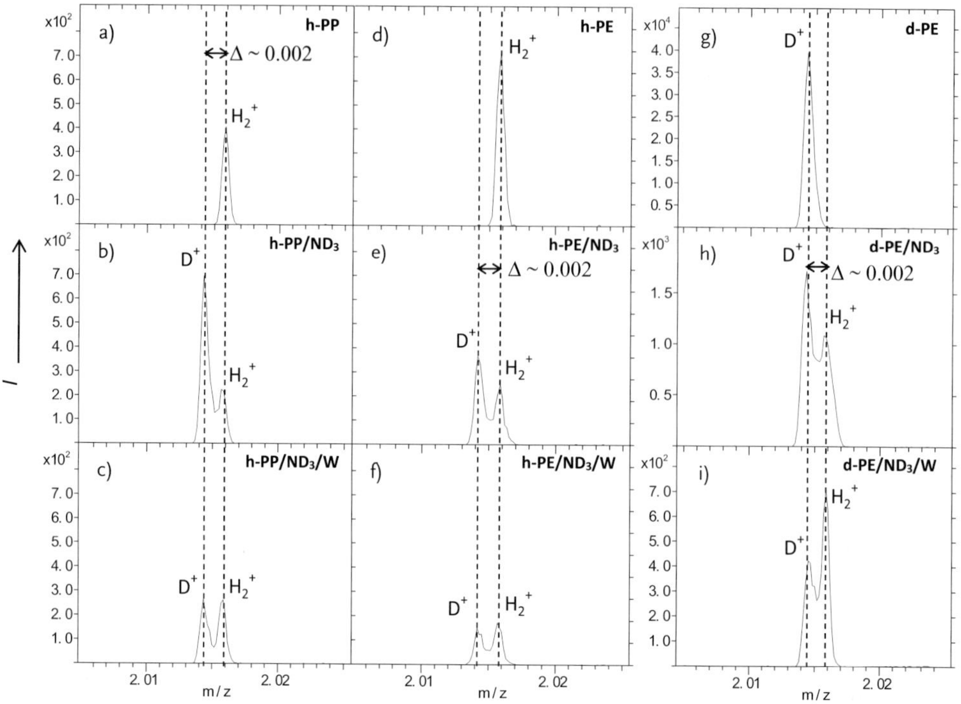

Figure 5.56 Positive ToF-SIMS spectra around m/z 2 of h-PP (a)–(c), h-PE (d)–(f), and d-PE (g)–(i) films untreated, ND_3, or NH_3 low-pressure rf plasma-treated, and post-plasma washed with THF. D^+ (m/z 2.014) and H_2^+ (m/z 2.016) peaks are clearly separated by using the high mass resolution ($M/\Delta M > 10000$) mode of ToF-SIMS.

5.3.6
ToF-SIMS Investigations

Specific information on H–D exchange in polypropylene (h-PP), polyethylene (h-PE), and deuterated polyethylene (d-PE) foils exposed to ammonia plasma was obtained by ToF-SIMS (time-of-flight secondary ion mass spectrometry) analysis (for details see Reference [68]). The introduction of N-functionalities was also studied by this approach. D–H and H–D exchange was principally proved in terms of D^+ and H_2^+ secondary ion yields. Figure 5.56 verifies that the positive deuterium secondary ion (D^+) peak was clearly separated from H_2^+ ion peak by a mass difference of 0.002 m/z at given SIMS acquisition parameters.

A wide range of characteristic $CH_xD_y^+$, NHD^+, and $CNH_xD_y^+$ secondary fragment ions were detected, which suggests complex reaction processes at the polymer–plasma surface. Specifically, the nitrogen incorporation has been found to be similar for NH_3 and ND_3 plasma treatment of PE and PP. ToF-SIMS spectra for ND_3 plasma treatment are, semi-quantitatively, rather similar for treated and washed h-PE and h-PP samples. Nevertheless, h-PP has been found to be more susceptible to ND_3 plasma modification than h-PE, as has been cross-checked also by XPS (cf. Figure 5.55). Deuterium in d-PE is more efficiently substituted by hydrogen from NH_3 plasma than the other way around (hydrogen in h-PE by D in the ND_3-plasma treatment). It is speculated that protons or hydrogen atoms play such a slightly more important role in the low-pressure ammonia plasma modification of polyolefins than deuterons or deuterium because of their lower atomic dimension, which may be also affect the mean free path and electron temperature in the plasma [69].

As shown by related secondary fragment ions in ToF-SIMS in Table 5.1 the polymer samples studied here are plasma functionalized with -NH_2, -NHD, or -ND_2 groups. The existence of mixed H-D fragments suggests complex reaction pathways in addition to a simple grafting of -ND_2 or -NH_2 moieties, which seems to be rather efficient. Interestingly, a ND_2^+ peak was not found in the spectra for both ND_3-plasma treated h-PP and h-PE samples but CD_2N^+ or CD_4N^+ peaks were present.

5.3.7 ATR-FTIR

The ammonia plasma treated HTC probes were immediately (0.2–1.0 h) measured after plasma processing using the ATR (attenuated total reflectance) technique with a diamond cell. The wavelength-dependent sampling depth is approximately 2500 nm (depending on wavenumber) and contrasted strongly to that of XPS (ca. 5 nm). Figure 5.57 presents results for the untreated h-HTC as well as the NH_3 and ND_3 treated h-HTC. As known from earlier investigations, nitrogen-containing groups cannot be detected, either in the deformation or in the stretching region.

H-D substitution of CH_2 groups produces a shift in $\nu_{as}CH_2 \rightarrow CD_2 = 727\,cm^{-1}$ and $\nu_sCH_2 \rightarrow CD_2 = 765\,cm^{-1}$ and, approximately, for the deformation signal $\delta = 383\,cm^{-1}$. Common signals in all spectra are rocking $\rho CH_2 = 724\,cm^{-1}$ and $\nu_{as}CH_3 = 2958\,cm^{-1}$. New bands for deuterated methyl groups are not clearly identified because of their low concentration.

The deuterated hexatriacontane (d-HTC) shows exclusively C–D vibrations, as documented in Figure 5.58. The ND_3 plasma treatment of d-HTC also showed features assigned to CH_x-structures, which might be affected through traces of hydrogen in the plasma chamber or by post-plasma reaction with humidity in the air. Indications for nitrile or acetylenic bonds in the region 2150–2250 cm^{-1}, which overlaps the C–D stretching bands, were not found (Figure 5.57).

In general, plasma exposure of h-HTC to ND_3 plasma and d-HTC to NH_3 plasma produces the same D-related signals in the IR spectrum.

Considering the 500-fold larger sampling depth of ATR/diamond in comparison to that of XPS, and also the relative strong absorbance of the CD_x-related signals, it can be concluded that the hydrogen/deuterium reactions in NH_3/ND_3 plasma have much more significance than the N-incorporation. Obviously, the range of plasma-induced reactions in the polyolefin materials is about one order of

Table 5.1 Positive low mass secondary fragment ions in ToF-SIMS and their proposed origins of detection for pristine and ND_3 plasma-treated h-PP and h-PE, and for NH_3 plasma-treated d-PE.

	Untreated	Plasma treated			
Ion (*m*/*z*)	C_2H_3 (27)	CH_2N^+ (28)	$CHDN^+$ (29)	CD_4N^+ (34)	$C_2H_2D_2N^+$ (44)
h-PP	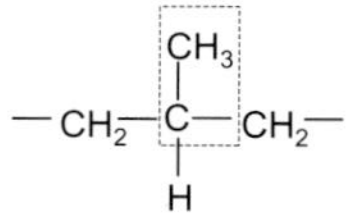	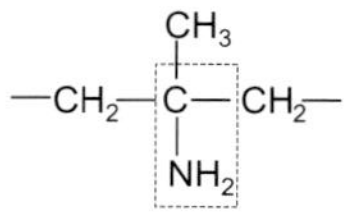	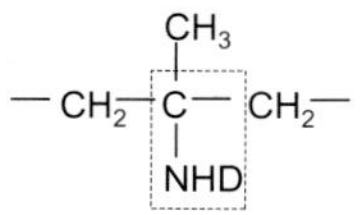	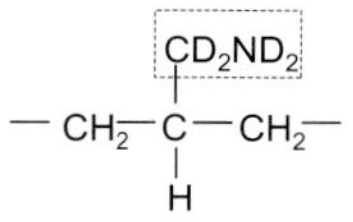	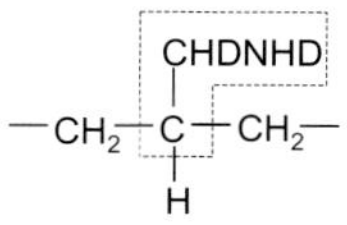
Ion (*m*/*z*)	C_2H_3 (27)	CH_2N^+ (28)	$CHDN^+$ (29)	CD_4N^+ (30)	$C_2H_2D_2N^+$ (44)
h-PE	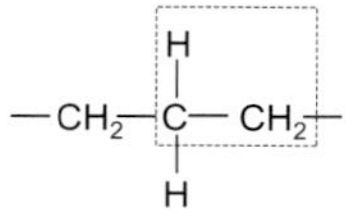	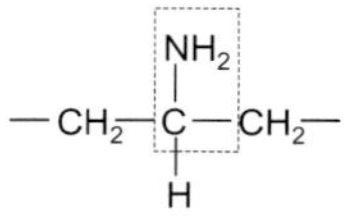	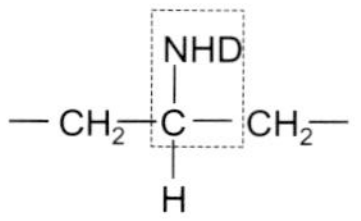	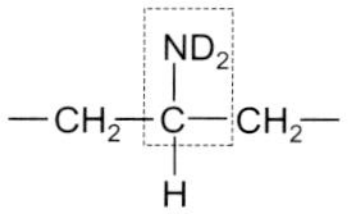	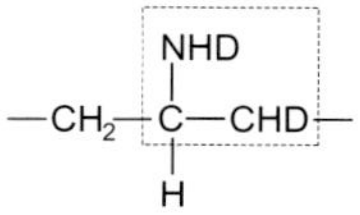
Ion (*m*/*z*)	$C_2D_3^+$ (30)	$C_2H_2N^+$ (42)	$C_2H_2D_2N^+$ (44)		
d-PE	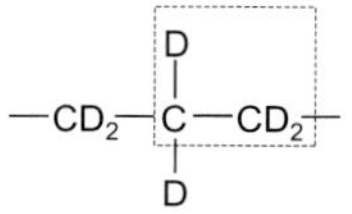	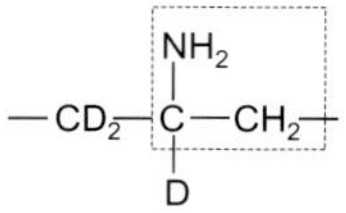	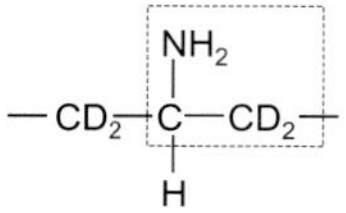		

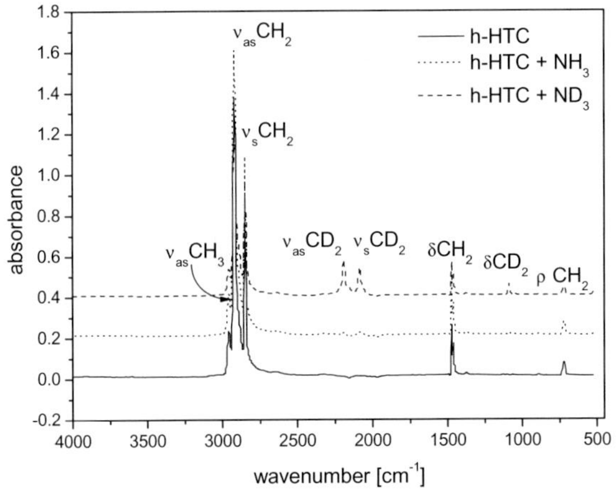

Figure 5.57 Hexatriacontane (h-HTC) exposed to NH_3 or ND_3 plasma.

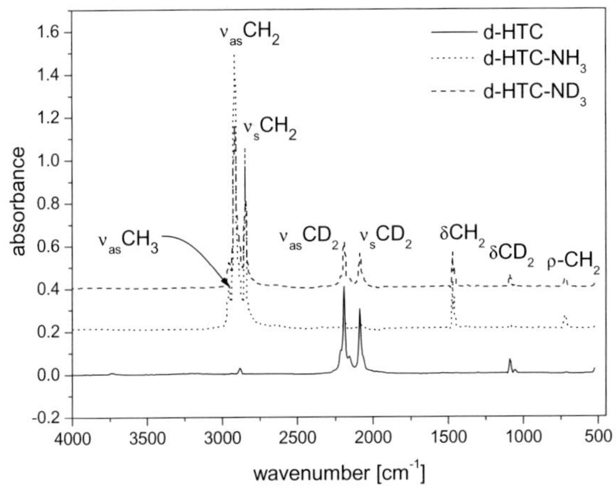

Figure 5.58 Deuterated hexatriacontane (d-HTC) exposed to NH_3 or ND_3 plasma.

magnitude higher for hydrogen/deuterium reactions than that for nitrogen. This might also be due to the smaller dimensions of H/D atoms in comparison to the N atom.

5.3.8 CHN Analysis

CHN-analysis of d- and h-hexatriacontane did not show any significant changes in stoichiometry after exposure to ammonia plasma. The nitrogen incorporation was within the standard deviation of N-analysis. The hydrogen percentage was minimally increased, but only slightly above the standard deviation range (±0.2 wt%). It must be considered that the HTC molecule ($C_{36}H_{74}$) is maximally saturated with hydrogen. Only plasma-induced chain scission may allow a marginal increase in hydrogen percentage by replacing of two CH_2-groups with two CH_3 (end) groups:

$$H_3C[CH_2]_{34}CH_3 + 2H\ (NH_3) \rightarrow 2H_3C[CH_2]_{17}CH_3$$

This chain scission increases the hydrogen percentage from 14.6 to 14.9 wt%. Actually, an increase of 0.5% was observed, indicating chain scissions. The H-NH_2 substitution also increases the hydrogen content (74 → 75%) but only marginally, from $C_{36}H_{74}$ to $C_{36}NH_{75}$:

$$H_3C[CH_2]_{34}CH_3 + NH_2\ (NH_3) \rightarrow H_3C[CH_2]_{33}CH(NH_2)CH_3$$

Following this equation, the amino group incorporation would decrease the hydrogen percentage from 14.6 to 14.4 wt% but increase the nitrogen percentage from 0 to 2.69 at.%. This N-concentration is far from the measured values of nearly zero. Therefore, it can be concluded that the ammonia plasma affects only the HTC surface. Hence, the only plausible mechanism for additional hydrogen incorporation in the bulk is chain scission and the formation of methyl groups. However, the more or less constant, or very slightly increased percentage of hydrogen is astonishing. It was assumed that short-wavelength vacuum UV radiation (VUV) of the ammonia plasma with its characteristic hydrogen lines produces dehydrogenation in analogy to observed dehydrogenation and formation of C=C double bonds on hydrogen plasma exposure to poly(vinyl chloride) [70] or to polyethylene [71]. If transferred to HTC the following reaction is proposed:

$$H_3C[CH_2]_{34}CH_3 + VUV\ (NH_3\ plasma) \rightarrow H_3C[CH_2]_{32}CH{=}CH{-}CH_3 + H_2$$

Moreover, observed di-, trimerization, and crosslinking of h-HTC on exposure to oxygen low-pressure plasma by means of MALDI-ToF mass spectrometry also produces a slight hydrogen deficiency [29, 72]. In contrast to this MALDI-analyzed O_2-plasma exposed HTC material, the ammonia plasma treated HTC could not be analyzed using MALDI, possibly due to crosslinking provoked by the ammonia plasma. A similar observation was made by Hudis, who has investigated the effect of hydrogen plasma, which may be acting in similar fashion to ammonia plasma [73].

5.3.9 NMR

Figure 5.59 displays the ^{13}C NMR spectrum of d-HTC (top spectrum). Direct excitation was required using a single 90° pulse because d-HTC does not contain any protons for a CPMAS (charge polarized magic angle spinning) experiment. Hence, extraordinarily long repetition times of about 6 h between each scan had to be used because of the extremely long T_1 times of the CD_2 groups in the middle section of chains. In Figure 5.59 the measured resonances at 13.9 ppm (CH_3), 23.7 ppm

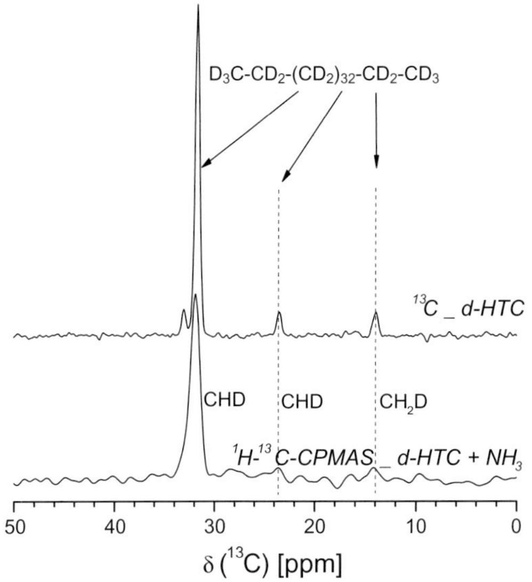

Figure 5.59 Comparison of the ^{13}C direct excitation spectrum of d-HTC (upper spectrum) with the ^{1}H–^{13}C CPMAS NMR spectrum of d-HTC + NH_3 (bottom spectrum), including line assignment (top right). The CPMAS spectrum unambiguously proves that NH_3 protons enter the d-HTC structure during the plasma process, forming CH_2D and CHD units as indicated.

(a-CH_2), and the strong signals around 32 ppm (residual CH_2) are assigned to the corresponding structural units. To verify whether NH_3 protons are incorporated into the d-HTC-structure during the plasma process with NH_3, ^{1}H–^{13}C cross polarization was used. There is a simple idea behind this: only if deuterons of the CD_2 and/or CD_3 units of d-HTC are exchanged by the NH_3 protons can these protons cause such CPMAS signals of the plasma modified d-HTC. This is in fact the case, as shown by the bottom spectrum in Figure 5.59, where a strong middle group signal was found. Hence, some protons enter the HTC structure, forming CDH units. Despite the limited signal-to-noise ratio it is also very likely that some CH_2D groups have been formed. The hatched lines in Figure 5.59 serve as a guide to the eye, and within the S/N-ratio a signal at 14 ppm seems to be present as well as the 23.7 ppm line for the adjacent CDH group. This means that the protons are incorporated in a statistical manner. There is no preferential substitution pattern.

The opposite case (exchange of HTC protons by ND_3 deuterons) was investigated using ^{2}H NMR (Figure 5.60). The static ^{2}H echo spectrum of d-HTC is displayed at the top of Figure 5.60 for convenience. This spectrum shows at a glance CD_3 and CD_2 signals. Owing to the three site jumps around the C3 axis, the quadrupole interaction is averaged down to $-\frac{1}{3}$ [74] and the corresponding peak pattern is narrower, as shown in Figure 5.60. Furthermore, as only very few deuterons enter the HTC structure during the plasma process, MAS (magic angle spinning) has been applied instead for signal-to-noise ratio reasons.

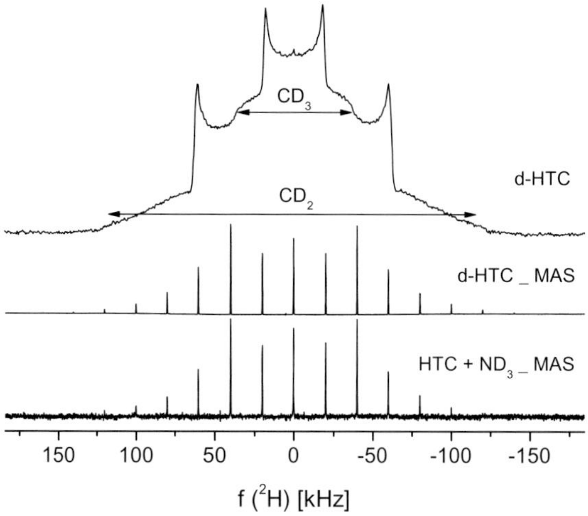

Figure 5.60 ^2H solid-echo (top) and corresponding ^2H MAS line shape for d-HTC, showing the presence of CD_3 and CD_2 units. The similar ^2H MAS line shape of the HTC + ND_3 sample allows us to conclude that ND_3 deuterons are incorporated into the HTC structure during the plasma process to form CH_2D and CHD units.

The corresponding MAS spectrum of d-HTC is shown in the middle of the figure. This time CD_3 and CD_2 MAS signals overlap. Nevertheless, a typical MAS sideband structure is found that ranges over the entire static echo frequency range, as expected. Again, ^2H signals of the HTC sample treated in the plasma process with ND_3 can only be observed if ND_3 deuterons are incorporated into HTC as CH_2D and CHD groups. This is the case, as shown by the bottom spectrum in Figure 5.60. The observed ^2H MAS sideband structure is very similar to that of d-HTC, confirming that these two units are at the surface.

5.3.10
Discussion of Hydrogenation and Amination of Polyolefins by Ammonia Plasma

As evidenced by XPS and ToF-SIMS, NH_x and ND_x groups were introduced onto the surface of paraffin (HTC) and polyolefins (PP, PE). However, on considering the different sampling depths of XPS and SIMS (ca. 1–10 nm) as well as those of IR and NMR (2.5 and 10 μm film thickness) and comparing the results, it is clear that nitrogen species are found only in the topmost nanometres of polymers. Thereby, ND_3 plasma achieves a slightly higher introduction of nitrogen than the NH_3 plasma and PP shows higher N-incorporation than PE. Whether the small differences between ND_3 and NH_3 plasma are realistic or within the deviations seen from experiment to experiment is not clearly indicated. However, the higher amount of incorporated N-species in polypropylene is obvious and can be confirmed also in principle.

The time-dependence of N-incorporation onto polypropylene surfaces shows differences between the efficiency of continuous-wave and pulsed plasma exposure. The pulsation of plasma obviously inhibits the formation of radicals, possibly by recombination in the plasma-off periods, as shown by the much lower post-plasma oxidation. Thus, the O-percentage is about twofold higher than the N-percentage using the cw-plasma and lower or equal when using the pulsed plasma mode. The N-introduction is slightly higher using NH_3 and $NH_3 + H_2$ plasmas than N_2 plasma. Note that NH_2 groups cannot be formed in N_2 plasma directly. Hydrogen must be removed from the polyolefin backbone, must react with N-species, and recombine with C-radical sites at polymer surface:

$$\text{-CH}_2\text{–CH}_2\text{-} + plasma \rightarrow \text{-CH=CH-} + 2\text{H}^\bullet\ (\text{H}_2)$$

or:

$$\text{-CH}_2\text{-CH}_{2\text{-}} + plasma \rightarrow \text{-CH}_2\text{-CH}^\bullet\text{-} + \text{H}^\bullet\ (0.5\text{H}_2)$$

$$\text{N}_2 + plasma \rightarrow 2\text{N}^\bullet$$

$$\text{N}^\bullet + 2\text{H}^\bullet \rightarrow {}^\bullet\text{NH}_2$$

$$\text{-CH}_2\text{-CH}^\bullet\text{-} \rightarrow \text{-CH}_2\text{-CH(NH}_2\text{)-}$$

The additional supply of hydrogen to ammonia plasma did not change significantly the N-introduction and the fraction of NH_2 groups among all N-containing groups. The mass action rule has no influence on this plasma reaction. The very low fraction of NH_2 groups among all N groups is obviously due to the energetically favored N and NH formation under the conditions of plasma and energy excess, as shown by standard bond dissociation values. In general, 1–4 NH_2 per 100 C is the maximum. Referenced to the total introduced nitrogen, a maximal 10% of all N-functional groups appears as primary amino groups. The assumed nitrile groups were only found using the surface-sensitive SEIRA-FTIR technique at the surface of ammonia plasma treated aliphatic LB films (traces) and polypropylene (minimal concentration). Extensive CH_3 group formation by chain scissions and hydrogenation are evident by its ν_{as} stretching vibration at $2960\,cm^{-1}$ in the IR spectrum.

ToF-SIMS-MS measurements of ammonia plasma exposed h-PP, h-PE, and d-PE show a wide range of differently composed N functional groups with a composition of NH_xD_y. These different N functional groups could not be clearly differentiated using XPS. Therefore, only the total N-introduction was determined by XPS. However, after derivatization of primary amino groups their concentration could be easily determined within the XPS sampling depths. FTIR did not show any nitrogen introduction because of its higher sampling depth.

The H or D modification of polyolefins and hexatriacontane achieved much deeper polymer layers and gave higher quantities. It can be concluded that the hydrogenation effect of ammonia plasma dominates the nitrogen introduction strongly. This is also evident by the appearance of strong D-related IR signals in the respective ATR-FTIR spectra and also by ToF-SIMS and NMR. The spectra suggest a statistic exchange of H↔D during plasma exposure. This exchange affects the whole bulk material as NMR results show and ATR-results confirm. The stoichiometry was not changed significantly, as shown by CHN analysis. Very small changes might be interpreted as a few chain scissions related to a very small increase in hydrogen percentage.

The polymer chains became substituted as well as end-groups, forming CD_3, CD_2H, CH_2D, and CHD groups, as shown by NMR. The IR spectra show the dominance of CD_2 groups among all other groups when h-HTC was exposed to the ND_3 plasma and, vice versa, CH_2 groups were found exclusively after exposure of d-HTC to NH_3 plasma. Hydrogen might be able to diffuse deeper into the bulk material than nitrogen, which was found only in the topmost surface layer. This behavior of hydrogen may be caused by the smaller collision cross section/atomic volume in comparison to nitrogen. NMR gave no hints of chain scissions. However, CHN analysis allowed us to assume a few chain scissions, as always observed for all plasma processes [66].

Thus, ammonia low-pressure plasma exposure of aliphatic structures produces a complex reaction mechanism with a wide range of structural changes into the bulk. Penetration of reactive (plasma produced) hydrogen or deuterium species into deeper polymer layers (>1 μm) than N-species and their reaction with the bulk material by preferential H↔D exchange was not expected. Highly reactive species react immediately, preferentially at the topmost surface, as observed for nitrogen and nitrogen-containing species (Figure 5.61).

As mentioned before, ammonia plasma emits strong hydrogen UV irradiation with lines in the vacuum UV range of the Lyman series from 91 to 121 nm [75–76]. Such wavelengths correspond to 10–12 eV, compared to about 3.7 eV necessary to dissociate C–H and C–C bonds. It may be assumed that this hard radiation also reaches deeper layers, where it produces C-radical sites. Nitrogen-related emissions are much softer and have wavelengths > 160 nm [77]. Aliphatic chains absorb UV radiation from ≤160 nm [78–80].

Lyman irradiation may be responsible for the formation of double and (possibly) triple C≡C and C≡N bonds [81]. Aromatic ring structures formed by dehydrogenation could not be identified as speculated above. In the presence of surplus hydrogen only isolated multiple bonds are formed, as shown by SEIRA. Under low-pressure plasma conditions it is also possible that the oxime dehydrates to nitrile as mentioned before:

$$\text{-CH=N-OH} \rightarrow \text{-C}\equiv\text{N} + H_2O$$

Post-plasma oxidation and rearrangement reactions such as Curtius, Beckmann, or Lossen may be a source of isocyanate groups, which possibly contribute to the IR band at 2140 cm^{-1}.

In general, for formation of hydrocyanic acid, nitrile, and acetylenic groups an excess of energy is demanded because dehydrogenation steps are necessary and it results in chain scissions:

$$\text{>CH-NH}_2 \rightarrow \text{-C}\equiv\text{N} + H_2\uparrow \text{ (may be HCN or RCN)}$$

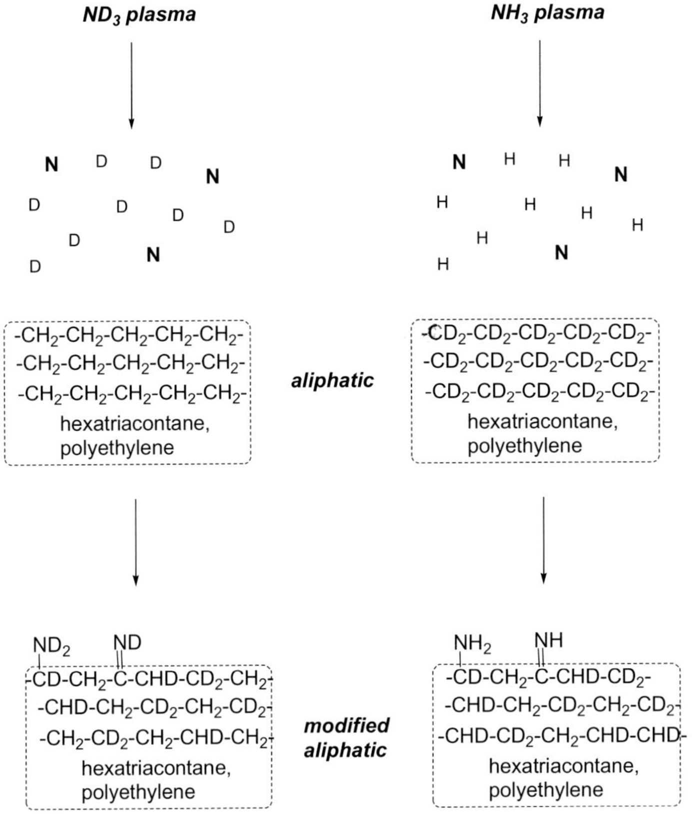

Figure 5.61 Sketch of ND_3 and NH_3 plasma treatment of hydrogen and deuterium-substituted aliphatic chains, respectively.

$$>C(CH_3){-}NH_2 \rightarrow \text{-}C{\equiv}N + H_2\uparrow \text{ (may be RCN)}$$

$$\text{-}CH_2{-}C(CH_3)H{-}CH_2\text{-}R \rightarrow \text{-}C{\equiv}C\text{-}CH_3 + CH_3{-}R + H_2\uparrow$$

5.4 Carbon Dioxide Plasmas

Carbon dioxide plasma oxidizes polymer surfaces. Preferential dissociation of carbon dioxide in the plasma gives [82]:

$$CO_2 + \text{plasma} \rightarrow CO + O$$

Thus, several different O-containing functional groups are formed. However, CO_2 and CO, as well as O, have ionization potentials in the range 13–14 eV. Therefore, the average electron energy should be high (of the same order). Thus, the electron energy exceeds the bond dissociation energies of C–C and C–H bonds. Therefore, the idea of softer plasma oxidation in the CO_2 plasma is not realistic.

On the other hand, attachment of the intact CO_2 molecule onto the polymer backbone should be also possible and should open a route for the introduction of carboxylic or carbonate structures onto a polymer surface [83]. Using low-pressure plasma and low wattage the attachment of electronically excited or ionic CO_2 (formal insertion into a C–H bond) to the polymer backbone dominates, as presented schematically here:

$$CO_2^* + \text{-}CH_2\text{-}CH_2\text{-} \rightarrow [\text{-}CH_2\text{-}CH(CO_2^\bullet)\text{-} + H^\bullet] \rightarrow \text{-}CH_2\text{-}CH(COOH)\text{-}$$

Because of the high energy of the first excitation state of CO_2 (12 eV) the above-described route may be too simple. More probably, CO_2 dissociates into atoms and these atoms form a new COOH group among other O-containing groups. Another route may be the formation of C-radical sites during the exposure to the oxygen plasma:

$$\text{-}[CH_2\text{-}CH_2]_n\text{-}\ \textit{plasma} \rightarrow \text{-}[CH_2\text{-}CH_2]_{n-1}\text{-}[CH_2\text{-}CH^\bullet]\text{-} + H^\bullet$$

$$\text{-}[CH_2\text{-}CH_2]_{n-1}\text{-}[CH_2\text{-}CH^\bullet]\text{-} + CO_2 \rightarrow \text{-}[CH_2\text{-}CH_2]_{n-1}\text{-}[CH_2\text{-}CH(CO_2^\bullet)]\text{-}$$

$$\text{-}[CH_2\text{-}CH_2]_{n-1}\text{-}[CH_2\text{-}CH(CO_2)^\bullet]\text{-} + RH \rightarrow\rightarrow \text{-}[CH_2\text{-}CH_2]_{n-1}\text{-}[CH_2\text{-}CH(COOH)]\text{-} + R^\bullet$$

Following the reported information, a concentration of 10–12 COOH groups per 100 C atoms were introduced using the CO_2 plasma [84]. Poll observed gravimetrically the introduction of reactive CO_2 species from the plasma into PET films [85]. Badyal applied the CO_2 plasma for the soft oxidation of polystyrene [86], Hoffman for polyethylene [87], and Gancarz for polysulfone membranes [88]. Inagaki also treated polyethylene films with the CO_2 plasma (and other plasmas such as CO, NO, NO_2) [89, 90]. He has used it to oxidize "softly" the polyethylene surface so as to increase the surface energy. Using the CO_2 plasma the surface free-energy of the polymer was maximally increased comparing to other plasma gases (Table 5.2).

Table 5.2 Surface energy of polyethylene after treatment with different plasma gases.

Plasma	XPS-measured O per 100 C	Total surface energy (γ_{total}) (mJ m^{-2})	Dispersive component of surface energy (γ_{disp}) (mJ m^{-2})	Polar component of surface energy (γ_{polar}) (mJ m^{-2})
As received	6	36	30	6
CO_2	36	65	18	47
CO	41	61	20	41
NO_2	21	59	23	36
NO	35	56	21	35
O_2	19	56	22	34

Table 5.3 Yield of COOH obtained using different plasma gases.

Plasma	Oxygen per 100 C		
	OH	>C=O	COOH
CO_2	0	3.1	5.5
CO	0	0.2	3.3
NO_2	2.5	2.1	1.1
NO	0.6	1.3	0.5
O_2	3.2	2.0	1.2

Surprisingly, Inagaki did not find a clear correlation between the percentage of introduced oxygen and the thus-produced polar contribution (γ_{polar}) to the surface energy (γ_{total}). The dispersive fraction of the surface energy reflects changes in polymer structure by plasma modification.

A more detailed inspection of the C1s signal by curve fitting and derivatization of the different O-functional groups (cf. Reference [91]) showed that about 5.5 COOH groups per 100 C were formed using the CO_2 plasma (Table 5.3) [92].

Inagaki also had another idea for increasing the yield in COOH groups. He developed a hybrid method by combination of plasma polymerization in the presence of CO_2, that is, acrylic acid was plasma polymerized in the presence of CO_2. Thus, it was possible to double the yield of COOH groups [93].

As mentioned above, COOH functional groups may be attached to macromolecular chains by substitution reactions, nucleophilic (S_N1 or S_N2) or radical (S_R), thus replacing a hydrogen atom from the polymer molecule. Alternatively, hydrogen is abstracted from polymer by plasma exposure, thus forming a C radical vacancy. The OH radical can recombine with the C radical and thus form the desired functional group. However, independently of the actual incorporation mechanism, COOH groups are not originally present in O_2 or CO_2 plasmas. Thus, they cannot be formed directly by attachment of plasma gas atoms or fragments. It is necessary to pick up hydrogen in plasma gas phase and form COOH (or OH) groups. The same is true for the formation of NH_2 groups in N_2 plasma. These functional-group forming intermediates need hydrogen for their formation, which is abstracted from the polymer chain. Traces of water in the gas phase or in the adsorption layer may be another source of hydrogen. The following reaction pathway is assumed:

$$CO_2 + {}^{\bullet}H \rightarrow {}^{\bullet}COOH$$

$$\text{Polymer}^{\bullet} + {}^{\bullet}COOH \rightarrow \text{polymer-COOH}$$

Insertion of CO_2 in unactivated C–H bonds of polyolefins by a nucleophilic substitution (S_N2) is improbable [61]:

$$C\text{–}H + CO_2 \rightarrow C\text{–}COOH?$$

CO_2 insertions are very rare but may be occur under plasma conditions.

5.5 SH-Forming Plasmas

Mercapto groups are of interest in biochemistry. Yanagihara [94] claimed the deposition of mercapto group-containing plasma polymers.

However, hydrogen sulfide plasmas or hydrogen plasmas in contact with solid sulfur are also able to modify polyolefins:

$$H_2 \text{ plasma} + S_8 \rightarrow H_2S$$

$$H_2S + \text{polymer-H} \rightarrow \text{polymer-SH} + H_2$$

Another way to produce SH groups at a polymer surface by plasma-assistance was described by Okusa *et al.* [95]. They equipped mica surfaces with OH groups by water vapor plasma followed by consumption with mercapto-silane:

$$\text{Mica} + H_2O\ \textit{plasma} \rightarrow \text{mica-OH}$$

$$\text{Mica-OH} + \text{HS-}(CH_3)_3\text{-Si}(OCH_3)_3 \rightarrow \text{Mica-O-Si}(OCH_3)_2\text{-}(CH_2)_3\text{-SH} + H_3C\text{-OH}$$

5.6 Fluorinating Plasmas

Plasma-chemical surface fluorination is usually performed with CF_4 or SF_6 plasmas [96]. Other fluorinating plasma gases are SiF_4, BF_3, NF_3, XeF_2, and $ClCF_3$ [97]. Measured fluorination degrees of PE and PP range from 1.1 to 1.8 F/C using the CF_4 or SF_6 plasma [98–100] and 1.82 on exposure to the NF_3 plasma (3 s). The highest fluorination ratio comes close to that of pure poly(tetrafluoroethylene) (PTFE, 2.0 F/C). The maximal fluorination ratio is also dependent on the plasma parameter and type of polymer: PE – 1.82, PP – 1.70, PS – 1.55, and PMMA [poly(methyl methacrylate)] – 1.14 [96]. The main disadvantage of plasma fluorination, and also of the chemical fluorination, is the undesired production of C-radical sites within the polymer due to the strong exothermic fluorination reaction and the interaction with plasma UV radiation. Thus, a post-plasma oxygen introduction ($C^{\cdot} + {}^{\cdot}O\text{-}O^{\cdot} \rightarrow C\text{-}O\text{-}O^{\cdot} \rightarrow$ reaction products) occurs over longer periods of storage on contact with ambient air [101]. Such undesired side-reactions produce oxidation, which results in strong polar groups because of the neighborhood of fluorinated sequences. Such a polar polyolefin surface is in contrast with hydrophobic surfaces produced by pure perfluorination. Therefore, oxygen addition during the fluorination is used to produce highly-adherent polar polyolefin surfaces. Perfluorination is used to produce thin barrier layers such as in polymer fuel tanks [102]. The idea behind the fluorination is to decrease the polymer free volume by H $\rightarrow$ F substitution. The more voluminous F atoms close the routes for solvent or fuel diffusion.

CF_4, SF_6, BF_3, NF_3, XeF_2, and so on produce fluorine atoms under plasma conditions by dissociation; thus silicon or silica can be attacked (plasma etching in microelectronics):

$$CF_4 \text{ plasma} \rightarrow (4-x)F + CF_{4-y}$$

$$nF + Si \rightarrow xSiF_4\uparrow$$

Using polymers as substrate, fluorination of the polymer backbone occurs. Hydrogen atoms are replaced by fluorines in an exothermic self-accelerating process occurring as chain propagation:

$$nF + H_n\text{-polymer} \rightarrow \text{fluorinated polymer} + nHF$$

$$\Delta_R H^O{}_{298} = \Delta_R H^O{}_{\text{H-polymer}} - \Delta_R H^O{}_{H-F} = 411 - 546 \text{ kJ mol}^{-1} = -135 \text{ kJ mol}^{-1}$$

Thus, chemical fluorination of the polymer backbone by fluorine atoms is strongly driven by thermodynamics and does not need any additional plasma energy. Only the dissociation of the fluorine precursor needs the plasma assistance to disrupt any bonds to fluorine. Because of the great excess in enthalpy the fluorination process shows low selectivity towards different polymer units and functional groups.

However, the ideal perfluorinated structure, for example, presenting the highest possible oxidation state, is not realized, possibly for statistical reasons, concurrent chain scissions, and local thermal effects (reaction enthalpy). These thermal effects lead also to chain scission and the formation of radicals, which are not completely saturated by fluorine atoms. Metastable C radical sites remain, were trapped, and react after contact with air to give peroxy radicals and rearrange to carbon acid fluorides as proposed:

$$\text{R-CF}^{\bullet} + {}^{\bullet}\text{O-O}^{\bullet} \rightarrow \text{R-CF-O-O}^{\bullet} \rightarrow \text{R-COF} + {}^{\bullet}\text{O}$$

The carbon acid fluorides hydrolyze rapidly to oxygen-containing products (carboxylic acids):

$$\text{-COF} + H_2O \rightarrow \text{-COOH} + HF$$

When neighbored by perfluorinated groups such carboxylic groups become strongly acidic, comparable to trifluoroacetic acid (CF_3COOH), because of the electron-withdrawing effect of the perfluorinated groups. Therefore, alcoholic OH in the neighborhood of CF_x groups has acidic properties comparable to those of phenols.

Fluorination proceeds to higher percentages when using the plasma technique (Figure 5.62).

The time-scale for the plasma fluorination process is much prolonged compared to those of oxygen or ammonia plasma exposure. Fluorination proceeds over longer periods as evidenced in Figure 5.62. This may be due to the small

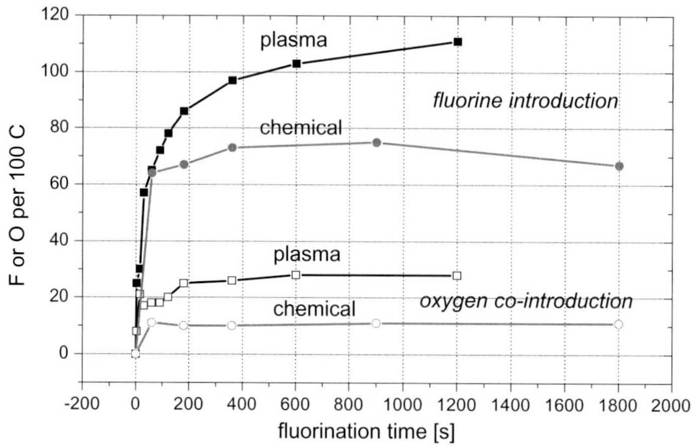

Figure 5.62 Introduction of fluorine into polypropylene surfaces on exposure to CF_4 plasma (100 W, 10 Pa) in comparison to chemical fluorination with a $2\%F_2/98\%N_2$ gas mixture.

Table 5.4 Electronegativity, dissociation energy, and ionization potential of several elements [24].

Element	Electronegativity (Pauling)	Electronegativity (Mulliken)	Dissociation energy ($kcal\,mol^{-1}$)	Ionization potential for X/X_2 (eV)
H	2.2	2.1	103 (CH_3–H)	13.6/15.4
C	2.6	2.5	90 (CH_3–CH_3)	13.0 (CH_4)
N	3.0	3.0	75 (CH_3–NH_2)	14.5/15.6
O	3.5	3.5	91 (CH_3–OH)	13.6/12.2
F	4.0	4.5	120 (CH_3–F)	17.4/15.8

dimensions of the fluorine atom, which can migrate within the full sampling depth of XPS and beyond.

However, the undesired post-plasma introduction of oxygen cannot be hindered, neither by chemical fluorination nor by plasma fluorination processing (Figure 5.62). At first, this reaction with oxygen in concurrence with fluorination is astonishing for a chemist. Fluorine is a stronger oxidizing species than oxygen; moreover, oxygen is not able to replace fluorine chemically. Only, the above given auto-oxidation process with C-radical sites and their reaction with molecular oxygen can explain the (post-plasma) co-introduction of oxygen-containing groups. This behavior is documented in Table 5.4, which presents the relative electronegativity after Pauling [103] and Mulliken [104].

Replacing fluorine with oxygen may be possible under plasma conditions, possibly by plasma particle bombardment; however, most oxygen is incorporated by

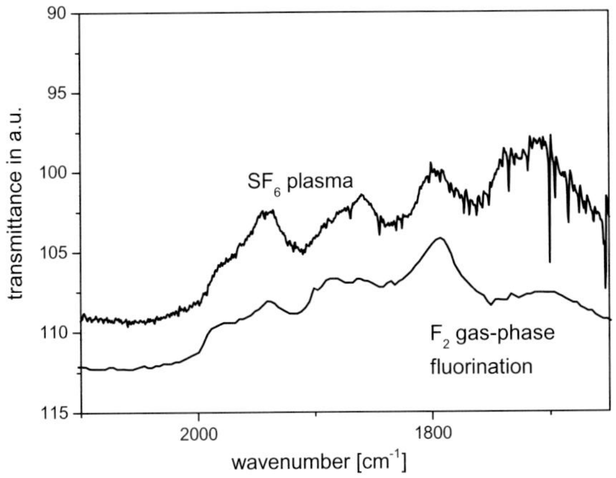

Figure 5.63 Section of the FTIR-ATR (Fourier-transform infrared–attenuated total reflectance) spectra of gas-phase fluorinated and plasma-fluorinated PP.

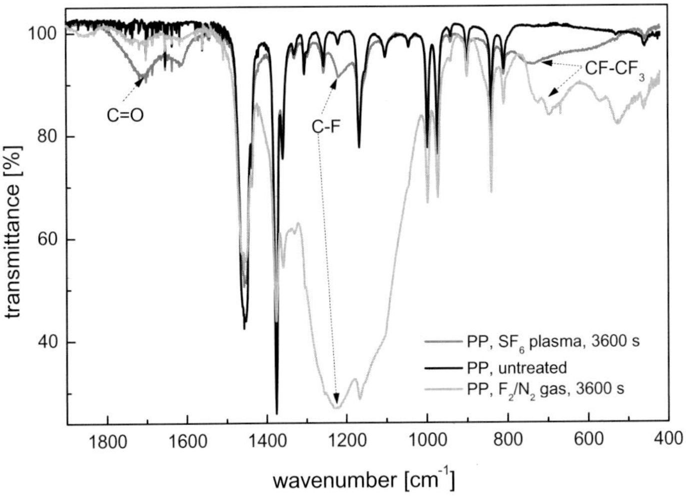

Figure 5.64 ATR spectra of plasma and chemically fluorinated PP foil.

the auto-oxidation process with oxygen from air and trapped C radical sites. The acid fluorides play the key role in forming new oxygen- and fluorine-containing polar groups as shown by the above-cited reactions and in the IR-spectra at about $1800\,cm^{-1}$ (Figure 5.63).

The higher concentration of fluorine in the topmost PP surface layer produced by plasma fluorination (Figure 5.62) and measured by XPS with a sampling depth of about 5 nm contrasts strongly with the intense absorption of the ν_{C-F} vibration near $1220\,cm^{-1}$ (Figure 5.64) after gas-phase fluorination or low-pressure plasma fluorination as measured by FTIR-ATR.

The reason for this different behavior and this different penetration depth is the different diffusivity of fluorinating species under atmospheric or elevated pressure and under vacuum conditions.

The fluorine precursor gases, such as CF_4 or SF_6, cannot react chemically in their neutral form with the polyolefin. Plasma-induced dissociation of the precursor is needed to produce F atoms, which can react with hydrocarbon molecules by $H \rightarrow F$ substitution or $C^{\bullet} + {}^{\bullet}F$ recombination. A (self-accelerating) radical chain-process (auto-fluorination) can only proceed with elemental fluorine, and not with CF_4. This is due to the difference in dissociation energies required to produce the needed fluorine atoms:

$$F{-}F \rightarrow 2{}^{\bullet}F \quad 155\ \text{kJ mol}^{-1}$$

$$CF_4 \rightarrow CF_3^{\bullet} + {}^{\bullet}F \quad 507\ \text{kJ mol}^{-1}$$

It must be considered that the dissociation energy of CF_4 is much higher when all four F atoms were removed from the C atom. The chemically reactive fluorine molecules and atoms are able to diffuse into the bulk, where fluorination also occurs. Fluorination with elemental fluorine is exothermic ($\Delta_R H^{O}{}_{298} = -135\ \text{kJ mol}^{-1}$), and is, thus, self-accelerating. Therefore, the chemical fluorination proceeds with time into the bulk (Figure 5.64).

Using different ATR crystals with different sampling depths the penetration depth of plasma and chemical fluorination can be appreciated (Figure 5.65).

The fluorination depth for the chemical fluorination amounts about 2 μm and contrasts with that of the plasma-assisted fluorination of a few hundreds of nanometers at maximum [102–107].

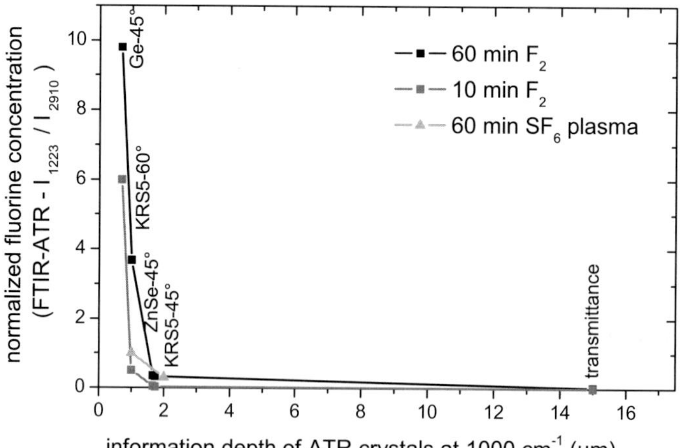

Figure 5.65 Plasma (SF_6) and chemically (F_2) fluorinated PP foil measured by ATR using different crystals, thus varying the information depth.

Other authors have also measured the fluorination depth by using various techniques such as angle-resolved XPS and measurement of (differential) etching rates of fluorinated/non-fluorinated polymers and so on for various polymers, as summarized by Egitto [108]. The plasma fluorination of polymers is an attractive technique by which to fluorinate, perfluorinate, or oxyfluorinate [109] the polymer surface to produce hydrophobic and water-repellent properties (perfluorination) [99, 100, 110–113] or hydrophilic surfaces (oxyfluorination, Figure 5.66) [107].

We need to distinguish between post-plasma oxygen implementation when the plasma-fluorinated polymer is exposed to ambient air (in the range of 5 to 25 O

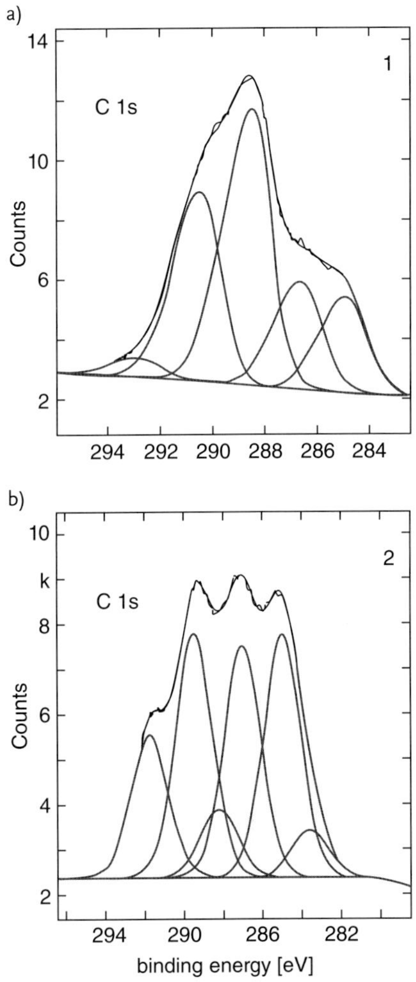

Figure 5.66 C1s signal of (a) plasma (SF_6) and (b) chemically (F_2) fluorinated PP.

per 100 C) and the purposeful addition of oxygen to the fluorinating plasma gas to form acid fluorides and other polar O-containing groups. The oxygen addition to the polymer surface is not clearly visible in the C1s signal because of the superposition of C-O_x-related peaks with dominant C-F_x-related peaks (Figure 5.66). However, the appearance of an intense O1s peak clearly indicates the co-introduction of oxygen (cf. Figure 5.62).

The plasma-chemical fluorination provokes the dominant formation of CF_2 and CF species beside CF_3, C–CF, and CO species. The chemical gas-phase fluorination with 2 vol.% fluorine gas shows an incomplete fluorination of polyethylene, manifest in the intense C1s subpeak at 285 eV, characteristic for C–H components, and a high concentration of CF_3 groups, indicating substantial degradation of polyolefin. The formation of CF_3 groups is only possible by chain scission and perfluorination of the newly formed end-group:

$$\text{-}[CH_2\text{-}CH_2]_n\text{-} + (n+1)\,F_2 \rightarrow \text{-}[CF_2\text{-}CF_2]_m\text{-}CF_3 + F_3C\text{-}[CF_2\text{-}CF_2]_{n-m}\text{-}$$

In CF_4 plasma-treated polymers CF_3 groups may also be formed by addition of CF_3 radicals, possibly by, for example:

$$\text{-}[CH_2\text{-}CH_2]_n\text{-} + CF_4 + \textit{plasma} \rightarrow \text{-}[CHF\text{-}CH(CF_3)]_n\text{-}$$

The F1s peak indicates the existence of covalently bonded C–F_x species (binding energy ≈ 689 eV) and traces of F^- anions (binding energy ≈ 686 eV).

The inevitable co-introduction of oxygen, causing a more hydrophilic surface, is often not observed, does not exist, or is not mentioned. Moreover, the presented high ratio of fluorination or the postulated poly(tetrafluoroethylene) (PTFE)-like perfluorination needs to be scrutinized. In terms of thermodynamics C–C chain scissions are obligatory.

Perfluorinated polyethylene is identical to PTFE and possesses after fluorination 200 F per 100 C without any C–C bond scission. However, the authors did not consider that such high values are indications of polymer degradation, such as is evident on comparing hexafluoroethane and PTFE with C_2F_6, C : F = 1 : 3 (C_2F_6) and 1 : 2 (-[CF_2–CF_2]$_n$-). Moreover, a proper calibration of the XPS is necessary.

Values near this theoretical limit of C : F = 1 : 2 are given by several workers [114, 115]. High F/C ratios have been presented by Yagi [97] using different inorganic fluorine precursors (Table 5.5).

As mentioned before, Badyal investigated systematically the incorporation of fluorine in different types of polymers on exposure to the CF_4 plasma [96]. Yagi compared the fluorination degree after NF_3 plasma treatment (Table 5.6) [97].

Strobel *et al.* also investigated the fluorination of polypropylene with different fluorinating plasma precursors. He measured much lower fluorine concentrations than Yagi (around 1.0 F/C) [100]. For the fluoroform plasma (CHF_3) the deposition of a fluorinated plasma polymer film was observed [111]. Thus, the authors differ between fluorinating plasmas, such as CF_4 or F_2, and film-forming plasmas such as fluoroform (CHF_3) [99]. Fluoroform split off H first because the dissociation energy of C–H is lower (440 kJ mol^{-1}) than that of C–F (502 kJ mol^{-1}). Similar observations were made by d'Agostino *et al.* [116]. Strobel also found in the C1s

Table 5.5 F/C ratio of plasma-fluorinated polyethylene before and after extraction with $CF_2ClCFCl_2$ [98].

Reactant	F/C ratio		
	Plasma	After extraction	Loss in F/C ratio (%)
NF_3	1.92	1.90	1
BF_3	1.12	0.93	17
SiF_4	1.12	0.92	19
CF_4	1.50	1.05	30
C_2F_4 (polymerizing)	1.44	0.63	57

Table 5.6 F/C ratio of NF_3 plasma-fluorinated polymers before and after extraction with $CF_2ClCFCl_2$ [98].

Polymer	F/C ratio		
	Plasma	After extraction	Loss in F/C ratio (%)
Polyethylene	2.00	1.82	9
Polypropylene	1.85	1.70	8
Polystyrene	1.90	1.55	18
Poly(methyl methacrylate)	2.06	1.14	45

signal of CHF_3 plasma exposed glass surfaces traces of a CH component in the deposited film [111]. The F/C ratio amounted to 1.24 and the O/C ratio 0.03; thus, a large number of carbonaceous, crosslinked, and unsaturated structures must occur.

Perfluorination (replacing of all hydrogen atoms by fluorines) should be minimize the surface free energy of polyethylene to $<20\,mN\,m^{-1}$; thus, values may be achieved similar to that of poly(tetrafluoroethylene). Such a perfluorinated polymer surface should be ideally hydrophobic. CF_3 and CF_2 groups are responsible for this hydrophobic character because of the complete shielding of all bonding forces by the F substituents [92]. However, this perfluorinated structure is not realized with exposure to fluorinating plasmas; therefore, the shielding is incomplete and a significant polar component in the surface energy is additionally produced besides the mentioned co-introduction of oxygen.

Why do these incompletely fluorinated and partially oxidized surfaces show very high contact angles towards water in some cases? Topography and particularly the roughening of surface also play an important role besides surface composition (Figure 5.67) [117].

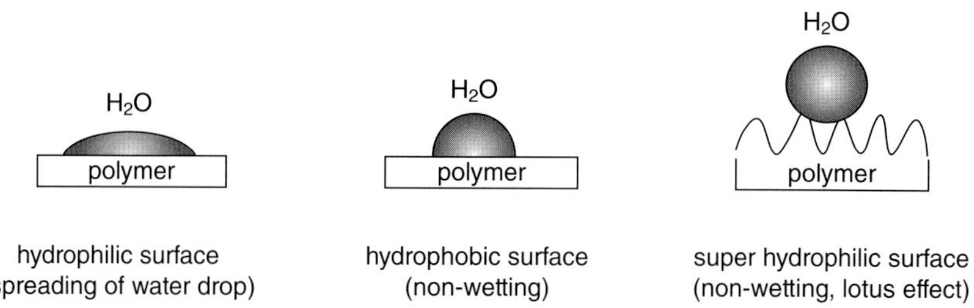

Figure 5.67 Wetting behavior of a water drop on the surface of differently modified polymers.

5.7 Chlorination

In contrast to the perfluorination of polyolefins the shielding of all residual binding forces by chlorine substituents is not complete. Therefore, chlorine functionalities can enhance the hydrophilicity [118–120]. CCl_4 plasma chlorination of polymers has been performed. Depending on the chlorination ratio and the type of activated hydrogen atom, which is replaced by chlorine, different contributions to the surface energy are produced [121, 122]. The largest increase in surface energy arises through the substitution of a hydrogen from a CH_3 group by Cl. Inagaki used the CCl_4 plasma for chlorination [92] and Strobel the CF_3Cl plasma [123, 124]. Inagaki comments that the tetrachloro-carbon plasma is one of the most efficient plasmas for increasing the surface energy. Under optimum conditions this plasma produces a surface tension of $67.7\,mJ\,m^{-2}$. This value is close to that of water ($72\,mJ\,m^{-2}$), which is also reflected in the advancing water contact angle of 7°. As reported for plasma activations generally the chlorination process is also accompanied by post-plasma incorporation of oxygen. Certainly, this oxygen introduction also contributes to the increase in surface energy. As mentioned before, in discussing plasma fluorination in the CF_4 plasma, considerable undesired co-introduction of oxygen is observed under different plasma conditions (<68 O per 100 C), exceeding the concentration of grafted Cl [122].

The split Cl 2p peak (near 200 eV) in Figure 5.68 shows the existence of covalently bonded C–Cl species and Cl^- anions after CCl_4 treatment of poly(ethylene terephthalate) (PET).

As is in also the case with fluorination, two possibilities for chlorination exist [118, 123, 125]:

1) direct chlorination of the macromolecules of the polymer substrate (CCl_4);
2) deposition of a Cl-containing plasma-polymer layer ($CHCl_3$, $C_2H_2Cl_2$, or $CHBr_3$).

Figure 5.69 shows the change in C1s signal of polyethylene after exposure to $CHCl_3$ plasma for different times. Notably, film deposition of plasma polymer is also possible using this plasma.

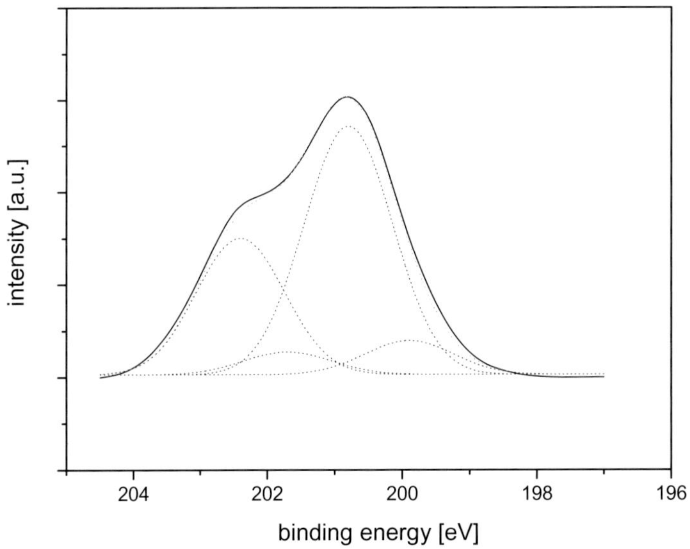

Figure 5.68 Cl 2p peak at PET surfaces after exposure to cw-rf CCl_4 plasma (60 s, 20 Pa, 150 W).

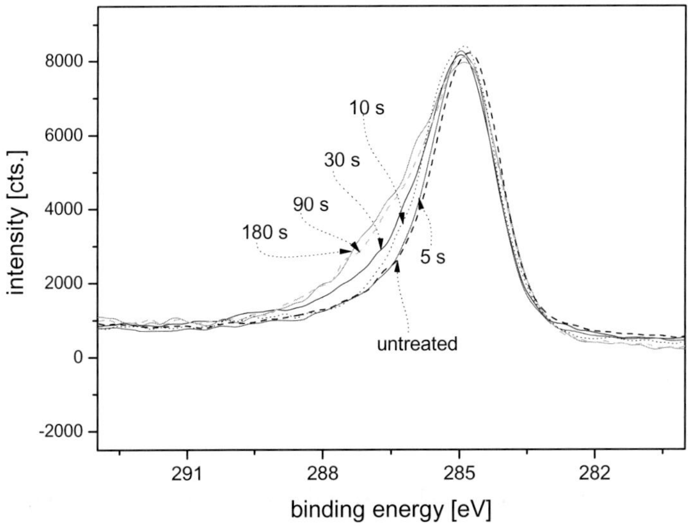

Figure 5.69 C1s signal of polyethylene after exposure to chloroform ($CHCl_3$) plasma for different times.

Other sources of surface chlorination were plasmas in vapors of $SOCl_2$, $POCl_3$, and PCl_3, which were also used to introduce sulfonic- and phosphonic acid groups.

Plasma chlorination has been applied to the modification of polyester fabrics [126].

5.8
Polymer Modification by Noble Gas Plasmas

As discussed in Section 4.1, low-pressure noble gas plasmas emit an intense plasma-vacuum radiation. Moreover, a particle shower of energy-rich and reactive species such as ions, electrons, and neutral metastable excited atoms bombards the polyolefin surface. These metastable states exist in high concentration and possess a long life-time [127]. Therefore, in remote/afterglow plasmas these metastable species travel long distances. These noble gas metastables are responsible for surface effects on polymer samples. On the other hand, the energy-rich plasma vacuum UV radiation causes crosslinking in surface-near polymer layers and in its neighboring bulk. This effect was originally attributed to the action of metastable species and, therefore, called crosslinking by activated species of noble gas plasmas (CASING) [128–129]. In 1972 Hudis attributed the crosslinking to the vacuum-UV irradiation from the (H_2) gas plasma [73].

The electron temperature (kinetic energy: $m/2v^2 = 1.5kT$) depends on the pressure and ionization potential of the noble gas (Figure 4.4).

Although the heavy particles (positive ions and neutral species) in non-isothermal low-pressure glow discharge plasmas have very low kinetic energy because of energy distribution via elastic collisions and energy transfer to the wall (temperature near room temperature) the electrons accumulate energy up to the threshold from where inelastic collision are possible. Inelastic collisions cause ionization, excitation, dissociation, and so on. Ions can also initiate chemical reactions at a polymer surface by ion bombardment (sputtering), adjusted by bias voltage, or ion–molecule reactions or neutralization as well as Penning ionization [130]. The penetration depths of ions and metastables into polymer surfaces are strongly dependent on their kinetic energy and electronic excitation or charge. The mean free path of these plasma species was experimentally determined for an oxide layer (PbO). The following order was found: He > Ne > Ar > Kr. Clark calculated the penetration depth of argon ions and metastables into a copolymer of ethylene and tetrafluoroethylene as 0.8 nm, based on a mean free path of 1 nm for electrons of about 960 eV kinetic energy [16]. While the electrons play the dominant role in the plasma, their role in interactions with the polymer is secondary. The mean free path of electrons in the plasma near zero kinetic energy is a few 10 nm. Thus, direct energy transfer in the surface region is likely to be relatively small, dominated by phonon excitation [131]. For electrons with a mean free path of 1.4 nm the penetration depth in the polymer has been estimated as 1.1 nm [132]. The

Table 5.7 Ionization potential and metastables energy of four noble gases [134–135].

Noble gas	Ionization potential (eV)	Metastables energy (eV)	Resonance radiation (eV)
He	24.58 ($^2S_{1/2}$)	19.82 (^1S); 20.61 (^3S)	21.2 (He^I); 40.8 (He^{II})
Ne	21.56 ($^2P_{3/2}$); 21.66 ($^2P_{1/2}$)	16.62 ($^3P_0$); 16.72 ($^3P_2$)	16.7, 16.8 (Ne^I); 26.8, 26.9 (Ne^{II})
Ar	15.76 ($^2P_{3/2}$); 15.94 ($^2P_{1/2}$)	11.55 ($^3P_0$); 11.72 ($^3P_2$)	11.6, 11.8 (Ar^I); 13.3, 13.5 (Ar^{II})
Kr	14.00 ($^2P_{3/2}$); 14.67 ($^2P_{1/2}$)	9.92($^3P_0$); 10.56 ($^3P_2$)	10.0, 10.6 (Kr^I); 12.8, 13.5 (Kr^{II})

penetration depths of all plasma particles are too low to be responsible for detected changes in polymer structure and composition in surface-near layers over the micrometer range. Only for the first few atomic monolayers is direct energy transfer from electrons, ions, and metastables of importance. Here, an ionized polymer chain produced in the topmost layer of the condensed polymer phase can undergo several structural transformations such as chain-scission or crosslinking.

Therefore, the multi-line and broad wavelength continuum radiation of the low-pressure plasma seems to be responsible for the detected effects over the μm range in polymers. Noble gas plasmas show strong resonance line emissions in the region 20–200 nm. Thus, singly charged noble gas plasmas emit in the range 58–110 (He), 74–100 (Ne), 105–155 (Ar), and 148–200 nm (Kr). Important (high-energetic) resonance lines are positioned at 105 and 107 nm (Ar) and 92 and 93 nm (Ar^+), 73 and 74 nm (Ne) and 46.1 and 46.2 nm (Ne^+), and 58 (He) and 30 nm (He^+) (Table 5.7) [133]. The total attenuation cross sections for the vacuum ultraviolet radiation passing into the polymer are dominated by the photo-ionization component [136], resulting in the production of polymer ions. At slightly longer wavelengths, components of the Rydberg transitions near the ionization limits also have substantial cross sections. Such states are also of importance in discussing the polymer–resonance radiation interactions. Modifications in the outermost few ångströms caused by the long-wavelength UV/visible component are small because of the low attenuation coefficients and low intensity compared with those of the vacuum ultraviolet component and its much higher intensity [137]. Thus, analyzing noble gas plasma treated ethylene–tetrafluoroethylene copolymers Clark could identify species that are characteristic for crosslinking of the copolymer [133]. The concentration loss for the $\underline{C}F_2$ component in the C1s signal and for the F1s peak decreases exponentially with time of exposure in the order He > Ne > Ar > Kr. However, for the changes in the subsurface and the bulk a different order was found caused by the different population of ions and metastables in the plasma (Kr > Ar > Ne > He) and other factors with Ne > He > Kr > Ar. Clark concludes that direct energy transfer at the surface produces $CF_2 \rightarrow CF$ or $CH_2 \rightarrow CH \rightarrow C$ and the radiative process $CF_2 \rightarrow CF \rightarrow C$, or $CH_2 \rightarrow CH \rightarrow C$. Thus, ultraviolet radiation has been demonstrated to have long-range effects in polymeric materials and in appropriate cases modifications in surface-near layers are detected to a depth of 10 μm [25, 73] or in a few cases up to the millimetre range [138–139].

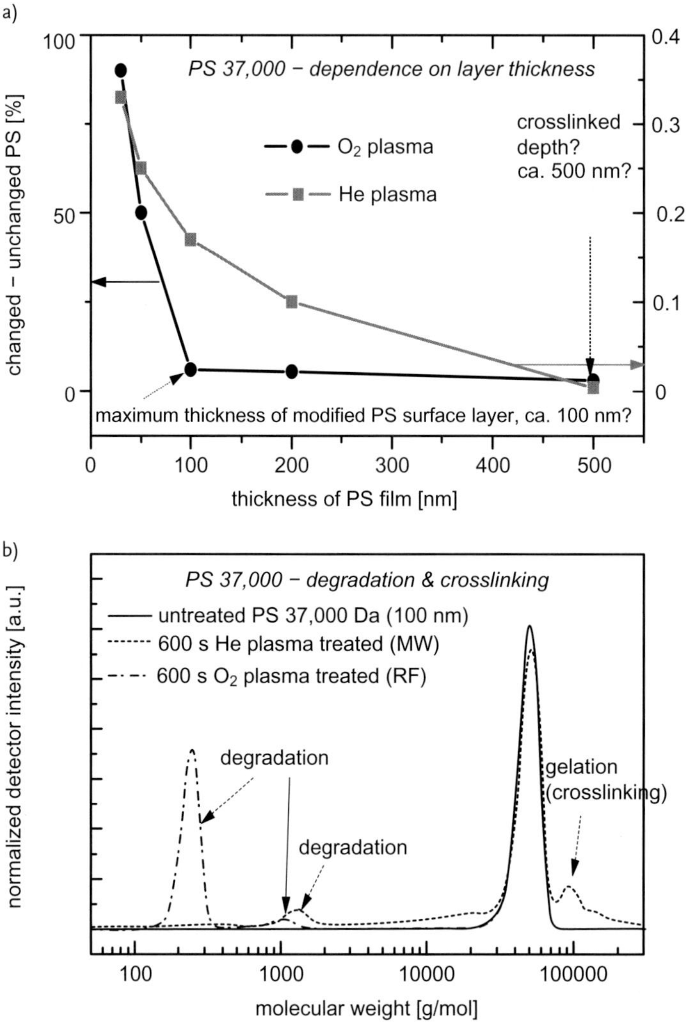

Figure 5.70 (a) Differences between oxygen and helium plasmas in terms of penetration depth of the crosslinking activity of the plasma UV radiation and (b) degradation and crosslinking influences on the molar mass distribution measured by gel-permeation chromatography.

The most suitable method to determine changes in the molar mass distribution (MWD), such as slight crosslinking, is gel-permeation chromatography (GPC), which is also known as size-exclusion chromatography (SEC). Figure 5.70b shows typical changes of MWD on exposure to helium or oxygen plasma. Both degradation and crosslinking products appear during the helium plasma treatment of polystyrene.

The thickness of plasma-modified surface-near layers differs between He- and O_2 plasma exposure (Figure 5.70a). After 180 s treatment time the He plasma modifies completely a polystyrene spin-coating film of about 500 nm [14]. In contrast, oxygen plasma also produced crosslinking, which was very surprising. It is limited to a thickness of <100 nm.

References

1 Kühn, G., Weidner, S., Decker, R., Ghode, A., and Friedrich, J. (1999) *Surf. Coat. Technol.*, **116–119**, 796–801.

2 Friedrich, J., Loeschcke, I., Gähde, J., and Richter, K. (1981) *Acta Polym.*, **32**, 337–343.

3 Kreidl, W.H. (1953) US-Patent 2632921.

4 Kreidl, W.H. (1959) *Kunststoffe*, **49**, 71.

5 Strobel, M., Branch, M.C., Ulsh, M., Kapaun, R.S., Kirk, S., and Lyons, C.S. (1996) *J. Adhesion Sci. Technol.*, **10**, 515–539.

6 Kim, C.Y., Evans, U., and Goring, D.A.I. (1971) *J. Appl. Polym. Sci.*, **15**, 1357.

7 Owens, D.K. (1975) *J. Appl. Polym. Sci.*, **19**, 265.

8 Owens, D.K. (1975) *J. Appl. Polym. Sci.*, **19**, 3315.

9 Friedrich, J., Unger, W., Lippitz, A., Gross, T., Rohrer, P., Saur, W., Erdmann, J., and Gorsler, H.-V. (1996) *Polymer Surface Modification: Relevance to Adhesion* (ed. K.L. Mittal), VSP, Utrecht, pp. 49–72.

10 Rossmann, K. (1956) *J. Polym. Sci.*, **19**, 141.

11 Friedrich, J., Geng, S., Unger, W., and Lippitz, A. (1998) *Surf. Coat. Technol.*, **98**, 1132–1141.

12 Unger, W.E.S., Lippitz, A., Friedrich, J.F., Wöll, C., and Nick, L. (1999) *Langmuir*, **15**, 1161–1166.

13 Geng, S., Friedrich, J., Gähde, J., and Guo, L. (1999) *J. Appl. Polym. Sci.*, **71**, 1231–1237.

14 Friedrich, J.F., Unger, W.E.S., Lippitz, A., Koprinarov, I., Weidner, S., Kühn, G., and Vogel, L. (1998) *Metallized Plastics 5&6: Fundamental and Applied Aspects* (eds K.L. Mittal), VSP, Utrecht, pp. 271–293.

15 Unger, W.E.S., Friedrich, J.F., Lippitz, A., Koprinarov, I., Weiss, K., and Wöll, C. (1998) *Metallized Plastics 5&6: Fundamental and Applied Aspects* (eds K.L. Mittal), VSP, Utrecht, pp. 147–168.

16 Clark, D.T. and Dilks, A. (1977) *J. Polym. Sci.: Polym. Chem. Ed.*, **15**, 2321.

17 Beamson, G. and Briggs, D. (1992) *High Resolution XPS of Organic Polymers-The Scienta ESCA300 Database*, John Wiley & Sons, Ltd, Chichester.

18 Friedrich, J., Mix, R., Schulze, R.-D., and Rau, A. (2010) *J. Adhesion Sci. Technol.*, **24**, 1329–1350.

19 Hare, L.O., Leadley, S., and Parbhoo, B. (2002) *Surf. Interface Anal.*, **33**, 335–342.

20 Friedrich, J., Unger, W., Lippitz, A., Gross, T., Rohrer, P., Saur, W., Erdmann, J., and Gorsler, H.-V. (1995) *J. Adhesion Sci. Technol.*, **9**, 575–598.

21 Weidner, S., Kühn, G., and Friedrich, J.F. (1998) *Rapid Commun. Mass Spectrom.*, **12**, 1373–1381.

22 Friedrich, J., Rohrer, P., Saur, W., Gross, T., Lippitz, A., and Unger, W. (1993) *Surf. Coat. Technol.*, **59**, 371.

23 Friedrich, J. (1981) *Contr. Plasma Phys.*, **21**, 261–277.

24 Wedenejew, W.J., Gurwitsch, K.W., Kondratjew, W.H., Medwedjew, W.A., and Frankewitsch, E.L. (1971) *Energien Chemischer Bindungen, Ionisationspotentiale Und Elektronenaffinitäten*, VEB Deutscher Verlag für Grundstoffindustrie, Leipzig.

25 Clark, D.T. and Feast, W.J. (1976) *Polymer Surfaces*, John Wiley & Sons, Ltd, Chichester, p. 185.

26 Clark, D.T. and Dilks, A. (1977) *J. Polym. Sci., Polym. Chem. Ed.*, **15**, 15.

27 Prohaska, G.W., Johnson, E.D., and Evans, J.F. (1984) *J. Polym. Sci., Polym. Chem. Ed.*, **22**, 2953–2972.
28 Friedrich, J.F., Unger, W.E.S., Lippitz, A., Weidner, S., Geng, S., Wigant, L., Wöll, C., Erdmann, J., and Gorsler, H.-V. (1996) *Applications of Surface and Interface Analysis* (eds H.J. Mathieu, B. Reihl, and D. Briggs), John Wiley & Sons, Ltd, Chichester, pp. 803–807.
29 Weidner, S., Kühn, G., Friedrich, J., Unger, W., and Lippitz, A. (1996) *Rapid Commun. Mass Spectrom.*, **10**, 727–737.
30 Weidner, S., Kühn, G., Decker, R., Roessner, D., and Friedrich, J. (1998) *J. Polym. Sci.*, **36**, 1639.
31 Friedrich, J., Unger, W., Lippitz, A., Koprinarov, I., Ghode, A., Geng, S., and Kühn, G. (2003) *Composite Interface*, **10**, 139–172.
32 Lippitz, A., Koprinarov, I., Friedrich, J.F., Unger, W.E.S., Weiss, K., and Wöll, C. (1996) *Polymer*, **37**, 3157.
33 Koprinarov, I., Lippitz, A., Friedrich, J.F., Unger, W.E.S., and Wöll, C. (1998) *Polymer*, **39**, 3001–3009.
34 Friedrich, J., Loeschcke, I., Frommelt, H., Reiner, H.-D., Zimmermann, H., and Lutgen, P. (1991) *Polym. Degrad. Stabil.*, **31**, 97.
35 Friedrich, J., Unger, W., and Lippitz, A. (1995) *J. Macromol. Sci.*, **100**, 111–115.
36 Koprinarov, I., Lippitz, A., Friedrich, J.F., Unger, W.E.S., and Wöll, C. (1998) *Polym. Commun.*, **38**, 2005–2010.
37 Lippitz, A., Friedrich, J., Unger, W., Schertel, A., and Wöll, C. (1996) *Polymer*, **37**, 3151–3156.
38 Friedrich, J., Unger, W., Lippitz, A., Koprinarov, I., Kühn, G., Weidner, S., and Vogel, L. (1999) *Surf. Coat. Technol.*, **116–119**, 772–782.
39 Friedrich, J.F., Unger, W.E.S., Lippitz, A., Giebler, R., Koprinarov, I., Weidner, S., and Kühn, G. (2000) *Polymer Surface Modification: Relevance to Adhesion*, vol. 2 (ed. K.L. Mittal), VSP, Utrecht, pp. 137–172.
40 Weidner, S., Kühn, G., Friedrich, J., and Schröder, H. (1996) *Rapid Commun. Mass Spectrom.*, **10**, 40–44.
41 Weidner, S., Kühn, G., and Friedrich, J. (1997) *Tagungsband, ECASIA 97, I. Olefjord* (eds L. Nyborg and D. Briggs), John Wiley & Sons, Ltd, Chichester, pp. 795–799.
42 Yasuda, H., Lamaze, C.E., and Sakaoku, K. (1973) *J. Appl. Polym. Sci.*, **17**, 137.
43 Hansen, R.H., Pascale, J.V., de Benedicts, T., and Rentzepis, P.M. (1965) *J. Polym. Sci. A*, **3**, 2205.
44 Throl, U., Gähde, J., and Friedrich, J. (1982) *Acta Polym.*, **33**, 561–566.
45 Friedrich, J., Kühn, G., and Gähde, J. (1979) *Acta Polym.*, **30**, 470–477.
46 Friedrich, J. and Gähde, J. (1980) *Acta Polym.*, **31**, 52.
47 Friedrich, J., Gähde, J., and Pohl, M. (1980) *Acta Polym.*, **31**, 310–315.
48 Friedrich, J. (1986) *Wissenschaft Fortschritt*, **36**, 12–15.
49 Krüger, P., Knes, R., and Friedrich, J. (1999) *Surf. Coat. Technol.*, **112**, 240–244.
50 Friedrich, J. (2005) Proceedings of 1st Thüringer Grenz- und Oberflächentage, September 15–16, 2005, Jena.
51 Miller, S.L. (1953) A production of amino acids under possible primitive earth conditions. *Science*, **117**, 528.
52 Miller, S.L. and Urey, H.C. (1959) Organic compound synthesis on the primitive earth. *Science*, **130**, 245.
53 Hollahan, J.R., Stafford, B.B., Falb, R.D., and Payne, S.T. (1969) *J. Appl. Polym. Sci.*, **13**, 807.
54 Lub, J., van Vroonhoven, F., Bruninx, E., and Benninghoven, A. (1989) *Polymer*, **30**, 40.
55 Friedrich, J., Ivanova-Mumeva, V.G., Gähde, J., Andreevskaya, G.D., and Loeschcke, I. (1983) *Acta Polym.*, **34**, 171.
56 d'Agostino, R., Stendardo, M., and Favia, P. (1995) *Proceedings of the 12th International Symposium on Plasma Chemistry*, vol. I (eds J.V. Heberlein, D.W. Ernie, and J.T. Roberts), University of Minneapolis, p. 27.
57 Oehr, C. and Müller, M. (1999) *Surf. Coat. Technol.*, **116–119**, 802.
58 Meyer-Plath, A., Mix, R., and Friedrich, J. (2007) *Adhesion Aspects of Thin Films*, vol. III (ed. K.L. Mittal), VSP, Leiden, Materials Research Society, Warrendale, PA, p. 177.

59 Boenig, H.V. (1988) *Fundamentals of Plasma Chemistry and Technology*, Technomic, Lancaster.

60 Meyer-Plath, A. (2002) Grafting of Amino and Nitrogen Groups on Polymers by Means of Plasma Functionalisation. PhD thesis. Ernst-Moritz-Arndt University of Greifswald.

61 Fanghänel, E. (2004) *Organikum*, 22nd edn, Wiley-VCH Verlag GmbH, Weinheim.

62 Hergenrother, P.M. and Levine, H.H. (1967) *J. Polym. Sci., A1*, **5**, 1453.

63 Oran, U., Swaraj, S., Friedrich, J., and Unger, W.E.S. (2005) *Surf. Coat. Technol.*, **200**, 463.

64 Swaraj, S., Oran, U., Lippitz, A., Friedrich, J., and Unger, W.E.S. (2005) *Surf. Coat. Technol.*, **200**, 494.

65 Holländer, A., Wilken, R., and Behnisch, J. (1999) *Surf. Coat. Technol.*, **116–119**, 991–995.

66 Strobel, J.M., Strobel, M., Lyons, C.S., Dunatov, C., and Perron, S. (1991) *J. Adhesion Sci. Technol.*, **5**, 119.

67 Friedrich, J., Wettmarshausen, S., and Hennecke, M. (2009) *Surf. Coat. Technol.*, **203**, 3647.

68 Min, H., Wettmarshausen, S., Friedrich, J., and Unger, W.E.S. (2011) *J. Anal. At. Spectrom.*, **26**, 1157–1165.

69 Wagner, D., Dikmen, B., and Döbele, H.F. (1988) *Plasma Sources Sci. Technol.*, 7, 462.

70 Behnisch, J., Zimmermann, H., and Friedrich, J. (1991) *Int. J. Mater. Sci.*, **16**, 139.

71 Vasilets, V.N., Kuznetsov, A.V., and Sevastianov, V.I. (2004) *J. Biomed. Mater. Res. Part A*, **69**, 428.

72 Murillo, R., Poncin-Epaillard, F., and Segui, Y. (2007) *Eur. Phys., J. Appl. Phys.*, **37**, 299.

73 Hudis, M.V. (1972) *J. Appl. Polym. Sci.*, **16**, 2397.

74 Jelinski, J. (1986) Deuterium NMR of solid polymers, in *High Resolution NMR Spectroscopy of Synthetic Polymers in Bulk* (ed. R.A. Komoroski), VCH Publishers, Deerfield Beach, FL.

75 Lyman, T. (1906) *Astrophys. J.*, **23**, 181.

76 Holländer, A., Klemberg-Sapieha, J.E., and Wertheimer, M.R. (1994) *Macromolecules*, **27**, 2893.

77 Liston, E.M., Martinu, L., and Wertheimer, M.R. (1993) *J. Adhesion Sci. Technol.*, **7**, 1091.

78 Painter, L.R., Arakawa, E.T., Williams, M.W., and Ashley, J.C. (1980) *Radiat. Res.*, **83**, 1.

79 Truica-Marasescu, F.E. and Wertheimer, M.R. (2005) *Macromol. Chem. Phys.*, **206**, 744.

80 Nakayama, M., Shibahara, M., Maruyama, K., and Kamat, K. (1993) *J. Mater. Sci. Lett.*, **12**, 1380.

81 Skurat, V. (2003) *Nucl. Instrum. Methods Phys. Res. B*, **208**, 27.

82 Klomp, A.J.A., Terlingen, J.G.A., Takens, G.A.J., Strikker, A., Engbers, G.H.M., and Feijen, J. (2000) *J. Appl. Polym. Sci.*, **75**, 480.

83 Shard, A.G. and Badyal, J.P.S. (1991) *J. Phys. Chem.*, **95**, 9438.

84 Medard, N., Soutif, J.-C., and Poncin-Epaillard, F. (2002) *Surf. Coat. Technol.*, **160**, 197.

85 Poll, H.-U. and Meichsner, J. (1980) *Acta Polym.*, **31**, 757.

86 Shard, A.G. and Badyal, J.P.S. (1991) *J. Phys. Chem.*, **95**, 9436.

87 Terlingen, J.G., Gerritsen, H.F., Hoffman, A.S., and Jen, F.J. (2002) *J. Appl. Polym. Sci.*, **40**, 719.

88 Gancarz, I., Pozniak, G., and Bryjak, M. (1999) *Eur. Polym. J.*, **35**, 1419.

89 Inagaki, N., Tasaka, S., Kawai, H., and Kimura, Y. (1990) *J. Adhesion Sci. Technol.*, **4**, 99.

90 Inagaki, N., Tasaka, S., Ohkubo, J., and Kawai, H. (1990) *J. Appl. Polym. Sci., Appl. Polym. Symp.*, **46**, 399.

91 Everhart, D.S. and Reilley, C.N. (1981) *Anal. Chem.*, **53**, 665.

92 Inagaki, N. (1996) *Plasma Surface Modification and Plasma Polymerization*, Technomic, Lancaster.

93 Inagaki, N. and Matsunaga, M. (1985) *Polym. Bull.*, **13**, 349.

94 Yanagihara, K., Kimura, M., Niinomi, M., Nishikawa, Y., and Mukaida, Y. (1986) US Patent 4,693,799.

95 Okusa, H., Kurihara, K., and Kunitake, T. (1994) *Langmuir*, **10**, 3577–3581.

96 Hopkins, J. and Badyal, J.P.S. (1995) *J. Phys. Chem.*, **99**, 4261.

97 Yagi, T. and Pavlath, A. (1984) *J. Appl. Polym. Sci.: Appl. Polym. Symp.*, **38**, 215.

98 Jama, C., Quensierre, J.-D., Gengembre, L., Moineau, V., Grimblot, J., Dessaux, O., and Goudmand, P. (1999) *Surf. Interface Anal.*, **27**, 653.
99 Anand, M., Cohen, R.E., and Baddour, R.F. (1981) *Polymer*, **22**, 361.
100 Strobel, M., Corn, S., Lyons, C., and Korba, G.A. (1985) *J. Polym. Sci., Polym. Chem. Ed.*, **23**, 1125.
101 Friedrich, J., Wigant, L., Unger, W., Lippitz, A., Erdmann, J., Gorsler, H.-V., Prescher, D., and Wittrich, H. (1995) *Surf. Coat. Technol.*, **74–75**, 910.
102 Friedrich, J., Wigant, L., Lippitz, A., Wittrich, H., Prescher, D., Erdmann, J., and Gorsler, H.-V. (1996) *Polymer Surface Modification: Relevance to Adhesion* (ed. K.L. Mittal), VSP, Utrecht, pp. 121–136.
103 Pauling, L. (1961) *J. Inorg. Nucl. Chem.*, **17**, 215.
104 Mulliken, R.S. (1976) Arithmetic average of electron affinity and ionization energy. *J. Am. Chem. Soc.*, **98**, 7869.
105 Friedrich, J., Wigant, L., Lippitz, A., Wittrich, H., Prescher, D., Erdmann, J., and Gorsler, H.-V. (1995) *J. Adhesion Sci. Technol.*, **9**, 1165–1180.
106 (a) Wigant, L. and Friedrich, J. (1996) *Wissenschaftliche Beiträge der TFH Wildau*, **1**, 40–48; (b) Friedrich J., Kühn G., Schulz U., Jansen K., and Möller B. (2002) *Vakuum*, **14**, 285–290.
107 Friedrich, J., Kühn, G., Schulz, U., Jansen, K., Bertus, A., Fischer, S., and Möller, B. (2003) *J. Adhesion Sci. Technol.*, **17**, 1127–1144.
108 Egitto, F.D. (1990) *Pure Appl. Chem.*, **62**, 1699.
109 Kim, M.T. (2003) *Appl. Surf. Sci.*, **211**, 285; (b) Volkmann T. and Widdecke H. (1989) *Kunststoffe*, **79**, 8.
110 Millard, M.M. (1972) *Anal. Chem.*, **44**, 828.
111 Strobel, M., Lyons, C.S., and Korba, G.A. (1987) *J. Polym. Sci., Polym. Chem. Ed.*, **25**, 1295.
112 Strobel, M., Thomas, P.A., and Lyons, C.S. (1987) *J. Polym. Sci., Polym. Chem. Ed.*, **25**, 3343.
113 Yagi, T., Pavlath, A.E., and Pittman, A.G. (1982) *J. Appl. Polym. Sci.*, **27**, 4019.
114 Sigurdsson, S. and Shishoo, R. (1997) *J. Appl. Polym. Sci.*, **66**, 1591.
115 Scott, P.M., Matienzo, L.J., and Babu, S.V. (1990) *J. Vac. Sci. Technol. A*, **8**, 2382.
116 d'Agostino, R., Cramarossa, F., and de Benedicts, S. (1982) *Plasma Chem. Plasma Process.*, **2**, 213.
117 Minko, S., Müller, M., Motornov, M., Nitschke, M., Grundke, K., and Stamm, M. (2003) *J. Am. Chem. Soc.*, **125**, 3896.
118 Upadhyay, D.J. and Bhat, N.V. (2003) *Plasma Polym.*, **8**, 237.
119 Tyczkowski, J., Krawczyk, I., and Wozniak, B. (2003) *Surf. Coat. Technol.*, **174–175**, 849–853.
120 Lu, W., Cao, T., Wang, Q., and Cheng, Y. (2011) *Plasma Process. Polym.*, **8**, 94–99.
121 van Krevelen, D.W. and Hoftyzer, P.J. (1974) *Properties of Polymers*, Elsevier, Amsterdam.
122 Inagaki, N., Tasaka, S., and Imai, M. (1993) *J. Appl. Polym. Sci.*, **48**, 1963.
123 Strobel, M., Vara, K.P., Corn, S., Lyons, C.S., and Morgan, M. (1990) *J. Appl. Polym. Sci., Appl. Polym. Symp.*, **56**, 61.
124 Corn, S., Vara, K.P., Strobel, M., and Lyons, C.S. (1991) *J. Adhesion Sci. Technol.*, **5**, 239.
125 Inagaki, N., Tasaka, S., and Suzuki, Y. (1994) *J. Appl. Polym. Sci.*, **51**, 2131.
126 Akovali, G. and Takrouri, F. (1991) *J. Appl. Polym. Sci.*, **42**, 2717.
127 McTaggart, F.K. (1967) *Plasma Chemistry in Electrical Discharges*, Elsevier, Amsterdam.
128 Hansen, R.H. and Schonhorn, H. (1966) *J. Polym. Sci., Polym. Lett. Ed.*, **4**, 209.
129 Schonhorn, H. and Hansen, R.H. (1967) *J. Appl. Polym. Sci.*, **11**, 1461.
130 Hertz, G. and Rompe, R. (1973) *Plasmaphysik*, Akademie-Verlag, Berlin.
131 Clark, D.T. and Feast, W.J. (1978) *Polymer Surfaces*, John Wiley & Sons, Ltd, Chichester.
132 Clark, D.T. and Thomas, H.R. (1977) *J. Polym. Sci., Polym. Phys. Ed.*, **15**, 15.
133 Clark, D.T. and Dilks, A. (1978) *J. Polym. Sci.*, **16**, 911.
134 Muschlitz, E.E. Jr (1968) *Science*, **159**, 599.
135 Delcroix, J.L., Ferreira, C.M., and Ricard, A. (1973) *Proceedings of 11th IGPIC*, vol. I, Prague, p. 301.

136 Partridge, R.H. (1966) *J. Chem. Phys.*, **45**, 1685.

137 Samson, J.A.R. (1977) *Techniques of Vacuum Ultraviolet Spectroscopy*, John Wiley & Sons, Inc., New York.

138 Friedrich, J., Gähde, J., and Nather, A. (1981) *Plaste Kautschuk*, **28**, 140.

139 Friedrich, J., Loeschcke, I., Reiner, H.-D., Frommelt, H., Raubach, H., Zimmermann, H., Elsner, T., Thiele, L., Hammer, L., and Merker, E. (1990) *Int. J. Mater. Sci.*, **13**, 43–56.

6
Bulk, Ablative, and Side Reactions

6.1
Changes in Supermolecular Structure of Polymers

Exposure to oxygen-plasma treatment introduces up to 30% oxygen but also changes the orientation of macromolecules at the top-most surface and in surface-near layers. It could be shown that in partially crystalline and/or oriented polymer systems any orientation diminishes completely [1–5]. In terms of polymer chemistry the orientation of molecules is responsible for the extraordinary mechanical properties of the systems, in particular if the bulk structure is affected. Changes near the surface demand surface-sensitive methods for identification of such orientation effects. By means of IR spectroscopy using polarized radiation such effects can be identified. Synchrotron-based X-ray absorption spectroscopy is also well-suited, for example, angle-resolved NEXAFS (near edge X-ray absorption fine structure) spectroscopy [6] (Figure 6.1).

This method is complementary to X-ray photoelectron spectroscopy (XPS). NEXAFS gives information on the orientation of groups, segments, and chains in the polymer surface layer, and also on the existence of π-electron-containing bonds such as olefinic, aromatic, or carbonyl species. For identification of changes in orientation NEXAFS spectra at different angles are recorded. Most often 20°, 55°, and 90° are used. A transition from the K-shell to an unoccupied molecular orbital has maximum intensity when the electric field vector ($\boldsymbol{E}$) of the linearly polarized X-ray beam points along the direction of maximum orbital amplitude. The transition is forbidden when $\boldsymbol{E}$ lies in the nodal plane of the orbital. In general, there is a $\cos 2\Theta$ dependence of the transition intensity where Θ is the angle between the $\boldsymbol{E}$ vector and the direction of maximum orbital amplitude. Note that for ethylene a resonance due to transitions to C–H antibonding MOs is observed in addition to the main π^* and σ^* resonances associated with the C=C double bond.

The 90–20° difference spectra reflect the orientation (anisotropy) of molecules [self-assembled or Langmuir–Blodgett layers] or macromolecules (polymers) within the analyzed layer (3–5 nm). The area of the difference spectrum on both sites of the zero line formed between the measured function and the zero line is the measure for the orientation (anisotropy). This can be also expressed as an orientation factor referenced to a defined orientation, which may be a Langmuir–Blodgett

The Plasma Chemistry of Polymer Surfaces: Advanced Techniques for Surface Design, First Edition.
Jörg Friedrich.

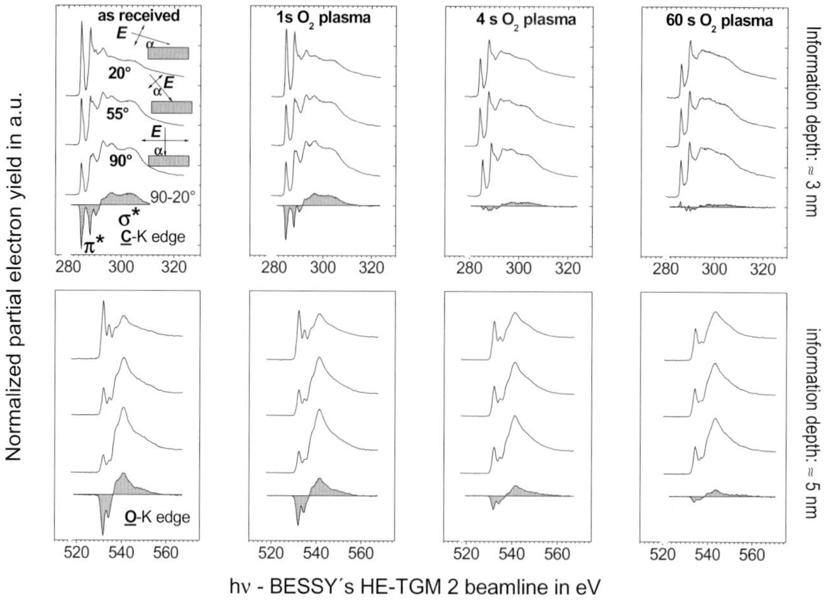

Figure 6.1 C_K- and O_K-edge spectra (NEXAFS) of poly(ethylene terephthalate) (PET) measured at different angles of incidence and the 90°–20° difference spectra according to the time of exposure to oxygen plasma.

layer or the highly-oriented original state in a (semi-)crystalline polymer. With amorphization of the surface layer by plasma exposure the orientation (possibly a supermolecular structure) becomes isotropic and the difference spectrum tends to zero, for example, the spectra of all angles of incidence are more or less identical (cf. Figure 6.1) [6].

In principle, such a near-edge photoelectron spectrum shows the structure of the element-specific absorption edge for each electron quantum number. Resonance transitions are superposed on this basic absorption as pre-edge sharp and intense π^*-resonances (C=C, C=O) and σ^* C–C and C–H (Rydberg) transitions (Figure 6.2). The π^*-resonances are an excellent analytical feature by which to detect any unsaturation within the topmost polymer layer of 3–5 nm. However, the different types of unsaturations are often not well-resolved, as is known from $\pi \rightarrow \pi^*$ transitions in UV/visible spectroscopy. Similar to XPS, cross-checking is possible by comparing the results for the absorption edges of elements involved in the bond, for example, for carbonyl feature information is available from the C_K and the O_K edge.

As can be seen from Figure 6.1 the C_K and O_K spectra become like each other the longer the PET was exposed to oxygen plasma. Therefore, complete amorphization has taken place after prolonged exposure. The original (biaxially stretched) orientation of macromolecules at the PET surface (cf. also Figure 6.2) diminishes completely. However, this structure withstood in the main the attack

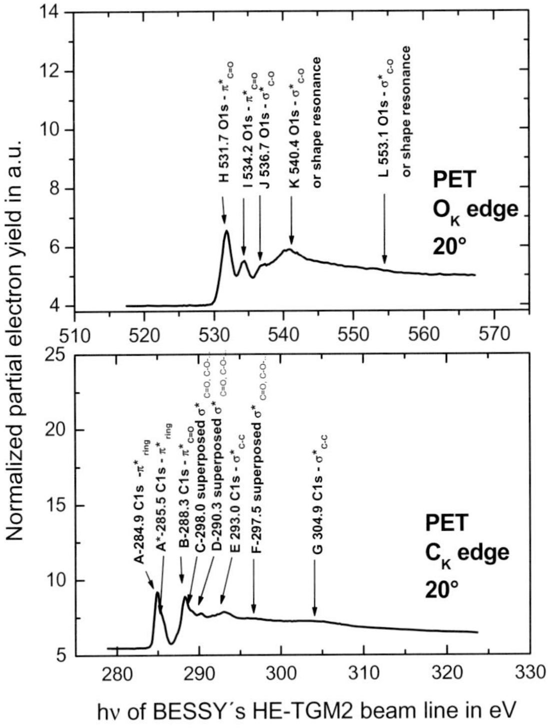

Figure 6.2 Assignation of π* and σ* transitions in the C_K and O_K spectra of poly(ethylene terephthalate) (PET) at an angle of incidence of 20°.

of the O_2 plasma particle bombardment and UV-induced rearrangements up to a maximal exposure time of 2 s (cf. Figures 6.3 and 6.4). PET does not possess well-suited points for introducing polar groups onto the backbone without its preliminary scission. Therefore, an increase of polar component of surface energy is strongly related with degradation and the formation of low-molecular weight oxidized material (LMWOM).

Oxidation of the ethylene gylcol unit produces decay of the ester group and chain scission. Chain scissions make the PET mechanically defective at the surface (weak boundary layer). Norrish I and II reactions produce such undesired chain scissions. Cracking of aromatic rings also makes chain scissions, which requires much more energy; however, the plasma involves sufficient energy. Formation of phenol and bezoic acid structures at aromatic rings by photo-Fries rearrangement also makes chain scissions. Thus, there is no defined way to equip the PET molecule with polar groups without chain scissions.

A similar response to oxygen plasma exposure is seen for highly ordered self-assembled aliphatic monolayers from octadecyltrichlorosilane (OTS), which is as an ideal model for polyethylene (PE) surfaces and polypropylene (PP) films

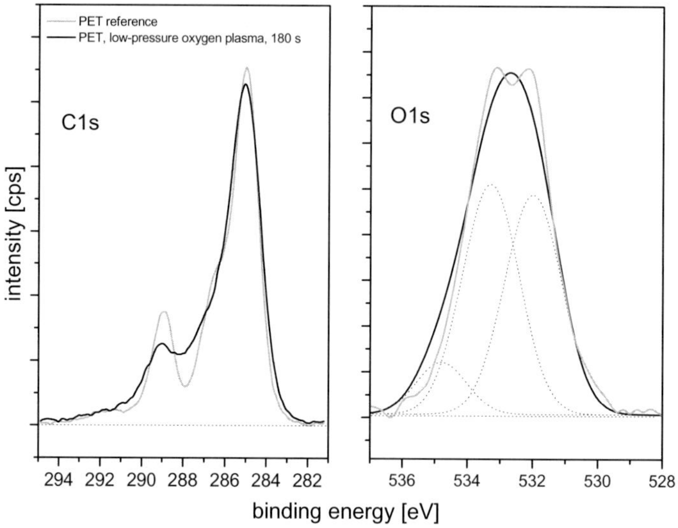

Figure 6.3 C1s and O1s peaks of PET before and after cw-rf plasma exposure for 180 s.

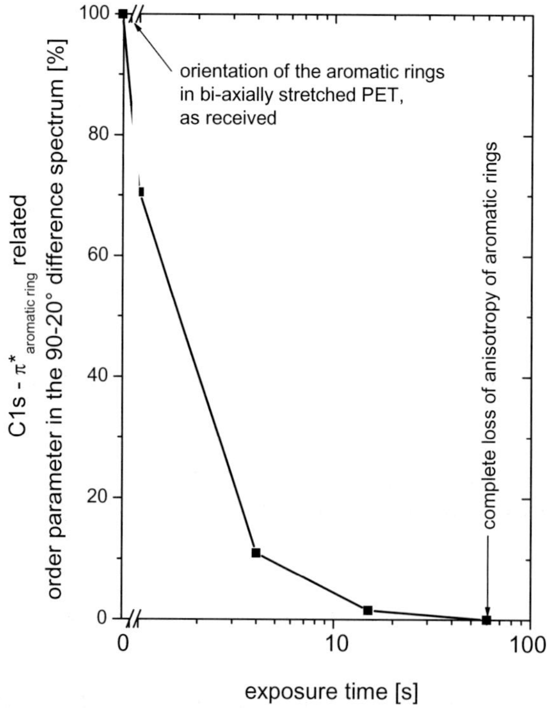

Figure 6.4 Loss in orientation of biaxially stretched PET films during the exposure to the DC oxygen plasma.

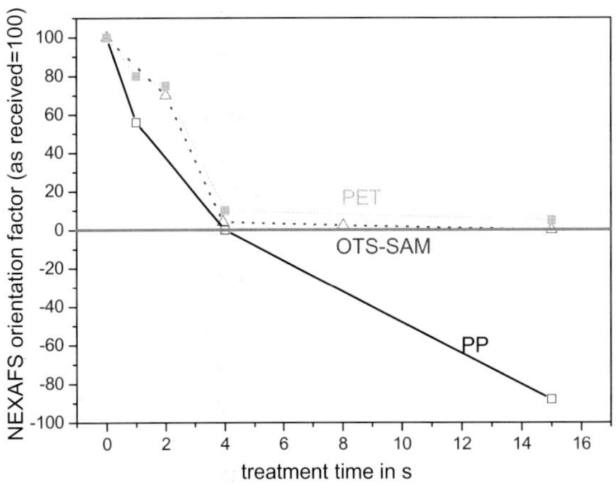

Figure 6.5 Comparison of NEXAFS measured surface-related orientation factors of PET, OTS, and PP with exposure time to the oxygen plasma (cw rf, 100 W, 6 Pa).

(Figure 6.5). In all cases, under the plasma conditions used in Reference [2], the orientation was roughly intact up to 2 s plasma exposure. Prolonged plasma treatment leads to very fast and complete loss in any orientation. Polypropylene shows surprising behavior. In addition in this case, the orientation of biaxially stretched polypropylene is diminished after 4 s exposure time but at prolonged exposure a completely new type of orientation appears (cf. Figure 6.5). This process is known as "chemi-crystallization" [7]. The high density of newly introduced O-functional groups into the surface-near layer enables the formation of new interactions between these groups by hydrogen bonds and other interactions. Polymer segments having such interacting groups undergo an (new) orientation by shortening the distances between interacting segments. Thus, the density increases.

More recently, this effect was also evidenced using X-ray wide-angle diffraction and density measurements [8] showing the same tendency although these methods are bulk methods. However, obviously, these changes of orientation at longer oxygen plasma exposure are not strictly limited to the upper few nanometers. Structural changes may also occur with some polymers in the μm range [8]. Using a X-ray wide-angle scan, the crystallinity of LDPE (low-density polyethylene) always increased, for example, from 0.69 to 0.75. In contrast, HDPE (high-density polyethylene) most often shows a slight decrease in crystallinity from 0.83 to 0.79. Thus, an average density of plasma-exposed polyethylene near 0.75–0.79 was observed [8]. The strong decrease in crystallinity of PET, on exposure to oxygen plasma, could be confirmed using a bulk method, namely, the density measurement. The density of polymers depends linearly on the crystallinity. All plasma treatments of PET show, more or less, a decrease of density (Table 6.1).

Another idea was that the amorphous phase of semi-crystalline polymers, such as PET, PP, or PE (polyethylene), is selectively etched. The remaining polymer

Table 6.1 Changes in the density of PET with the type of plasma employed.

Plasma	Density (g cm^{-3})
As received	1.396
O_2	1.394
Ar	1.394
H_2O	1.393
H_2	1.392
NH_3	1.391
N_2	1.390

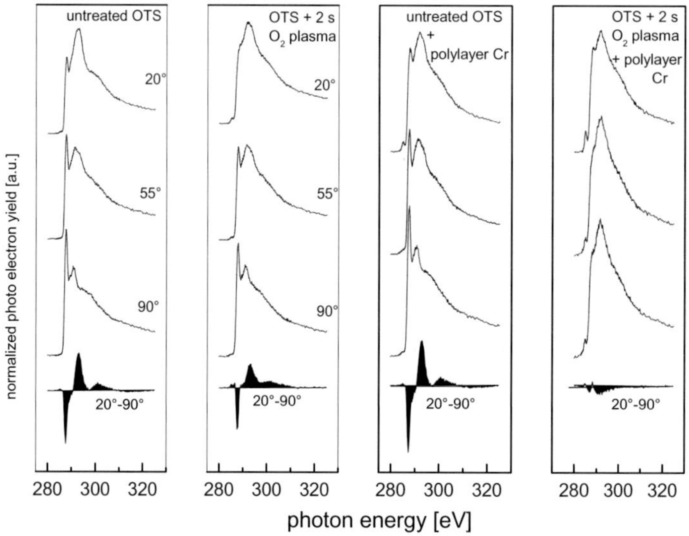

Figure 6.6 NEXAFS C_K edge spectra of octadecyltrichlorosilane (OTS) monolayers on Si wafers and also with a subsequent Cr deposition of about four Cr monolayers, both before and after exposure to cw-rf oxygen plasma.

surface-near layer should consist of pure crystalline polymer, thus increasing the resulting crystallinity degree (and polymer density) of the whole polymer sample.

The amorphization of the polymer subsurface correlates closely with the oxygen introduction during oxygen plasma exposure and is enhanced by additional deposition of a few chromium layers (Figure 6.6).

When a steady-state of oxygen introduction is achieved the orientation is zero (Figure 6.7). It is assumed that at this point a layer of LMWOM [9] is formed that acts as a weak boundary layer (WBL) and thus strongly weakens the adhesion between metal and polymer (Figure 6.7). The loss in any adhesion to metal coatings is shown below.

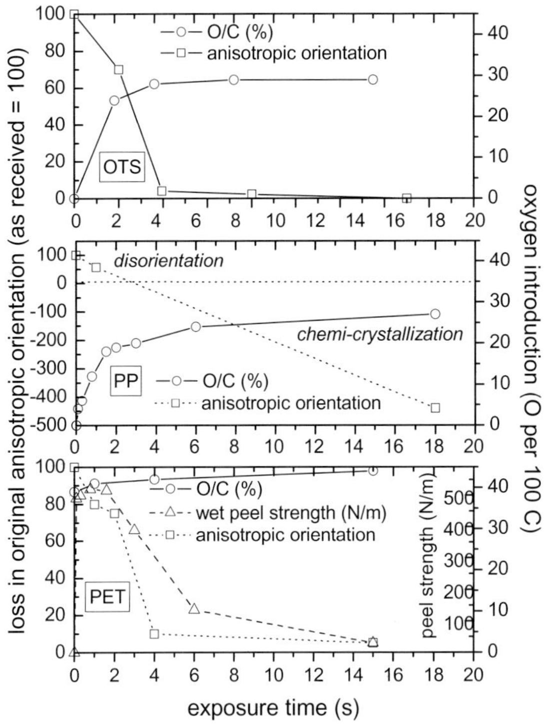

Figure 6.7 Oxygen introduction, orientation loss, and peel strengths of thermally evaporated 200 nm thick Al layers on octadecyltrichlorosilane (OTS), polypropylene (PP), and poly(ethylene terephthalate) (PET) on exposure to the oxygen plasma (cw rf, 100 W, 6 Pa).

To gain an impression of the processes a rough and generalizing scheme of possible processes on exposure of self-assembled monolayers to oxygen plasma were designed as shown in Section 6.7.

6.2 Polymer Etching

Polymers without layered structures were uniformly etched under exposure to the oxygen plasma [10]. Using other etching gases a strong dependence of the etching behavior on the type of plasma is obvious (Table 6.2) [11]. Thus, the etching rates are strongly dependent on the chemical reactivity of (dissociated) plasma gas, the plasma conditions, the reactor geometry, and type of polymer and its sensitivity towards etching. Therefore, the etching rates of polymers vary strongly with the above-mentioned influencing factors. Consequently, the measured etching rates are relative values with strong uncertainty. Nevertheless, the succession of etching

Table 6.2 Etching rates of poly(ethylene terephthalate) foils (PET) in different gas plasmas under comparable plasma conditions (cw-rf plasma, 100 W, 13 Pa).

Plasma gas (300 W, cw-rf plasma, 13 Pa)	$10^5 \times$ Etching rate ($mg\,cm^{-2}\,s^{-1}$)
O_2	6.4
H_2O	3.5
He	1.8
Ar	1.4
H_2	0.9
N_2	0.8

Table 6.3 Etching rates of different polymers on exposure to oxygen plasma (cw-rf plasma, 10 Pa, 300 W).

Polymer	$10^5 \times$ Etching rates ($mg\,cm^{-2}\,s^{-1}$)
Poly(vinyl acetate) (PVAc)	12
Poly(methyl methacrylate) (PMMA)	11
Low-density polyethylene (LDPE)	6
High-density polyethylene (HDPE)	6
Poly(ethylene terephthalate) (PET)	6
Poly(butylene terephthalate) (PBT)	6
Polystyrene (PS)	2
Natural rubber	1.5

rates can also vary according to the influencing factors. Thus, groups of polymers can be defined in terms etching behavior (Table 6.2).

Comparing the etching rates of poly(ethylene terephthalate) in oxygen and in argon plasma the percentage of the chemical contribution to the etching can be roughly appreciated ($6.4–1.4 \times 10^{-5}\,mg\,cm^{-2}\,s^{-1}$); for example, it can be concluded that roughly four-fifths of the polymer removal is caused by chemical processes. Nitrogen, hydrogen, and water vapor plasmas certainly have chemical contributions to the plasma etching process; however, this factor seems to be negligible within the material removal.

This type of polymer also plays a dominant role, as shown in Table 6.3, which lists the etching rates of several polymers [8].

H. Yasuda has measured the etching rates of different polymers in pure low-pressure helium plasma (Table 6.4) [12].

The etching rates of polymers reflect the sensitivity of polymer units to degradation under exposure to the O_2 or He plasma. Moreover, the lowering of etching rate by aromatic rings and by the tendency of several polymers to undergo crosslinking (polyethylene, polybutadiene, etc.) is obvious. Especially, alkane and

Table 6.4 Etching rates of different polymers on exposure to helium plasma (cw-rf plasma, 30 W, 13.6 Pa).

Polymer	**10^5 × Etching rates ($mg\,cm^{-2}\,min^{-1}$)**
Poly(oxymethylene) (POM)	17.0
Poly(acrylic acid) (PAA)	16.2
Poly(methyl methacrylate) (PMMA)	15.4
Poly(vinyl pyrrolidone) (PVP)	11.9
Poly(vinyl alcohol) (PVA)	9.4
Poly(ethylene terephthalate) (PET)	1.7
High-density polyethylene (HDPE)	1.2
Polyamide-6 (PA-6)	1.1
Polypropylene (PP)	0.8

rubber-like chains tend to crosslink, thus stabilizing the polymer towards the etch attack from the plasma [8]. Moss *et al.* have given a survey of the plasma etching rates of homo- and copolymers [13].

Hydrogen, water, carbon dioxide, and carbon monoxide are the most prominent degradation products (as expected) when polypropylene is exposed to the oxygen plasma. The released neutral degradation products in the off-gas were measured by means of mass spectrometry. Using gas chromatography and a thermal conductivity detector, on exposure of polypropylene to argon plasma the dominant degradation product was hydrogen (Section 6.5). Methane and C_2 products such as acetylene, ethylene, and ethane were also detected in significant concentrations; C_4 products such as butadiene, cyclobutadiene, butane, and butane were present as well as traces of C_6, C_3, and C_5 products (Friedrich, J., unpublished results, 1976) [14].

The etching progress can be also followed by IR spectroscopy. As discussed in Section 5.11 and demonstrated in Figure 5.11 for the functionalization with O-containing functional groups in an oxygen plasma and the transition to a steady-state of plasma etching after a few seconds a constant relation between the different oxygen products was observed. IR spectroscopy confirms these results when looking at the etching of a self-assembled monolayer of stearic acid on a silver target, that is, the lowering of CH_2/CH_3 stretching intensities at 2900 cm^{-1} (Section 6.5).

The increasing oxidation is also obvious if comparing the intensity ratio between the stretching regions of CH_x and C-O_x around 2900 and 1100 cm^{-1}. The decrease of intensity of the CH_2/CH_3 stretching region reflects the progressive etching process.

Using gravimetric measurements (micro balance) a linear loss in weight was measured for polypropylene exposed to oxygen plasma (Figure 6.8).

Etching of polymers begins with chain scissions of the backbone. These fragments are oxidized and degraded to LMWOM and ultimately to CO_2, CO, and H_2O.

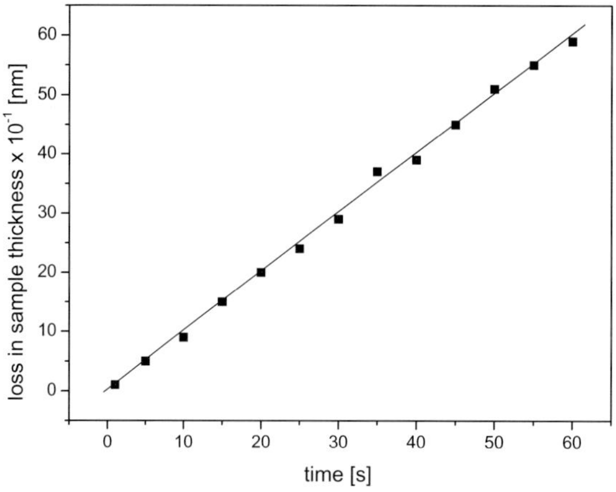

Figure 6.8 Weight loss of polypropylene on exposure to cw-rf plasma (frequency = 13.56 MHz, pressure = 10 Pa, wattage = 100 W).

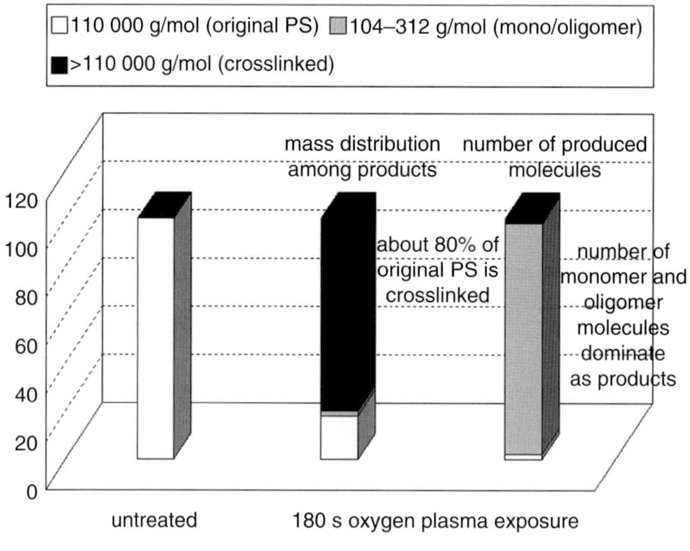

Figure 6.9 Overview of simultaneous degradation and crosslinking of PS bulk (200 nm layers of PS 110 000 g mol^{-1}) on different exposure times to cw-rf oxygen plasma (wattage = 100 W, pressure = 10 Pa, 200 nm layer thickness).

The change of molar mass distribution can be determined using gel-permeation chromatography (GPC) (also known as size-exclusion chromatography, SEC) as well as by different types of field-flow fractionation (FFF). Both degradation (SEC) and crosslinking (FFF) are present if polystyrene is exposed to the low-pressure oxygen plasma (Figure 6.9). This broadening of the original molar mass distribution is obvious from a combination of SEC and FFF spectra, as shown schematically in Figure 5.19.

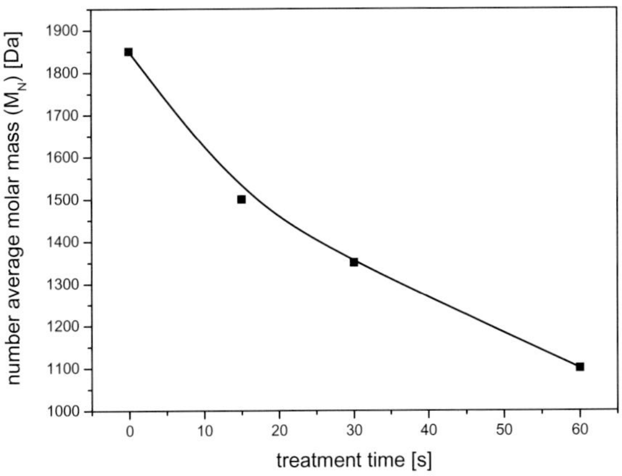

Figure 6.10 Loss in molar mass of PMMA 1900 on exposure to cw-rf low-pressure oxygen plasma (power = 100 W, pressure = 10 W) measured by matrix-assisted laser desorption/ionization time-of-flight (MALDI-ToF) mass spectrometry.

The molar mass distribution function was extremely broadened also at a short time of exposure to oxygen plasma, evidencing the susceptibility of polymer bulk and surface-near layers to plasma-induced changes in structure and composition.

Poly(methyl methacrylate) (PMMA) shows a nearly linear decrease in molar mass on exposure of 300 nm thick layers to the oxygen cw-rf plasma. No indications of crosslinking were found, which agrees with the general chemical tendency of PMMA on exposure to high-energy irradiation to decompose by depolymerization (Figure 6.10).

A line-scan (Figure 6.11) of oxygen plasma exposed polyesterurethane clearly shows the formation of C=C double bonds by dehydrogenation up to sampling depths of several micrometres. It is strongly argued that the dehydrogenation, crosslinking, formation of C-radical sites, and formation of C=C-double bonds is caused by the irradiation with plasma-vacuum UV radiation [8, 15].

Summarizing the obtained information (cf. Section 5.2) the course of plasma functionalization, crosslinking, and etching is as shown in Figure 6.12.

6.3 Changes in Surface Topology

Layered structures in polymers – such as quenched or amorphous surface layers, trans-crystalline structures at the interface to metals, polymer or paint coated surfaces, contaminated surfaces, laminated sheets, porous membrane structures, filled polymer composites, and so on – provoke changes in the etching rate (Figure 6.13).

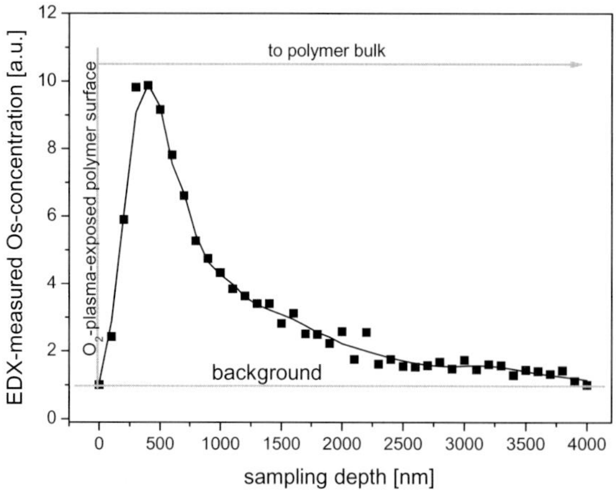

Figure 6.11 Exposure of polyesterurethane to oxygen cw-rf plasma (wattage = 100 W, pressure = 10 Pa) and consumption of plasma-formed C=C double bonds with OsO_4 by measuring the Os-concentration along a line by means of EDX (energy-dispersive X-ray spectroscopy).

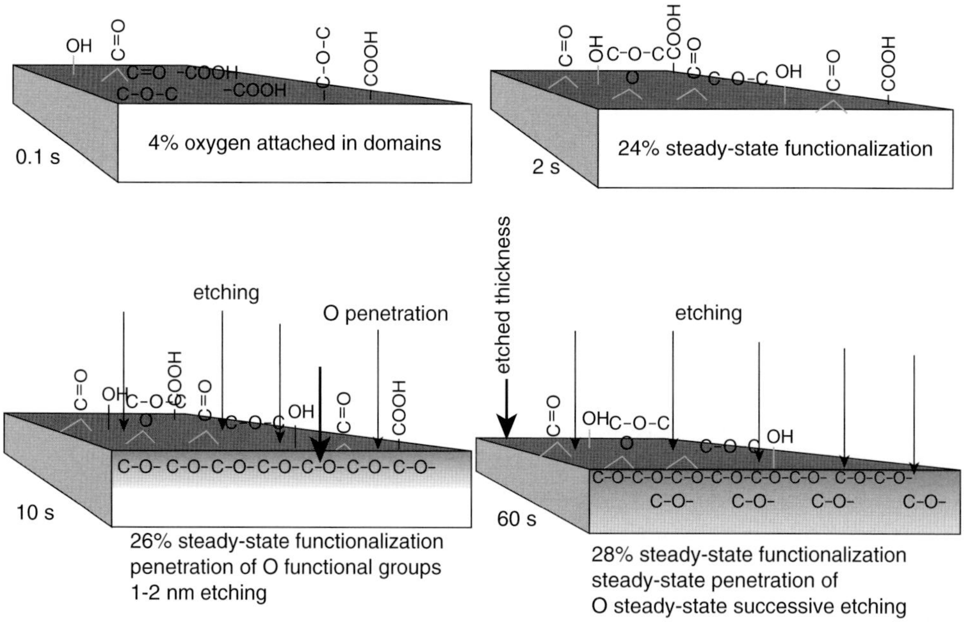

Figure 6.12 Schematics of plasma-induced effects on polymers.

By means of *plasma etch gravimetry* differential changes of the etching rate (dR/dt) can be estimated for the detection of such layered structures [16]. The changes in sample mass were detected by electronic microbalances, quartz syringes, or quartz microbalances [17]. In the vicinity of inorganic or metallic surfaces the etching rate of polymers is often strongly reduced. This effect is

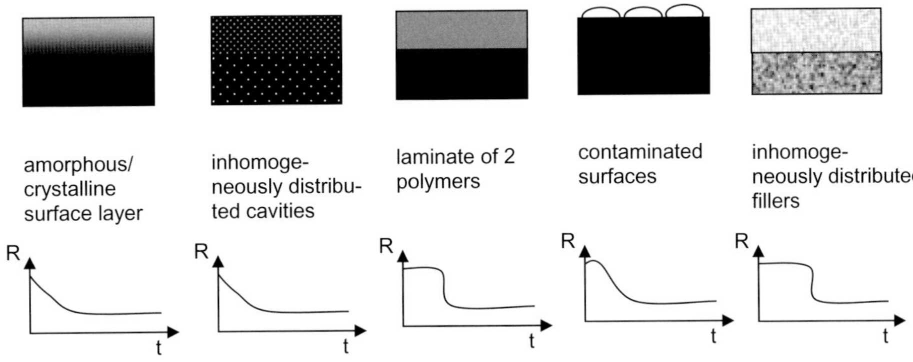

Figure 6.13 Variants of layered structures (gradient layers) in surface-near layers and the principal course of etching rates with exposure time to the oxygen plasma.

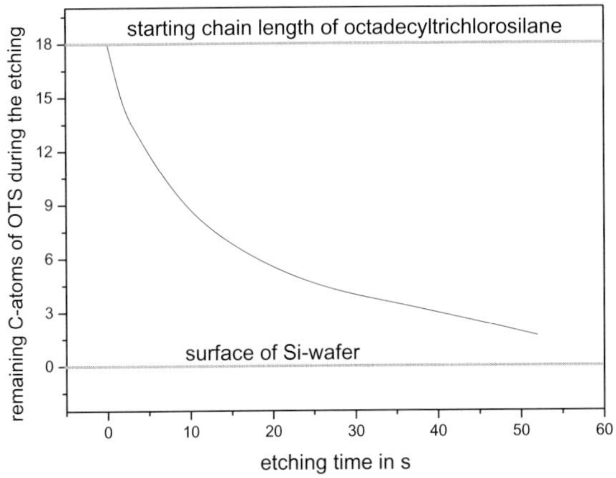

Figure 6.14 Dependence on etching time of the etching of C_{18}-OTS chains deposited onto Si-wafer surfaces by exposure to oxygen cw-rf plasma (power = 100 W, pressure = 13 Pa).

caused by the more etch-resistant polymer structures formed through the interaction between metal or inorganics surface and the polymer or by the absence of a homogeneous closed polymer layer and the remaining isolated polymer islands on the target surface. An example is the etching of an octadecyltrichlorosilane (OTS) monolayer deposited on a Si-wafer on exposure to oxygen plasma [4]. Because of the extremely flat Si-surface islands of remaining OTS clusters can be more or less excluded. In this case the special influence of the wafer surface and the linkage to the wafer may cause the decrease in etching rate. In Figure 6.14 the change in etching rate of the OTS layer as it approaches to the surface is plotted, as measured by using the IRRAS technique [18].

6.4 Plasma Susceptibility of Polymer Building Blocks

As mentioned before, the most fragile polymer building blocks involve a precursor or preformed structure for the elimination of a gaseous plasma degradation product such as CO_2, CO, and H_2O (cf. Section 5.2.8). Another indicator is a high oxygen concentration in the polymer molecule. One example is cellulose, an isotactic β-1,4-polyacetale of cellobiose, $(C_6H_{10}O_5)_n$. Units with such a high oxidation degree (83%) are predestined for fast degradation on exposure to oxygen plasma. This is reflected in high etching rates and the preferred destruction of such units or groups [19]. Other easily degradable and therefore already highly oxidized groups that decay to CO_2, CO, or H_2O are carbonates [20], peroxyacids, acids, and esters [21].

An overview of a few polymers and their degradation behavior on exposure to oxygen plasma is plotted as a set of NEXAFS spectra in Figure 6.15.

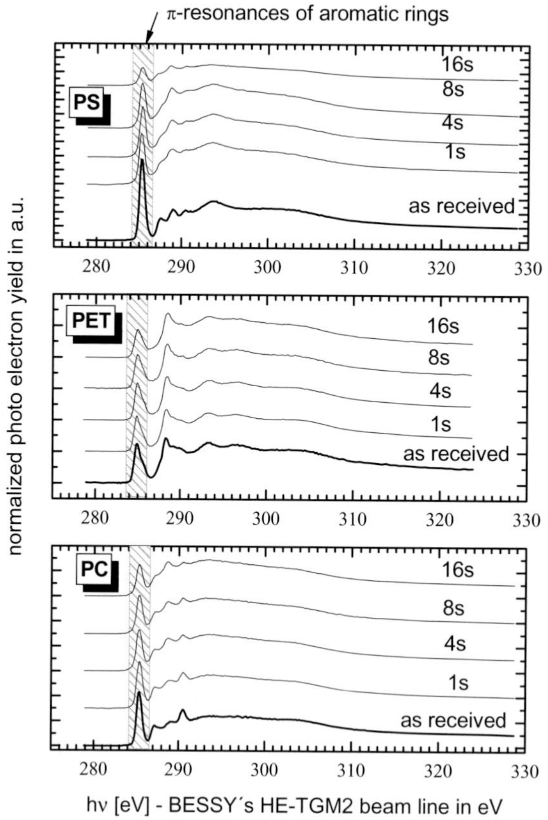

Figure 6.15 Set of C_K edge absorption spectra of polystyrene (PS), poly(ethylene terephthalate) (PET), and poly(bisphenol-A carbonate) (PC) exposed to a DC oxygen plasma (6 Pa).

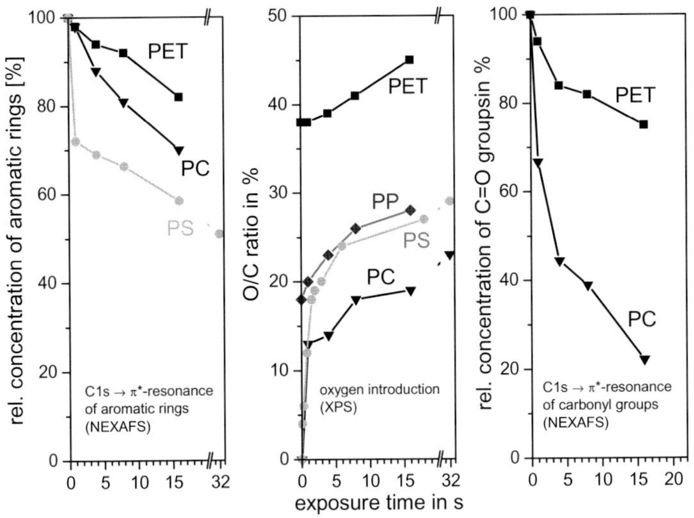

Figure 6.16 Changes in the concentration of aromatic rings and carbonyl groups as well as of total oxygen concentration with exposure time to an oxygen cw-rf plasma (pressure = 6 Pa, power = 100 W).

Evidently, all π-resonances strongly decrease in intensity with prolonged exposure to plasma. This effect is maximal for polystyrene, which is also reflected in Figure 6.16.

Here, the NEXAFS gives information on the (existence and) concentration of π- and n-electron (n = free electron pairs) groups in polymers as measured by the appropriate π-electron resonances in the C and O K-edges, indicating the aromatic ring systems and manifold carbonyl features in the corresponding functional groups or units. The changes in concentration were also measured by XPS using the $\pi \rightarrow \pi^*$ shake-up signal. However, this signal is not as sensitive the π-resonances in the absorption spectra. To complete the picture of oxidation and degradation of polymers on exposure to oxygen plasma the oxygen uptake of polymers is plotted again (Figure 6.16).

The splitting-off (removal) of aromatic rings from the polymer backbone depends strongly on the number of chemical bonds to the polymer backbone. Singly-bonded side groups, known as phenyl groups, are eliminated very easily, as seen for polystyrene. Rings doubly bonded to the backbone, such as para-bonded rings in PET, are more difficult to remove. Here, it is more probable that the aromatic ring becomes destroyed (cracked) than that the intact ring can be removed. The carbonyl feature within ester or carbonate groups degrades strongly. As expected, the carbonate group is most sensitive towards the oxygen plasma exposure. π-Resonance producing groups were affected by their chemical neighborhood. Thus, plasma exposure produces, generally, broadening of all resonances.

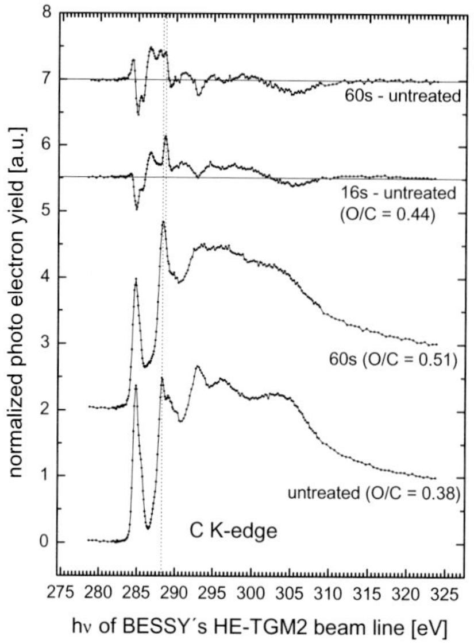

Figure 6.17 NEXAFS C_K edge spectra of poly(ethylene terephthalate) exposed to oxygen plasma (cw-rf, power = 100 W, pressure = 6 Pa). Dotted lines mark the original carbonyl and the newly formed carbonyl feature.

However, in NEXAFS and XPS spectra the distinction must be made between the original carbonyl features in the untreated polymer, such as ester or carboxylic groups, and newly formed carbonyl-containing groups, such as aldehydes or ketones (Figure 6.17).

As was shown in Figure 5.29 further new O functional groups are formed at the surface of poly(ethylene terephthalate).

In addition, the time-of-flight secondary ion mass spectrometry (ToF-SIMS) is a useful tool for investigating the degradation behavior of the polymer building blocks.

6.5
Plasma UV Irradiation

The plasma vacuum UV (VUV) radiation consists of *line* (*resonance*) and *continuum* (*recombination* and *Bremsstrahlung*) radiation (Figure 6.18).

According to the electron energy distribution in the plasma the corresponding electron excitations and radiative transitions to the ground state occur. For example, the helium plasma emits VUV radiation in the range of ca. 5 to 200 nm [22–24]. The emitted radiation is produced if an electron switches from an excited singlet state to the singlet ground state (resonance radiation, fluorescence radiation). This resonance radiation is associated with discrete emission lines of high energy (low

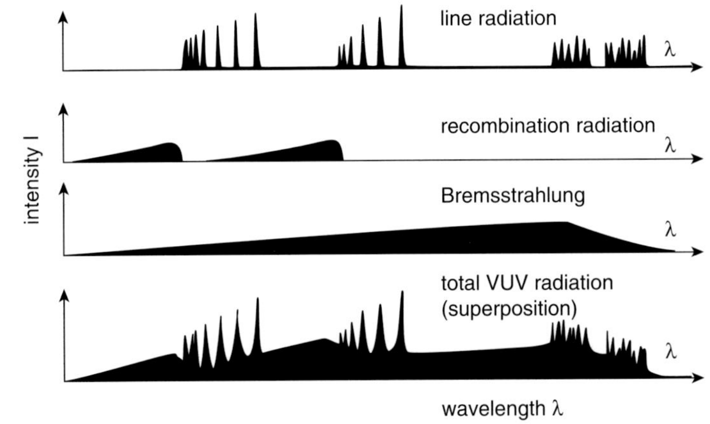

Figure 6.18 Schematic presentation of plasma-emitted radiation.

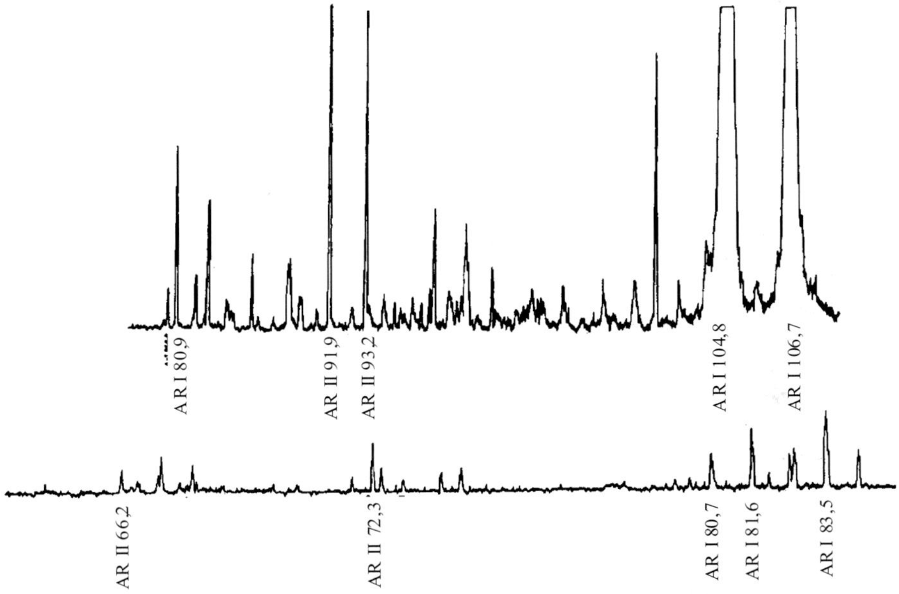

Figure 6.19 Ar plasma emission in the range 60–160 nm.

wavelength) and high intensity [25]. Figure 6.19 presents an example of Ar plasma VUV radiation (60–110 nm) [23].

Single ionization of atoms and molecules dominates in low-pressure DC, rf, or microwave-excited glow discharges. Twofold ionized species are seldom found under glow discharge conditions. Thus, VUV radiation of the wavelength range 100–200 nm is most intense in contrast to the range 10–100 nm, which is associated with twofold ionized species (extreme ultraviolet = EUV) and energies in the range 120–12 eV. The energy of radiation amounts 6.2 eV for the wavelength (λ) = 200 nm and 1239.8 eV at $\lambda = 1$ nm, which must be correlated to the ordinary C–C and C–H binding and dissociation energies in polymers of about 3.7 eV.

Low-pressure glow discharge plasmas in noble gases, hydrogen, oxygen, or metal vapors are usually line radiation emitting plasmas (resonance radiation) [26]. Continuum radiation-emitting plasmas need a high-power input such as arc discharges, but molecular gas plasmas of low-power density also have a background of continuum radiation.

6.6 Absorption of Radiation by Polymers

Organic molecules and polymers exhibit similar absorption behavior with ultraviolet radiation, which is summarized in the simplified Jablonski diagram in Figure 6.20.

At first an energetic photon is absorbed and one electron of the bond occupies a singlet states (e.g., $S_0 \rightarrow S_2$). Within S_2, vibrational and rotational excited states are also occupied (Franck–Condon principle). The electron excitation is followed by an internal conversion characterized by transition to the vibrational ground state of the excited electronic state (S_1) under release of heat. In a few cases, transition to the triplet system by spin conversion (intersystem crossing) occurs:

$$S_0 + h\nu \rightarrow S_2{}^* \text{ radiation absorption}$$

$$S_2{}^* \rightarrow S_1{}^0 + \Delta E \text{ internal conversion}$$

$$S_1{}^0 \rightarrow S_0 + h\nu \text{ resonance radiation (fluorescence radiation)}$$

$$S_1{}^0 \rightarrow T_2{}^* \text{ intersystem crossing}$$

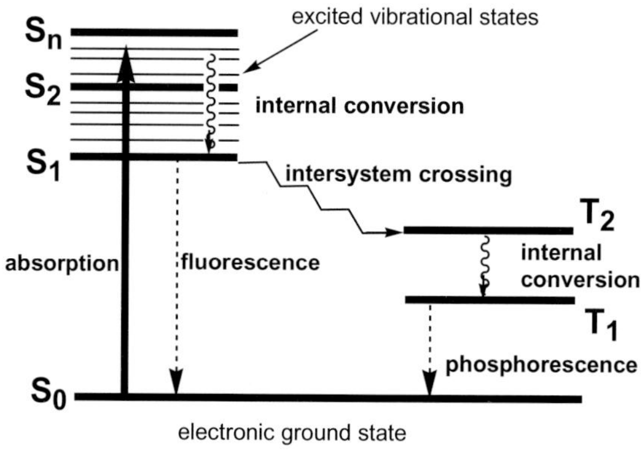

Figure 6.20 Simplified Jablonski diagram (S = singlet state, T = triplet state).

$T_2^* \rightarrow T_0 + \Delta E$ internal conversion

$T_0 \rightarrow S_0 + h\nu$ phosphorescence radiation

Figure 6.21 shows a typical absorption–fluorescence spectrum. The emission is shifted bathochromically by 123 nm.

The phosphorescence transition ($T_1 \rightarrow S_0$) is forbidden by known selection rules and shows therefore a time-lag. From S_1 or T_1 states radical formation occurs or energy transfer from the excited donor to the acceptor takes place. Moreover, complexation of phenyl rings leads to the formation of excimers and excimer singlets and, therefore, to emission of excimer radiation, as shown in Scheme 6.1.

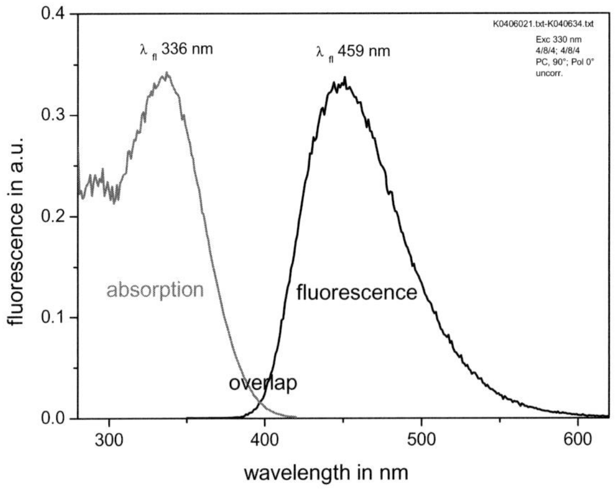

Figure 6.21 Absorption–fluorescence spectra of dansyl hydrazine grafted onto surface amino groups at PP surfaces.

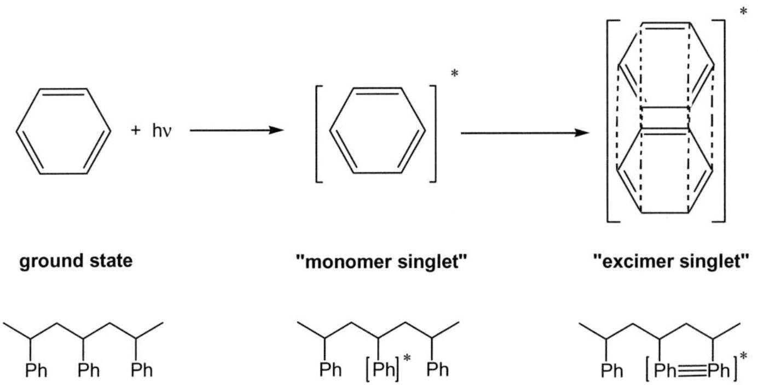

Scheme 6.1 Excimer formation.

Carbonyls play the most important role in light emission of polymers. Absorption of photons can produce singlet S* (n → π* transition) or triplet T* (π → π* transition).

The excited triplet state exists as a biradical:

$$>C=O + h\nu \rightarrow {}^{3}[>C^{\bullet}\text{-}O^{\bullet}]$$

Figure 6.22 shows the chemiluminescence resulting from an oxygen-plasma treated piece of rubber that was heated.

Typical photochemical reactions of carbonyl-containing polymers based on such radiation-excited triplet states are Norrish I and Norrish II rearrangements (Scheme 6.2).

At $\lambda < 200$ nm (>6.2 eV) bond scissions are potentially able to occur ($C\text{–}C_{PE}$ = 3.6 eV and $C\text{–}H_{PE}$ = 3.7 eV) in the polymer, including the scission of backbone

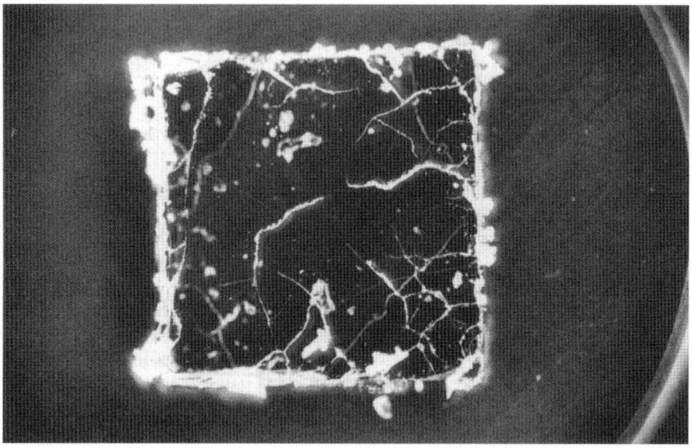

Figure 6.22 Chemiluminescence of a piece of rubber exposed to oxygen plasma.

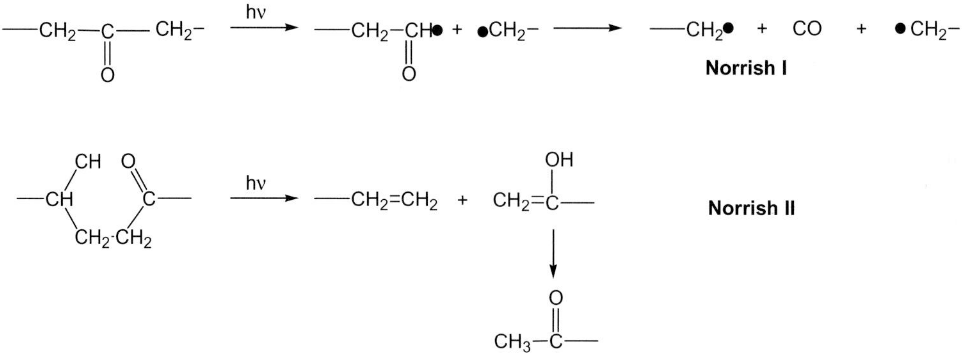

Scheme 6.2 Norrish I and II rearrangements.

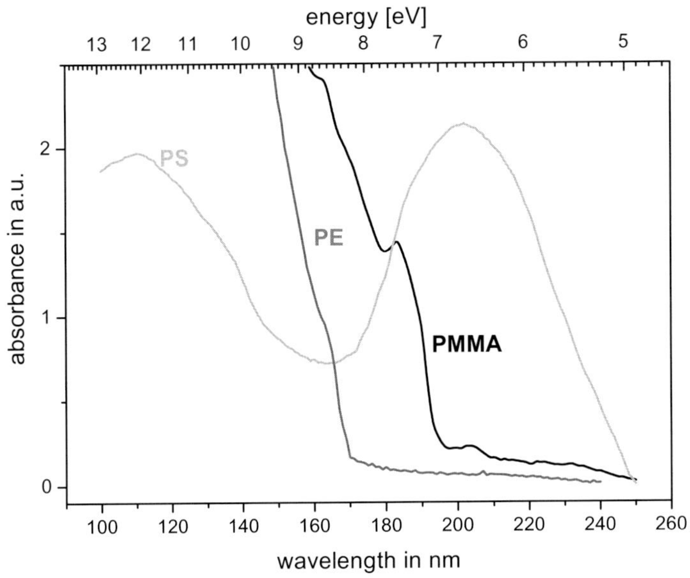

Figure 6.23 VUV absorption spectra of PS, PE, and PMMA.

and the formation of radical sites by hydrogen abstraction or C–C bond dissociation, as well as formation of C=C double bonds [27]. Last but not least these radiation-induced $\sigma \rightarrow \sigma^*$-transitions also contribute to the crosslinking of the polymer [15]. Figure 6.23 presents a few absorption spectra of polymers. Werthe-imer and other authors investigated systematically the absorption behavior of polymers [28–31].

Figure 6.24 presents the electron distribution within the MO scheme for most important bonds in polymers.

6.7 Formation of Unsaturations

Polymers tend to hydrogen abstraction, formation of radicals and double bonds, and crosslinking if exposed to oxygen plasma. Such behavior was found with polyolefins [32], poly(vinyl chloride) (PVC) [33] and other polymers (Figure 6.25). PVC shows an intense brown–violet discoloration after exposure to oxygen-free plasmas due to the formation of a topmost thin conjugated polyene layer:

$$\text{-[CH}_2\text{–CHCl]}_n\text{-} + h\nu \rightarrow \text{-[CH=CH]}_m\text{-[CH}_2\text{–CHCl]}_{n-m}\text{-} + m\text{HCl}$$

The degree m of polyene conjugation amounts to 10, as identified by the longest wavelength of superposed fine structure [33–35]. Using the oxygen plasma these polyenes are successively oxidized similar to their oxidation on exposure to air [36].

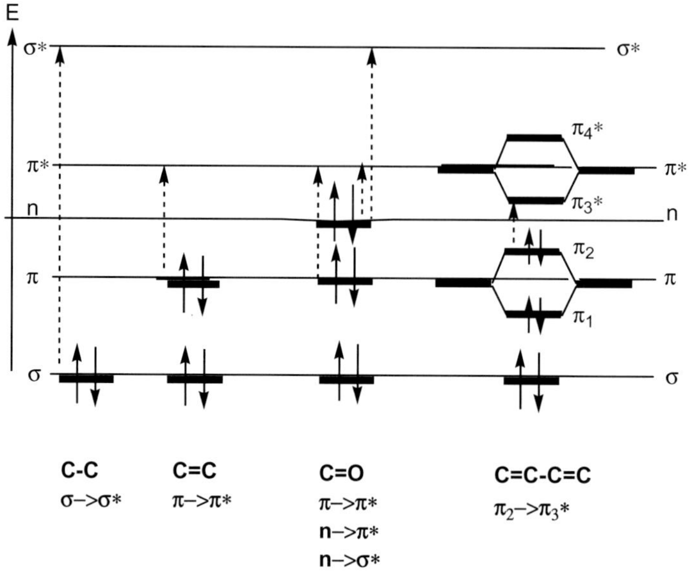

Figure 6.24 MO scheme of C–C, C=C, C=O, and C=C–C=C bonds.

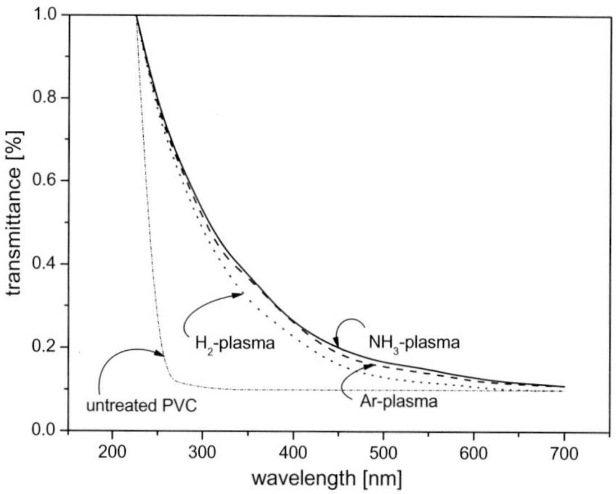

Figure 6.25 Changes in UV-spectra of poly(vinyl chloride) after exposure to different types of plasmas (each 60 min).

Polyurethanes show an intensive brown discoloration up to hundreds of microns or even millimetres [35, 37]. Octadecyltrichlorosilane (OTS) self-assembled layers (SAM) also show the formation of C=C double bonds (Figure 6.26).

Changes to polypropylene after oxygen plasma exposure can be seen using NEXAFS spectroscopy (Figure 6.27). NEXAFS (ca. 3 nm for the C_K edge and ca.

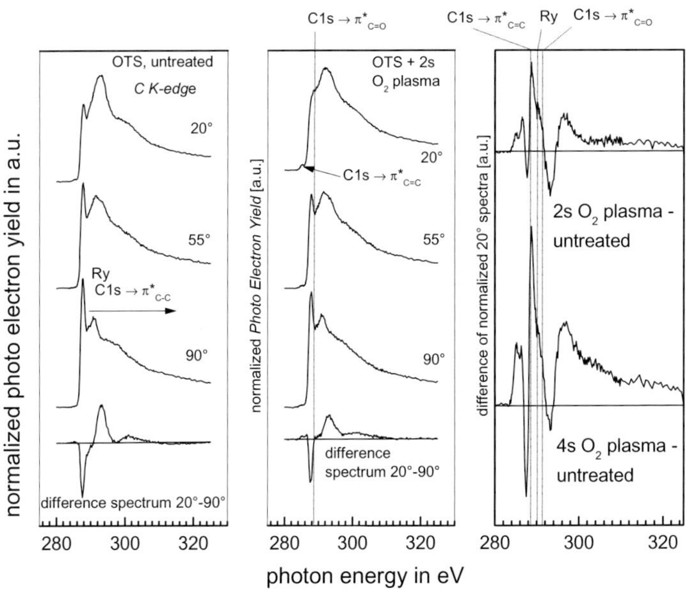

Figure 6.26 NEXAFS spectra of OTS monolayers exposed to oxygen plasma.

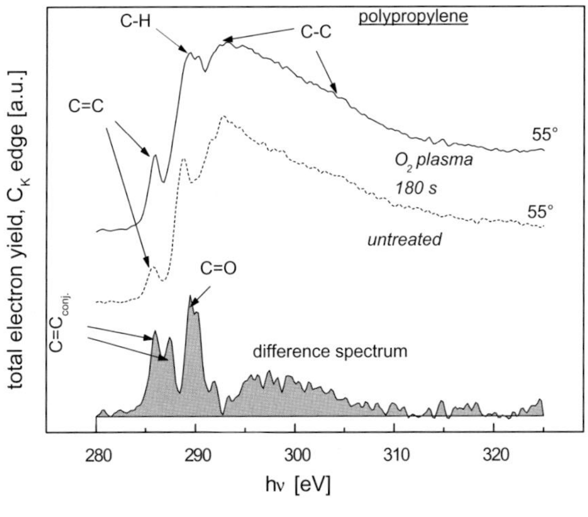

Figure 6.27 NEXAFS C_K-edge spectra of polypropylene exposed to O_2 plasma (cw-rf plasma, exposure time = 180 s, pressure = 6 Pa, power = 100 W).

5 nm for the O_K edge) clearly identified π-electron containing bonds such as C=C double bonds and C=O carbonyl features for polypropylene after its exposure to the low-pressure oxygen-plasma (Figure 6.27).

For this purpose C_K edge spectra were recorded at the glancing angle (ca. 55°) and the difference spectra of the polymer sample before and after plasma exposure were calculated [1, 5, 38, 39].

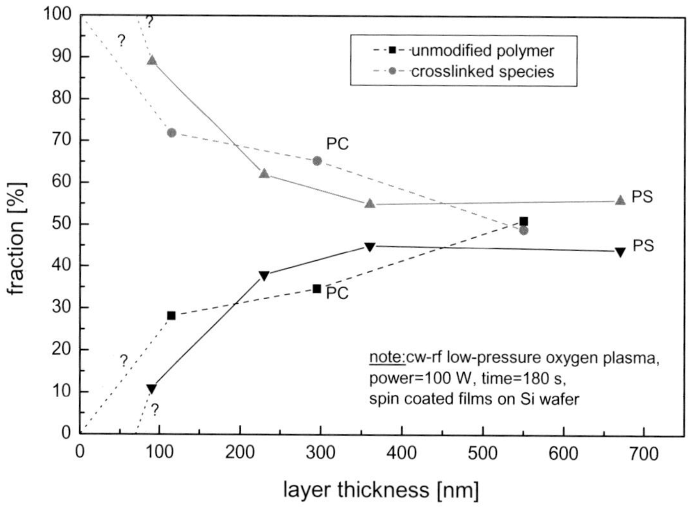

Figure 6.28 Thermal field flow fractionation (ThFFF) of oxygen plasma-treated polymers (cw rf plasma, 100 W, 180 s, 6 Pa).

In Figure 6.27 this difference spectrum of polypropylene as received and after 180 s exposure to the oxygen plasma shows two new features, which are assigned to the plasma-induced formation of C=C and C=O bonds. The splitting of the C1s $\rightarrow \pi^*_{C=C}$ resonance indicates the existence of conjugated double bond systems [6].

The formation of double bonds and the occurrence of crosslinking are the most prominent structural changes in near-surface layers and in the bulk of polymers. Considering the absorption property of polyethylene in the VUV region and also the spectral distribution of the emitted plasma radiation Holländer calculated the depth of radiation-induced defects in the polymer to ca. 100 nm [40].

Using the thermal field flow fractionation (ThFFF) results of polymer films with different thickness, each exposed for 180 s to an oxygen plasma, the crosslinking depth in poly(bisphenol-A carbonate) and polystyrene was estimated roughly to at least 300 nm. Thicker polymer layers did not show an increasing fraction of (crosslinked) gel fraction of ultrahigh molar masses (Figure 6.28).

After prolonging the exposure time more crosslinking was found (Figure 6.29).

After intensifying and prolonging the plasma conditions deeper zones in the bulk of polymers were branched or crosslinked as shown in Figure 6.30 for oxygen-plasma exposed polystyrene and using polystyrene standards of different molar mass.

It can be concluded from this plot that "harder" plasma conditions influence near-surface layers of the polymers up to the micrometer range, as exemplified with PS (further details of the crosslinking and its detection are given in Section 6.10).

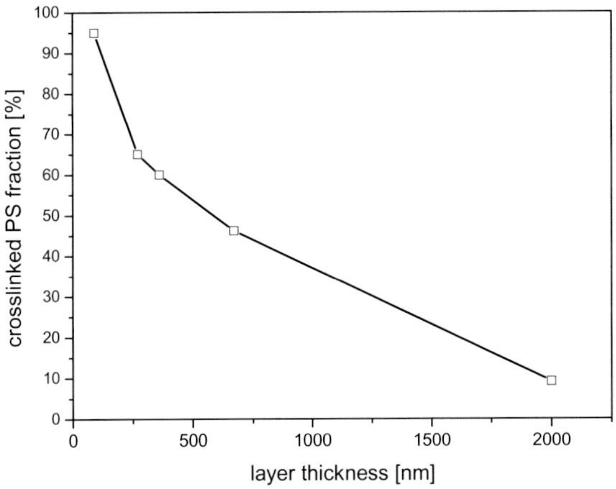

Figure 6.29 ThFFF of oxygen plasma-treated polystyrene (cw rf plasma, power = 300 W, time = 600 s, pressure = 6 Pa).

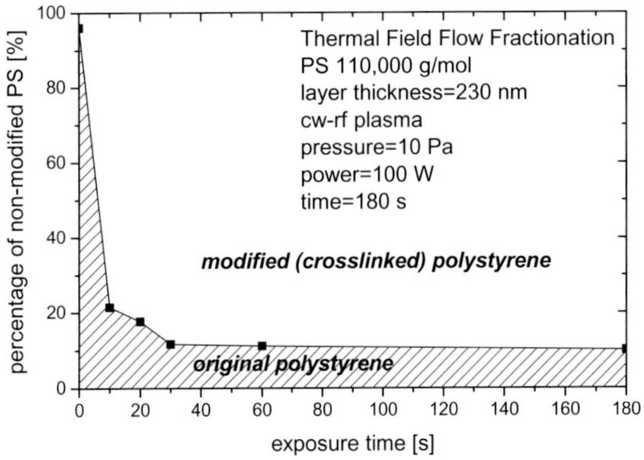

Figure 6.30 ThFFF of oxygen plasma-treated polystyrene (cw rf plasma, power = 100 W, time = 180 s, pressure = 10 Pa).

6.8
Formation of Macrocycles

The formation of macrocycles in polymers during their exposure to plasma is a very important process, but it was not considered up to now. A distinction must be made between linear chains, with perhaps 100 or 1000 monomer units, and two end-groups and macrocycles with approximately 3–20 monomer units without end-groups, cyclic polymers with infinite polymer cycles, and other polymer architectures (Figure 6.31).

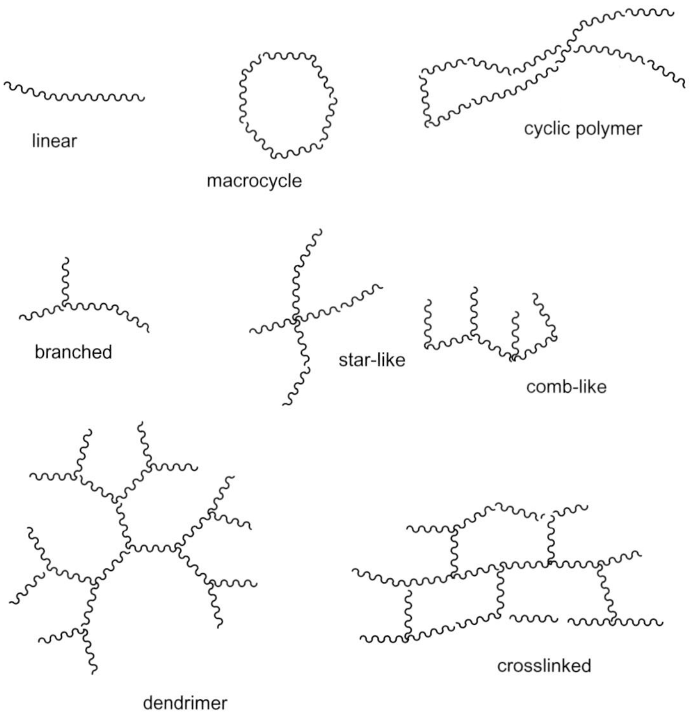

Figure 6.31 Examples of molecular architecture in polymer chemistry.

The formation of macrocycles is a thermodynamic problem. The shorter the chain the more bending of all participating bonds is necessary to form a cyclic compound. Therefore, the formation of three- or four-membered cycles is not very probable. At the other extreme, the coming together of end-groups of infinitely long chains has no probability. Therefore, the above-mentioned maximum exists, which, however, depends on the type of polymer. Macrocycles possess no polar end-groups and have therefore different properties to the linear species. Moreover, the average molar masses of macrocycles are much lower than those of the linear species. Therefore, the more hydrophobic macrocycles tend to migrate to the polymer surface. This macrocycle-enriched surface has insufficient wettability and the loosely bonded low-molecular-weight macrocycles exhibit as a weak boundary layer.

Because of the thermodynamic equilibrium between linear and cyclic species, macrocycle formation occurs when activation energy is available to adjust this equilibrium. Therefore, macrocycle-free oligo-ethylene terephthalate (before chromatographic separation of macrocycles) undergoes a rapid reformation of macrocycles when exposed to oxygen plasma (Figure 6.32) [41].

The poly(ethylene terephthalate) forms different ring structures and also several different end-group combinations in the linear species.

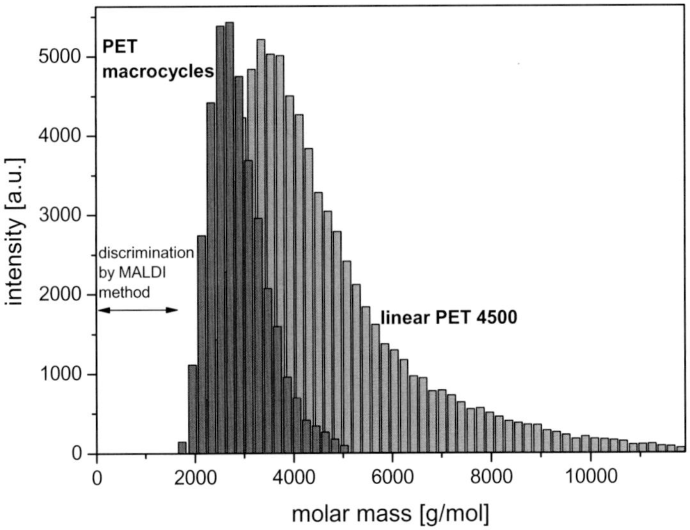

Figure 6.32 MALDI-MS measured bimodal molar mass distribution of poly(ethylene terephthalate) (4500 g mol^{-1}) exposed to an oxygen plasma (cw-rf plasma, 100 W, 6 Pa, 600 s).

Polyoxymethylene shows similar behavior. Among linear species with different end group combinations, macrocycles with a different number of monomer units are also found, again using MALDI-ToF MS (Figure 6.33). Most other polymers also show the formation of macrocycles [3, 42].

6.9 Polymer Degradation and Supermolecular Structure of Polymers

The etching rate of polymers also depends on the supermolecular structure. The supermolecular structure is formed from crystalline, semi-crystalline, transition zones, and amorphous regions. However, in the crystalline regions the macromolecules may be folded as regular crystals but the loops at the margins of such a "crystal" are points of preferred plasma attack. These folds may be assigned to the amorphous region and they are preferably oxidized when exposed to oxygen plasma. The etching rate of a polymer depends on the supermolecular structure because of the different susceptibility towards the plasma (Figure 6.34).

Such a supermolecular structure is a feature with increased orientation degree (cf. Figure 6.34 and other supermolecular and molecular structures), for example, the polymer chains have a near molar mass distribution and arrange themselves in microcrystallites, liquid crystalline, or supermolecular structures. In contrast to them amorphous regions also coexist in such semi-crystalline

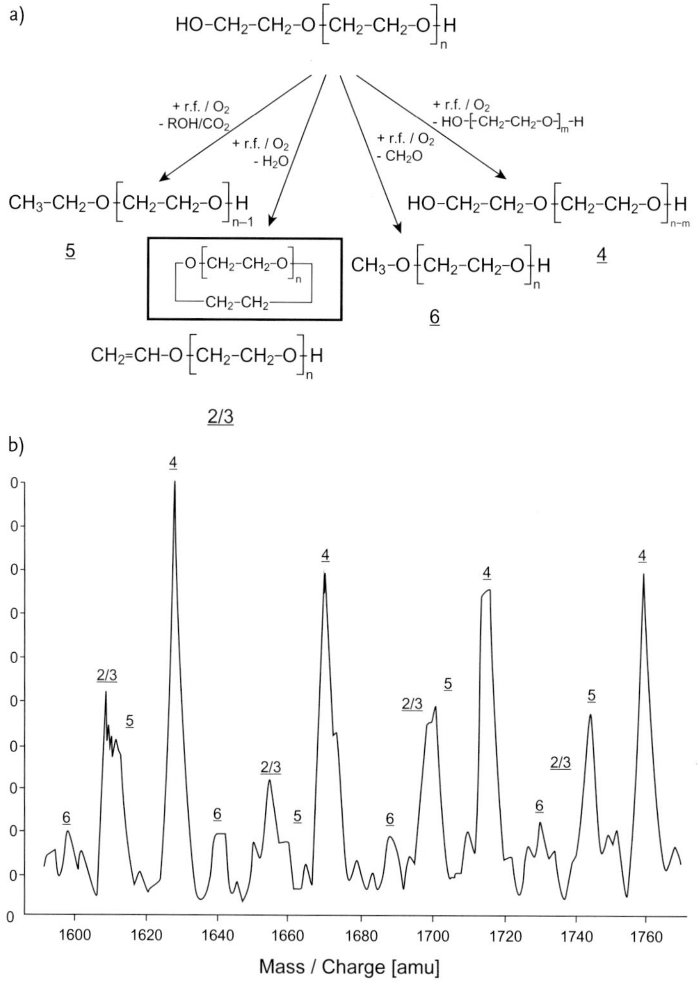

Figure 6.33 (a) Formation of macrocycles within all reactions of poly(ethylene glycol) (PEG 4000) on exposure to O_2 plasma and (b) the corresponding MALDI-ToF spectrum.

polymers, such as in polyethylene, polypropylene, poly(ethylene terephthalate), or poly(tetrafluoroethylene). There are also several completely (100%) amorphous polymers characterized by transparency [polystyrene, polycarbonate, poly(methyl methacrylate)]. The random folded polymer chains that form a coil are loosely packed and possess a much lower density than the crystalline fractions, which are more densely packed with strong inter- and intra-actions. In such a way, HDPE (high-density polyethylene) shows differences in density ranging from 1.0 to 1.1 $g\,cm^{-3}$. The crystallinity can be determined by IR, XRD (X-ray diffraction),

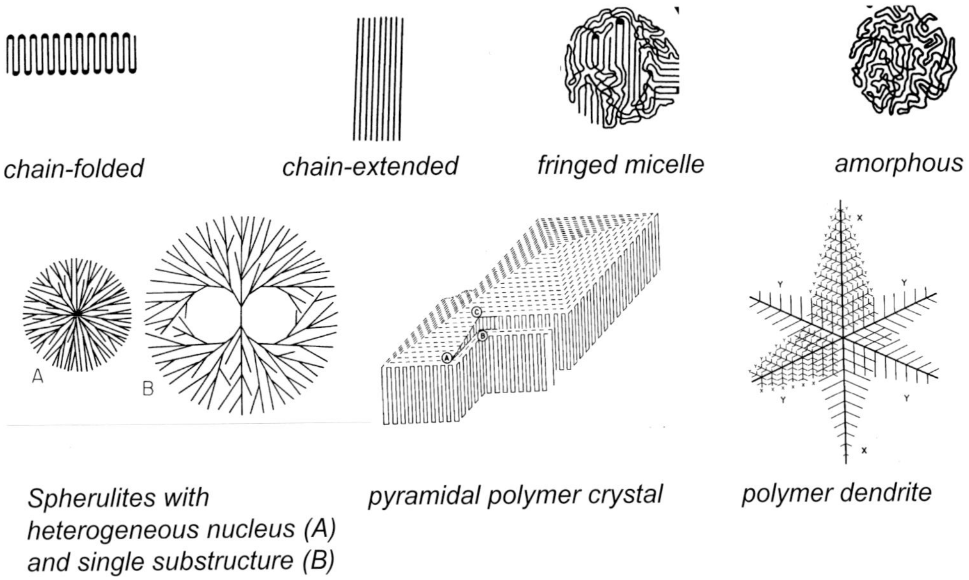

Figure 6.34 Highly ordered structures in polymers, which influence the degradation and aging behavior.

density measurement, DSC (differential scanning calorimetry), and NMR (nuclear magnetic resonance) spectroscopy [8].

Thus, the co-existing structures have different etching rates [43–45]. Therefore, the oxygen plasma etching of such polymers allows selective removal of amorphous phases or phases with lower orientation. The crystalline and the ordered structures should survive and remain. Thus, such an etching process should be an appropriate preparation method for excavation of buried supermolecular structures of polymers used for electron microscopy [46–48]. Figure 6.35 shows a typical example. Impact-resistant polystyrene (10% polybutadiene) was etched in a oxygen low-pressure cw-rf plasma.

The spheres in the micrograph can be assigned to poly(butadiene). However, the plasma etching produces many artifacts, and thus this method could not achieve acceptance.

It should be clearly stated that the establishing of plasma etching as a preparation method for specimens investigated by electron microscopy was not successful. The specimens were changed in structure during the plasma etching process. The plasma particle bombardment of the topmost layer, unspecific oxidation, and the secondary changes on a few 100 nm scale by energetic VUV radiation were responsible. It could be shown that plasma etched reliefs are formed by secondary plasma effects and are not produced by simple excavation of crystalline regions (Figure 6.36).

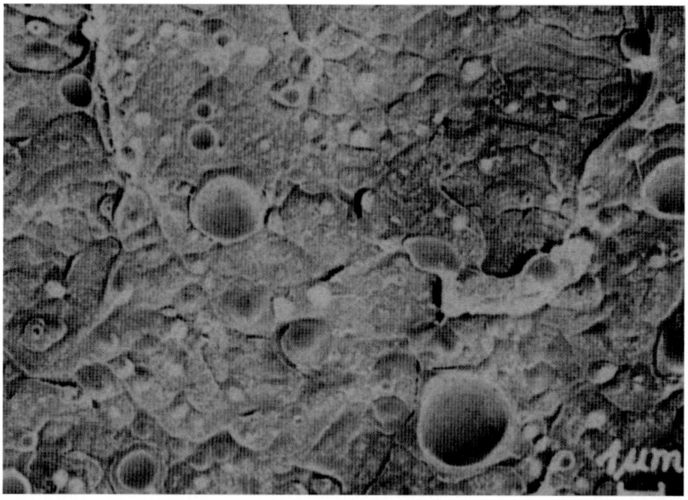

Figure 6.35 Micrograph of impact-resistant polystyrene after 300 s exposure to O_2-plasma (1 : 8600).

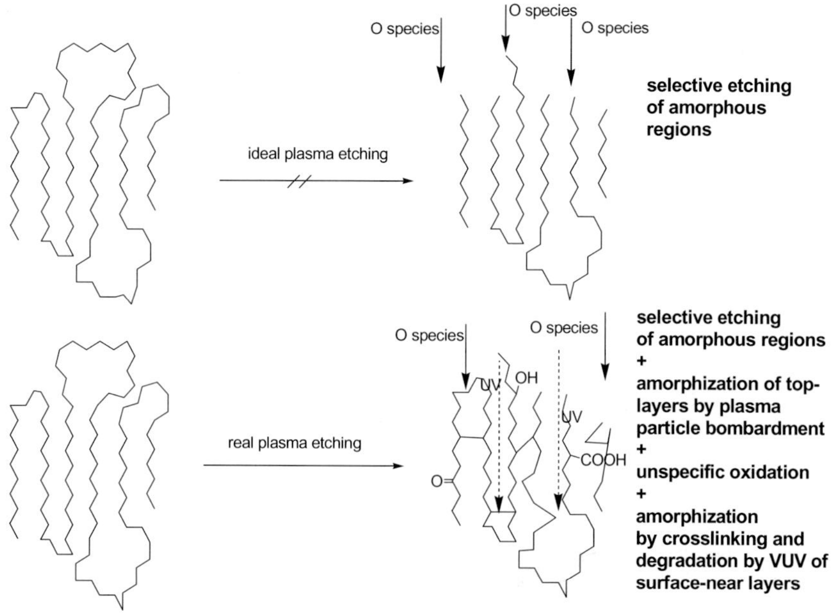

Figure 6.36 Secondary amorphization of crystalline structures in polymers by particle shower, oxidation, and VUV irradiation from a plasma.

6.10 Crosslinking versus Degradation of Molar Masses

As mentioned before, any plasma treatment of polymers is associated with a broadening of the molecular weight distribution (MWD), which is specific for each polymer and its synthesis (Figure 6.37). This means, on the one hand, lower molecular weight polymer molecules and newly formed low-molecular-weight polymer derivates appear and, on the other hand, chain-extended as well as random crosslinked and branched polymers occur [49].

As can be seen in Figure 6.37 He-plasma produces crosslinking of the polystyrene (37 000 g mol^{-1}) up to molar masses of at least 100 000 g mol^{-1}. Molar masses > 100 000 g mol^{-1} were removed by filtering, which is necessary to avoid the obstruction of chromatographic columns. Small fractions of low molecular-weight PS appear also after helium plasma treatment, in the range of a few 1000 g mol^{-1}.

The oxygen plasma produces a significant degradation of the polymer. However, dimers, trimers, and other fragments occupy the region of a few 100 g mol^{-1}. Therefore, it was very surprising to find by means of MALDI-ToF mass spectrometry that crosslinking also takes place during the often applied oxygen plasma treatment of polymers. It could be shown that the archetype of aliphatic polymers, hexatriacontane (HTC, C_{36}), characterized by the defined molecular weight, exhibits the formation of dimers and trimers when exposed to the oxygen plasma, as shown by MALDI mass spectrometry (MALDI MS) (Figure 6.38) [32]. It was therefore proposed that oxygen plasma exclusively produces degradation (fragmentation) and

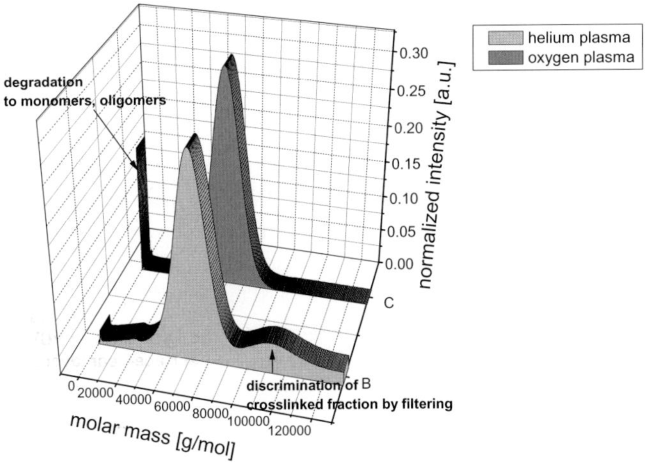

Figure 6.37 Size exclusion chromatography (SEC; also known as GPC, gel permeation chromatography) chromatogram of PS (37 000 Da) after exposure to O_2- and to He-plasma (cw-rf plasma, pressure = 10 Pa, wattage = 100 W).

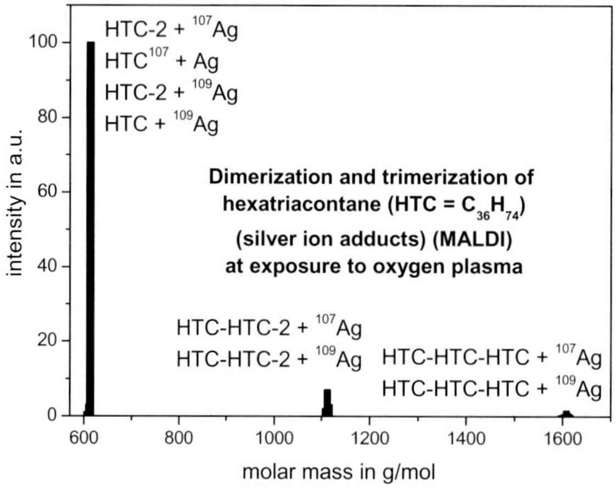

Figure 6.38 MALDI-MS spectrum of hexatriacontane products formed on exposure to O_2 plasma (time = 180 s, power = 100 W, pressure = 10 Pa).

not any dimerization or crosslinking. The unexpected crosslinking was investigated in detail by field flow fractionation (FFF) and is discussed below.

The polymer degradation was measured using SEC, which produces the mantle curve of the molecular weight distribution (MWD). MALDI MS detects single molar masses up to a few 10 000 g mol^{-1}. At higher molar masses only the mantle curve of the MWD is recorded. It must be added that both SEC and MALDI MS are only applicable to soluble polymers, consequently even only slightly crosslinked polymers cannot be measured. Moreover, MALDI has a cut-off for masses <500 g mol^{-1}. Lower masses were superposed by matrix-related signals. A further precondition for MALDI is a low polydispersity (Q) (M_W/M_N, small distribution) of the investigated polymer (<1.3), for example, the quotient of the weight average and the number average of the polymer must be low ($Q = M_W/M_N < 1.3$).

The polymer degradation is dependent on the type of polymer, as shown in Figure 6.39. Poly(ethylene terephthalate) (PET), polystyrene (PS), and bisphenol A-polycarbonate (PC) were exposed to the oxygen rf and microwave low-pressure plasma and to an atmospheric-pressure plasma of a dielectric barrier discharge (DBD). The degradation of polymers to lower molar masses shows differences. PET was quite stable and showed the occurrence of discrete low-molecular-weight species. This behavior was concisely pronounced for PS. It tends to form monomer, dimer, and trimer when exposed to oxygen plasma. In contrast, polycarbonate shows a continuous shift in its MWD to lower molar masses.

The distribution functions of ultrahigh molar masses of linear, branched, and weakly crosslinked polymers were measured by FFF. Thermal field flow fractionation (ThFFF) was often used and also asymmetric (cross-)flow-FFF (AFFF or F^4) [49]. FFF is related to chromatography but, however, works without separation phase. It is a simple channel. The polymer solution flows through the channel. Perpendicularly to the flow direction the thermal field or the cross-flow of the solvent to a membrane acts. Using the signal of multi-angle laser light scattering

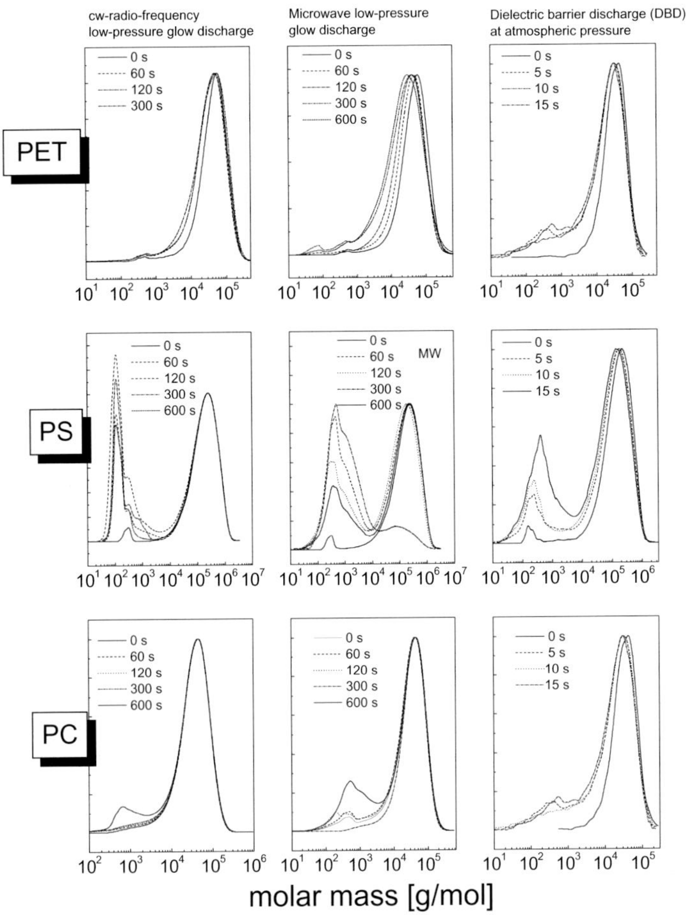

Figure 6.39 GPC (SEC) chromatograms of PET, PS, and PC (250 nm thick spin-coating films) after exposure to low-pressure O_2 plasma and to atmospheric DBD plasma.

(MALLS) as detection signal the significance of the crosslinking process during the oxygen plasma treatment of polymers becomes obvious. Surprisingly, the oxygen plasma also causes crosslinking and not only degradation [41]. As mentioned several times before, the reason for this is the short-wavelength UV from the plasma (and is not due to crosslinking by activated species of inert gases – CASING as originally postulated by Hansen and Schonhorn) [50]. This crosslinking effect was used to promote the adhesion and sticking ability of polymer surfaces [51].

6.11
Radicals and Auto-oxidation

As noted in Section 6.10, irradiation with UV radiation from the plasma leads to strong changes in polymer structure. The result of this interaction between radiation and macromolecules is the degradation and/or crosslinking processes and the post-plasma (auto-)oxidation. However, the first step of these processes is the abstraction of hydrogen from the polymer chain or the scission of C–C bonds (irreversible $\sigma \rightarrow \sigma^*$-electronic transitions), as shown below for the most probable processes of polyethylene, polytetrafluoroethylene, polypropylene, and polystyrene:

Polyethylene:

$$\text{-CH}_2\text{-CH}_2\text{-} + \textit{plasma-VUV} \rightarrow \text{-CH}_2\text{-CH}^{\bullet}\text{-} + {}^{\bullet}\text{H (0.5H}_2\text{) H-abstraction}$$

$$\text{-CH}_2\text{-CH}_2\text{-} + \textit{plasma-VUV} \rightarrow \text{-CH}_2^{\bullet} + {}^{\bullet}\text{CH}_2\text{- C–C scission}$$

$$\text{-CH}_2\bullet + ({}^{\bullet}\textit{O-O}^{\bullet}\ \textit{from air}) \rightarrow \text{-CH}_2\text{-O-O}^{\bullet}\text{ post-plasma peroxide formation}$$

Polytetrafluoroethylene:

$$\text{-CF}_2\text{-CF}_2\text{-} + \textit{plasma-VUV} \rightarrow \text{-CF}_2\text{-CF}^{\bullet}\text{-} + {}^{\bullet}\text{F (0.5F2) F-abstraction}$$

$$\text{-CF}_2\text{-CF}_2\text{-} + \textit{plasma-VUV} \rightarrow \text{-CF2}^{\bullet} + {}^{\bullet}\text{CF}_2\text{- C-C scission}$$

$$\text{-CF}_2\bullet + ({}^{\bullet}\textit{O-O}^{\bullet}\ \textit{from air}) \rightarrow \text{-CF}_2\text{-O-O}^{\bullet}\text{ post-plasma peroxide formation}$$

Polypropylene:

$$\text{-CH}_2\text{-CH(CH}_3\text{)-} + \textit{plasma-VUV} \rightarrow \text{-CH}_2\text{-C}^{\bullet}\text{(CH}_3\text{)-} + {}^{\bullet}\text{H (0.5H}_2\text{) H-abstraction}$$

Polystyrene:

$$\text{-CH}_2\text{-CH(phenyl)-} + \textit{plasma-VUV} \rightarrow \text{-CH}_2\text{-C}^{\bullet}\text{(phenyl)-} + {}^{\bullet}\text{H (0.5H}_2\text{) H-abstraction}$$

$$\text{-CH}_2\text{-CH(phenyl)-} + \textit{plasma-VUV} \rightarrow \text{-CH}_2\text{-CH}^{\bullet}\text{-} + {}^{\bullet}\text{phenyl (0.5 biphenyl) phenyl-abstraction}$$

In all cases primarily radicals are formed, which undergo post-plasma auto-oxidations, molar mass degradation, crosslinking, formation of double bonds, and so on. Nevertheless, in deeper layers of the polymer (up to a few micrometres [37]) a significant concentration of free but confined and therefore metastable radicals remains. Over the course of some weeks or months these radicals react with interdiffusing oxygen from the air to peroxy radicals and further products. An example of long-term reactivity is the exposure of PP to low-pressure ammonia plasma and its oxygen-uptake during 24 h exposure to the air (Figure 6.40) [8].

Figure 6.41 presents the post-plasma attachment of oxygen and also the incorporation of the radical scavengers Br and NO:

$$\text{C}^{\bullet} + {}^{\bullet}\text{O-O}^{\bullet} \rightarrow \text{C-O-O}^{\bullet}\text{ (+RH)} \rightarrow \text{C-O-OH}$$

$$\text{C}^{\bullet} + \text{Br}_2 \rightarrow \text{C-Br} + \text{Br}^{\bullet}$$

$$\text{C}^{\bullet} + {}^{\bullet}\text{NO} \rightarrow \text{C-NO (+O}_2\text{)} \rightarrow \text{C-NO}_2$$

Kuzuya has investigated the formation of radicals in different polymers under plasma exposure by means of electron spin resonance (ESR) [52, 53]. These peroxy

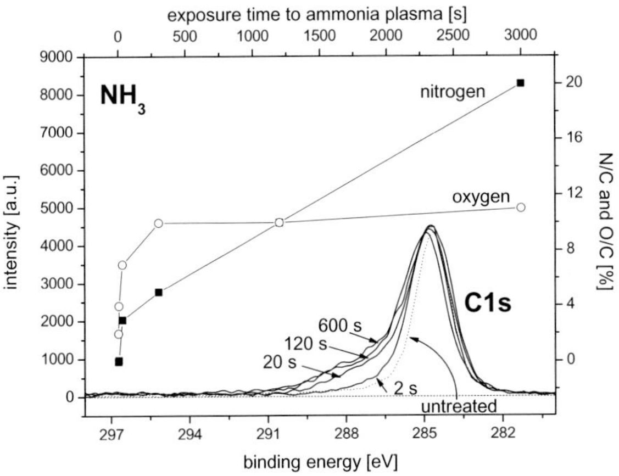

Figure 6.40 Treatment time of PP with ammonia plasma (cw rf plasma, 300 W, 10 Pa) versus elemental composition (XPS) after subsequent exposure to air for 24 h as well as changes of the C1s peak.

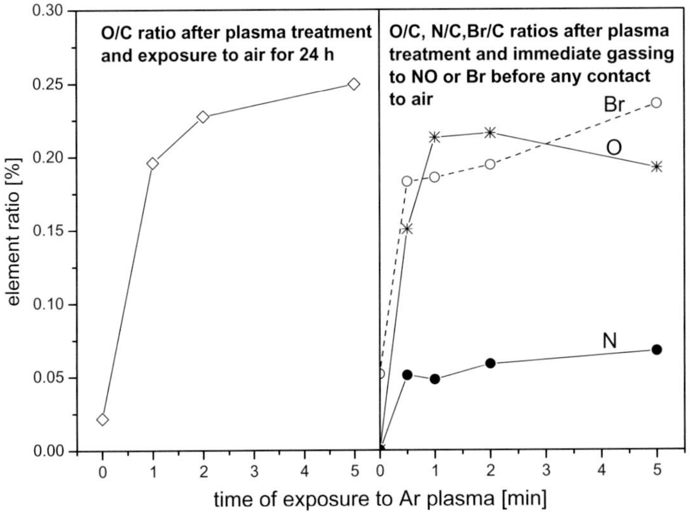

Figure 6.41 Secondary oxygen (air), Br (Br_2), and NO incorporation into PP after argon plasma treatment.

radicals are formed preferentially in the amorphous phase of semi-crystalline polyethylene because the oxygen diffuses into the amorphous phase ten-times faster than into the crystalline phase [35]. Radicals at dangling bond sites in the amorphous phase react faster with oxygen to give peroxy radicals than do alkyl- and allyl radicals in the crystalline phase of polyethylene. These three radical types were also detected in PTFE; however, the formation rate of radicals was much slower than in PE [52]. This was explained by the much higher dissociation energy of the C–F bond compared to C–C and C–H bonds [53]. Two different radical types were identified in polystyrene, one of the cyclohexadienyl type formed by the attachment of hydrogen to the aromatic ring and the other radical species was unspecific for radicals at branching points [53]. Using the reaction of radicals with DPPH (diphenylpicrylhydrazil), the (post-plasma) formation of peroxy radicals in polyethylene exposed to argon plasma was determined by Suzuki (Figure 6.42) [54]:

$$C^{\bullet} + (\text{aryl})_2\text{N-N}^{\bullet}\text{-(aryl)(-NO}_2)_3 \rightarrow (\text{aryl})_2\text{N-N(C-)-(aryl)(-NO}_2)_3$$

Poll *et al.* also used the DPPH-method to evaluate the efficiency in radical formation of several different gas plasmas on PE [55]. They found the following succession: $N_2 < Kr < Ne < He \approx H_2 <$ air. This sequence lets us again assume that the energy of UV lines plays an important role. Especially, the He, Ne, and H_2 plasmas exhibit (line) radiation of high energy content (20–120 nm). The tendency to enhance the yield in radical formation seems to depend on the energy of radiation as the most important factor. The special position of air in this succession is due to the stabilizing action of peroxide formation upon the DPPH derivatization.

Radicals were also determined by gasification with nitric oxide (NO) [56]. Nitric oxide reacts with radicals to give nitroso groups and higher oxidized products. Nitrogen can easily be determined by XPS analysis [40]. NO is a permanent radical and can be also written as $^{\bullet}$NO. Therefore, it reacts preferentially and rapidly with C-radical sites as shown before and also with peroxy radicals to give nitrates:

$$\text{C-O-O}^{\bullet} + {}^{\bullet}\text{NO} \rightarrow \text{CO-O-NO} \rightarrow \text{CONO}_2$$

NO reacts also with hydroperoxides and hydroxyl groups [35]:

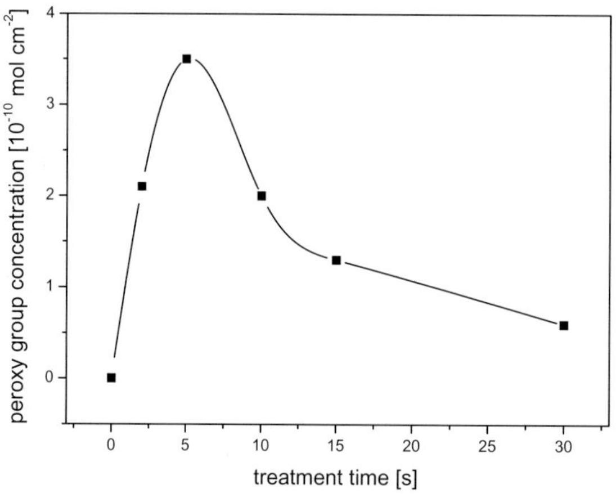

Figure 6.42 Variation of peroxy radical concentration at PE surfaces, as determined by reaction with DPPH, with exposure time to an argon plasma.

$$\text{C-O-OH} + {}^{\bullet}\text{NO} \rightarrow \text{CO-O}^{\bullet} + \text{HNO}$$

$$\text{C-OH} + 2{}^{\bullet}\text{NO} \rightarrow \text{C-ONO} + \text{HNO}$$

The NO molecule is very small and is able to easily penetrate into polymer layers far from the surface and react with radicals within the bulk.

Bromine acts more efficiently, as demonstrated in Figure 6.41. However, Br also reacts with other groups in the polymer (olefinic double bonds, possibly substitution to aromatic rings, etc.).

In contrast, the above-mentioned labeling with DPPH affects only the radical sites directly located at the polymer surface. This is true because of the larger volume of the DPPH molecule. The product of any auto-oxidation, the hydroperoxides, may be quantified by reactions using iodine, thiocyanate, SO_2/Fe^{2+}, or triphenylphosphine [35]. Gassing with NO is also used for permanent stabilization of any plasma-treated polymer against auto-oxidation.

6.12
Plasma-Induced Photo-oxidations of Polymers

The plasma-emitted VUV radiation also initiates several ordinary photochemical processes [35] such as Norrish II rearrangements or photo-Fries reactions [57]. Among all polymers, poly(ethylene terephthalate) especially undergoes on oxygen plasma exposure a chemically defined degradation following a mechanism known from photochemistry (Scheme 6.3) [41].

Chain scissions were observed with corresponding loss in molar masses due to Norrish I and II rearrangements, decarbonylation (CO↑), decarboxylation (CO_2↑) of the ester group, formation of phenols (aryl-OH, photo-Fries rearrangement), branching, side-group scission, and formation of polyenes [21]. These ordinary classic reactions dominate. However, a few unspecific plasma products are also detected that are caused by the excess of energy in the plasma. Such plasma-specific reactions are the cracking of aromatic rings (producing polyenes or branched/crosslinked polyenes) and the formation of peroxyacids [C(O)–O–OH]. Another plasma-specific reaction is the formation of different types of macrocycles (cf. Section 6.8).

Using MALDI-ToF MS these products could be detected together with all possible end group combinations (Figure 6.43).

6.13
Different Degradation Behavior of Polymers on Exposure to Oxygen Plasma

If polymers are irradiated with ionizing radiation three categories of degradation behavior are known, depolymerization, pyrolysis, and crosslinking [58, 59]. Surprisingly, the behavior of polymers on exposure to the oxygen plasma could also be divided into categories. All polymer behaviors could be assigned to one of these

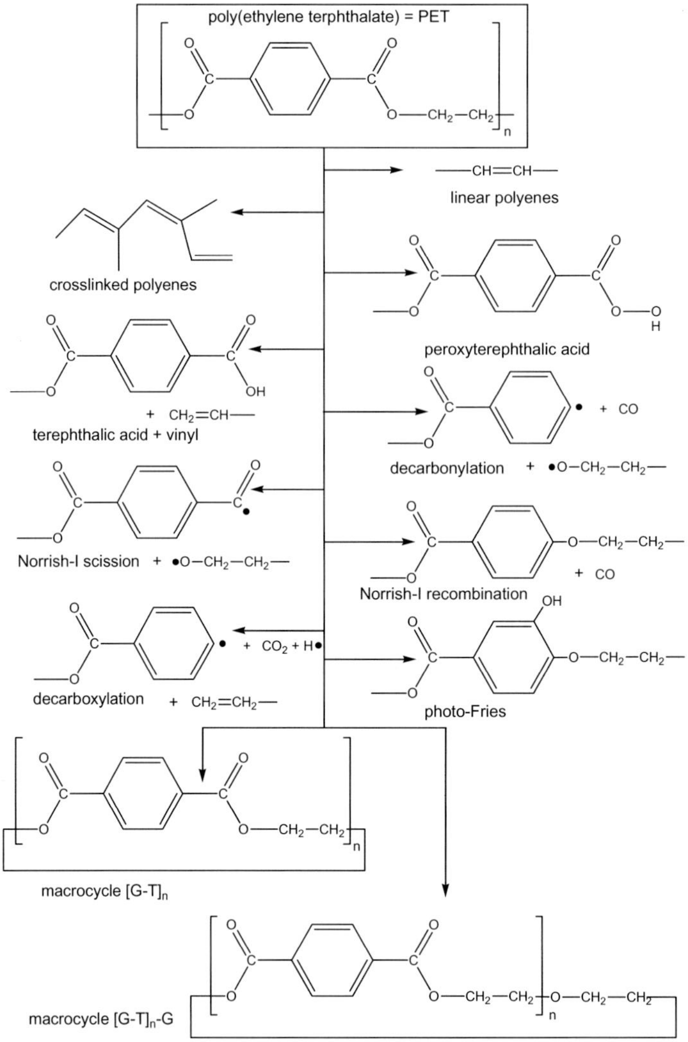

Scheme 6.3 Reaction products of PET when exposed to oxygen plasma (cw-rf plasma, power = 300 W, pressure = 6 Pa).

categories [60]. The sensitivity towards the oxygen plasma is not an integral polymer property but differs from unit to unit and group to group. The oxidation and crack stability of the backbone, the tendency to branch and crosslink, the supermolecular structure including crystallinity, stretching, orientation, the molar mass, the polydispersity, and so on lead to the specific behavior of each polymer. An integral measure of this behavior in the oxygen plasma is the etching rate. Yasuda measured this etching rate for several polymers. He found that POM [poly(oxymethylene)] and PET [poly(ethylene terephthalate)] have high etching

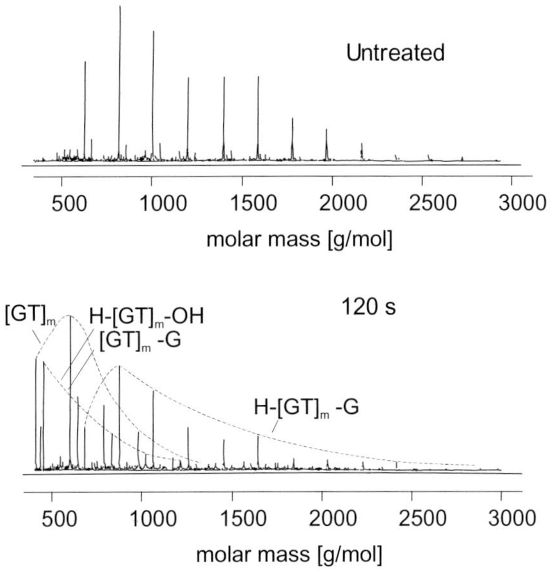

Figure 6.43 MALDI-ToF MS of poly(ethylene terephthalate) before and after O_2 plasma treatment (cw-rf plasma, power = 100 W, pressure = 6 Pa, abbreviations = G–glycol unit, T–terephthalate unit, H–hydrogen).

rates. It must be added that the etching rates of polymers are strongly dependent on the type of plasma reactor, plasma parameters, type of plasma, and so on and thus also the order of polymer etching rates can be changed. Chemically, non-active plasmas from inert gases, such as the helium plasma, act only by their vacuum UV radiation and the particle bombardment (sputtering) from the plasma. Chemical reactions are limited to hydrogen abstraction, chain scissions, rearrangements, and reaction with traces of leak gases and adsorption layers in the reactor. Therefore, the inert gas plasmas also weakly degrade polymers. In this case, POM, poly(acrylic acid) (PAA), and poly(methyl methacrylic acid) (PMAA) possess the highest etching rates [12].

The general behavior of polymers in the oxygen plasma has been studied by MALDI-ToF MS [3, 49, 61]. For this purpose low-molecular-weight polymers were used (2000–6000 g mol^{-1}), such as poly(ethylene glycol) (PEG, Figure 6.44), PET (Figure 6.45), and PMMA (Figure 6.46). They were exposed to the oxygen plasma and subsequently analyzed.

Poly(ethylene glycol) shows a random degradation of molar masses. During oxygen plasma exposure a second irregular and broad molecular weight distribution occurs. The peak-to-peak distances are randomly distributed and positioned close together in this second distribution. This type of polymer degradation on exposure to the oxygen plasma was called *random degradation*.

As all the MALDI spectra are well-resolved, single masses can be unambiguously identified and assigned. PET shows the above-mentioned *photo-oxidative*

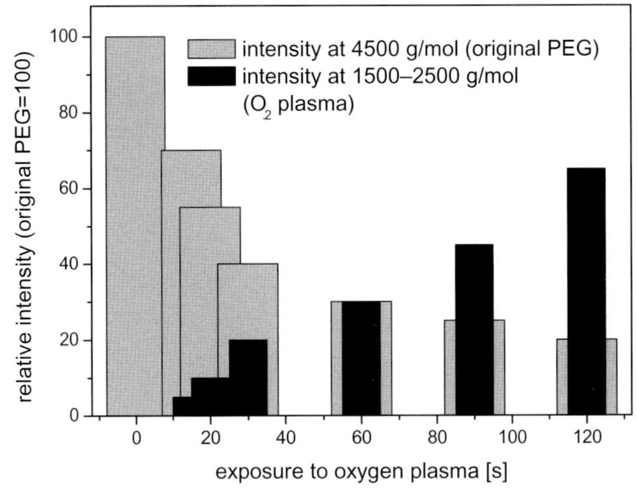

Figure 6.44 Changes in intensity of MALDI detected maxima of a bimodal mass distribution of PEG after oxygen plasma exposure (15–120 s).

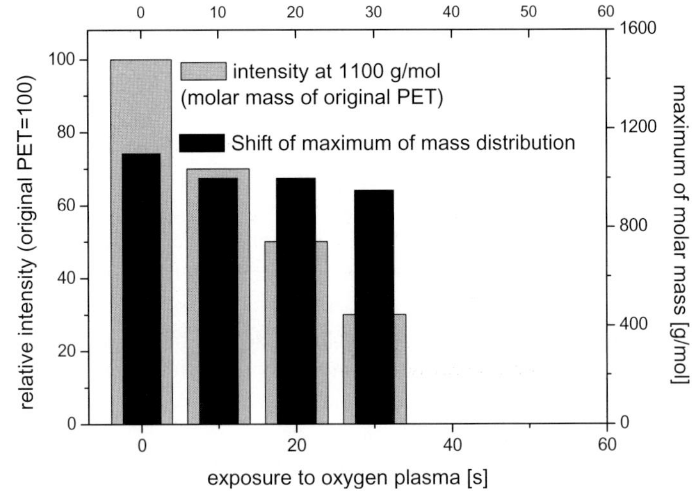

Figure 6.45 Changes in intensity of MALDI detected maxima of PET mass distribution after oxygen plasma exposure (10–30 s).

degradation behavior. Thus, all peaks of degradation products could be assigned to defined products (cf. Figure 6.45).

For PMMA the peak-to-peak distances remain constant at $\Delta = 100\,\mathrm{g\,mol^{-1}}$ (Da) after oxygen plasma exposure. However, the molecular weight distribution function shifts successively to lower molar masses during oxygen plasma treatment. Such behavior is typical for a polymer *depolymerization,* for example, the polymer chain loses monomer units step by step.

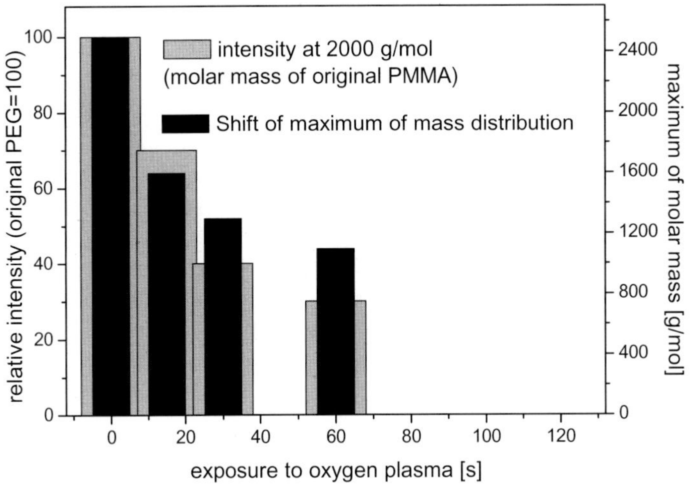

Figure 6.46 Intensity loss and shift of molar mass distribution in MALDI mass spectra of low-molecular-weight PMMA after oxygen plasma exposure (15–60 s).

Table 6.5 Categories of polymers that can be distinguished following exposure to oxygen plasma.

Category	Polymers
Crosslinked	PE, PP, polybutadiene
Randomly degraded	PEG, PPG, cellulose
Photo-oxidatively degraded	PET, PBT, BPA-PC
Depolymerized	PMMA, PAA, PS

A fourth class of polymers is those that tend to preferentially *crosslink* upon exposure to the oxygen plasma. Polyolefins (polyethylene and polypropylene) and polybutadienes show such a tendency.

Summarizing the results, especially those obtained from MALDI MS, Table 6.5 shows the prototypes of polymer response that can be distinguished following exposure to oxygen plasma.

6.14 Derivatization of Functional Groups for XPS

The derivatization of functional groups, bonds, or radicals by markers or labels allows their unambiguous identification on measuring the labeled sample using XPS. Several functional groups show the same analysis pattern, which hinders the exact determination of their existence and their concentration. Groups, bonds, or radicals have a specific new signal presented by the newly introduced atoms or groups of the label [56, 62, 63]. Most often the bond or functional group (X) is derivatized by a label (Y) that includes one or more fluorine atoms [64], such as:

$$C{=}X + YF_3 \rightarrow C\text{-}X\text{-}YF_3$$

Fluorine is easily and unambiguously measurable using XPS as the F1s peak in the photoelectron spectrum. On introducing more than one fluorine atom in place of one bond or group the XPS signal becomes additionally increased. Quantitative analysis can be attained when three to five fluorine atoms substitute one bond or group. Thus, the specific signal is also amplified three- to fivefold in comparison to the original (unspecific) signal. However, replacing a functional group by attaching the marker molecule changes the stoichiometry at the top-most surface layer [65]. The following preconditions must be fulfilled:

- selective reaction (100%);
- replacing the group by a marker with a new element;
- fluorine-containing labels are used because they are often small, can achieve a high sampling depth, and do not disturb the polymer structure too much;
- the group or bond should be marked but not the polymer structure unspecifically, such as by enrichment of markers at interfaces, phase boundaries, and so on;
- the marker should possess more than one F, possibly three or five;
- reaction must be complete (yield = 100%) and stoichiometric;
- rapid derivatization process;
- gas-phase reactions may be favored;
- no other reaction products; if there are such products they must be (easily) removable;
- the label should not possess too many carbon atoms because these carbon atoms go into the calculation;
- the sampling depth of XPS should be, preferably, homogeneously derivatized (no gradient);
- the derivatized product should be stable to XPS measurement and also during X-ray exposure;
- low-molecular-weight material may disturb the analysis because of removal by solvents or condensed solvent vapors during the derivatization process;
- low-molecular-weight material generally adulterates the result because it is not clear what is important, the concentration on a degraded unstable surface or the concentration on intact macromolecules.

Derivatization is necessary for distinguishing different oxidation or binding states of the investigated element that otherwise cannot be differentiated by XPS. Carbon

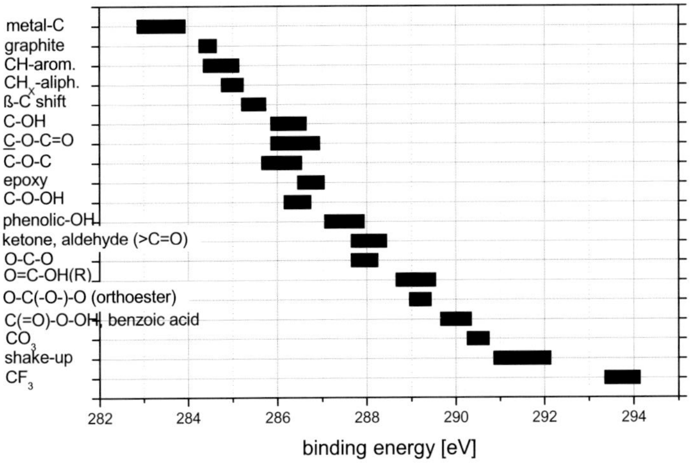

Figure 6.47 Binding energies of groups that contain carbon in the same oxidation state but have different functional groups.

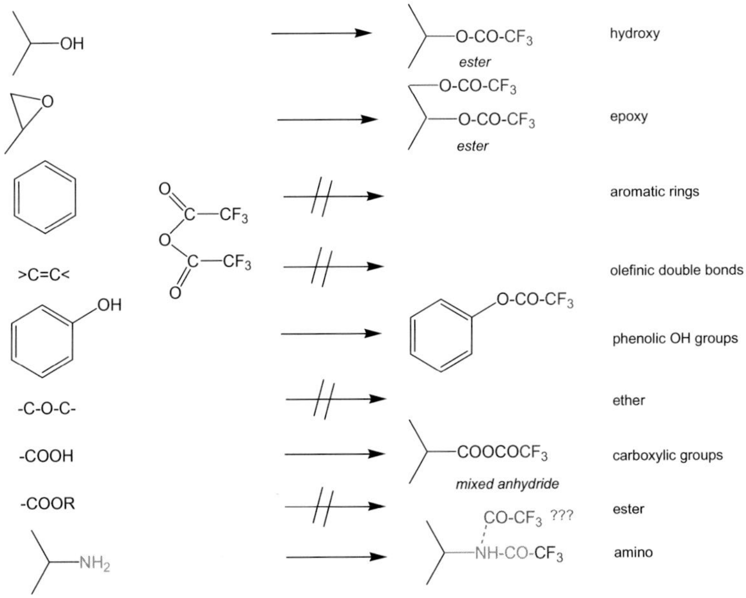

Scheme 6.4 Derivatization reactions of OH, NH_2, and COOH groups using trifluoroacetic anhydride (TFAA), for the application of XPS.

possesses nine oxidation states. In polymers, these oxidation states are realized by different numbers of bonds to oxygen or fluorine as well as to metals. Figure 6.47 lists characteristic groups of bonds that cannot be measured in isolation by XPS. Here, the labeling helps to distinguish groups with the same oxidation state (same number of bonds to oxygen):

Scheme 6.4 lists typical derivatization reactions for labeling OH groups.

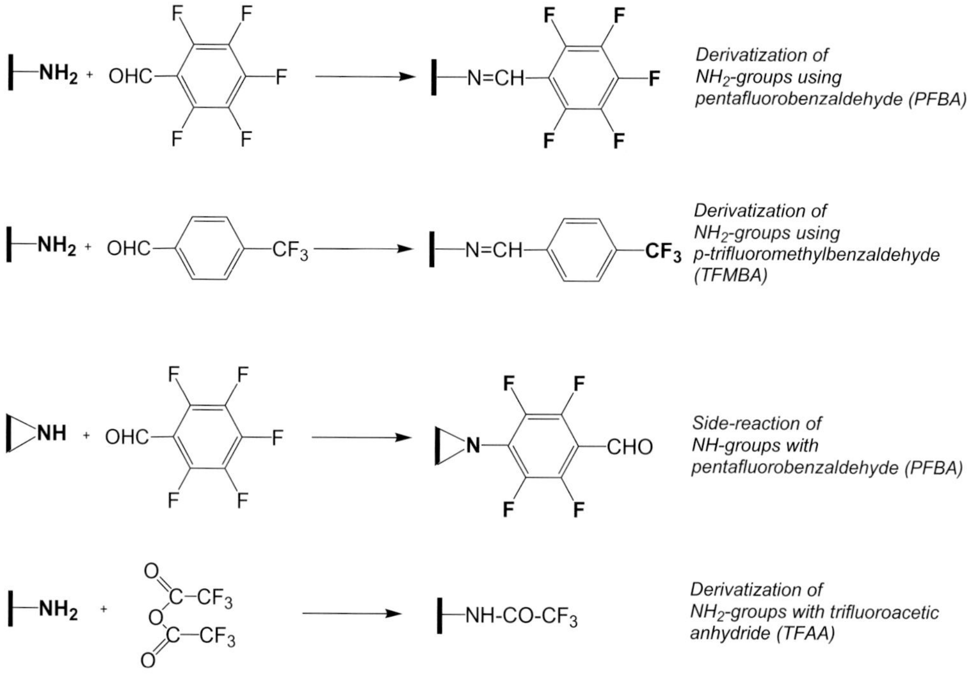

Scheme 6.5 Derivatization reactions of NH_2 and NH groups using benzaldehydes and trifluoroacetic anhydride (TFAA), for the application of XPS.

For example, poly(vinyl alcohol) has an elemental composition of C_2H_4O, for example, for the C/O ratio it is 67 at.% C and 33 at.% O (hydrogen cannot be measured by means of XPS) or referenced to 100 carbon atoms 50 O per 100 C. Notably, the presentation of values referenced to 100 C or % O/C is the most exciting presentation for statistical processes. Atomic percent may be preferred for characterization of defined stoichiometries of low-molecular-weight species. After consumption of the OH group with trifluoroacetic anhydride (TFAA, cf. Scheme 6.4):

$$[CH_2\text{-}CH(OH)]_n\!- + C_2O_3(CF_3)_2 \rightarrow \text{-}[CH_2\text{-}CH(O\text{-}CO\text{-}CF_3)]_n\!- + CF_3\text{-}COOH.$$

Thus, from 50 O per 100 C the elemental composition was changed to 50 O and 75 F per 100 C.

More problems exist with NH_2 groups. The plasma-produced amino groups undergo oxidation, side- and post-plasma reactions. Scheme 6.5 lists the labeling reactions [56].

Ketones or aldehydes are labeled by the reactions shown in Scheme 6.6 [66].

Carboxylic acids are consumed to give esters by reaction with trifluoroethanol (TFE) (Scheme 6.7).

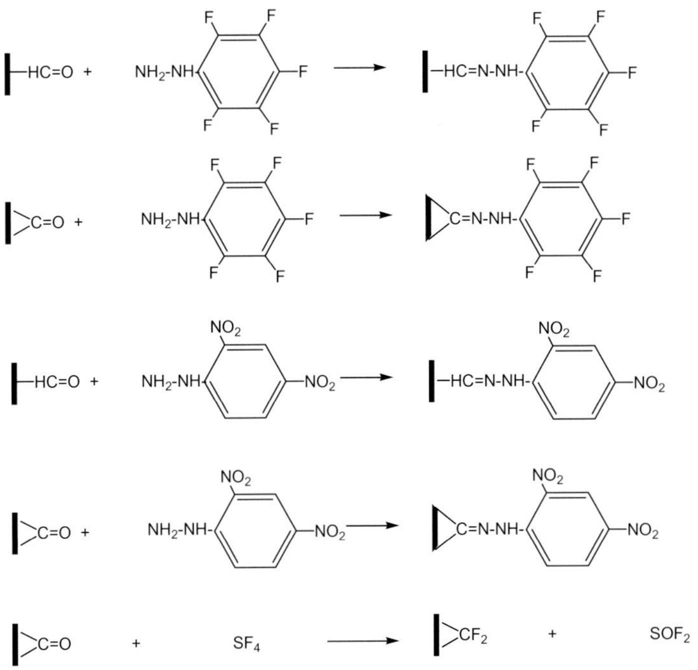

Scheme 6.6 Derivatization reactions of C=O and CHO groups using hydrazine derivates, for the application of XPS.

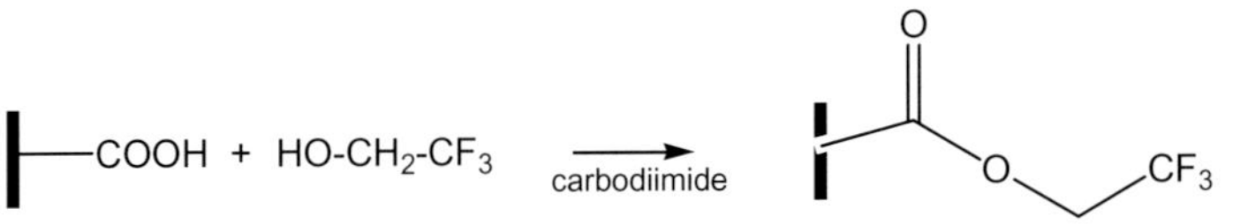

Scheme 6.7 Derivatization reactions of COOH using trifluoroethanol, for the application of XPS.

As mentioned before, radicals can also be derivatized and, thus, analyzed by XPS (Scheme 6.8).

To detect very low concentrations of functional groups, derivatization with fluorescence labels is advantageous [63, 65, 67–69]. Mix *et al.* tested several different labels attached to OH groups of plasma-modified polymer surfaces (Scheme 6.9) [67, 69]

Then, the fluorescence labels were grafted (Scheme 6.10).

Everhart and Reilley [64], Briggs and Seah [56] as well as Holländer [65] have published summaries of different derivatization techniques of functional groups at polymer surfaces. For all surface functional groups specific derivatization labels and reactions exist.

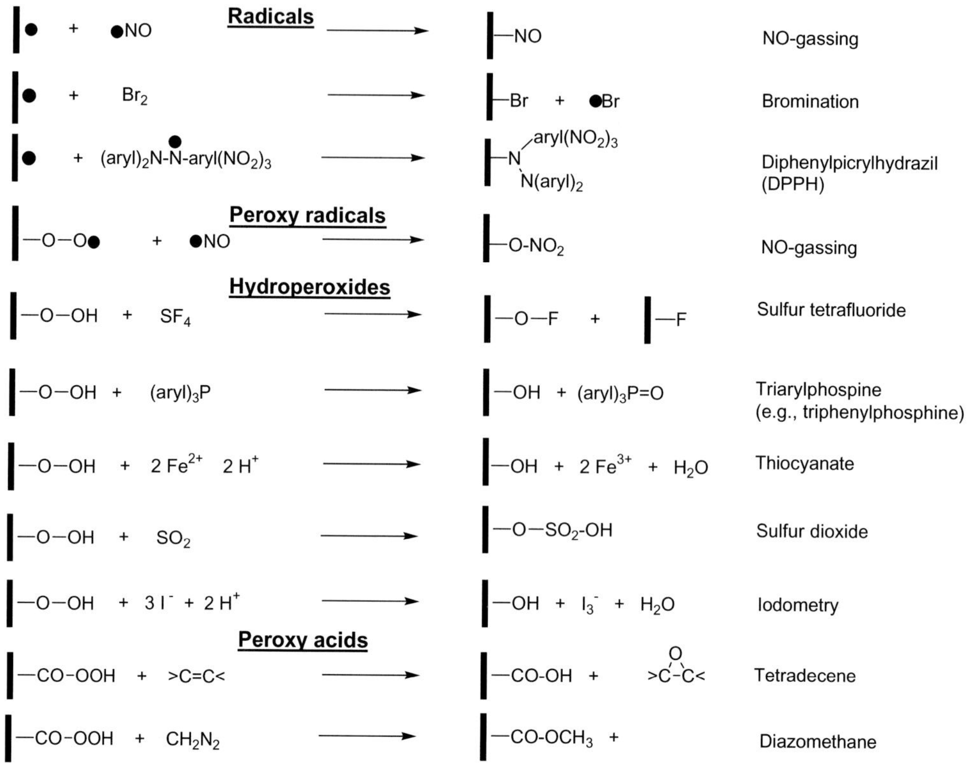

Scheme 6.8 Derivatization of various radicals and oxygenated groups, for the application of XPS.

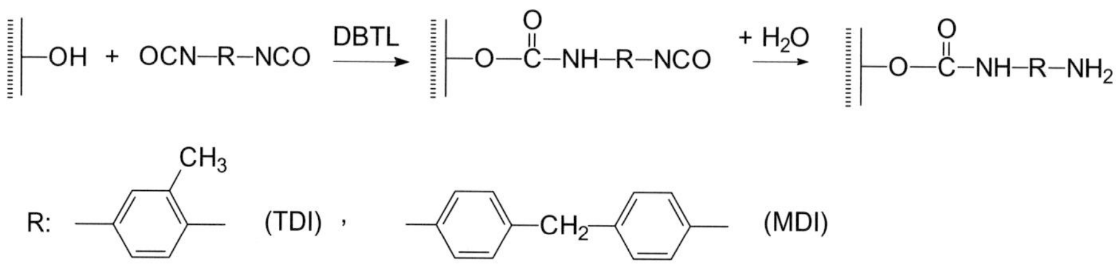

Scheme 6.9 Attachment of labels to OH groups; abbreviations: toluene-2,4-diisocyanate (TDI), (80/20 isomer mixture, methylene-di-*p*-phenylene isocyanate (MDI), and dibutyltin dilaurate (DBTL).

For non-defined statistical surface functionalizations the derivatization reactions must be calibrated using a well-defined reference material. Such reference materials may be poly(vinyl alcohol) for OH groups and poly(allylamine), 4,4′-diaminodiphenylmethane for NH_2 groups as well as poly(acrylic acid) for COOH groups. However, such reference materials are not as well-defined as desired because of the enrich-

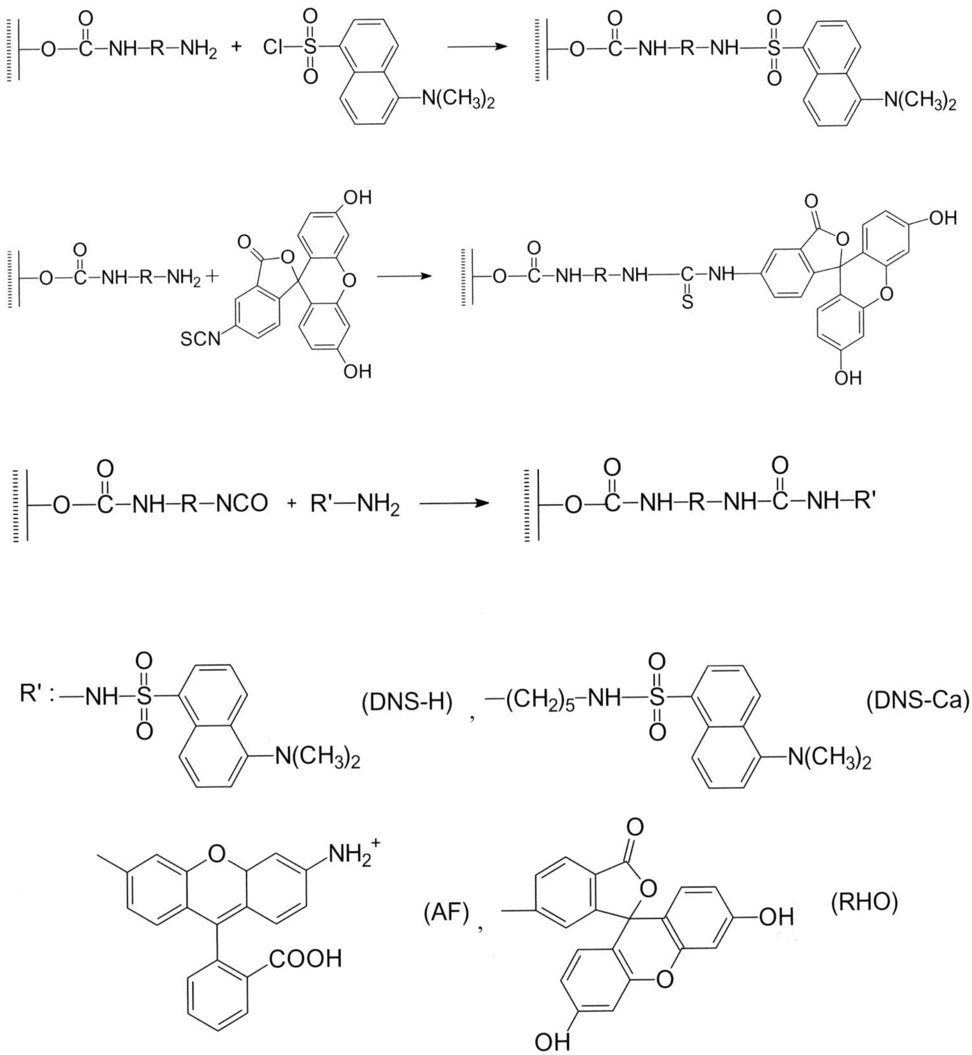

Scheme 6.10 Grafting of fluorescence labels; abbreviations: dansyl chloride (DNS), dansyl cadaverine (DNS-Ca), dansyl hydrazine (DNS-H), 5(6)-aminofluorescein (AF), and rhodamine110 (RHO).

ment or deficiency of functional groups by diffusion or reaction (acetal, N-oxide, amide, anhydride formation, etc.).

In addition, derivatization not only introduces new elements such as fluorine, which can be detected and quantified via the F1s signal, but it is also possible to reference internally changes in the C1s signal (Figure 6.48).

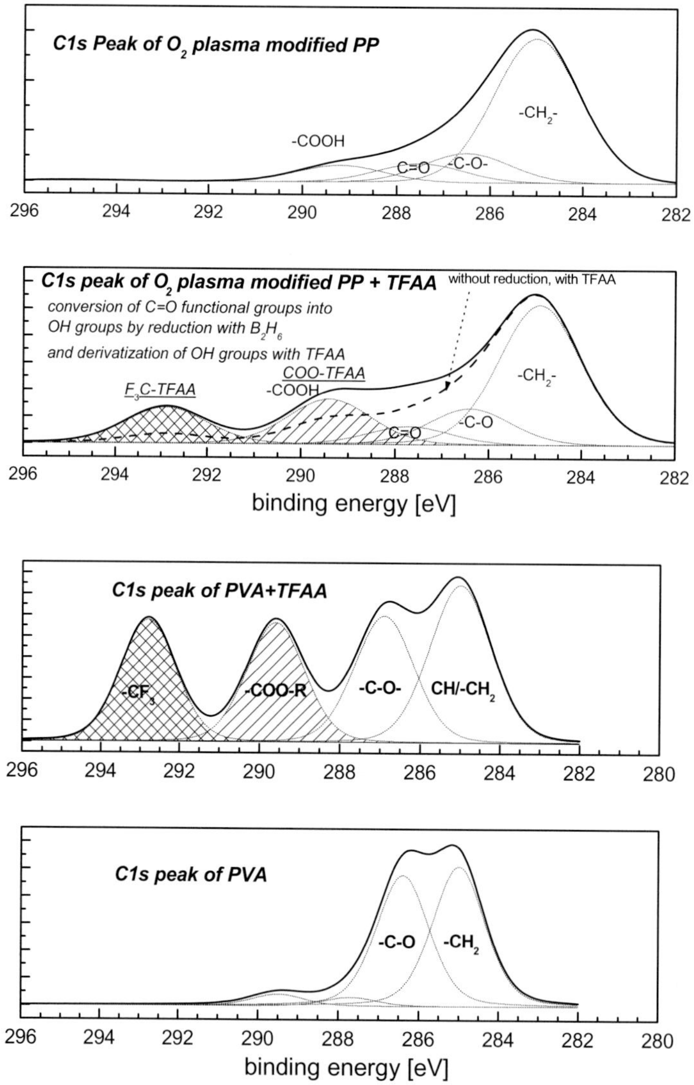

Figure 6.48 Derivatization of oxygen-plasma modified polypropylene and poly(vinyl alcohol) reference material derivatized with trifluoroacetic anhydride (TFAA).

Considering the derivatization technique for the estimation of the concentration of functional groups, one, two, or three gradients from polymer surface towards the polymer bulk exist or are superposed:

1) A gradient density of functional groups is present after plasma surface modification but it is not present after deposition of homogeneous layers carrying functional groups.

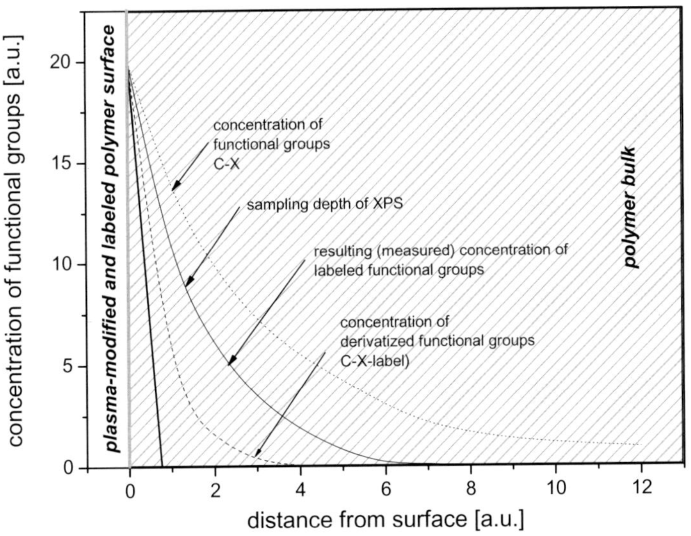

Figure 6.49 Schematic of gradients from polymer surface to bulk caused by enrichment of functional groups at polymer surface, along with sampling depths of XPS and labeling depth.

2) The sampling depth function of the XPS method shows that more than 60% of the signal is produced by incidences within the first 1 nm.

3) Gradient of marker fixing from surface towards bulk; voluminous labels are attached exclusively onto the topmost surface, small markers also in surface-near layers.

Figure 6.49 depicts schematically a superposition of all gradients. The resulting gradient should be very strong; thus, only the first atom layers are detected. A homogeneous distribution of functional groups at the surface and in the bulk eliminates one gradient, which is the case for homogeneous layers or polymers.

References

1 Lippitz, A., Friedrich, J., Unger, W., Schertel, A., and Wöll, C. (1996) *Polymer*, **37**, 3151–3156.

2 Koprinarov, I., Lippitz, A., Friedrich, J.F., Unger, W.E.S., and Wöll, C. (1996) *Polymer*, **37**, 2005–2010.

3 Weidner, S., Kühn, G., Friedrich, J., Unger, W., and Lippitz, A. (1996) *Rapid Commun. Mass Spectrom.*, **10**, 727–737.

4 Friedrich, J., Geng, S., Unger, W., and Lippitz, A. (1998) *Surf. Coat. Technol.*, **98**, 1132–1141.

5 Unger, W.E.S., Lippitz, A., Friedrich, J.F., Wöll, C., and Nick, L. (1999) *Langmuir*, **15**, 1161–1166.

6 Stöhr, J. (1992) *NEXAFS Spectroscopy*, *Springer Series in Surface Science*, vol. 25, Springer Verlag, Berlin.

7 Chalykh, A.B., Petrova, I.I., Vasilenko, Z., Gerasimov, V.I., and Brusentsova, V.G. (1974) *Vysokomol. Soedin.*, **6**, 1289.

8 Friedrich, J., Kühn, G., and Gähde, J. (1979) *Acta Polym.*, **30**, 470.

9 Strobel, M., Corn, S., Lyons, C.S., and Korba, G.A. (1987) *J. Polym. Sci., Part A1*, **25**, 129.

10 Friedrich, J., Gähde, J., and Nather, A. (1981) *Plaste Kautsch.*, **28**, 620–625.

11 Friedrich, J., Gähde, J., and Pohl, M. (1980) *Acta Polym.*, **31**, 312–315.

12 Yasuda, H., Lamaze, C.E., and Sakaouku, K. (1973) *J. Appl. Polym. Sci.*, **17**, 137.

13 Moss, S.J., Jolly, A.M., and Tighe, B.J. (1986) *Plasma Chem. Plasma Process.*, **6**, 401.

14 Rosskamp, G. (1972) Organische Reaktionen im Plasma von Glimmentladungen. PhD thesis. University of Tübingen.

15 Hudis, M. (1972) *J. Appl. Polym. Sci.*, **16**, 2397.

16 Friedrich, J. (1981) *Contr. Plasma Phys.*, **21**, 261.

17 Possart, W. and Friedrich, J. (1986) *Plaste Kautschuk*, **33**, 273.

18 Geng, S., Friedrich, J., Gähde, J., and Guo, L. (1999) *J. Appl. Polym. Sci.*, **71**, 1231.

19 Friedrich, J., Throl, U., Gähde, J., and Schierhorn, E. (1982) *Acta Polym.*, **33**, 405.

20 Lippitz, A., Koprinarov, I., Friedrich, J.F., Unger, W.E.S., Weiss, K., and Wöll, C. (1996) *Polymer*, **37**, 3157.

21 Friedrich, J., Loeschcke, I., Frommelt, H., Reiner, H.-D., Zimmermann, H., and Lutgen, P. (1991) *J. Polym. Degrad. Stabil*, **31**, 97–119.

22 Lianos, L., Hoc, T.Q., and Duc, T.M. (1994) *J. Vac. Sci. Technol., A*, **12**, 2491.

23 Clark, D.T. and Dilks, A. (1980) *J. Polym. Sci., Polym. Chem. Ed.*, **18**, 1233.

24 Friedrich, J. (1986) *Wissenschaft Fortschritt*, **36**, 311–319.

25 Enzyklopädie, B.-K. (1970) *Atom-Struktur Der Materie* (eds C. Weissmantel, R. Lenk, W. Forker, R. Ludloff, and J. Hoppe), VEB Bibliographisches Institut, Leipzig.

26 Skurat, V. (2003) *Nucl. Instrum. Methods Phys. Res. B*, **208**, 27.

27 Rutscher, A. and Deutsch, H. (1983) *Wissensspeicher Plasmatechnik*, VEB Fachbuchverlag Leipzig.

28 Fozza, A.C., Klemberg-Sapieha, J.E., and Wertheimer, M.R. (1999) *Plasmas Polym.*, **4**, 183.

29 Holländer, A., Klemberg-Sapieha, J.E., and Wertheimer, M.R. (1995) *J. Polym. Sci. A: Polym. Chem.*, **33**, 2013.

30 Babucke, G., Lange, H., and Ohl, A. (2001) Quantitative analyse der VUV-strahlung in mikrowellenplasmen, in *10. Bundesdeutsche Fachtagung Plasmatechnologie*, Deutsche Gesellschaft für Plasmatechnologie, Greifswald, conference held in Greifswald, organized by DGPT, 64 pp.

31 Fozza, A.C., Kruse, A., Holländer, A., Ricard, A., and Wertheimer, M.R. (1988) *J. Vac. Sci. Technol., A*, **16**, 72.

32 Friedrich, J.F., Unger, W.E.S., Lippitz, A., Giebler, R., Koprinarov, I., Weidner, S., and Kühn, G. (2000) *Polymer Surface Modification: Relevance to Adhesion*, vol. 2 (ed. K.L. Mittal), VSP, Utrecht, pp. 137–172.

33 Loeschcke, I., Friedrich, J., and Lutgen, P. (1990) *Acta Polym.*, **41**, 553.

34 Behnisch, J., Friedrich, J., and Zimmermann, H. (1991) *Acta Polym.*, **42**, 51–53.

35 Rabek, J.F. (1995) *Polymer Photodegradation – Mechanisms and Experimental Methods*, Chapman & Hall, London.

36 Friedrich, J., Pohl, M., Elsner, T., and Altrichter, B. (1988) *Acta Polym.*, **39**, 544.

37 Friedrich, J. and Frommelt, H. (1988) *Acta Chim. Hung.*, **125**, 165.

38 Friedrich, J.F., Koprinarov, I., Giebler, R., Lippitz, A., and Unger, W.E.S. (1999) *J. Adhesion*, **71**, 297.

39 Friedrich, J.F., Unger, W.E.S., Lippitz, A., Koprinarov, I., Weidner, S., Kühn, G., and Vogel, L. (1998) *Metallized Plastics 5&6: Fundamental and Applied Aspects* (ed. K.L. Mittal), VSP, Utrecht, p. 271.

40 Holländer, A., Wilken, R., and Behnisch, J. (1999) *Surf. Coat. Technol.*, **116–119**, 788.

41 Weidner, S., Kühn, G., Friedrich, J., and Schröder, H. (1996) *Rapid Commun. Mass Spectrom.*, **10**, 40–44.

42 Elias, H.G. (1990) *Macromolecules*, Hüthig & Wepf, Basel.

43 Grasenick, F. (1957) *Radex-Rundschau*, **5/6**, 843.

44 Spit, B.J. (1967) *Faserforsch. Textiltechnik*, **18**, 161.

45 Lipatow, J.S., Bezruk, L.I., Lebedew, E.V., and Gomza, J.P. (1974) *Vysokomol. Soedin.*, **5**, 328.

46 Friedrich, J., Gähde, J., and Pohl, M. (1980) *Acta Polym.*, **31**, 310.

47 Friedrich, J., Pohl, M., and Gähde, J. (1981) *Acta Polym.*, **32**, 48–56.

48 Friedrich, J., Gähde, J., and Pohl, M. (1982) *Acta Polym.*, **33**, 209–216.

49 Weidner, S., Kühn, G., Decker, R., Roessner, D., and Friedrich, J. (1998) *J. Polym. Sci.*, **36**, 1639.

50 Schonhorn, H. and Hansen, R.H. (1967) *J. Appl. Polym. Sci.*, **11**, 1461.

51 Hall, J.R., Westerdahl, C.A.L., Bodnar, M.J., and Levi, D.W. (1972) *J. Appl. Polym. Sci.*, **16**, 1465.

52 Kuzuya, M., Kondo, S., Sugito, M., and Yamashiro, T. (1998) *Macromolecules*, **31**, 3230.

53 Kuzuya, M., Ito, H., Kondo, S., Noda, N., and Noguchi, A. (1991) *Macromolecules*, **24**, 6612.

54 Suzuki, M., Kishida, A., Iwata, H., and Ikada, Y. (1986) *Macromolecules*, **19**, 1804.

55 Poll, H.-U., Kleemann, R., and Meichsner, J. (1981) *Acta Polym.*, **32**, 139–142.

56 Briggs, D. and Seah, M.P. (eds) (1994) *Practical Surface Analysis*, vol. 1, John Wiley & Sons, Ltd, Chichester.

57 Makhlis, J. (1975) *Radiation Physics and Chemistry of Polymers*, John Wiley & Sons, Inc., New York.

58 Chapiro, A. (1964) *Radiat. Res. Suppl.*, **4**, 197–1919.

59 Henglein, A., Schnabel, W., and Wendenburg, J. (1969) *Einführung in Die Strahlenchemie*, Verlag Chemie, Weinheim.

60 Friedrich, J., Unger, W., Lippitz, A., Koprinarov, I., Kühn, G., Weidner, S., and Vogel, L. (1999) *Surf. Coat. Technol.*, **116–119**, 772–782.

61 Weidner, S., Kühn, G., and Friedrich, J.F. (1998) *Rapid Commun. Mass Spectrom.*, **12**, 1373–1381.

62 Chilkoti, A. and Ratner, B.D. (1996) *Surface Characterization of Advanced Polymers* (eds L. Sabbattini and P.G. Zambonin), VCH, Weinheim, p. 221.

63 Friedrich, J., Kühn, G., and Mix, R. (2006) *Prog. Colloid Polym. Sci.*, **132**, 62–71.

64 Everhart, D.E. and Reilley, C.N. (1981) *Anal. Chem.*, **53**, 665.

65 Holländer, A. (2004) *Surf. Interface Anal.*, **36**, 1023.

66 Altmann, K. (2011) Analytische Bestimmung von oberflächengebundenen Aldehyd- und Ketogruppen sowie phenolischen Hydroxygruppen [Analytical quantification of aldehyde, keto and phenolic hydroxy groups bonded to polymer surfaces]. Master's thesis. Freie Universität Berlin.

67 Hoffmann, K., Resch-Genger, U., Mix, R., and Friedrich, J. (2006) *J. Fluoresc.*, **16**, 441–448.

68 Hoffmann, K., Mix, R., Hoffmann, K., Buschmann, H.-J., Friedrich, J.F., and Resch-Genger, U. (2009) *J. Fluoresc.*, **19**, 229–237.

69 (a) Resch-Genger, U., Hoffmann, K., Mix, R., and Friedrich, J.F. (2007) *Langmuir*, **23**, 8411–8416; (b) Mix, R., Hoffmann, K., Resch-Genger, U., Decker, R., and Friedrich, J. F. (2007) *Polymer Surface Modification: Relevance to Adhesion*, vol. IV (ed. K.L. Mittal), VSP, pp. 171–192.

7
Metallization of Plasma-Modified Polymers

7.1
Background

A prominent application of polymer surface modification of polymers by exposure to plasma is the production of highly adherent metal–polymer composites. The adhesion of metal layers deposited onto polymer surfaces is determined by the concentration and the bond strength of chemical and physical interactions between the metal atoms and the functional (polar) groups at the polymer surfaces. Each type of functional group produces individual metal–polymer interactions, making a specific contribution that depends on its concentration to the interfacial adhesion and, consequentially, to the related shear or peel strength of metal–polymer systems (Figure 7.1).

Each type of functional group makes an individual contribution in terms of interaction with the metal. The interfacial adhesion is the sum of all specific metal-group contributions, depending exclusively on the respective functional group concentration as seen above (Figure 7.1). There are many different functional groups and therefore many different types of metal–groups interactions with different bond strengths, each interaction contributing its bond strength and partial concentration to the total adhesion of the metal–polymer composite. However, additional unspecific contributions to the adhesion strength due to van der Waals interactions, hydrogen bonding, and interdiffusion processes must be also considered [1]. Thus for each type x of metal–functional group interaction (σ_x), the work of adhesion ($W^x{}_a$) is calculated by:

$$W^x{}_a = ALc_i\sigma_x$$

where

A = area,
L = Loschmidt constant,
c_i = concentration.

The total work of adhesion (W_a) follows by summation over all types of interaction:

The Plasma Chemistry of Polymer Surfaces: Advanced Techniques for Surface Design, First Edition.
Jörg Friedrich.

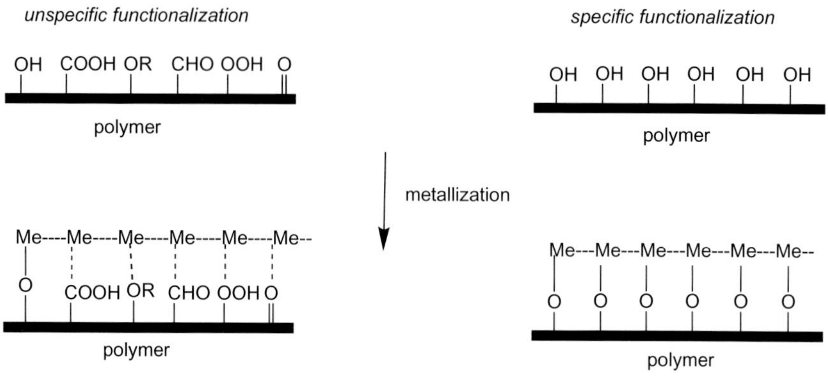

Figure 7.1 Schematics of metal–polymer interactions.

$$W_a = AL\sum_{i=1}^{m} c_i\sigma_i$$

which presents a correlation to the macroscopically measured adhesive bond strength, which might be the peel strength.

7.2 Polymer Plasma Pretreatment for Well Adherent Metal–Polymer Composites

Adhesion improvement of metal–polymer composites by plasma pretreatment of the polymer component was often announced in the past [2–7]. The most important effect of plasma treatment is the incorporation of functional groups, which can interact physically or chemically with metal atoms [8–9]. Another possibility is the roughening of a polymer surface by plasma exposure [10] or by crosslinking and therefore by changing the surface properties, which is known as CASING (crosslinking by activated species of inert gases) (cf. Figure 7.2) [11].

Modification of metal substrates by plasma pretreatment is also possible. It could be shown that cleaning, oxide conditioning, and formation as well as plasma-assisted primer deposition can improve the adhesion of polymer coatings on metal substrates [12]. Reductive as well as oxidative acting plasmas could enhance the peel strength between steel and polyurethane coatings. Soft copper oxidation by O_2 plasma improves strongly the adhesion of poly(phenylquinoxaline)s to copper foils [13]. Often, plasma polymers were used as adhesion-promoting interlayers in metal–polymer composites [14–19]. A special problem is the well-adherent metallization of polytetrafluoroethylene (PTFE) using plasma surface modification [20–21].

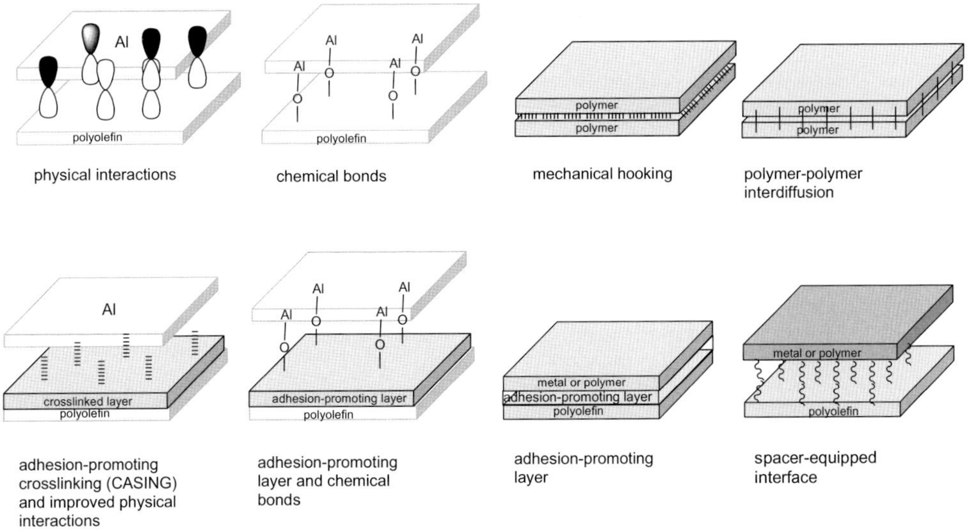

Figure 7.2 Schematics of different adhesion mechanisms.

7.2.1
Surface Cleaning by Plasma for Improving Adhesion

Two types of contaminations are found at polymer surfaces, organic and inorganic impurities. Using a plasma process, the nature of contaminants must be considered. Inorganic components often remain after exposure to plasma. Most often oxygen or other oxidizing plasmas are applied. Metal oxides or silica-like residues remain. Therefore, two different plasma processes must be used to remove all contaminations. Silicones are partially oxidized and form siloxane-like compounds:

$$\text{-[Si-O(CH}_3)_2]_x\text{-} + \text{O-}\textit{plasma} \rightarrow \text{-[Si-O}_2]_x\text{-} + 2x\text{CO}_2\uparrow + 3x\text{H}_2\text{O}\uparrow$$

$$\text{-[Si-O}_2]_x\text{-} + \text{CF}_4\text{-}\textit{plasma} \rightarrow x\text{SiF}_4\uparrow + x\text{CO}_2\uparrow$$

However, it must be considered that intense plasma exposure also affects the properties of polymers by etching, degradation, radical formation, oxidation, and crosslinking. Therefore, the thickness, thickness distribution, chemical nature, and topography of contamination layers play an important role. Sources of contaminations are fingerprints, lubricants, oils, mold release agents, air pollutions, vapor precipitates, traces of foods, beverages, blood, tissue, and so on. Thick contamination layers should be removed as much as possible by mechanical abrasion or by solvent extraction. Remaining inhomogeneously distributed contaminations (agglomerates) are difficult to remove by plasma etching. The duration of plasma exposure must be sufficient to remove the thickest spots of contamination;

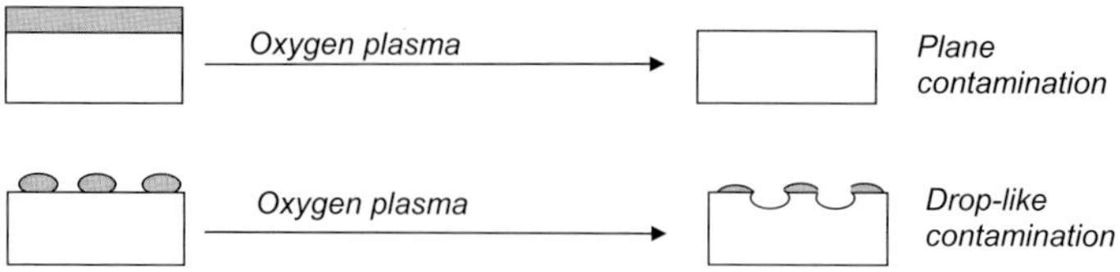

Figure 7.3 Homogeneous and inhomogeneous contaminations at polymer surfaces both before and after exposure to oxidative plasma treatment.

however, uncontaminated and uncovered areas are then exposed intensely to the plasma without any need. Indeed, these areas are "over-etched" and the polymer material becomes degraded (Figure 7.3).

Contamination films act also as anti-adhesives and adulterate the adhesion. Low-pressure oxygen plasma was the most efficient for removing organic contaminations [22]. Removing or reducing inorganic contaminations, such as oxide layers, hydrogen, and tetrafluoromethane plasmas have proved to be best achieved as:

$$\mathrm{Fe_3O_4} + H_2\ plasma \rightarrow \mathrm{3Fe + 2H_2O}\uparrow$$

$$\mathrm{SiO_2} + CF_4\ plasma \rightarrow \mathrm{SiF_4}\uparrow + \mathrm{CO_2}\uparrow$$

Fluorine-releasing plasmas for cleaning surfaces (CF_4, BF_3, XeF_2, NF_3, SF_6, OF_2, etc.), however, also etch the material and the reactor strongly. The waste gases are toxic. Acidic components in waste gases can be easily and cheaply neutralized ($CaCO_3$, CaO), while other toxic gases can be removed by use of catalysts. Such waste gas processing is found in the microelectronic industry. Strong toxic and/or etching products are hydrofluoric acid, perfluoroisobutene, hexafluoroacetone, perfluoroacetic acid, and 1,1-difluoroethene.

The main reaction pathways during removal of organic contaminations to the above-mentioned products are:

$$O_2\ plasma \rightarrow \mathrm{H_2O + CO + CO_2}$$

$$H_2\ plasma \rightarrow \mathrm{CH_4 + H_2O}$$

$$N_2\ plasma \rightarrow \mathrm{NH_3, HCN, NO}\ldots$$

$$CF_4\ plasma \rightarrow \mathrm{HF, R\text{-}CFO, NF_3}\ldots$$

Figure 7.4 shows the influence of various cleaning procedures of stainless steel with different efficiencies on lap-shear strength of thick layers (10 mm) of polyurethane (PU) [8, 23].

Figure 7.5 presents the variation in cleaning of glass bulbs of lamps by exposure to low-pressure oxygen plasma with the time of the process.

After cleaning with aqueous solutions a glass surface of an electrical bulb has about 50 ng cm^{-2} residual hydrocarbon contaminations. After 360-s exposure to the low-pressure oxygen plasma only 2 ng cm^{-2} was found. Then, a thin lacquer film can be deposited with sufficient adhesion.

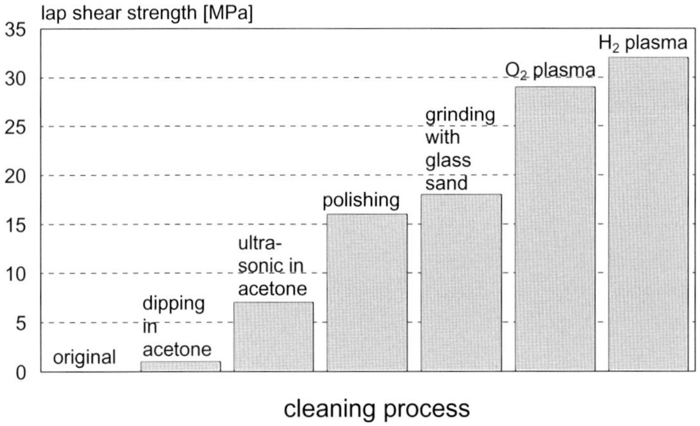

Figure 7.4 Dependence of lap-shear strength, as expressed in the XPS C1s/Fe2p intensity coefficient, of steel coated with polyurethane on the cleaning process employed.

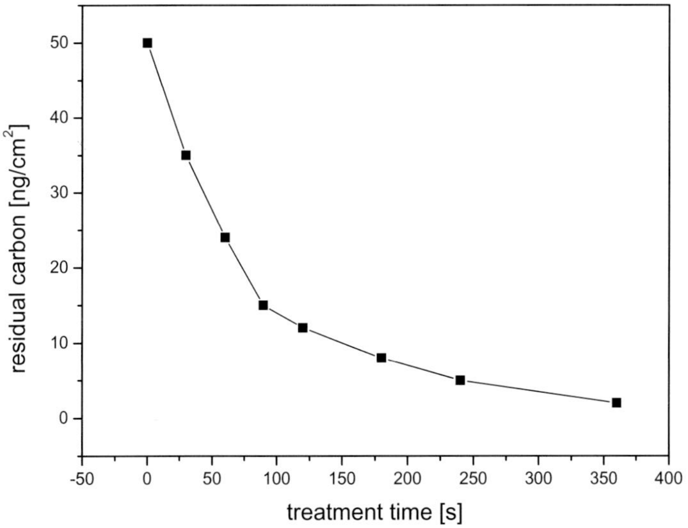

Figure 7.5 Residual carbon contaminations at glass surfaces during O_2 plasma etching.

Polymer surfaces are also often contaminated with additives, catalysts, fingerprints, dust, oil, and water from ambient air pollutions. Such a contamination layer blocks the interactions between metal and polymer and must, therefore, be avoided or if present removed by solvent washing, sonication, or plasma etching. The contamination of an industrial polyethylene foil was analyzed and then removed by solvent extraction because the prior metallization cleaning did not

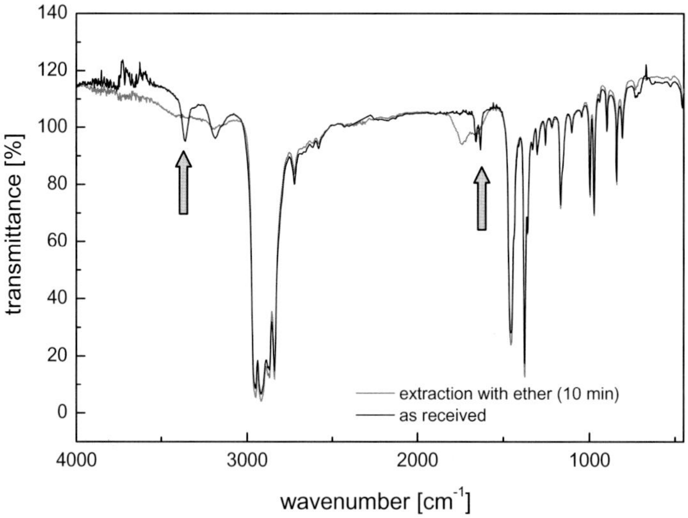

Figure 7.6 Contamination of polyethylene foils with eruca acid amide measured by FTIR before and after solvent washing.

produce significant adhesion. The covering with a slip agent [eruca acid amide = $H_3CCH_2)_7CH{=}CH(CH_2)_{11}CONH_2$] was determined as the reason for poor adhesion (Figure 7.6).

7.2.2
Oxidative Plasma Pretreatment of Polymers for Adhesion Improvement

Oxidation of polyolefin surfaces by low-pressure oxygen plasma produces a broad variety of different O-functional groups at concentrations near the steady-state limit between maximal surface functionalization with O-functional groups and the etching of the polymer and production of gaseous etching products. The maximal oxygen percentage fixed as functional groups at polyethylene or polypropylene surfaces amounts about 28 O per 100 C, and for poly(ethylene terephthalate) (PET) about 42 O/100 C. While PE (polyethylene) and PP (polypropylene) can incorporate high concentrations of oxygen at their surface without complete degradation and rapid etching, the basic percentage of oxygen in untreated PET is 40 O/100 C. Marginally exceeding this value by 2% accelerates the etching and formation of gaseous degradation products. Higher O-concentrations are not realistic and were affected by simultaneous deposition of metal oxides from electrodes or from similar undesired reactions.

High oxygen concentrations and therefore the presence of O-containing polar groups at polyolefin surfaces increase the surface energy and make the polyolefins wettable by polar liquids and also by metal layers. High surface energy and high polar contributions to surface energy are preconditions for strong adhesion.

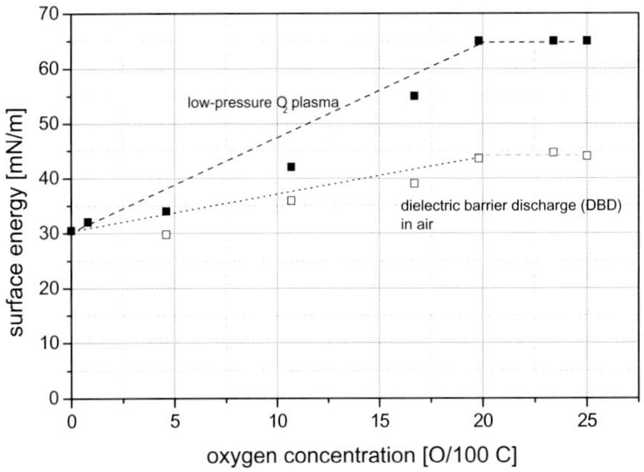

Figure 7.7 Surface energy of polyethylene as a function of oxygen concentration for low- and atmospheric-pressure plasma.

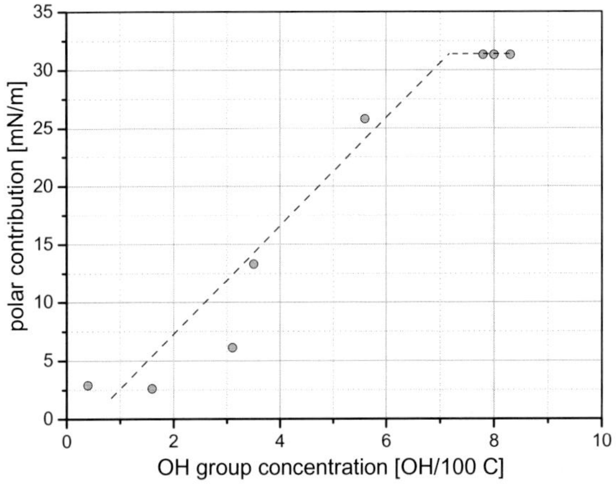

Figure 7.8 Polar contribution of surface energy versus concentration of OH groups of ethylene–allyl alcohol copolymers produced under low-pressure plasma exposure.

Polyolefins are characterized by the absence of any functional groups and, therefore, by low surface energy and low or zero polar contribution to it. Oxidation of the surface increases the surface energy linearly to 20 O/100 C; it then reaches a plateau with constant surface energy (Figure 7.7).

The same type of dependence also exists for the polar contribution of OH groups to the surface energy at the surface of plasma-produced ethylene–allyl alcohol of varied composition (Figure 7.8). The OH concentration was measured by XPS (X-ray photoelectron spectroscopy) and derivatization with trifluoroacetic anhydride (TFAA).

These dependencies of surface energy (σ), (σ_{polar}) = f(O or OH), do not consider the existence of low-molecular-weight oxidized material (LMWOM) at the polymer surface [24]. These degradation layers affect high polarity but are only weakly bonded to the non- or weakly modified polymer substrate. Moreover, Bikerman has found that such degraded molecules at the surface form a weak boundary layer (WBL), which decreases the adhesion to metals because of the weakened bonding of these molecular fragments to the intact macromolecules of the polymer bulk [25]. Such layers of polymer degradation products may also be partially dissolved in the drops of test liquids used, thus influencing the measurement.

The surface energies of metals and inorganics are higher by about two orders of magnitude than those of polymers. This strong difference reflects the weak interactions between metals and polymers, such as between aluminium and polyethylene or polypropylene. It is a common approach to increase the surface energy of polyolefins by oxidation for improved interactions between metal and polymers along their interface and for stronger adherent composites. However, after polyolefin oxidation the low surface energy of the untreated material is only marginally increased, by 50–100% (Figure 7.7). Therefore, the large difference in surface energies of metals and polymers remains nearly unchanged (28 $mN\,m^{-1} \leftrightarrow \leq 1000$ Al and after pretreatment $\rightarrow$ 38 $mN\,m^{-1} \leftrightarrow \leq 1000$ Al). This nearly unchanged difference is the driving force for further (redox) reactions as reported below. Therefore, the interactions between metals and oxidized polyolefin surfaces have to be characterized as metastable. The chemical bonds formed (covalent bonds with high polar contribution) between metal and polymer are not the final reaction product; they are only an intermediate (metastable) step. Therefore, they undergo further reactions when additional activation energy is present. In conclusion:

- The desired (most efficient) chemical bonds must be stabilized to preserve them from hydrolysis, because these bonds are polar covalent ones, which tend to hydrolyze.
- They are, however, unstable for long runs because of the thermodynamics involved, which requires full balancing of surface energy at interface by redox reactions.

The peel strength of aluminium on polyolefins and poly(ethylene terephthalate) increases very strongly with the introduction of only low concentrations of O-functional groups. Further increase in peel strength was moderate. Thus, the peel strength versus plasma exposure shows another characteristic than the surface energy versus exposure time. It was argued that in contrast to test liquids for contact angle measurements (surface energy) low-molecular weight products (LMWOM) at polymer surface produced by the plasma exposure cannot be tolerated by metal layers. Such metal deposits are solid phases, which cannot dissipate, dissolve, or distribute such LMWOM in the metallic phase as is possible by the test liquid. The low-molecular weight products remain at metal–polymer interface as weak boundary interlayer and thus weaken the peel strength. For this comparison, the measured peel strength of Al–polyolefin and Al–poly(ethylene terephtha-

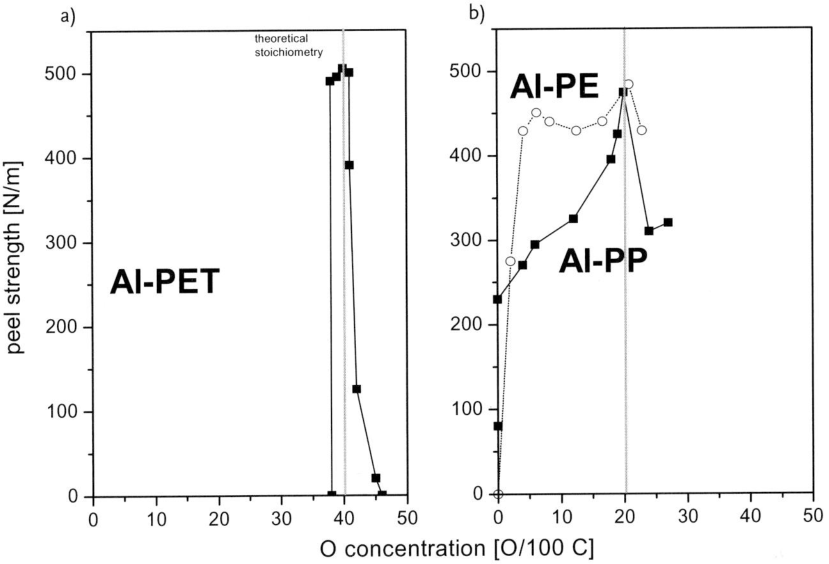

Figure 7.9 Dependence of Al peel strength of Al–polymers on XPS-measured O-concentration introduced by low-pressure oxygen plasma exposure (rf plasma, 6 Pa, 100 W).

late) composites [18] as well as tensile shear strengths of polypropylene-polyurethane composites [19] were also referenced to oxygen percentage at the surface of oxygen-plasma exposed polymers (Figure 7.9).

Indeed, differences between surface energy and peel strength are obvious. A strong deviation from the theoretical elemental composition of PET ($C_{10}O_4 = 40$ O/100 C) produced low peel strength, which may be also due to destruction of the original PET structure and formation of a weak boundary layer at the interface.

Oxygen plasma exposure of polyolefins may also produce extensively LMWOM, for oxidation degrees >20 O/100 C, that is not bonded to the polymer bulk. Therefore, metal deposits are not linked to the polymer substrate and the measured peel strength becomes lower for >20 O/100 C. However, this LMWOM can be removed before metal deposition by washing with solvents. The thus cleaned polymer surface shows a much lower concentration of functional groups. Comparison of both functions, the surface energy of PE and the Al peel strength of Al–PE versus oxygen concentration (Figure 7.9b), shows characteristic differences.

The removal of oxygen-plasma produced LMWOM from a polypropylene surface by diffusion and dissolving into the viscous phase of a polyurethane adhesive was expected for polyurethane–polyolefin composites. Indeed, such composites show another type of adhesion dependence on oxygen percentage (Figure 7.10).

Not until 20–25 O/100 C at polypropylene surface was a decrease in lap shear strength observed. The peel strength increases rapidly with O-concentration and achieves a maximal plateau from 5 to 25 O/100 C. The dielectric barrier discharge

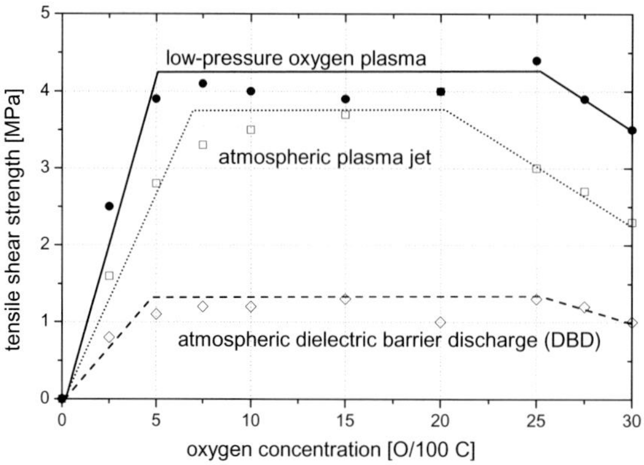

Figure 7.10 Tensile shear strengths of PU–PP specimens as a function of oxygen percentage at polypropylene surfaces introduced by atmospheric and low-pressure plasmas.

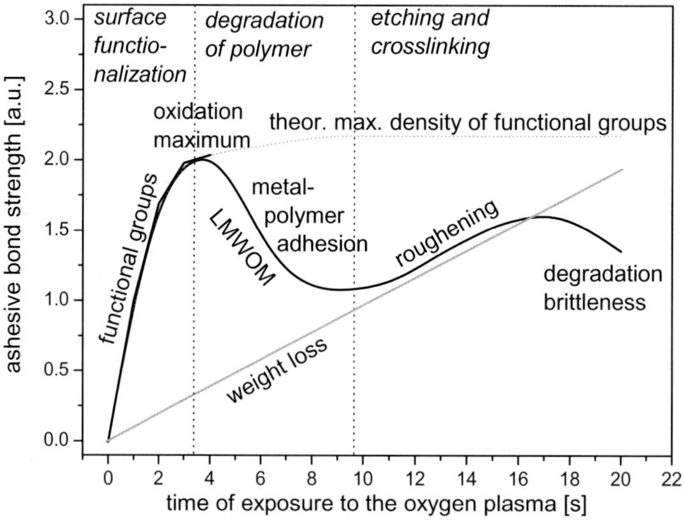

Figure 7.11 Dependence of the general behavior of metal–polymer adhesion on the duration/intensity of exposure to low-pressure glow discharge plasma.

(DBD) treatment produced lowest tensile shear strength. It is assumed that much degradation to LMWOM occurs using this type of atmospheric-pressure plasma, thus forming a pronounced WBL, as demonstrated by Strobel [24].

Figure 7.11 presents a general trend for plasma surface treatment of polymers with oxygen low-pressure plasma [10].

An increase in adhesive bond strength with concentration of O-functional groups is compensated or weakened by increasing oxidation and formation of loosely bonded LMWOM. This slight decrease in adhesive bond strength can be partially compensated by plasma etching and subsequent polymer roughening, thus allowing mechanical interlocking.

7.2.3
Reductive Plasma Pretreatment of Perfluorinated Polymers

Perfluorinated polymers cannot be oxidized by any plasma exposure chemically because fluorine is the strongest oxidation agent. Possible ways of removing fluorine are by sputtering or, more efficiently, by reducing the system followed by introduction of adhesion-promoting functional groups. Thus, fluorine-containing polymers can be roughened by exposure to noble gas plasmas (He, Ar) or can be reduced by plasmas (Na, K, H_2, NH_3) to carbon or hydrocarbons. The main disadvantage of all chemical and physical modification is the degradation of polymer backbones by C–C scissions:

$$\text{-CF}_2\text{-CF}_2\text{-} + Ar\ plasma \rightarrow 2\text{C} + 2\text{F}_2$$

$$\text{-CF}_2\text{-CF}_2\text{-} + H_2\ plasma \rightarrow \text{-CH}_2\text{-CH}_2\text{-} + 4\text{HF}$$

or:

$$\text{-CF}_2\text{-CF}_2\text{-} + H_2\ plasma \rightarrow 2\text{CH}_4 + 4\text{HF}$$

or:

$$\text{-CF}_2\text{-CF}_2\text{-} + Na\ plasma \rightarrow 2\text{C} + 4\text{NaF}$$

A few pretreatments of polytetrafluoroethylene (PTFE) for producing well-adherent Al–PTFE composites have been tested (Figure 7.12) [26]: (i) deposition of mono-type functional group-carrying plasma polymers as adhesion promoters on untreated PTFE, (ii) H_2 plasma pretreatment of PTFE followed by deposition of an adhesion-promoting plasma polymer layer, and (iii) H_2 plasma pretreatment of PTFE alone. Process (i) did not give any Al–PTFE bond strength because of the poor adhesion of the plasma polymers to the PTFE surface. Processes (ii) provided

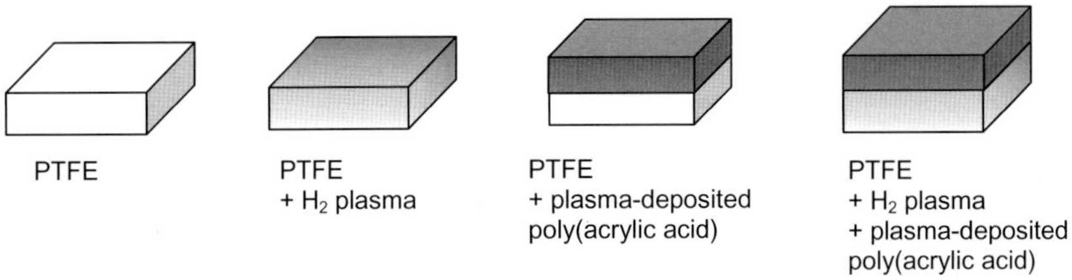

Figure 7.12 Different plasma pretreatments of polytetrafluoroethylene (PTFE) for coating with aluminium.

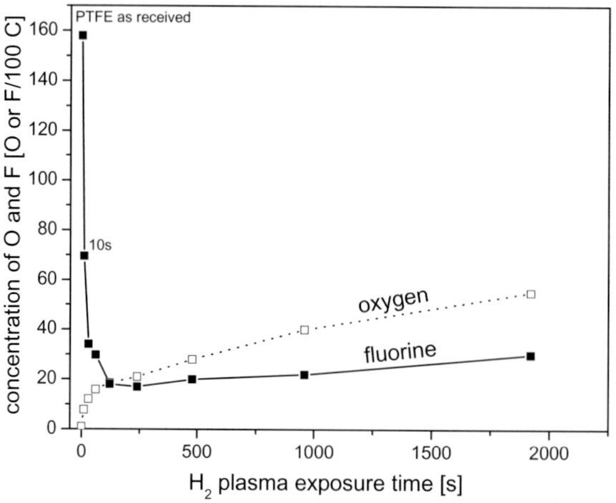

Figure 7.13 Fluorine removal and oxygen incorporation as a function of exposure time to H_2 plasma (power = 100 W, pressure = 6 Pa, mode = cw).

adequate and strongly adherent Al–PTFE systems, whereas process (iii) produced moderately to highly adherent Al–PTFE systems.

Similar to the results described in References [27–29], defluorination in a continuous-wave or pulsed (0.1 ms plasma "on," 0.9 ms plasma "off," 1000 Hz pulse frequency) hydrogen rf plasma leads to a minimum of fluorine atoms at the PTFE surface of 17 F/100 C (35 F/100 C in Reference [30]) and to 55 O/100 C (6.4 O/100C in Reference [31]) (Figure 7.13).

The introduction of oxygen is most likely due to post-plasma reactions of plasma-produced C radical sites at the PTFE surface with molecular oxygen from air during sample transfer from the plasma reactor to the XPS spectrometer [31, 32]. This post-plasma reaction cannot be avoided. The O/C ratio (*ex situ* XPS measurements) also increases with exposure time to the plasma, in particular for treatment times longer than 120 s (Figure 7.13). An explanation for the residual fluorine after prolonged plasma pretreatment may be the increasing temperature and, therefore, enhanced hydrophobic recovery [33]. On the other hand, a simple abrasion test (wiping with a cotton cloth) and the solvent stability test (6 h rinsing in tetrahydrofuran–THF) of the defluorinated PTFE surface show that this near-surface layer became unstable when exposed for longer than 10 s to hydrogen plasma. The respective XP spectra became very similar to the spectrum of untreated PTFE. Therefore, for further plasma-chemical processing and metal deposition a H_2 plasma treatment time of 10 s was chosen.

Characteristic changes in the bimodal shape of the C1s signals (CF_2 and CH_x) can be observed by prolonging the time of exposure to the hydrogen plasma (Figure 7.14). Indications of C–N related bonds are also found (10N/100 C).

The C1s peak shift to lower binding energy is characteristic for hydrocarbon and O bonded C species. This interpretation was also given in Reference [1, 34–37].

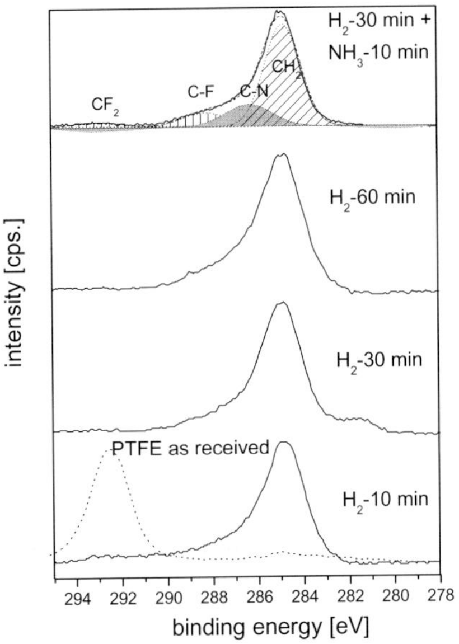

Figure 7.14 C1s peaks of PTFE after different exposures to hydrogen (and ammonia) plasma using the cw-rf mode (pressure = 6 Pa, wattage = 100 W).

Hydrogen plasma reduction of PTFE produces preferably a CH_x component with a C1s electron binding energy (BE) of 285 eV. A graphite-like species at BE = 284.5 eV or lower binding energies could not be detected. Therefore, it was concluded that a complete reduction of the PTFE backbone to loosely bonded graphite did not take place as sometimes is the case with alkaline metal reduction. It can be speculated that the hydrogen reduction at short treatment times (<10 s) is much gentler and, therefore, avoids the formation of a pronounced weak boundary layer.

The Al peel strength to hydrogen plasma pretreated and then metalized PTFE is shown as a function of plasma treatment time in Figure 7.15.

In comparison to metalized untreated PTFE the peel strength is significantly increased. Only 10 s pretreatment is sufficient to increase the Al peel strength to the maximum. However, very prolonged hydrogen plasma pretreatment (>1000 s) results in slightly lower peel strength. Then, failure was observed between Al and PTFE.

The introduction of an adhesion-promoting pulsed-plasma polymer interlayer onto the H_2-plasma pretreated (each 10 s) PTFE substrate improved the peel strength further to a range of 350–400 N m^{-1}, limited by the adhesion measurement conditions. The real bond strength is much higher.

The locus of failure during peeling was identified by analyzing the peeled 200 nm thick Al layer and the peeled PTFE surfaces using XPS. At maximum peel strength, the peeled PTFE surface showed an XPS C1s signal at a binding energy

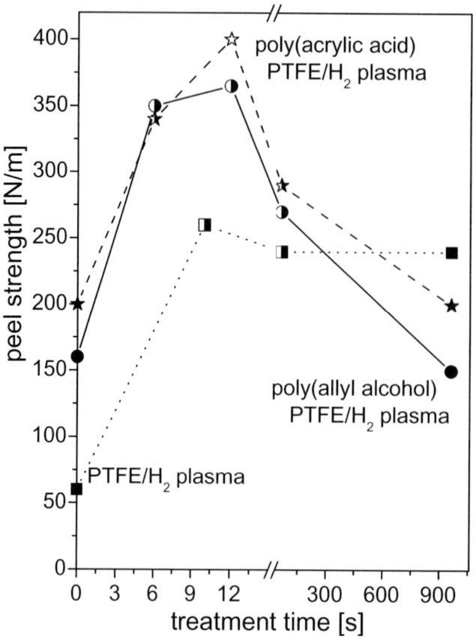

Figure 7.15 Aluminium peel strength as a function of pretreatment time and type of modification; filled symbols = interface failure between Al and PTFE, open symbols = interface failure between tape and Al, and half-filled symbols = mixed failure.

of 291 eV, which is characteristic for unmodified PTFE, and the peeled metal surface exhibits a signal near 285 eV, which was assigned to CH_x components. Thus, the peel front should have propagated along the interface between unmodified PTFE and a H_2 plasma modified PTFE near-surface layer. Adhesion-promoting poly(acrylic acid) interlayers between Al and hydrogen plasma pretreated PTFE also provoke peel propagation along the plasma polymer and the unmodified PTFE (Figure 7.16).

7.2.4
Adhesion Improvement Using Homo- and Copolymer Interlayers

The pulsed plasma-initiated radical polymerization of functional group bearing monomers was used to produce such a model surface. Table 7.1 presents the peel strengths of Al-plasma homopolymer–PP systems.

It was confirmed that layers of pulsed-plasma polymerized ethylene homopolymer did not promote adhesion to Al when applied in Al–PP systems as an adhesion-promoting interlayer. With NH_2 group-containing layers, very weak peel forces were produced, in contrast to the high peel strength using OH and especially COOH group-containing interlayers (Table 7.1 and Figure 7.17).

In the case of CH_2 and NH_2 groups, failures occur at plasma polymer–PP interface, as was shown by XPS of peeled surfaces. With OH groups the peel front propagates within the allyl alcohol plasma polymer. The adhesion of acrylic acid

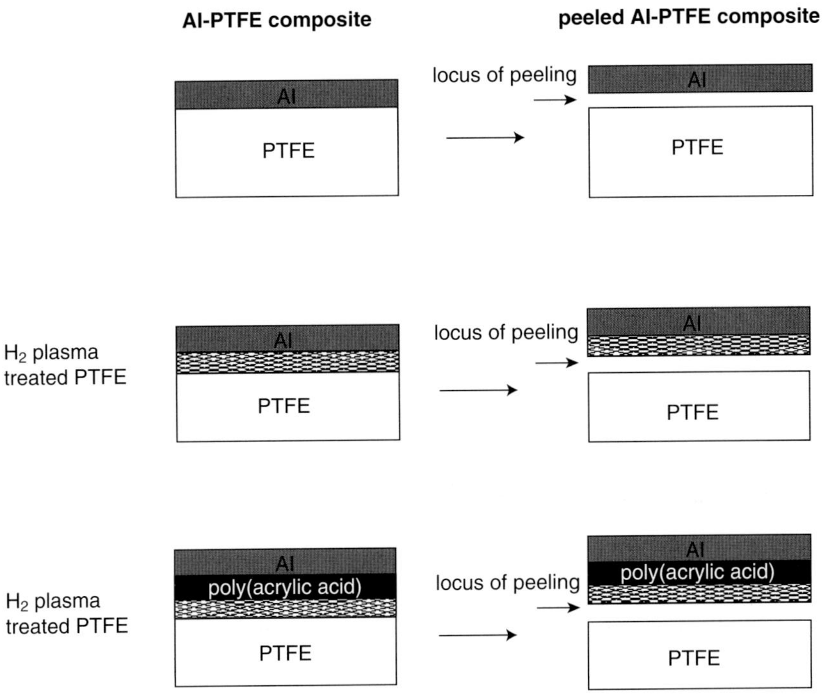

Figure 7.16 Schematics of the locus of failure during peeling of aluminium in different Al–PTFE composites.

Table 7.1 Variation of Al peel strength at PP with the nature of the adhesion-promoting plasma homopolymer interlayer.

Functional group	Peel strength ($N\,m^{-1}$)	Concentration	Type of failure	Location
CH_2	0–10	–	Interface	Plasma polymer–PP
NH_2	≈30	18 NH_2/100 C	Interface	Plasma polymer–PP
OH	≈85	31 OH/100 C	Cohesive	Inside plasma polymer
COOH	≈650	24 COOH/100 C	Cohesive	Inside PP

layers to PP as well as to Al was so strong that peeling occurs within the PP substrate, as demonstrated by XPS analysis of both peeled surfaces. Thus, the interactions between aluminium and the monotype functional groups of homopolymers depend strongly on the type of functional group in the order COOH >> OH >> NH_2 > CH_2.

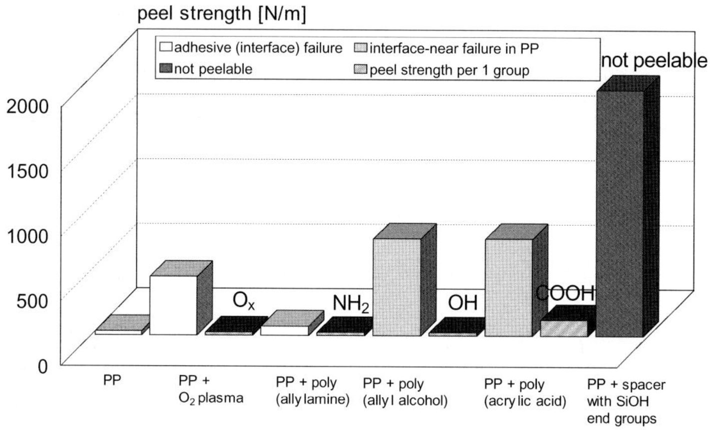

Figure 7.17 Dependence of Al–PP peel strength on plasma pretreatment and the increment realized by the contribution of functional groups/100 C.

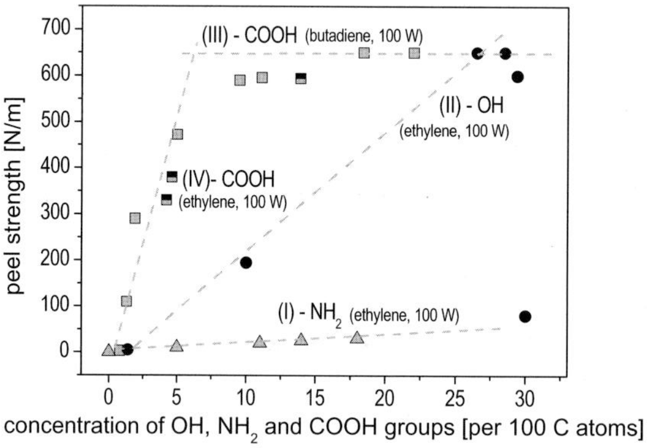

Figure 7.18 Aluminium peel strength of Al–plasma copolymer–PP composites as a function of functional group density at the plasma-polymer surface (gray filled squares: copolymers of acrylic acid and butadiene; half-filled squares: acrylic acid and ethylene; filled circles: allyl alcohol and ethylene; gray filled triangles: allylamine and ethylene).

Copolymers made from co-monomers with and without functional groups allow us to vary the concentration of functional groups in accordance with the co-monomer ratio. Allyl alcohol–ethylene plasma copolymers show the maximum in adhesion (650 N m^{-1}) at a concentration of 27–29 OH per 100 C atoms. Therewith, this copolymer produces a significantly higher peel strength than the allyl alcohol homopolymer (85 N m^{-1}) (Figure 7.18).

A nearly linear dependence of peel strength on the density of OH, NH_2, and COOH functional groups was observed in the range of 0–27 OH, 0–15 NH_2, and 0–5 COOH/100 C (Figure 7.18). Using poly(allyl alcohol) copolymers, the peel front propagates along the interface plasma-polymer–PP (interfacial failure) at OH concentrations of 1–25 OH/100 C. A plateau of maximum peel strength (650 N m^{-1}) at concentrations of 27–29 OH groups/100 C atoms was observed. At these concentrations, partially or complete cohesive failures (inside the plasma copolymer) occurred at peeling. The pure allyl alcohol plasma polymer is tacky and weak and shows a low cohesive strength, resulting in low peel strength and a pure cohesive failure within the allyl alcohol homopolymer layer. NH_2 groups also showed a pure interfacial failure at the Al–plasma polymer interface and the lowest maximum peel strength to Al.

COOH groups produce the highest peel strength to Al (Figure 7.18). The Al peel strength depends linearly on the concentration of functional groups in the range 0–5 COOH/100 C. Higher concentrations of COOH groups (>5 COOH/100 C) do not increase the peel strength further. This is characterized by pure cohesive failures within the polypropylene substrate [38]. This was also confirmed by producing copolymers of different composition made from poly(vinyl chloride) and maleic anhydride. These copolymers show a characteristic dependence of peel strength of evaporated aluminium topcoats on the concentration of carboxylic groups in each copolymer [1]. However, such bulk-modified copolymers of different composition also vary in their bulk properties, which also influence the measured peel strength. In contrast, polymer surface functionalization of polyolefins or deposition of ultrathin adhesion-promoting layers on them does not change the bulk properties of the polymer substrate.

Remember that in the case of O_2 plasma treatment OH, COOH, and other O species are produced. There is a difference between peel strength using O_2 plasma treatment (400–450 N m^{-1}) and OH-group containing plasma polymer interlayers (600–650 N m^{-1}). Moreover, increasing the yield of OH groups of the O_2 plasma treatment by post-plasma reduction using diborane to a maximal 14 OH/100 C (after O_2 plasma treatment 3 OH/100 C) could be realized, which is much lower the 30 OH/100 C of the allyl alcohol copolymer. However, the maximum peel strength was not further improved.

It can be assumed that excessive O_2 plasma modification causes polymer degradation at the surface, thus forming a weak boundary layer. This was concluded by the abrupt decrease in peel strength when the O concentration exceeded 20 O/100 C. Polymer degradation starts with the beginning of plasma exposure [10] and is caused by irradiation with vacuum UV (VUV) radiation ($\lambda < 200$ nm, photo-oxidation) and to a much lower extent by plasma particle bombardment [38]. Consequently, the adhesion-improving effect of introducing functional groups and the polymer degradation superpose. Plasma polymer deposition avoids more or less such degradation.

7.3 New Adhesion Concept

Adhesion problems between metal layers and polymers exist in various electronic or decorative assemblies or other types of metal–polymer systems. The reasons

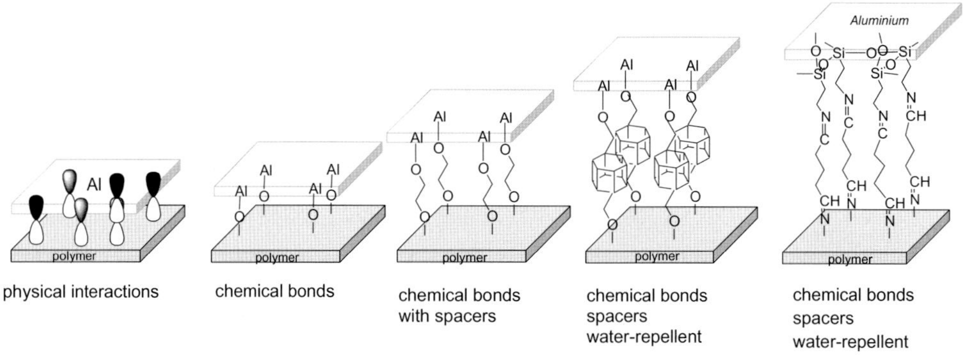

Figure 7.19 Schematics of adhesion promotion in metal–polymer systems.

are the different natures of materials and the absence of adhesion-promoting functional groups, the very different thermal expansion coefficients of polymers and metals, and the water ingress along the interface. The dominance of weak physical interactions and the deficiency of strong covalent bonds often lead to interface failures during mechanical loading (shearing or peeling). Here, the introduction of covalently bonded organic spacers of different chain lengths, stiffness, and end-groups into the metal–polymer interface was realized to improve the adhesion bond strength. The spacer molecules compensate the mechanical stress along the interface and flexibilize and lower the migration of moisture along the interface by introduction of hydrophobic, water-repellent spacers such as pure aliphatic chains or siloxane units.

To introduce spacers the polymer surface was plasma-chemically functionalized and then grafted under wet-chemical conditions. To establish chemical bonds between metals, such as aluminium, and non-polar polymer (polyolefin) surfaces, such as polypropylene (Figure 7.19), chemically reactive groups must be introduced onto the polymer surface. Such groups for enhancing the adhesion to aluminium were found to be COOH and OH groups, which can react with metallic aluminium by formation of covalent (bidentate) or ionic bonds [39].

Even unmodified PP and PE surfaces react with evaporated Al atoms by charge transfer from Al to carbon (Al–C) [40–42]. The role of chemical interactions between Al and primary amino groups is more controversial. Amino groups form a σ complex $Al \rightarrow NH_2^{\cdot}$, which may re-arrange to Al–N [43]. OH groups in poly(vinyl alcohol) (PVA) react with $(CH_3)_3Al$ and also with metallic Al to form Al–O–C complexes [44–46]. It could be shown that all OH groups present at the PVA surface were consumed by Al–O–C formation. In terms of metal–organic chemistry Al–O–C bonds are alcoholates. Other authors interpret the electron transfer from Al to different functional groups present at a polymer surface as acid–base interactions [47]. Moreover, Al undergoes a redox reaction with O functional groups present at O_2 plasma-modified polypropylene surfaces and forms Al_2O_3

and thus reduces the O functional groups along the metal–polymer interface [48]. Aluminium adheres very strongly to COOH group-modified polymers as Al–O–CO groups [38, 42, 49–51] in mono or bidentate form [42, 49–51]:

$$\mathrm{Al + HOOC{-}R \rightarrow Al{-}O{-}CO{-}R} \quad \text{monodentate}$$

$$\mathrm{2Al + HOOC{-}R \rightarrow (Al{-}O)_2CO{-}R} \quad \text{bidentate}$$

$$\mathrm{2Al + HOOC{-}R \rightarrow Al(O)_2CO{-}R} \quad \text{chelating bidentate}$$

At low pressure, evaporated Al atoms (*in statu nascendi*) are very reactive towards the monotype functional groups. Therefore, the goal was to replace weak physical interactions with strong chemical interactions but hinder any extensive redox reaction or hydrolysis of formed metal–polymer bonds.

Metal–polymer interfaces are characterized by the abrupt transition of the metal phase with it specific properties (surface energy, thermal expansion coefficient) to that of the polymer as mentioned before. Now, all mechanical stress under mechanical load or caused by the strongly different thermal expansion coefficients of metals and polymers (two orders of magnitude) is focused on the chemical bonds across the interface and neighboring polymer layers. Direct linking of metal to polymer by covalent bonds makes the interface inelastic, stiff, and, therefore, brittle (cf. Figure 7.19). All mechanical stress is transferred to the interface-near polymer molecules. In addition, these interface-near molecules are often degraded by pretreatment or redox reactions. Obviously, therefore, this stress should be distributed and compensated by inserting long and flexible spacer molecules instead of direct chemical bonding (Figures 7.19 and 7.20).

By using flexible spacer molecules along the metal–polymer interface a mechanical failure caused by thermal expansion or shrinking can be partially avoided.

With growing length of spacer molecules, the adhesion was increased (as demonstrated later). Simultaneous attachment of spacers with two different chain

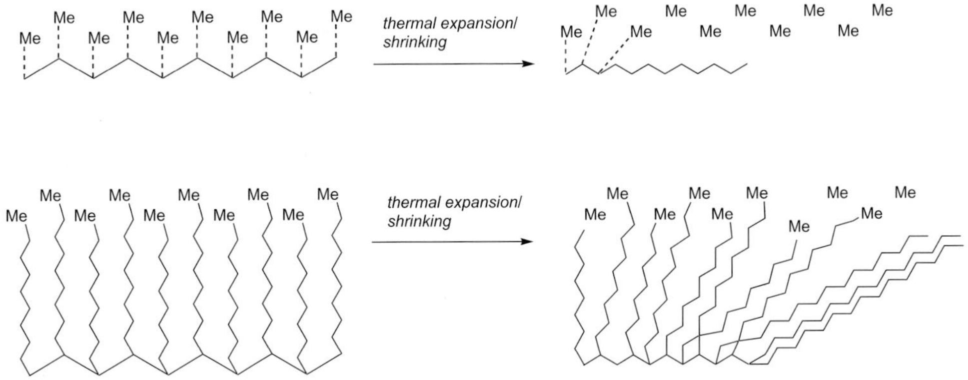

Figure 7.20 Shrinking–expanding of a metal–polymer at the interface, with or without spacers.

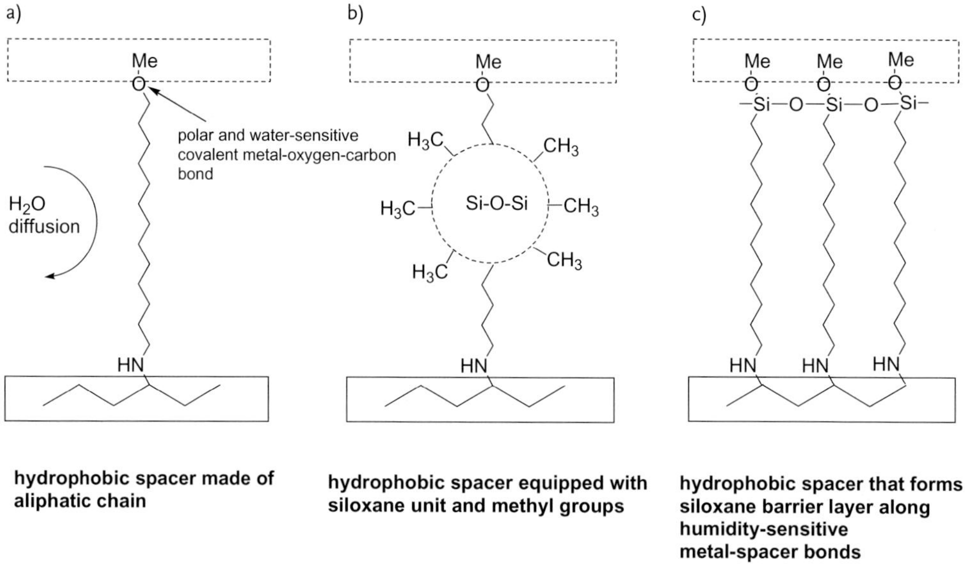

Figure 7.21 Variants of water-repelling elements placed within the interface by insertion of tailor-made spacer molecules.

lengths may produce a two-step breaking under mechanical load, thus, hindering abrupt failure.

Introduction of additional water-repellent elements within the flexible spacer hinders the hydrolysis of metal–polymer bonds, which is essential for the long-term durability of metal–polymer composites (Figures 7.19 and 7.21).

These water-repellent units should preferably consist of siloxane units equipped with methyl, isopropyl, or *tert*-butyl groups. Fluorine-containing segments may undergo redox reactions with metal atoms:

$$\text{-CF}_2\text{-CF}_2\text{-} + \text{Al} \rightarrow \text{AlF}_3 + \text{C}_2\text{F}$$

Since redox reactions are associated with ion transportation from metal to polymer (across the interface) an intermolecular linked siloxane-containing barrier layer was introduced to hinder such ion diffusion across the interface (Figure 7.21b,c). This architecture was achieved by coupling of aminosilane onto aldehyde-modified polymer surfaces or by coupling of aminosilane to the brominated polyolefin surface as described later.

Different types of spacer molecules were grafted onto the polyolefin surface. The plasma-introduced monosort functional groups such as NH_2 or Br groups were used as anchoring points. Ethylene glycols, diols and alcohols, and diamines and amines with different end groups, chain length, and branching degree were introduced as spacers. Thus, the number of CH_2- or CH_2-CH_2-O- units in ethylene glycols were varied from 1 to about 100, and those of diols and diamines from 2 to 10 [52]. Different spacer end-groups were used (NH_2, OH, COOH, CHO, SH, phenol, methoxy) and the spacer structure was varied from stiff to flexible or branched (Figure 7.22).

It should be remembered that C-Br groups can be converted into other functional groups, such as OH, NH_2, CN, NO_2, and so on, in a simple chemical way. Azide groups (CN_3) introduced by consumption of C-Br with NaN_3 have served as the starting point for click-chemistry [53–54]. Moreover, atomic transfer radical polymerization (ATRP) was achieved and substitution by organic–inorganic POSS (polyhedral oligomers of silsesquioxanes) molecules in noticeable quantities (cf. Figure 7.22).

All spacer molecules were introduced by nucleophilic substitutions onto polyolefin surfaces or by electrophilic addition onto (fully substituted) aromatic double bonds of graphene (absence of H, which can be substituted).

At most, each fifth carbon atom of polyolefin surface is spacer-substituted with small molecules and large-volume molecules are linked to each 100th carbon. The largest molecules linked to the surface were poly(ethylene glycol) 5000, octaaminophenylene-T8-POSS, and PAMAM [poly(amido amine)–amino group-carrying dendrimers of third, fourth, and fifth generation].

POSS was introduced for water repellence. However, the bonding of aminosilane (3-aminopropyltrimethoxysilane) was simpler and more efficient than POSS introduction. For this purpose the polypropylene was coated with a thin layer (50 nm) of plasma-deposited poly(allylamine). The primary amino groups could be grafted by glutaraldehyde:

$$\blacksquare + \text{allylamine} + \textit{plasma polymerization} \rightarrow \blacksquare\!\boxdot\text{-NH}_2$$

$$\blacksquare\!\boxdot\text{-NH}_2 + \text{OHC-(CH}_2)_3\text{-CHO} \rightarrow \blacksquare\!\boxdot\text{-N=CH-CH}_2)_3\text{-CHO}$$

$$\blacksquare\!\boxdot\text{-N=CH-(CH}_2)_3\text{-CHO} + \text{H}_2\text{N-R-Si(OCH}_3)_3$$
$$\rightarrow \blacksquare\!\boxdot\text{-N=CH-(CH}_2)_3\text{-CH=N-(CH}_2)_3\text{-Si(OH)}_3$$

$$\blacksquare\!\boxdot\text{-N=CH-(CH}_2)_3\text{-CH=N-(CH}_2)_3\text{-Si(OH)}_3 + \text{Al}$$
$$\rightarrow \blacksquare\!\boxdot\text{-N=CH-(CH}_2)_3\text{-CH=N-(CH}_2)_3\text{-Si(OH)}_2\text{-O-Al}$$

Intermolecular condensation of the Si(OH) groups can also occur:

$$\text{Si(OH)}_2 \rightarrow \text{Si-O-Si} + \text{H}_2\text{O}$$

Alternatively, the plasma bromination of polypropylene is simpler and produces the same result:

$$\blacksquare\text{-Br} + \text{H}_2\text{N-CH}_2)_3\text{-Si(OCH}_3)_3 \rightarrow \blacksquare\text{-NH-(CH}_2)_3\text{-Si(OCH}_3)_3$$

$$\blacksquare\text{-NH-(CH}_2)_3\text{-Si(OCH}_3)_3 + \text{H}_2\text{O} \rightarrow \blacksquare\text{-NH-(CH}_2)_3\text{-Si(OH)}_3 \rightarrow$$

$$\blacksquare\text{-NH-R-Si(OH)}_2\text{-O-Al}$$

The formed silanol groups may be responsible for extra strong adhesion-promotion to aluminium coatings [55]. The remaining silanol groups can react intermolecularly to form Si–O–Si bonds. The thus-modified Al–PP composites could not be

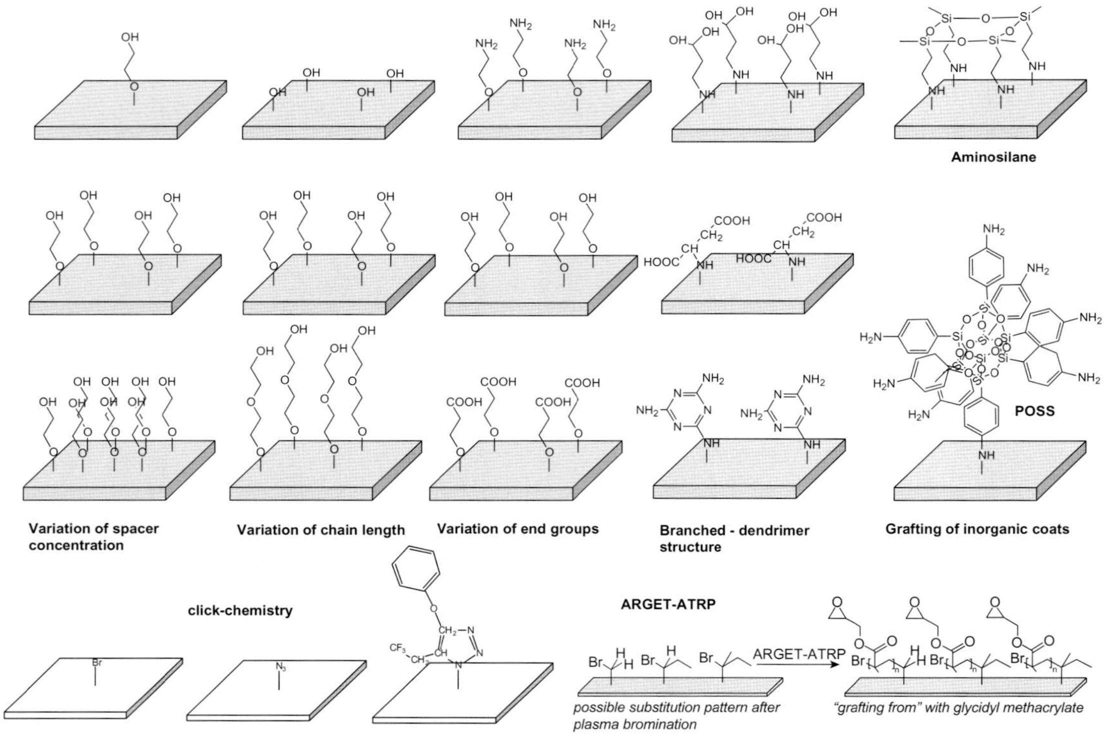

Figure 7.22 Realized architectures produced by grafting of spacer molecules onto plasma-brominated polypropylene surfaces (or allylamine plasma polymer coated surface).

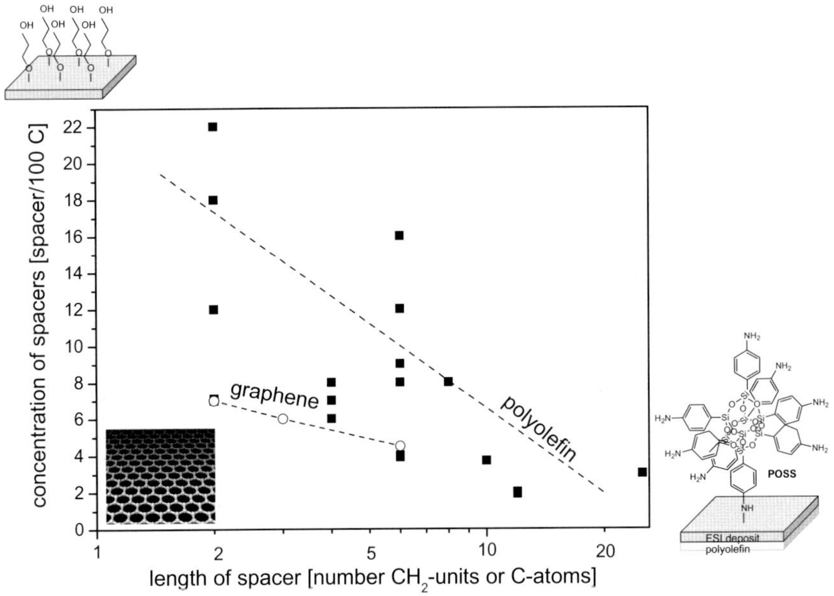

Figure 7.23 Maximal achieved spacer densities (concentration) at polyolefin and graphene surfaces by nucleophilic or electrophilic substitution.

peeled because of the strong adhesion. Moreover, the adhesion was durable for 6 months on exposure to ambient air with 60% humidity. It is assumed that after hydrolysis the formed silanol groups are responsible for interactions with the metal (possibly Al–O–Si bonds) and for barrier formation by partial condensation (Figure 7.21c).

The concentration (graft density) of wet-chemically grafted organic spacer molecules on brominated polypropylene or polyethylene surfaces shows a variance. Averaging gives a linear dependence of spacer concentration/density on spacer length/molecular volume as depicted in Figure 7.23.

The peel strength of aluminium from spacer-equipped polypropylene surfaces was maximal for COOH end-groups (Figures 7.24 and 7.25). About two C_{10} spacers with COOH end groups per 100 C on a polypropylene surface produced high peel strength and cohesive failure, when aluminium layers were peeled off. Linear, branched, and dendrimer structures show the same results if equipped with COOH end groups (Figure 7.24). OH groups containing spacers also promote adhesion but are best in long-chain spacers.

Moreover, the stiffness of spacer molecules were changed from flexible aliphatic chains to stiff molecules, thus showing the advantage in measured peel strength of using flexible molecules as spacers in Al–polyolefin composites [52]. The concentration of grafted POSS and PAMAM-dendrimer molecules as well as PEG oligomers (500–5000 g mol^{-1}) amounts to 1–2 molecules per 100 C or 0.4–0.8

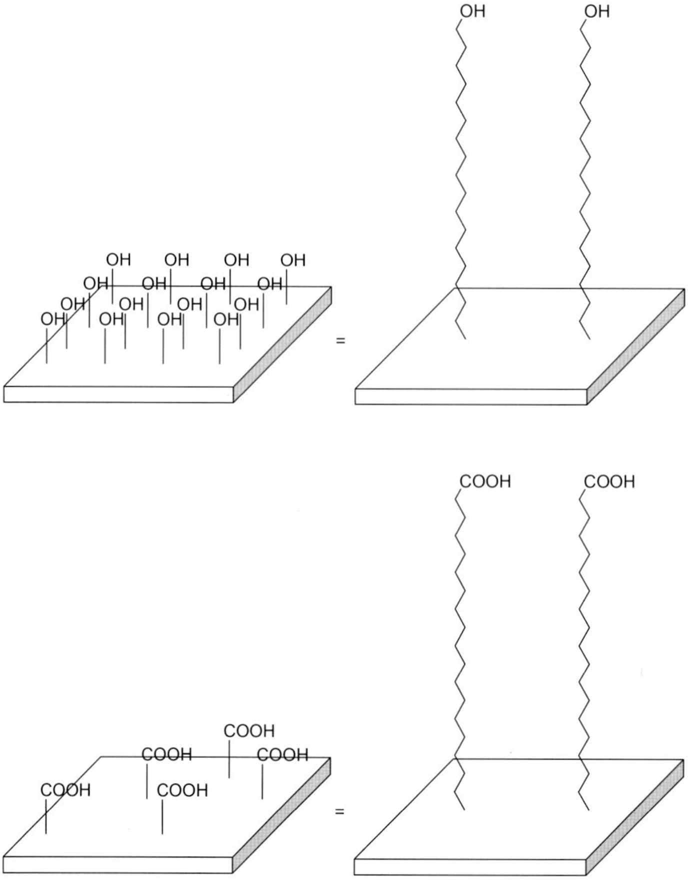

Figure 7.24 Schematics of the density of functional groups or spacers at a polypropylene surface that is necessary to produce maximal peel strength of thermally evaporated aluminium thin films (cohesive failure during peeling).

molecules nm^{-2}. Incorporation of SiO_x structures by grafting of POSS enhanced the durability of Al–PP composites. POSS molecules covalently bonded to the polyolefin surface are water-repellent, in particular when the SiO_x cage is substituted by isopropyl or *tert*-butyl groups, as shown by enhanced durability towards moisture.

7.4 Redox Reactions along the Interface

Evaporating metal atoms in vacuum and depositing them onto the polymer surface produces direct contact between metal atoms and functional groups or building

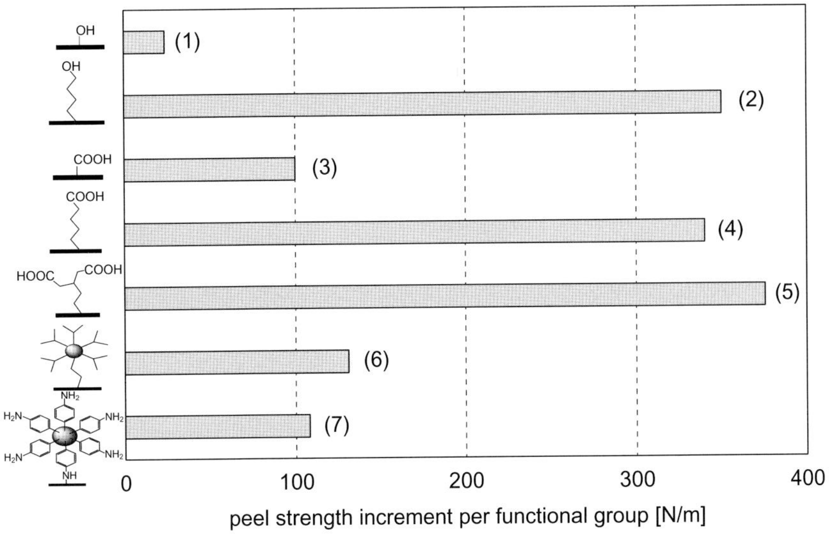

Figure 7.25 Contribution of spacer molecules with the shown architecture and end group to the measured peel strength; PAMAM dendrimers (5) and octaaminophenylene-T8-POSS (6).

blocks of the polymer, as already shown for hydrogen plasma exposure of polytetrafluoroethylene (PTFE). Strong electronegative metals, such as aluminium, magnesium, or alkalines, are very reactive towards functional groups and tend to form M–O–C or M–Hal bonds or oxides (M_xO_y) or halides (M_xHal_y). In this way, functional groups and building blocks were destroyed [13, 48, 52, 56]. Redox reactions are often accompanied with scission of covalent bonds of the polymer backbone. A further example of redox reactions at metal–polymer interface is that between alkaline metals and PTFE:

$$\text{-[CF}_2\text{-CF}_2]_n\text{-} + 4n\text{K} \rightarrow 2n\text{C} + 4n\text{KF}$$

This can be seen from the C1s XPS signal (Figure 7.26) [56, 57].

As mentioned before aluminium picks up oxygen from the polymer to form Al_2O_3, and also attacks PTFE to form AlF_3 (Figure 7.27) [48].

Conversely, noble metals with positive redox potential have the tendency to convert their oxide or sulfide layers into inferior oxides or to metal under oxidation of the polymer interface, such as CuO on Cu in contact with PE [57–59]:

$$2\text{CuO} + \text{polyethylene} \rightarrow \text{Cu}_2\text{O} + \text{polyethylene-O or } 2\text{Cu} + \text{polyethylene-O}_2$$

Typical oxidation products contain carboxylic groups, such as copper carboxylates [58]. CuO from printed circuit boards also had oxidized polyimide [59] and poly(phenylquinoxaline) [13], including its aromatic rings, at the interface.

Aromatic rings and other π-electron-containing bond systems in polymers can be attacked oxidatively [13] or by transition metals under preliminary formation of

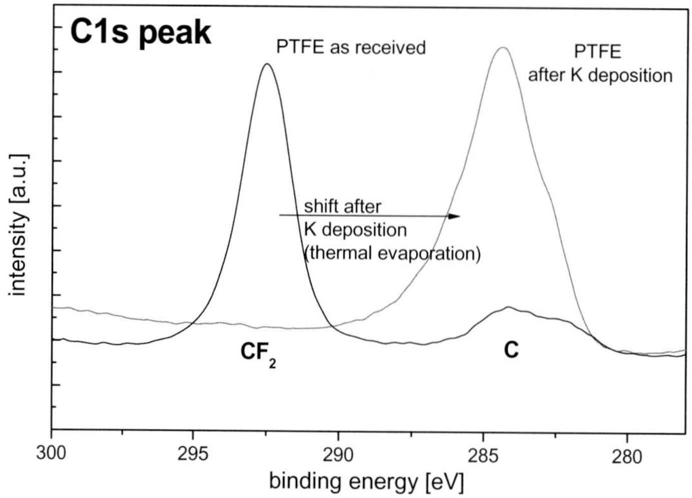

Figure 7.26 C1s signals before and after thermal K evaporation onto PTFE [18].

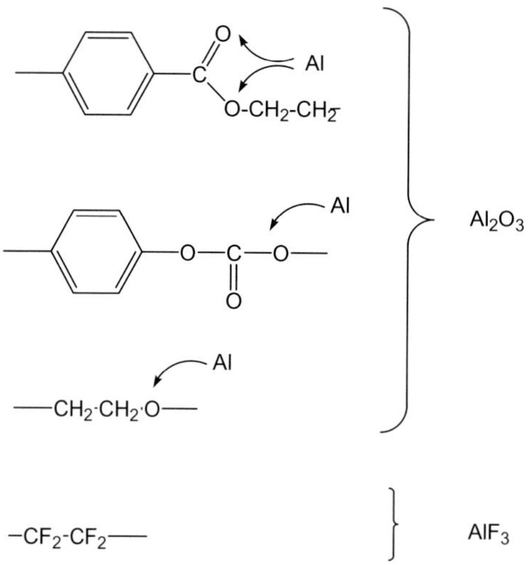

Figure 7.27 Examples of reduction reactions along the aluminium–polymer interface, evaporating thin films of aluminium.

d_{π}–p_{π} interactions [56]. Transition metals, such as chromium, form such complexes. Poly(ethylene terephthalate) formed such complexes, which were converted into σ-bonded Cr to the ring and finally converted into chromium carbide (Scheme 7.1) [56]. Additionally, a redox reaction between PET and the Cr was also observed (Scheme 7.1).

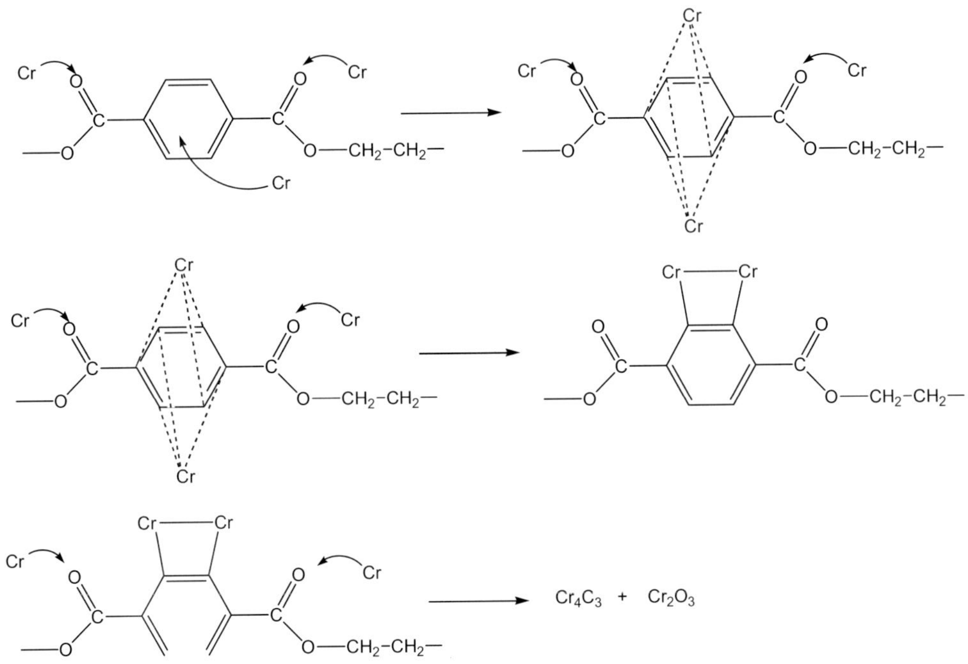

Scheme 7.1 Proposed reactions between Cr and aromatic rings of poly(ethylene terephthalate).

Using polystyrene the formation of sandwich-complexes with aromatic phenyl rings was evidenced on employing commercial Cr complexes as reference and near-edge X-ray absorption fine structure (NEXAFS) spectroscopy for their identification (Scheme 7.1) [60, 61].

Moreover, conversion of the Cr-complex into chromium covalently bonded to aromatic rings at two positions to afford a two-ring system is shown schematically in Scheme 7.1 for PET and in Figure 7.28 through NEXAFS spectra for polystyrene [56].

The appearance of a new C1s peak of Cr-evaporated polystyrene at a binding energy of 283.1 eV as well as 283.5 eV for bisphenol-A polycarbonate and 283.4 eV for poly(ethylene terephthalate) (not shown) were assigned to the formation of chromium carbide, the end-product of all Cr attacks on aromatic systems [62].

The methyl side-groups at the aromatic ring hinder the attachment of Cr as side-gated second cycle. This was evidenced by blocking such groups using poly(2,6-dimethyl-1,4-phenylene oxide) (PDMPO; Figure 7.29).

Independently of these d_π–p_π interactions redox reactions also occur, as seen by the pick-up of oxygen from the polymer associated with Cr(III) oxide formation (observed in the O1s peak) (Scheme 7.1).

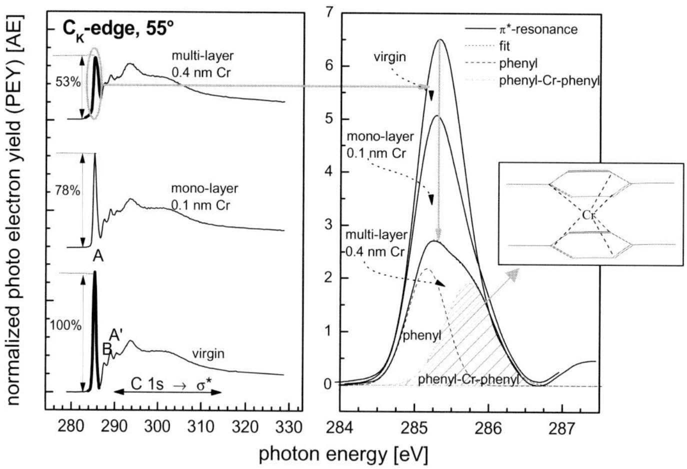

Figure 7.28 NEXAFS spectra C1s→π* transitions of virgin, mono-, and poly-atom Cr layers at the surface of polystyrene.

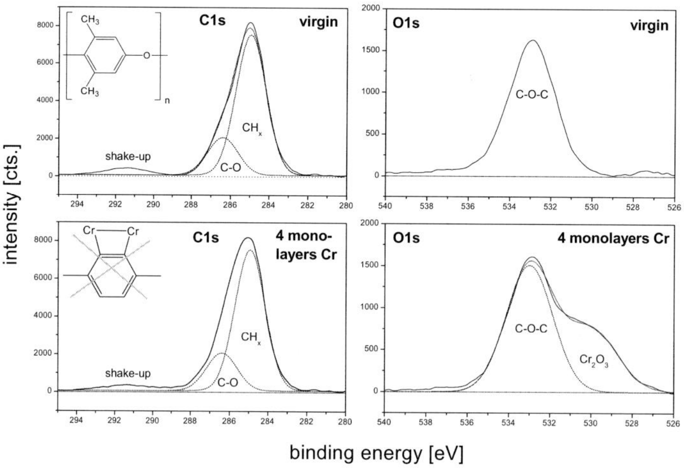

Figure 7.29 XPS C1s and O1s signals of poly(2,6-dimethyl-1,4-phenylene oxide) (PDMPO), with and without Cr monolayers.

7.5
Influence of Metal–Polymer Interactions on Interface-Neighbored Polymer Interphases

Polymer surfaces in contact with metals show often a new supermolecular structure that differs from both the original and the remaining and unaffected bulk

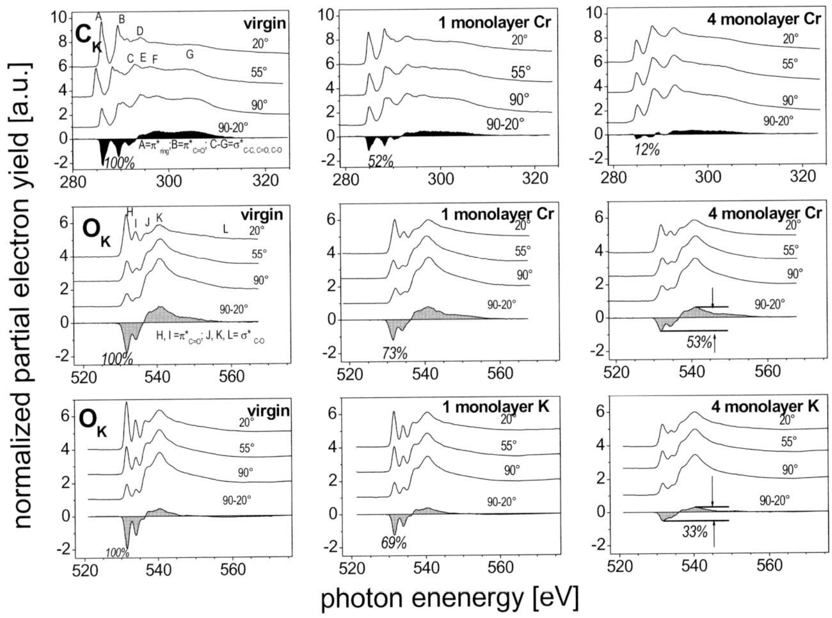

Figure 7.30 Variation of NEXAFS C_K and O_K edge spectra with chromium mono- and poly-layers.

structure. A prominent example is the "trans-crystalline structure" of polymers along the metal–polymer interface, as assigned by Schonhorn and coworkers and occurring on the micrometer scale [63, 64]. Such a special surface orientation in micrometer dimensions, provoked by orientation affected by the interface, was also detected using "etch gravimetry" by manifesting more etch resistance (lower etch rate) towards oxygen plasma than the unaffected polymer bulk material [65].

However, sometimes more dramatic changes in molecular orientation occur on the nanometer scale, caused by interfacial reactions, as discussed already when looking at interfacial redox reactions. Thus, it could be shown by means of NEXAFS spectroscopy that biaxially stretched poly(ethylene terephthalate) (PET) loses macromolecular orientation in the approximately 3 nm thick contact zone with chromium or potassium (Figures 7.30 and 7.31) [56].

The difference spectrum 20°–90° characterizes the molecular orientation and the greater the difference is (area in difference spectrum) the higher is the anisotropy. It should be considered that the information depths are about 3 nm for the C_K-edge and 5 nm for the O_K-edge (cf. Figure 7.31).

Interestingly, the loss in orientation near the interface (nm range) to metals is similar to that of oxygen-plasma exposure used for increasing the adhesion property (polarity) of a polymer surface. Both processes produce fast and complete loss of polymer orientation in the direct neighborhood of the interface (Figure 7.31).

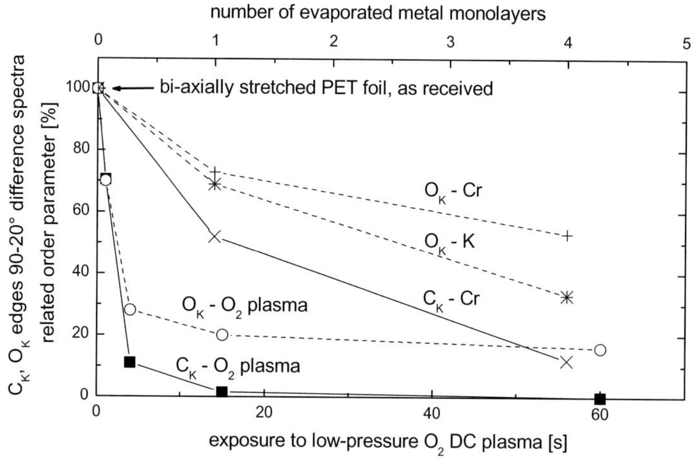

Figure 7.31 Variation of NEXAFS-determined order parameter of biaxially stretched PET with metal thickness and oxygen plasma exposure.

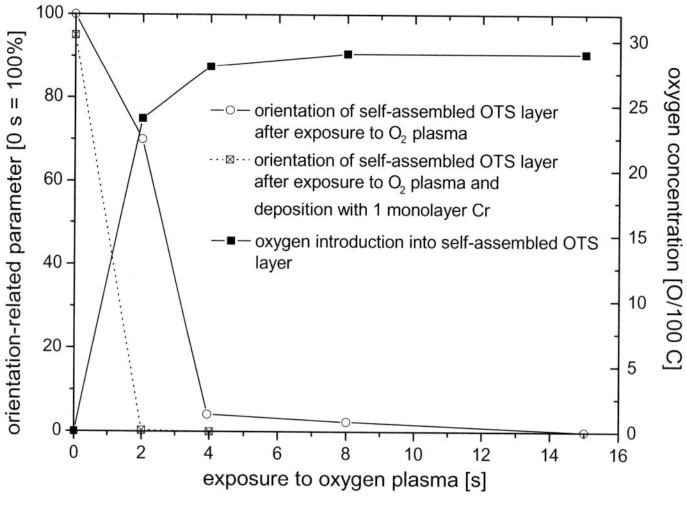

Figure 7.32 Loss of orientation and oxygen introduction of self-assembled monolayers of octadecyltrichlorosilane exposed to oxygen plasma and Cr deposition.

The changes in structure and orientation become maximal when the oxygen plasma pretreated polymer surface meets aluminium atoms during thermal evaporation under low-pressure conditions [66].

Using self-assembled monolayers (SAMs) made from octadecyltrichlorosilane (OTS) as models of paraffin or polyolefin surfaces strong disorientation is caused by short-time exposure to low-pressure oxygen plasma (Figure 7.32).

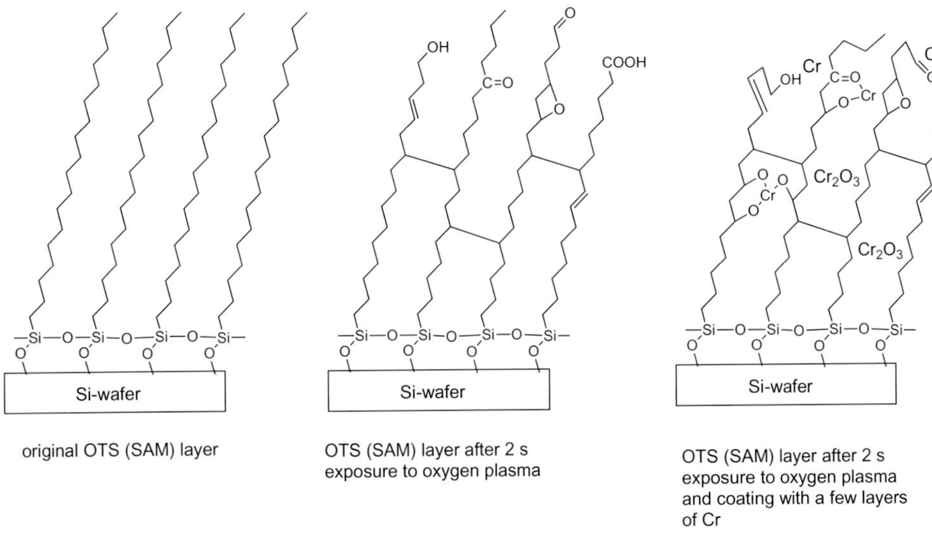

Figure 7.33 C_{18}-self assembled monolayers (OTS) on Si-wafer surface before and after exposure to oxygen plasma and chromium vapor.

Much more enhanced disorientation of the OTS-SAM film is observed after additional evaporation of Cr, which causes redox reactions with the oxygen plasma-introduced O-functional groups. This was demonstrated by evaluation of angle-dependent NEXAFS C_K-edge spectra (Figure 7.32) [67].

The NEXAFS spectra show the complete loss of orientation of OTS molecules after combination of both 2 s oxygen plasma exposure + 1 monolayer Cr-coating. Thus, extraordinarily strong redox reactions were initiated between plasma-introduced O-containing groups and metallic chromium to give Cr_2O_3 and reduction of plasma-introduced O-functional groups (Figures 7.32 and 7.33) [68].

7.6 Metal-Containing Plasma Polymers

Metal-containing plasma polymers have been produced by simultaneous thermal evaporation of metals and deposition of plasma polymers, thus simultaneously embedding metal clusters into the plasma polymer [69–71]. In this way plasma polymers with special properties were produced, such as highly unsaturated conducting plasma polymers with simultaneous embedding of dopants. Another variant of producing metal-containing plasma polymers consists of plasma polymerization of metal–organic precursors [72–76]. Kay and Dilks [77] or J. W. Coburn [78] may have been the first to describe metal-containing plasma polymers produced by sputtering of metal from the inner electrode of the plasma system. One example is the plasma polymerization of the C_2 hydrocarbons in the presence of

$Fe(CO)_5$ vapor, which can lead to the formation of soluble cluster complexes (I) and/or high oxidation state iron both in the form of the oxide (II) and associated with the carboxylate or β-diketonate iron (III). The relative yield of (I) with respect to (II + III) can be controlled, at will, by variation of the Yasuda factor W/FM [79]. In another example the metals tellurium and bismuth, dispersed in plasma polymers derived from CS_2 and styrene, were incorporated in the amorphous state [80]. An overview on metal-nanoparticles in polymers has been presented by Heilmann [81]. Munro *et al.* had investigated the structure of such plasma deposits of metal–polymer combinations by means of XPS [82, 83].

Another way to produce metal-containing plasma polymers is the combination of polymer and metal sputtering [84, 85].

7.7
Plasma-Initiated Deposition of Metal Layers

PECVD (plasma chemically enhanced vapor deposition) was used to deposit metal layers from metal–organic precursors. The first reported deposition was that of Ni from nickel tetracarbonyl [86, 87]. The $Ni(CO)_4$ was plasma chemically dissociated to deposit a Ni metal layer:

$$Ni(CO)_4 + \textit{plasma} \rightarrow Ni\downarrow + 4CO\uparrow$$

Silver- [88], boron- [89], copper- [90], barium-, yttrium, erbium-, and europium- [91], high-temperature superconductors [92], and tin-containing layers [93] were prepared in Suhr's group at the University of Tübingen by synthesis of numerous volatile metal–organic precursors and their decomposition under plasma exposure to metal films.

7.8
Inspection of Peeled Surfaces

After peeling, the peeled polymer and metal surfaces were analyzed to recognize the locus of peeling. Figure 7.34 shows a schematic sketch of interface analysis.

Figure 7.35 shows the peeling of two composites made of Al top layer–plasma-poly(allyl alcohol)–polypropylene substrate and Al top layer–plasma-poly(allyl alcohol)–ethylene copolymer–polypropylene substrate.

The peel failure in the composite Al–copolymer–PP propagates along the copolymer–PP interface. Using pure plasma poly(allyl alcohol) as primer the peel front propagates within the primer layer.

Using pulsed-plasma polymerized poly(acrylic acid) (pp-PAA) as adhesion-promoting interlayer the composite fails within the polypropylene substrate (Figure 7.36). Both the PP surface and the Al surface of peeled composite show the C1s peak of pure PP and no indications for the occurrence of the characteristic C1s signal of poly(acrylic acid).

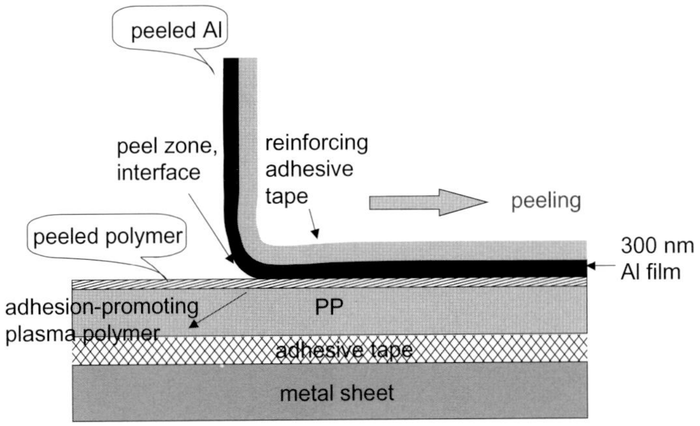

Figure 7.34 Schematic of XPS inspected peeled polymer and metal surfaces.

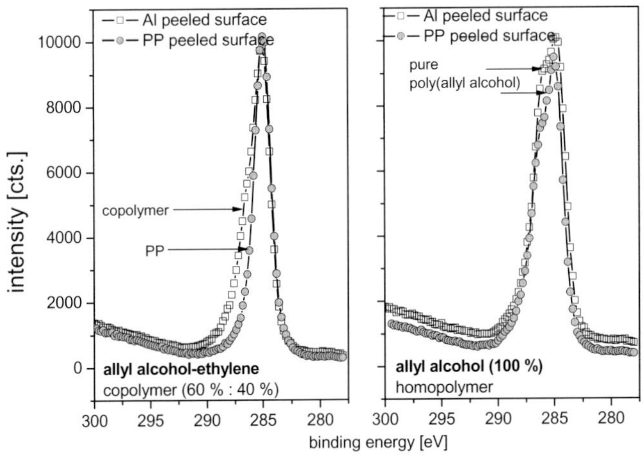

Figure 7.35 Locus of peeling in Al–copolymer (homopolymer)–PP composites.

7.9
Life Time of Plasma Activation

In particular, plasma polymers undergo extensive aging when exposed to ambient air. This problem has been mentioned in different sections of this book. Therefore, only one example is presented here, namely, the incorporation of oxygen from air by exposure of pulsed-plasma polymerized allylamine to air (Figure 7.37) [94].

The surface energy often decreases during storage and the contact angle re-increases. This unexpected behavior is in contrast to the post-plasma oxidation,

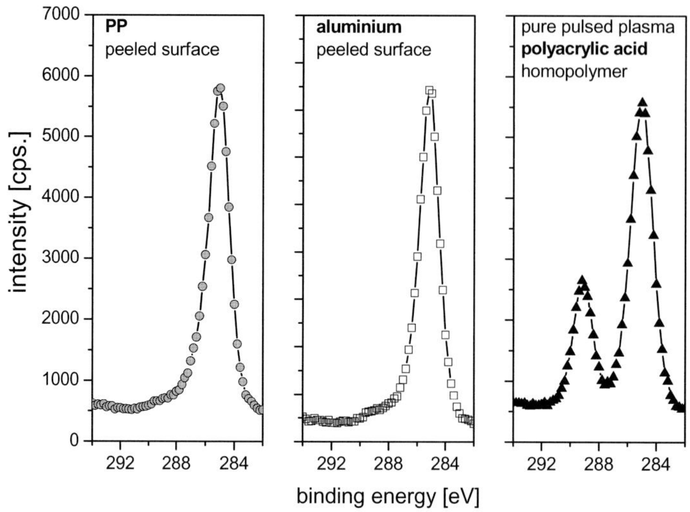

Figure 7.36 Locus of peeling in Al-pulsed-plasma polymerized acrylic acid–PP composites.

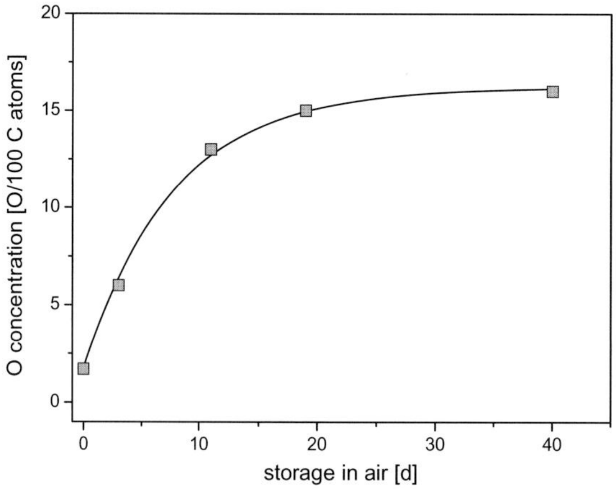

Figure 7.37 Oxygen incorporation of plasma-polymerized allylamine during exposure to air.

which produces additional polar groups. This phenomenon, as stated above, is attributed to "hydrophobic recovery." Figure 7.38 shows that this characteristic "hydrophobic recovery" was not found on exposing a special PP film to the dielectric barrier discharge in air.

Short exposure to discharge shows a slight decrease in water contact angle during exposure to air, while long-time exposure shows a very slight increase in contact angle.

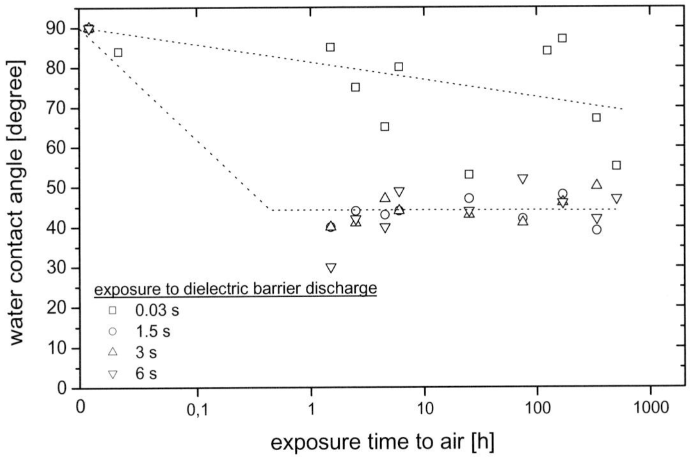

Figure 7.38 Re-increase of water contact angle of polypropylene on exposure to air after treatment with atmospheric dielectric barrier discharge.

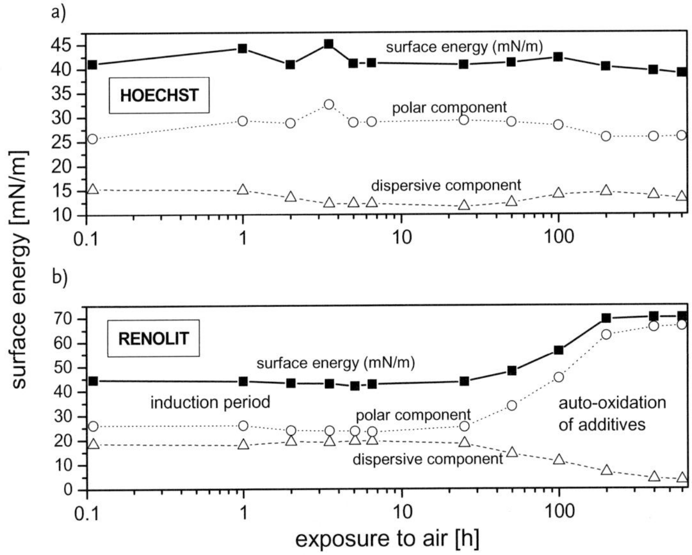

Figure 7.39 Different behavior of surface energies of two different polypropylene foils that were exposed to dielectric barrier discharge (0.1 s) at atmospheric pressure.

The missing re-increase of water contact angle was ascribed to the presence of an unsaturated slip agent at the polypropylene surface and its post-plasma auto-oxidation, which compensates for the "hydrophobic recovery."

The nearly constant remaining surface energy during exposure to air as shown in Figure 7.39a was confirmed for low-pressure cw-rf oxygen plasma

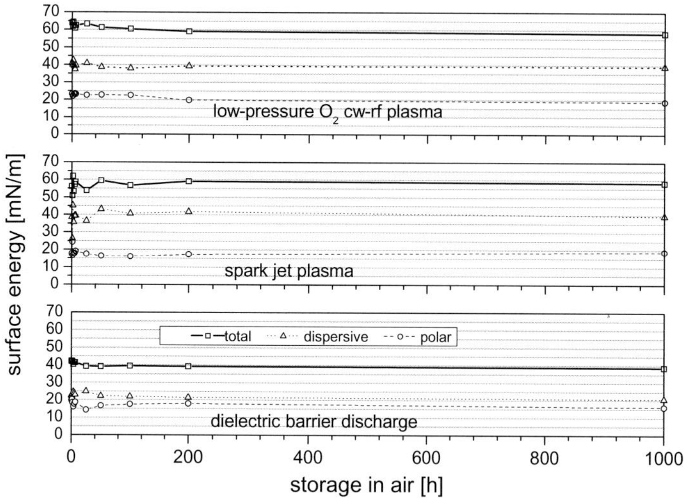

Figure 7.40 Comparison of surface energy during storage of polypropylene treated by low-pressure oxygen plasma, atmospheric-pressure plasma-jet, and dielectric barrier discharge plasma.

treatment as well as exposure to plasma jet and dielectric barrier discharge in air (Figure 7.40).

The PP film in Figure 7.39a shows a very low decrease in surface energy (increase in water contact angle) during exposure to air. The other film (Figure 7.39b) produces a strong auto-oxidation after long-time storage and a strong increase in surface energy (decrease in water contact angle). Here, auto-oxidation of the slip agent was responsible for the extraordinary increase in surface energy. Nevertheless, the oxidized slip agent shows high surface energy but no adhesion to aluminium because it works as a weak boundary layer.

Figure 7.40 compares the changes in surface energy during exposure to ambient air of polypropylene films treated in low-pressure oxygen plasma, atmospheric low-temperature plasma jet, and atmospheric dielectric barrier discharge. In all cases a slow decrease in surface energy was determined for all treatments.

In most cases the surface energy decreases slightly or sometimes remains nearly constant; nevertheless, the adhesion is strongly decreased after very short storage. Thus, the surface energy does not give decisive information on expected adhesive bond strength. This can be evidenced as shown in Figure 7.41. PP was pretreated with an atmospheric plasma jet. A rapid loss in lap shear strength of polyurethane–polypropylene composites is evident, when the pretreated PP was exposed to air for more than a few minutes before gluing it with polyurethane (Figure 7.41) [95].

Low-pressure oxygen plasma treatment of PP shows only a slight increase of the O/C ratio during exposure to air for 1 year, that is, the post-plasma auto-oxidation

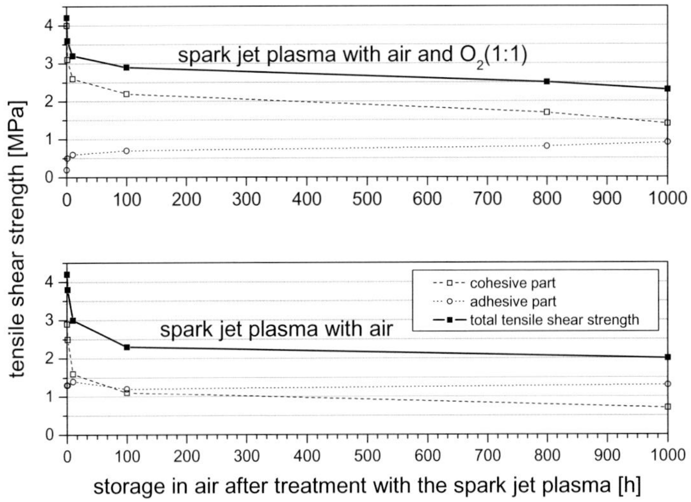

Figure 7.41 Loss in lap-shear strength of polyurethane–polypropylene composites with time between plasma treating and gluing.

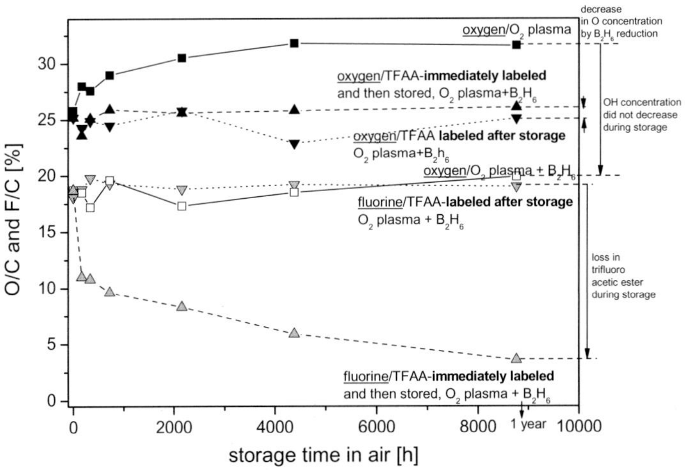

Figure 7.42 Oxygen plasma treatment of PP and reduction with B_2H_6 and the dependence on storage in air.

is minimal. The same oxygen plasma treatment combined with the post-plasma chemical reduction (B_2H_6) of carbonyl species to OH groups shows a constant O/C ratio also after one year exposure to ambient air. This means that the post-plasma wet-chemical reduction has quenched all trapped radicals in the polymer surface layer. Trifluoroacetic anhydride (TFAA) labeling of OH groups also shows a constant OH concentration over a period of one year. However, immediate labeling with TFAA after plasma pretreatment and storing the labeled sample for one year shows a strong decrease in F/C ratio due to hydrolysis of the ester during storage and the re-elimination of TFAA (Figure 7.42).

References

1 Wu, S. (1982) *Polymer Interface and Adhesion*, Marcel Dekker, New York.
2 Hall, J.R. and Westerdahl, C.A.L. (1969) *ACS Div. Org. Coat. Plast. Chem.*, **29**, 472.
3 Lerner, R.M. (1969) *Adhesion Age*, **12**, 35.
4 Rauhut, H.W. (1969) *Adhesion Age*, **12**, 28.
5 Dynes, P.J. and Kaelble, D.H. (1976) *J. Macromol. Sci.*, **A-10**, 535.
6 Hamermesh, C.L. and Crane, L.W. (1978) *J. Appl. Polym. Sci.*, **22**, 2395.
7 Burkstrand, J.M. (1978) *J. Vac. Sci. Technol., A*, **15**, 223.
8 Gähde, J., Friedrich, J., Loeschcke, I., Gehrke, R., and Sachse, J. (1992) *J. Adhesion Sci. Technol.*, **6/5**, 569.
9 Boenig, H.V. (1988) *Fundamentals of Plasma Chemistry and Technology*, Technomic, Lancaster.
10 Friedrich, J. (1986) *Wissenschaft Fortschritt*, **36**, 311–319.
11 Vogel, S.L. and Schonhorn, H. (1979) *J. Appl. Polym. Sci.*, **23**, 495.
12 Gähde, J., Friedrich, J., Fischer, Th., Unger, W., Lippitz, A., and Falkenhagen, J. (1996) *Prog. Colloid Polym. Sci.*, **101**, 194–198.
13 Friedrich, J., Falk, B., Loeschcke, I., Rutsch, B., Richter, Kh., Reiner, H.-D., Throl, U., and Raubach, H. (1985) *Acta Polym.*, **36**, 310–320.
14 Friedrich, J., Gähde, J., Frommelt, H., and Wittrich, H. (1976) *Faserforsch. Textiltechn./Z. Polymerenforsch.*, **27**, 599.
15 Novis, Y., Chtaib, M., Caudano, R., Lutgen, P., and Feyder, G. (1989) *Br. Polym. J.*, **21**, 171.
16 Nichols, M.F., Hahn, A.W., James, W.J., Sharma, K., and Yasuda, H.K. (1984) *J. Appl. Polym. Sci. Appl. Polym. Symp.*, **38**, 21.
17 Inagaki, N. and Yasuda, H. (1981) *J. Appl. Polym. Sci.*, **26**, 3333.
18 Sharma, A.K., Millich, F., and Hellmuth, E.W. (1981) *J. Appl. Polym. Sci.*, **26**, 2197.
19 Dorn, L. and Rasche, M. (1982) *Kunststoffberater*, **27**, 35.
20 Vilenskij, A.J., Wirlitsch, E.E., Stefanowitsch, N.N., and Krotowa, N.N. (1973) *Plast. Massy*, **5**, 60.
21 Collins, G.C.S., Lowe, A.C., and Nicholas, D. (1973) *Eur. Polym. J.*, **9**, 1173.
22 Krüger, P. and Friedrich, J. (2010) Feinreinigung von Oberflächen, in *Vakuum-Plasma-Technologien Beschichtung und Modifizierung von Oberflächen, Teil I und II* (eds G. Blasek and G. Bräuer), Eugen-G-Leuze-Verlag, Saulgau.
23 Krüger, P., Knes, R., and Friedrich, J. (1999) *Surf. Coat. Technol.*, **112**, 240.
24 Strobel, M., Corn, S., Lyons, C.S., and Korba, G.A. (1987) *J. Polym. Sci., Part A: Polym. Chem.*, **25**, 129–142.
25 Bikerman, J.J. (1968) *The Science of Adhesion Joints*, Academic Press, New York.
26 Friedrich, J., Mix, R., Schulze, R.-D., and Kühn, G. (2005) in *Adhesion* (ed. W. Possart), Wiley-VCH Verlag GmbH, Weinheim, pp. 265–288.
27 Shi, M.K., Selmani, A., Martinu, L., Sacher, E., Wertheimer, M.R., and Yelon, A. (1996) in *Polymer Surface Modification: Relevance to Adhesion* (ed. K.L. Mittal), VSP, Utrecht, p. 73.
28 Vargo, T.G., Gardella, J.A., Jr., Meye, A.E., and Baier, R.E. (1991) *J. Polym. Sci., Part A: Polym. Chem.*, **29**, 555.
29 Charbonnier, M., Romand, M., Alami, M., and Duc, T.M. (2000) in *Polymer Surface Modification: Relevance to Adhesion*, vol. 2 (ed. K.L. Mittal), VSP, Utrecht, p. 3.
30 Yagi, T. and Pavlath, A.E. (1984) *J. Appl. Polym. Sci. Appl. Polym. Symp.*, **38**, 215.

31 Friedrich, J., Kühn, G., Schulz, U., Jansen, K., Bertus, A., Fischer, S., and Möller, B. (2003) *J. Adhesion Sci. Technol.*, **17**, 1127.

32 Friedrich, J.F., Unger, W., Lippitz, A., Gross, Th., Rohrer, P., Saur, W., Erdmann, J., and Gorsler, H.-V. (1996) in *Polymer Surface Modification: Relevance to Adhesion* (ed. K.L. Mittal), VSP, Utrecht, pp. 49–72.

33 Yasuda, T., Okuno, T., Yoshida, K., and Yasuda, H. (1988) *J. Polym. Sci., Part B: Polym. Phys.*, **26**, 1781.

34 DeLollis, N.J. and Montoya, O. (1969) *Adhesion Age*, **12**, 36.

35 Miller, M.L., Postal, R.H., Sawyer, P.N., Martin, J.G., and Kaplit, M.J. (1970) *J. Appl. Polym. Sci.*, **14**, 257.

36 Dwight, D.W. and Riggs, W.M. (1974) *J. Colloid Interface Sci.*, **47**, 650.

37 Hooper, A. and Allara, D.L. (1998) in *Metallized Plastics 5&6: Fundamentals and Applied Aspects* (ed. K.L. Mittal), VSP, Utrecht, p. 203.

38 Friedrich, J.F., Unger, W.E.S., Lippitz, A., Giebler, R., Koprinarov, I., Weidner, St., and Kühn, G. (2000) in *Polymer Surface Modification: Relevance to Adhesion*, vol. 2 (ed. K.L. Mittal), VSP, Utrecht, p. 137.

39 Kühn, G., Ghode, A., Weidner, St., Retzko, I., Unger, W.E.S., and Friedrich, J.F. (2000) in *Polymer Surface Modification: Relevance to Adhesion*, vol. 2 (ed. K.L. Mittal), VSP, Utrecht, pp. 45–64.

40 Bou, M., Martin, J.M., Le Mogne, Th., and Vovelle, L. (1991) *Appl. Surf. Sci.*, **47**, 149.

41 Marcus, P., Hinnen, C., Imbert, D., and Siffre, J.M. (1992) *Surf. Interface Anal.*, **19**, 127.

42 DeKoven, B. and Hagans, P.L. (1986) *Appl. Surf. Sci.*, **27**, 199.

43 Klabunde, K.J. (1980) *Chemistry of Free Atoms and Particles*, Academic Press, New York.

44 Fisher, L., Hooper, A., Opila, R.L., Jung, D.R., Allara, D.L., and Winograd, N. (1999) *J. Electron. Spectrosc. Relat. Phenom.*, **98–99**, 139.

45 Akhter, S., Zhou, X.-L., and White, J.M. (1989) *Appl. Surf. Sci.*, **37**, 201.

46 Cotton, A. and Wilkinson, G. (1966) *Advanced Inorganic Chemistry*, John Wiley& Sons, Ltd, Chichester.

47 Droulas, J.L., Jugnet, Y., Duc, T.M. (1992) in *Metallized Plastics 3: Fundamental and Applied Aspects* (ed. K.L. Mittal), Plenum Press, New York, pp. 123–140.

48 Friedrich, J., Loeschcke, I., and Gähde, J. (1986) *Acta Polym.*, **37**, 687.

49 Rancourt, J.D., Hollenhead, J.B., and Taylor, L.T. (1993) *J. Adhesion*, **40**, 267.

50 Ho, P.S., Hahn, P.O., Bartha, J.W., Rubloff, G.W., LeGoues, F.K., and Silvermann, B.D. (1985) *J. Vac. Sci. Technol., A*, **3**, 739.

51 Alexander, M.R., Beamson, G., Blomfield, C.J., Leggett, G., and Duc, T.M. (2001) *J. Electron. Spectrosc. Relat. Phenom.*, **121**, 19–32.

52 Friedrich, J., Mix, R., and Wettmarshausen, S. (2008) *J. Adhesion Sci. Technol.*, **22**, 1123–1143.

53 Wettmarshausen, S., Mittmann, H.-U., Kühn, G., Hidde, G., and Friedrich, J.F. (2007) *Plasma Process. Polym.*, **4**, 832–839.

54 Chen, R.T., Muir, B.W., Such, G.K., Postma, A., Evans, R.A., Pereira, S.M., McLean, K.M., and Caruso, F. (2010) *Langmuir*, **26**, 3388–3393.

55 Huajie, Y., Mix, R., and Friedrich, J. (2011) *J. Adhesion Sci. Technol.*, **25**, 799–818.

56 Friedrich, J.F., Koprinarov, I., Giebler, R., Lippitz, A., and Unger, W.E.S. (1999) *J. Adhesion*, **71**, 297–310.

57 Chan, M.G. and Allara, D.L. (1974) *J. Colloid Interface Sci.*, **47**, 697–704.

58 Evans, R.G. and Packham, D.E. (1978) *J. Adhesion*, **9**, 267.

59 Evans, J.R.G. and Packham, D.E. (1979) *J. Adhesion*, **10**, 39–47.

60 Koprinarov, I., Lippitz, A., Friedrich, J.F., Unger, W.E.S., and Wöll, Ch. (1998) *Polymer*, **39**, 3001–3009.

61 van Order, N., Geiger, W.E., Bitterwolf, T.E., and Rheingolds, A.L. (1987) *J. Am. Chem. Soc.*, **109**, 5680–5690.

62 Friedrich, J., Unger, W., Lippitz, A., Koprinarov, I., Kühn, G., Weidner, St., and Vogel, L. (1999) *Surf. Coat. Technol.*, **116–119**, 772–782.

63 Hansen, R.H. and Schonhorn, H. (1966) *J. Polym. Sci., B*, **4**, 203.
64 Kwei, T.K., Schonhorn, H., and Frisch, L. (1967) *J. Appl. Phys.*, **38**, 2512.
65 Possart, W. and Friedrich, J. (1986) *Plaste Kautsch.*, **33**, 273–279.
66 Friedrich, J., Geng, Sh., Unger, W., and Lippitz, A. (1998) *Surf. Coat. Technol.*, **98**, 1132–1141.
67 Unger, W.E.S., Lippitz, A., Friedrich, J.F., Wöll, Ch., and Nick, L. (1999) *Langmuir*, **15**, 1161–1166.
68 Unger, W.E.S., Friedrich, J.F., Lippitz, A., Koprinarov, I., Weiss, K., and Wöll, C. (1998) in *Metallized Plastics 5&6: Fundamental and Applied Aspects* (ed. K.L. Mittal), VSP, Utrecht, pp. 147–168.
69 Holzinger, H. and Tiller, H.-J. (1972) *Plaste Kautsch.*, **9**, 641.
70 Friedrich, J., Retzko, I., Kühn, G., Unger, W., and Lippitz, A. (2001) in *Metallized Plastics 7: Fundamental and Applied Aspects* (ed. K.L. Mittal), VSP, Utrecht, p. 117.
71 Retzko, I., Friedrich, J.F., Lippitz, A., and Unger, W.E.S. (2001) *J. Electron. Spectrosc. Relat. Phenom.*, **121**, 111.
72 Gross, Th., Retzko, I., Friedrich, J., and Unger, W. (2003) *Appl. Surf. Sci.*, **203–204**, 575.
73 Kay, E. and Dilks, A. (1978) *ACS Polym. Prep.*, **19**, 511.
74 Suhr, H., Etspüler, A., Feurer, E., and Oehr, Ch. (1988) *Plasma Chem. Plasma Process.*, **8**, 9.
75 Chen, X., Rajeshwar, K., Timmons, R.B., Chen, J.-J., and Chyan, O.M.R. (1996) *Chem. Mater.*, **8**, 1067.
76 Biswas, M., and Mukherjee, A. (1994) *Photoconducting Polymers/Metal-Containing Polymers, Synthesis and Evaluation of Metal-Containing Polymers*, pp. 89–123, Advances in Polymer Science, vol. 115, Springer.
77 Kay, E. and Dilks, A. (1981) *Thin Solid Films*, **78**, 309–318.
78 Coburn, W.C. (1980) Plasma method for forming a metal containing polymer. United States Patent 4226896.
79 Morosoff, N., Haque, R., Clymer, S.D., and Crumbliss, A.L. (1985) *J. Vac. Sci. Technol., A*, **3**, 2098.
80 Asano, Y. (1983) *Thin Solid Films*, **105**, 1–8.
81 Heilmann, A. (2003) *Polymer Films with Embedded Metal Nanoparticles*, Springer Series in Materials Science, vol. 52, Springer.
82 Munro, H.S. and Till, C. (1984) *J. Polym. Sci. Polym. Ed.*, **22**, 3933–3942.
83 Munro, H.S. and Eaves, J.G. (1985) *J. Polym. Sci. Polym. Ed.*, **23**, 507–515.
84 Gritsenko, K., Grynko, D., Lozovski, V., Friedrich, J., Schulze, R.-D., Jurga, J., Convertino, A., and Kotko, A. (2004), in *Nowe Kierunki Modyfikacji I Zastosowan Tworzyw Sztucznych* (eds J. Jurgi, B. Jurkowskiego, and T. Sterzynskiego), Wydanictwo Politechniki Poznanskiej, Poznan.
85 Grytsenko, K.P., Capobianchi, A., Convertino, A., Friedrich, J., Schulze, R.-D., Ksenov, V., and Schrader, S. (2005) in *Polymer Surface Modification and Polymer Coating by Dry Process Technologies* (ed. S. Iwamoto), Research Signpost, Trivandrum, pp. 85–109.
86 Wittrich, H., Friedrich, J., Gähde, J., and Loeschcke, I. (1975) DDR-patent 120 473, July 17, 1975.
87 Friedrich, J., Wittrich, H., and Gähde, J. (1980) *Acta Polym.*, **31**, 59–62.
88 Oehr, Ch. and Suhr, H. (1989) *Appl. Phys. A*, **49**, 691.
89 Hegemann, D., Riedel, R., Dreßler, W., Oehr, C., Schindler, B., and Brunner, H. (1997) *Chem. Vap. Deposition*, **35**, 257.
90 Oehr, Ch. and Suhr, H. (1988) *Appl. Phys. A*, **45**, 151.
91 Holzschuh, H., Oehr, C., Suhr, H., and Weber, A. (1988) *Mod. Phys. Lett. B*, **2**, 1253.
92 Suhr, H., Oehr, C., Holzschuh, H., Schmaderer, F., Wahl, G., Kruck, T., and Kinnen, A. (1988) *Physica C*, **153–155**, 784.
93 Oehr, Ch. and Suhr, H. (1987) *Thin Solid Films*, **155**, 65.
94 Friedrich, J., Kühn, G., Mix, R., Retzko, I., Gerstung, V., Weidner, St., Schulze, R.-D., and Unger, W. (2003) in

Polyimides and Other High Temperature Polymers: Synthesis, Characterization and Applications (ed. K.L. Mittal), VSP, Utrecht, pp. 359–388.

95 Friedrich, J., Unger, W., Lippitz, A., Gross, Th., Rohrer, P., Saur, W., Erdmann, J., and Gorsler, H.-V. (1995) *J. Adhesion Sci. Technol.*, **9**, 575–598.

8
Accelerated Plasma-Aging of Polymers

8.1
Polymer Response to Long-Time Exposure to Plasmas

In contrast to plasma particle bombardment with electrons, ions, excited states, and other reactive neutrals the vacuum UV (VUV) radiation of the (oxygen) plasma modifies a few tens of nanometres up to a few micrometres or sometimes millimetres of polymer top-layers depending on the polymer type and plasma conditions [1, 2]. The usual plasma conditions with short exposure time do not influence the bulk properties of polymers. Only a very thin surface layer of a few nanometres becomes modified during short plasma exposure time. To produce long-range effects in the polymer bulk by plasma exposure, irradiation for several hours is needed (4–48 h). Moreover, using the oxygen plasma treatment a penetration of plasma-induced modification into the bulk cannot be expected because of simultaneous propagation of the irradiation-modified layer into the bulk as well as the etching front. A steady-state between radiation modification and etching is established, resulting in a thin radiation-modified layer of thickness (λ), which is permanently produced during plasma exposure (Figure 8.1).

Much greater λ is to be expected using hydrogen or noble gas plasmas because they emit more energetic VUV radiation [3] and have lower etching ability than oxygen plasma (Figure 8.2). However, one important ingredient for polymer aging is then missing, namely, the presence of oxygen to simulate oxidative, photo-oxidative, and thermo-oxidative processes.

Changes in bulk and in the surface top-layer have been followed by both surface (XPS = X-ray photoelectron spectroscopy, ATR = attenuated total reflectance IR) and bulk-sensitive (ESR = electron spin resonance, transmission FTIR, chromatography, thermogravimetry, etc.) analytical methods [1].

Very surprisingly, the analytical indications for polymer outdoor weathering, artificial and accelerated aging, thermo-oxidative degradation, and photo-oxidation are the same or are similar to those produced by exposing the polymer to plasmas for very long exposure times. The most important characteristics for polymer aging are the increase in carbonyl concentration (carbonyl index), occurrence of photo-oxidations, yellowing, molar mass degradation, formation of polyenes, and so on [4].

The Plasma Chemistry of Polymer Surfaces: Advanced Techniques for Surface Design, First Edition.
Jörg Friedrich.

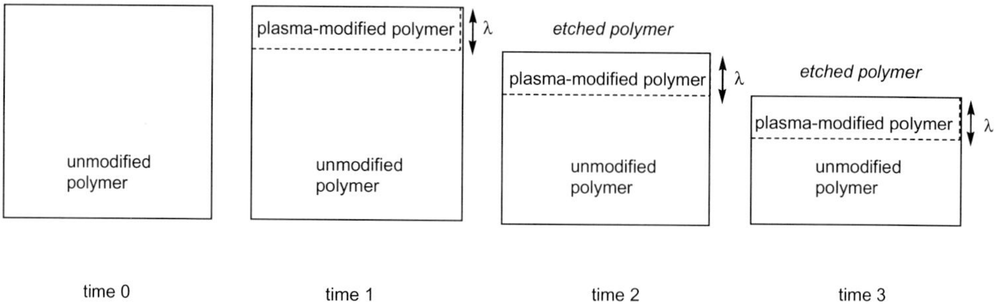

Figure 8.1 Steady-state behavior during oxygen plasma exposure.

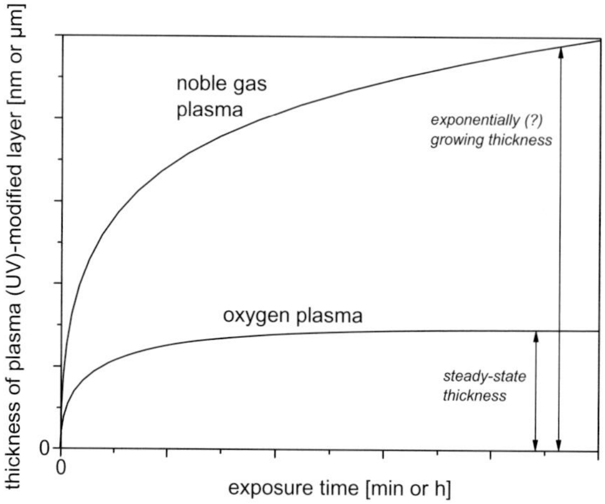

Figure 8.2 Proposed growth of a plasma-UV modified polymer layer during plasma exposure schematically presented for oxygen and noble gas plasmas.

The most important indications for polymer aging are loss in mechanical strength and, possibly, tensile strength and the elongation at break as well as worsening of surface properties such as gloss, haze, and yellowing. A strongly accelerated simulation of these surface-related aging factors is easy to achieve by plasma treatment [5], such as yellowing by hydrogen plasma exposure of poly(vinyl chloride) [6] or an increase of haze and loss in glance [7].

The mechanical properties of some polymers were also significantly decreased during plasma exposure [8–10].

However, to develop a new artificial aging process with a high acceleration factor it is necessary to have the same characteristic dependence on exposure time for mechanical strength (and surface properties) as found for outdoor weathering. The loss in mechanical strength as a function of exposure to oxygen plasma should (must) be the same as using the outdoor weathering or more intensified to aggressive climates such as present in Florida. Photo-oxidative artificial aging simulating such a climate can further shorten the exposure time. UV-C irradiation, higher temperature, high humidity, and the presence of aggressive gases such as SO_2 or NO_2 can further increase the polymer aging. However, there is the danger of initiating new degradation mechanisms that are not present during the usual application conditions of polymers. The oxygen plasma exposure of polymers can be regarded as an *ultra-accelerated artificial aging* of polymers. Especially, aromatic polyesters such as poly(ethylene terephthalate) or poly(butylene terephthalate) show the same characteristic course of mechanical strength with aging time as found for outdoor weathering. Looking to the outdoor-analog aging of PET foils (200 μm) the exposure to low-pressure oxygen plasma leads to the same characteristic aging behavior as found for outdoor weathering, as exemplified by the tensile strength (Figure 8.3).

The acceleration factor of plasma exposure compared to the outdoor exposure was about 311. This behavior is also comparable to the outdoor aging found for polycarbonate, polyurethane, epoxy resins, polystyrene, and polyacrylates as schematically presented in Figure 8.4 for the top row of polymer etching. In contrast, aliphatic polyolefins and poly(vinyl chloride) show inadequate behavior (Figure 8.4, bottom row). They tend to strong crosslinking and the formation of

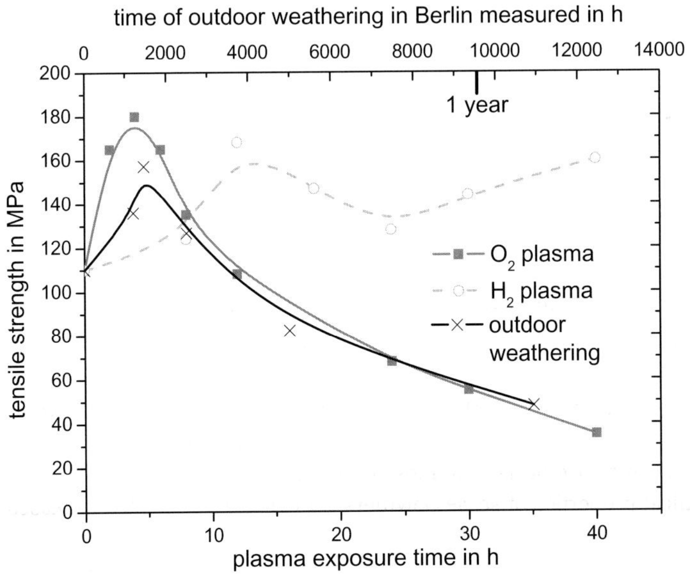

Figure 8.3 Comparison of changes in tensile strength of PET foils (200 μm thick) versus time of exposure to oxygen and to hydrogen plasma (cw-rf plasma, 500 W, 10 Pa) and on outdoor weathering.

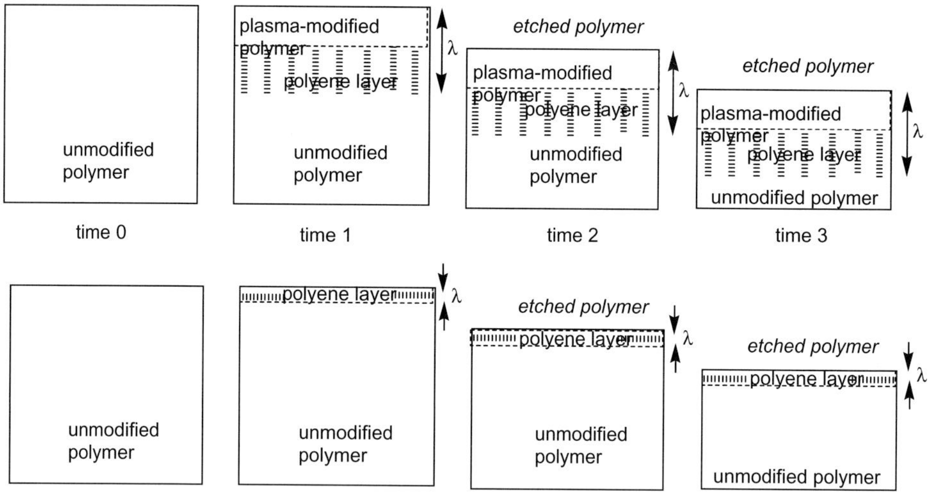

Figure 8.4 Two types of steady-state behavior during oxygen plasma exposure and polyene formation.

a surface-near polyene layer, which shields the bulk against further UV irradiation. The consequence is the strong degradation and etching of a very thin surface-near layer and survival of the nearly unchanged polymer bulk structure (Figure 8.4) [2].

Other investigators have observed similar changes in mechanical strength during plasma exposure. For example, Arefi-Khonsari and Amouroux detected after less than 10 s exposure of a 8 μm thick polypropylene film to oxygen plasma (kHz glow discharge) a loss of about 60% in elongation at break [11].

The mechanism of accelerated (oxygen) plasma weathering is much different from those of outdoor-weathering, UV-induced photo-oxidation, Xenotest irradiation, and so on. Pure weathering uses chromophores, natural "defects" in polymers, arising from contaminations. Accelerated aging processes uses radiation sources with a spectral distribution of radiation similar to that of the Sun's irradiation on Earth (Figure 8.5) [4].

These artificial plasma-radiation sources emit superposed resonance (line) and continuous recombination irradiation with additions of Bremsstrahlung, and thus more or less strongly differ from the spectrum of sun.

UV-A radiation and visible light (violet–blue) are only able to excite reversible $\pi \rightarrow \pi^*$ and $n \rightarrow \pi^*$ transitions of conjugated double and carbonyl bonds, which most often arise from defects or contaminations (chromophores). The transition from reversible excitation processes to irreversible chain scissions is indirect, follows a special mechanism, and is slow. Two general possibilities exist, C–H or C–C bond scission by radiation of sufficient energy or excitation of chromophores such as a Norrish-I reaction (α-scission):

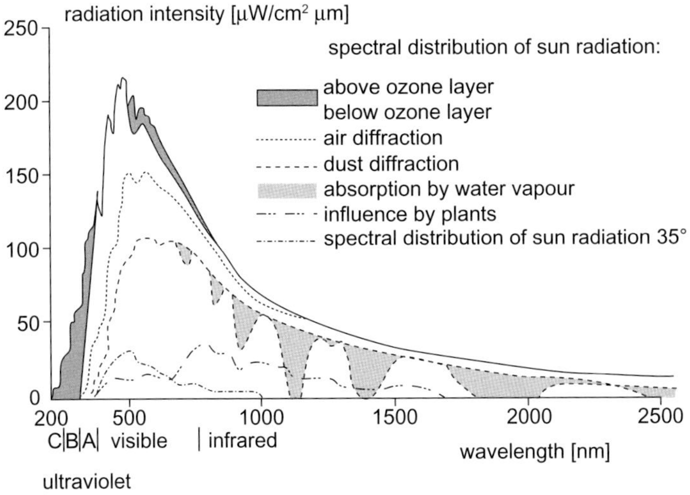

Figure 8.5 Spectral distribution of the Sun's radiation reaching the Earth.

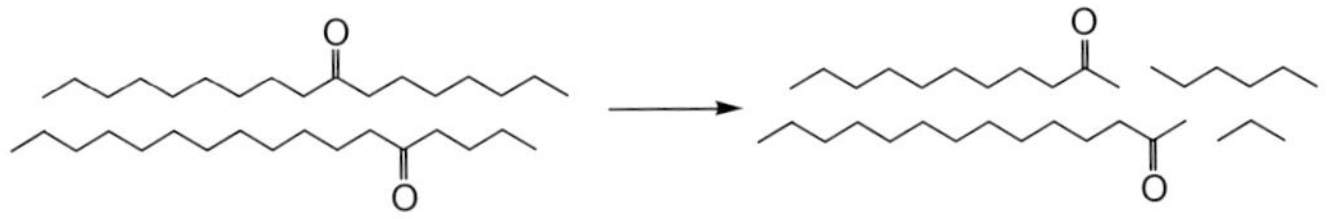

$$R\text{-}H + h\nu \rightarrow R^{\bullet} + 0.5H_2 \quad \text{-}CH_2\text{-}CO\text{-}R + h\nu \rightarrow \text{-}CH_2\text{-}CO^{*}\text{–}R$$

$$R^{\bullet} + {}^{\bullet}O\text{-}O^{\bullet} \rightarrow R\text{-}O\text{-}O^{\bullet} \quad \text{-}CH_2\text{-}CO^{*}\text{-}R \rightarrow \text{-}CH_2^{\bullet} + {}^{\bullet}CO\text{-}R$$

$$R\text{-}O\text{-}O^{\bullet} + RH \rightarrow R\text{-}O\text{-}OH + {}^{\bullet}R \quad \text{-}CH_2\text{-}CO^{*}\text{-}R \rightarrow \text{-}CH_2\text{-}CO^{\bullet} + {}^{\bullet}R$$

$$R\text{-}O^{\bullet} + RH \rightarrow R\text{-}OH + {}^{\bullet}R$$

$$R\text{-}O\text{-}OH \rightarrow RO^{\bullet} + {}^{\bullet}OH$$

$$2R\text{-}O\text{-}OH \rightarrow R\text{-}O\text{-}O^{\bullet} + R\text{-}O^{\bullet} + H_2O$$

$$2R^{\bullet} \rightarrow R\text{-}R$$

$$R\text{-}O\text{-}O^{\bullet} + {}^{\bullet}R \rightarrow R\text{-}O\text{-}O\text{-}R$$

Auto-oxidation *Norrish-I*

The material-changing process is the subsequent auto-oxidation [12, 13].

Plasma-induced polymer aging attacks directly C–H and C–C bonds in the polymer. The energy dose in plasmas is generally high and numerous covalent bonds can be broken at the same time. Figure 8.6 presents the possible energetic effects in polymers, considering simple processes that depend on the wavelength of UV radiation.

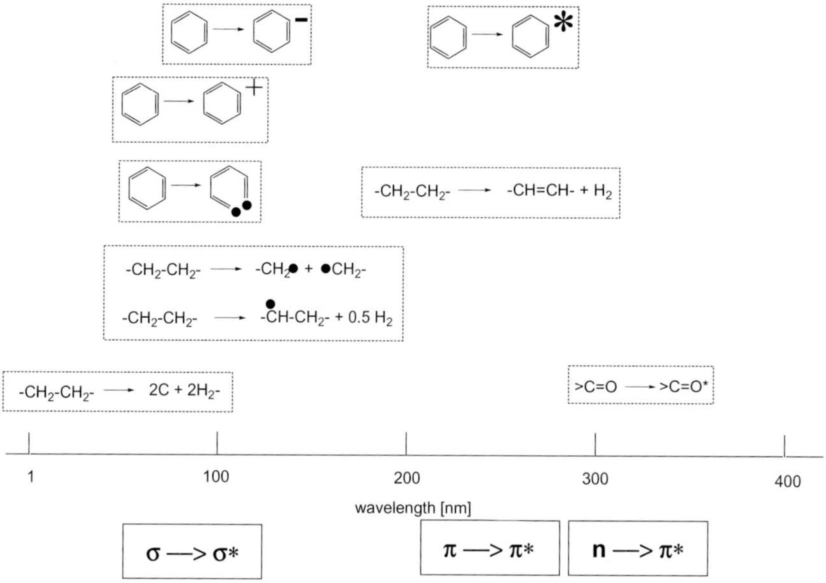

Figure 8.6 Energetic location of the most important chemical reactions according to the causative radiation.

The so-produced fragments are radicals, which can also continue the polymer degradation by auto-oxidation (Scheme 8.1).

8.2 Hydrogen Plasma Exposure

The hydrogen plasma modifies polymers differently. Chemically, it should be an ideal hydrogenation/reducing medium for any chemical processes because the plasma should be the source of H atoms (*in statu nascendi*):

$$H_2 + plasma \rightarrow 2H^{\bullet}$$

The high concentration of hydrogen atoms in the plasma predestines it as a chemical tool for reducing oxidized functional groups and hydrogenating any C=C double bonds. First, the bombardment with energetic particles from the (hydrogen) plasma removes functional groups and also hydrogen from the backbone. In such a way crosslinking is possible:

$$H^{\bullet} + RH \rightarrow R^{\bullet} + H_2$$

$$R_1^{\bullet} + R_2^{\bullet} \rightarrow R_1\text{-}R_2$$

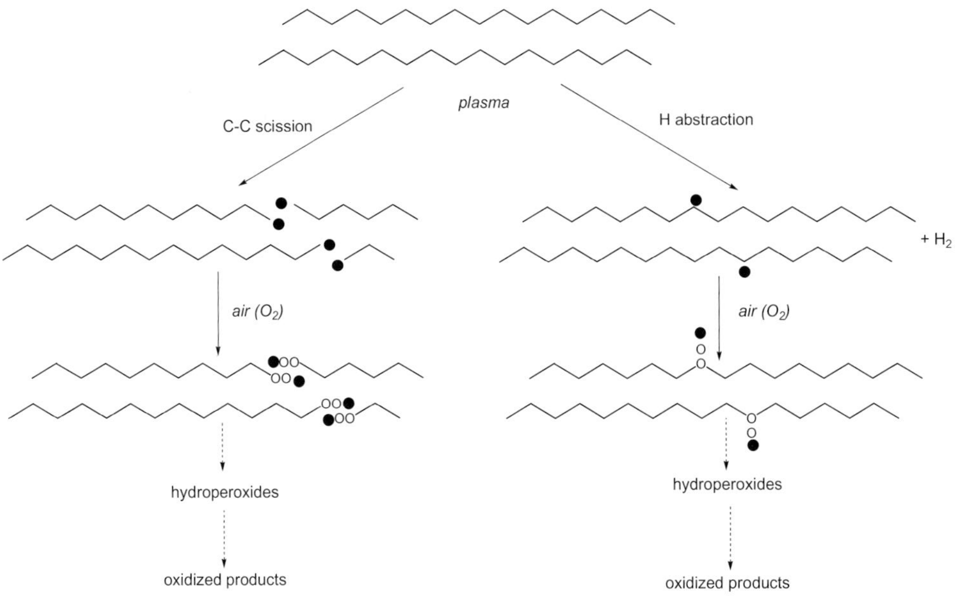

Scheme 8.1 Direct scission of covalent bonds by energetic UV radiation emitted from plasma.

or, which was not undoubtedly found, formation of double bonds:

$$2H^{\bullet} + \text{-}CH_2\text{-}CH_2\text{-} \neq \text{-}CH{=}CH\text{-} + 2H_2$$

However, any hydrogen deficiency in top-most polymer layers was not found after exposure to hydrogen plasma.

Such dehydrogenation is possible, though, by irradiation of deeper polymer layers:

$$\text{-}CH_2{-}CH_2\text{-} + UV \rightarrow \text{-}CH{=}CH\text{-} + H_2$$

Therefore, the maximal possible hydrogen concentration in polymers is a steady-state between its hydrogenation and dehydrogenation. The kinetic energy of hydrogen atoms and the released energy at recombination with a radical site hinder the achievement of complete hydrogenation because of the concurrence of re-dissociation, of crosslinking, or the formation of C=C double bonds. The energy-rich UV radiation of the hydrogen plasma reaches a few μm of the bulk in many polymers and thus it produces unsaturations, radical sites, and crosslinking. Using a deuterium precursor H–D exchange in the range of 2 μm was found in polyethylene [14, 15].

The radiation-induced formation of double bonds depends on the wavelength of irradiation (from plasma) and the wavelength-dependent absorption properties of the polymer. Hudis first found this strong polymer-changing effect of hydrogen plasma and measured the gelation (crosslinking) of polymers by size exclusion chromatography (SEC) [3, 16]. Looking to the well-known UV spectrum of the hydrogen plasma with its Lyman, Balmer, Paschen, Brackett, and Pfund series, enough intense and energy-rich lines are present in the spectrum to produce the above-mentioned effects. The Lyman series covers energies from about 10.2 to 12.8 eV [17]. The transition from an excited state (E^*) to the ground state (E^0) possesses the highest energy (resonance energy):

$$h\nu = E^* - E^0$$

Considering the basics of spectroscopy the wavenumbers of all lines of an atom can be described by:

$$\nu/c = 1/\lambda = T^0 - T^*$$

This results in:

$$E^* = -hcT^*$$

For hydrogen, the Rydberg equation is valid:

$$T^* = R/n^2$$

and therefore the energy is:

$$E^* = -hcR/n^2$$

where R = Rydberg constant and $n = \infty \rightarrow E = 0$.

It should be added that noble gas plasmas also emit line radiation of high energy content. Under plasma conditions all possible electron transitions were excited and resonance radiation is emitted. Organic contributions to the plasma emit fluorescence (resonance) radiation that is most important because of its high energy (Jablonski diagram).

Electronic excitations occur by inelastic collision with electrons. Electrons have kinetic energy (E_{kin}), which can be expressed in terms of the exposed voltage (U) and the elemental charge of an electron (e) as $E_{\mathrm{kin}} = Ue$. Considering the Maxwell equation it follows:

$$Ue = m/2v^2$$

For a voltage of 1 V the kinetic energy is defined as 1 eV.

Considering the series limit or the ionization energy of hydrogen (13.5 eV) all emitted radiation has energy lower than that value. Thus, the resonance (Lyman) radiation has lines from 10.2 to ca. 12.8 eV. The intensity of the emitted radiation or the probability of a transition depends on a few parameters. Apart from the pressure in the plasma and the density of particles in the volume the probability of an exciting inelastic collision depends also on the size of the atom. Hydrogen is the smallest atom and, therefore, the probability of collision with an electron is generally low (mean free path is high). Therefore, the kinetic energies of electrons are relatively high because they have more time for acceleration in the electrical field of the plasma, thus exciting the Lyman transitions. In summary, the hydrogen plasma emits very intense and energetic radiation that is responsible for several effects in surface-near layers or in the bulk of polymers.

Table 8.1 Energy of ionization ($E_{ionization}$), lowest excited state (E_{min}), triplet state (E_{tripl}), and lowest excited state of the single ion ($E_{ion\ 1}$) of noble gases.

Plasma gas	$E_{ionization}$ (eV)	E_{min} (eV)	E_{tripl} (eV)	$E_{ion\ 1}$ (eV)
Helium	24.5	19.8	21.1	54.2
Neon	21.5	16.6	17.0	40.9
Argon	15.7	11.5	11.8	27.8
Krypton	14.0	9.9	9.9	26.4
Xenon	12.1	8.3	8.4	21.2

8.3 Noble Gas Plasma Exposure, CASING

Noble gas plasmas cannot react chemically with polymers. Their action is pure physical (sputtering). Ions, electrons, and excited atoms are the acting species. Special attention was focused on the metastable states of noble gas atoms. For plasma UV radiation and for modification of polymers the energy of ionization ($E_{ionization}$), the lowest excited state of the atom (E_{min}), the metastable (E_{tripl}), and the lowest excited state of singly charged ($E_{ion\ 1}$) ion are of special importance (Table 8.1).

All species listed in Table 8.1 are present in the plasma, for example, He^+, He^{++}, H^{*1}, He^m, He^{+*}, and so on [18]. Schonhorn and Hansen had assigned metastable species as responsible for the crosslinking effect because of their frequent occurrence [19, 20]. Therefore, they called this effect crosslinking by activated species of inert gas plasmas (CASING).

A few years later Hudis found that the particle stream from the plasma only influences the topmost nanometres of the polymer surface and, therefore, the particle bombardment or diffusion of energetic particles could not be the source of crosslinking over the dimensions of micrometres. He could show that the VUV radiation is the reason for the crosslinking of polymers [3, 21]. However, the efficiency of the plasma-UV irradiation was 100–500 times higher than conventional UV irradiation. It was claimed that the plasma radiation is of polychromatic nature and therefore much more efficient (cf. Section 6.5).

Wertheimer *et al.* have investigated in detail the role of plasma UV irradiation in polymer surface modification during plasma exposure [22].

References

1 Friedrich, J., Gähde, J., and Nather, A. (1981) *Plaste Kautsch.*, **28**, 140.

2 Friedrich, J., Loeschcke, I., Reiner, H.-D., Frommelt, H., Raubach, H., Zimmermann, H., Elsner, Th., Thiele, L., Hammer, L., and Merker, E. (1990) *Int. J. Mater. Sci.*, **13**, 43–56.

3 Hudis, M. (1972) *J. Appl. Polym. Sci.*, **16**, 2397.

4 Rabek, J.F. (1995) *Polymer Photodegradation – Mechanisms and Experimental Methods*, Chapman & Hall, London.
5 Friedrich, J., Pohl, M., Elsner, T., and Altrichter, B. (1988) *Acta Polym.*, **39**, 544–549.
6 Behnisch, J., Friedrich, J., and Zimmermann, H. (1991) *Acta Polym.*, **42/1**, 51–53.
7 Friedrich, J., Kühn, G., Schulz, U., Erdmann, J., and Möller, B. (1999) *Mater. Test.*, **41**, 375.
8 Friedrich, J. and Gähde, J. (1980) DDR-patent 224 824, October 10, 1980.
9 Friedrich, J., Gähde, J., and Nather, A. (1981) *Plaste Kautsch.*, **28**, 420–425.
10 Friedrich, J. and Frommelt, H. (1988) *Acta Chim. Hung.*, **125**, 165–175.
11 Kurdi, J., Arefi-Khonsari, F., Tatoulian, M., and Amouroux, J. (1998) in *Metallized Plastics 5&6: Fundamental and Applied Aspects* (ed. K.L. Mittal), VSP, Utrecht, p. 295.
12 Jellinek, H.H.G. (1978) *Aspects of Degradation and Stabilization of Polymers*, Elsevier, Amsterdam.
13 Doležel, B. (1978) *Die Beständigkeit von Kunststoffen und Gummi*, Hanser, Munich.
14 Min, H., Wettmarshausen, S., Friedrich, J., and Unger, W.E.S. (2011) *J. Anal. At. Spectrom.*, **26**, 1157–1165.
15 Wettmarshausen, S., Min, H., Unger, W., Jäger, C., Hidde, G., and Friedrich, J. (2011) *Plasma Chem. Plasma Proc.*, **31** 551–572.
16 Friedrich, J., Kühn, G., and Gähde, J. (1979) *Acta Polym.*, **30**, 470–477.
17 Grotrian, W.R.W. (1928) Graphische Darstellung der Spektren von Atomen und Ionen, Bd. II, p. 3.
18 Egitto, F.D. (1990) *Pure Appl. Chem.*, **62**, 1699.
19 Hansen, R.H. and Schonhorn, H. (1966) *J. Polym. Sci., Polym. Lett. Ed.*, **4**, 209.
20 Schonhorn, H. and Hansen, R.H. (1967) *J. Appl. Polym. Sci.*, **11**, 1461.
21 Hudis, M. and Prescott, L.E. (1972) *J. Polym. Sci.*, **10**, 179.
22 Wertheimer, M.R., Fozza, A.C., and Holländer, A. (1999) *Nucl. Instrum. Methods Phys. Res., Sect. B*, **151**, 65–75.

9
Polymer Surface Modifications with Monosort Functional Groups

9.1
Various Ways of Producing Monosort Functional Groups at Polyolefin Surfaces

If a polyolefin surface is covered with monotype-functional groups chemical synthesis can be performed on them. In this way the polyolefin surface structure and functionality can be tailored and specified by grafting of molecules of different chain length, complexity, and function. Such surfaces are applied in biochemistry and medicine and used for biochips, microfluidics, or for enhancing the printability, gluability, or adhesion to metal coatings. A precondition is the existence of chemically reactive anchoring points for such graft synthesis, or radical sites and grafting via peroxy radicals (Figure 9.1).

These anchoring points for the more reproducible grafting via functional groups are monotype (monosort) functional groups, which must occur in sufficient density at the surface. Typical densities of such monotype functional groups for graft synthesis range from 3 to 30 groups per 100 C atoms. To produce these functional groups two different methods are used:

1) direct bonding of the monosort functional group onto the polymer backbones;

2) deposition of a well-adherent thin plasma polymer layer, which carries the desired monotype functional group. Four types of plasma-enhanced polymer surface functionalizations are in focus here (Scheme 9.1 and Table 9.1).

To fulfill the technical demands a simple unspecific plasma treatment of polyolefin surface is often not sufficient, as explained by the co-existence of different types of O-functional groups at the surface. Therefore, plasma treatments or plasma-assisted processes must be developed to produce a high density of monosort functionalizations at polyolefin surfaces, such as listed here:

1) Oxygen plasma treatment → reduction of C=O groups → OH-groups;

2) selective plasma treatment producing monosort functional groups (unknown, exception → plasma bromination) → C–Br bonds;

3) chemical conversion of C–Br groups into OH, NH_2, CN, and so on;

The Plasma Chemistry of Polymer Surfaces: Advanced Techniques for Surface Design, First Edition.
Jörg Friedrich.

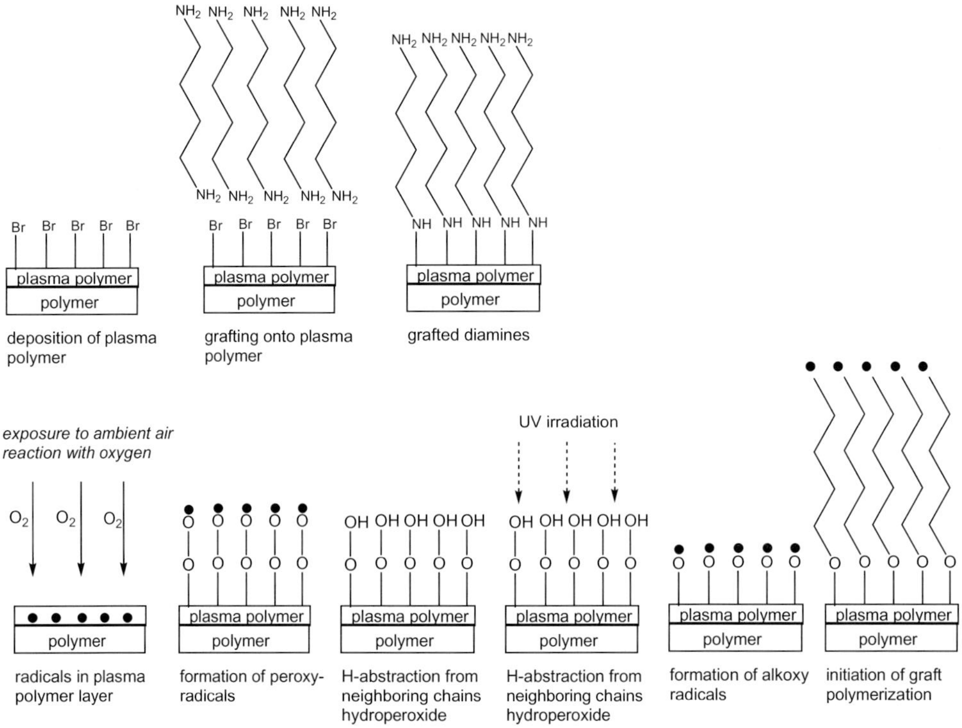

Figure 9.1 Grafting onto monosort functional groups at polyolefin surfaces or onto peroxide/alkoxy radical sites.

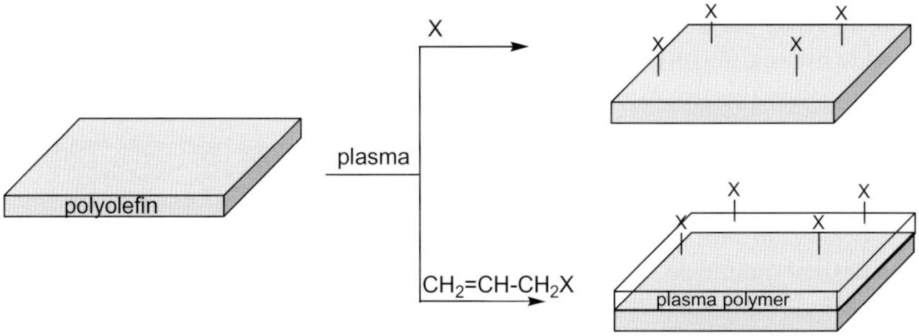

Scheme 9.1 Two routes for producing a bromine-rich polyolefin surface.

Table 9.1 Examples of processes for producing monotype functional groups.

Type of plasma functionalization	Plasma treatment	Chemical treatment	Type of functional group	Groups per 100 C
Direct attachment	Oxygen	Reduction with B_2H_6	OH	10–14
Direct attachment	Bromoform, bromine		Br	30–100
Deposition	Allyl bromide		Br	24
Deposition	Allyl bromide + Br_2		Br	50–60
Deposition	Allylamine		NH_2	8–16
Deposition	Allyl alcohol		OH	30
Deposition	Acrylic acid		COOH	27

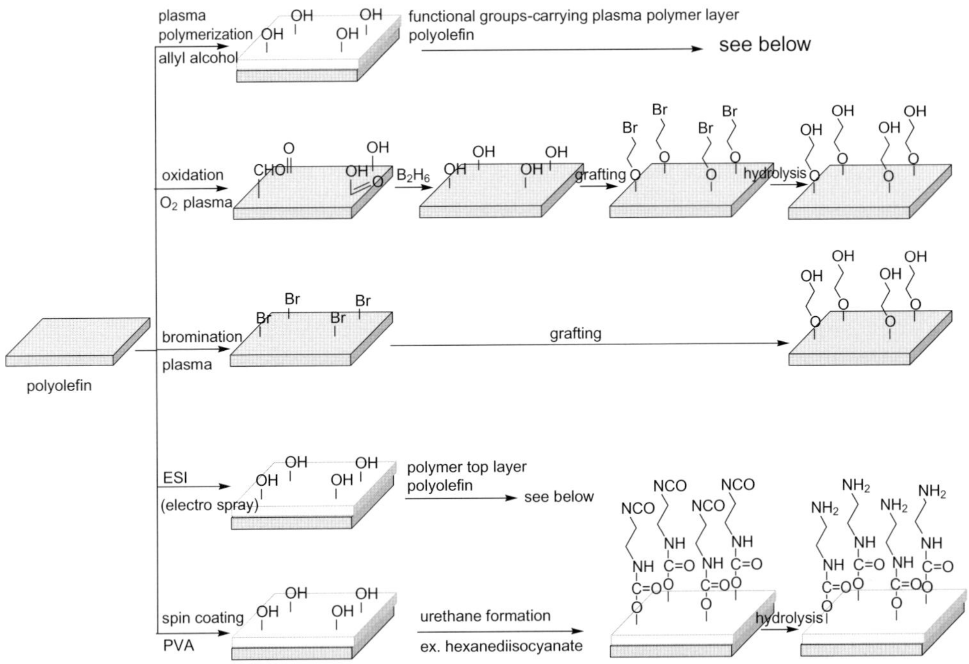

Scheme 9.2 Various means of polyolefin modification with monosort functional groups and chemical grafting via functional groups.

4) grafting of vinyl or acrylic monomers carrying functional groups (OH, COOH, NH_2. . . .) onto C radical sites or peroxy radicals at the polymer surface;

5) deposition of plasma polymer layers made of functional group carrying monomers (allyl alcohol, acrylic acid, allylamine. . . .); variation of density is possible by plasma-initiated copolymerization of functional group bearing co-monomers and "chain-extending, neutral" co-monomers [1].

In turn, these possibilities of producing monosort-functionalized polymer surfaces and chemical grafting onto these functional groups are discussed by reference to a few examples.

The above-listed variants of polyolefin surface modification with monotype-functional groups are presented schematically in Scheme 9.2.

9.2 Oxygen Plasma Exposure and Post-Plasma Chemical Treatment for Producing OH Groups

Polyolefins do not possess functional groups. As described in Chapter 5 oxidation processes can introduce polar O-functional groups onto the surface. On exposure of polymers to an oxygen plasma a broad variety of O-functional groups is produced (C–O–C, C–OH, CHO, >C=O, COOR, COOH, CO_3, etc.), which is a hindrance for defined chemical processing [2, 3]. For subsequent chemical reactions, accompanied by formation of azomethine, ether, ester, urethane, or urea bonds, only one sort of functional group is needed at polyolefin surface. Such monosort functionalities may be hydroxy, amino, epoxy, or carboxy groups. However, if different O-functional groups are present it is not possible to consume all functionalities. The O-functional groups can be assigned to different kinds of C–O bonds, namely, one, two, three, or four bonds between carbon and oxygen, as it can be differentiated by XPS (X-ray photoelectron spectroscopy) for groups located at the polymer surface.

In general, there is the possibility of applying oxidative or reductive wet-chemical processes for transferring all the bonds to, respectively, the highest oxidation ratio (C^{4+}, O–CO–O) or to the lowest (C^{2-} = CH_3–OH; C^-=-CH_2–OH, C^0 = >CH–OH, C^+ = C–OH) state of oxidation ratio. These oxidation states are assigned the following the oxidation rules in organic chemistry [4]. Other medium oxidation ratios are difficult to obtain and, also, the C^{4+} ratio is characterized by degradation (carbonate or CO_2 formation), auto-oxidation, and etching, that is, the process cannot be stopped. Moderate oxidations, such as using SeO_2, also do not have enough selectivity to produce exclusively one oxidation state or one type of C–O bond. Therefore, selective oxidation, such as to COOH (carboxylic) groups, is not realistic.

Such chemical oxidation processes often produce decay of the functional group or scission of the polymer backbone. The dissociation energies of C–H and C–C bonds in polyolefins are nearly the same ($\Delta H_{HC-H} = 396\,kJ\,mol^{-1}$, $\Delta H_{CH2-CH2} = 370\,kJ\,mol^{-1}$), as are the entropy terms $T\Delta S$. Therefore, chain scissions should occur very often. However, in fact, such chain scissions do not occur very often because the C–C backbone is shielded by its H-substituents [4].

Reduction can be limited to the C^{2-}-C^+ oxidation ratios, that is, to the formation of OH groups, to $CH_{(x)}$-OH. Such chemical reductions are more selective and easier to handle; however, a total reduction of the O-containing functional group to C–H species (C^{2-} to C^{4-}, =CH_2 to CH_4) is possible as a side-reaction. By using strong reduction agents such as $LiAlH_4$ or Zn/HCl, C–O–C ether structures were also broken and transformed into OH groups. However, a few polymers possess these ether bonds in their polymer backbone, such as polyethers (POM, PEG), and may therefore be degraded during chemical reduction [5]:

$$\text{-C-O-C-} + LiAlH_4 \rightarrow \text{-C-OH} + \text{HC-}$$

HI and HCl/Zn also attack C–O–C bonds quantitatively [4]:

$$\text{-C-O-C-} + 2HI \rightarrow 2\text{C-I} + H_2O$$

$$\text{C-I} + OH^- \rightarrow \text{C-OH}$$

The process using diborane (B_2H_6) reduces only carbonyl features and does not attack ether structures, therefore chain scissions are minimized [6–8]. The diborane reduction was developed by Brown [9]:

$$▮\text{-COOR} + B_2H_6 \rightarrow ▮\text{-CHOBH}_2\ (H_2SO_4) \rightarrow ▮\text{-CH}_2\text{–OH}\ \textit{slow}$$

$$▮\text{-COOH} + B_2H_6 \rightarrow ▮\text{-COBO} + H_2\ (H_2SO_4) \rightarrow ▮\text{-CH}_2\text{–OH}\ \textit{fast}$$

$$▮\text{-C=O} + B_2H_6 \rightarrow ▮\text{-CHOBH}_2\ (H_2SO_4) \rightarrow ▮\text{-CH}_2\text{–OH}\ \textit{fast}$$

A similar reduction process of carbonyl features to OH groups can be performed using Vitride® (Na-bis [2-methoxyethoxy]aluminium hydride) [10]. By employing diborane in combination with hydrogen peroxide, as well as special hydrolysis all olefinic double bonds in the polymer can be additionally hydroborated, that is, changed to OH groups [11]:

$$>C=C< + B_2H_6 \rightarrow \text{adduct} + H_2O_2 \rightarrow >CH–C(OH)<\ \textit{very fast}$$

Scheme 9.3 and Figure 9.2 present the above-mentioned reduction and the corresponding labeling reaction of OH groups with trifluoroacetic anhydride (TFAA) schematically and by respective XPS spectra. Polypropylene was exposed to oxygen plasma for 2 s. In this way about 22 oxygen atoms per 100 C-atoms were introduced to its surface. Labeling with trifluoroacetic anhydride (TFAA) and XPS measurement following [8] the introduction of fluorine showed that less than 10% of all O functional groups were OH groups after this oxygen plasma treatment [5, 6].

The formation of OH groups demands prior hydrogen abstraction from the polymer, for example:

$$\text{-CH}_2\text{–CH}_2\text{-} + \textit{plasma} \rightarrow >CH=CH< + 2^{\bullet}H\ (H_2)$$

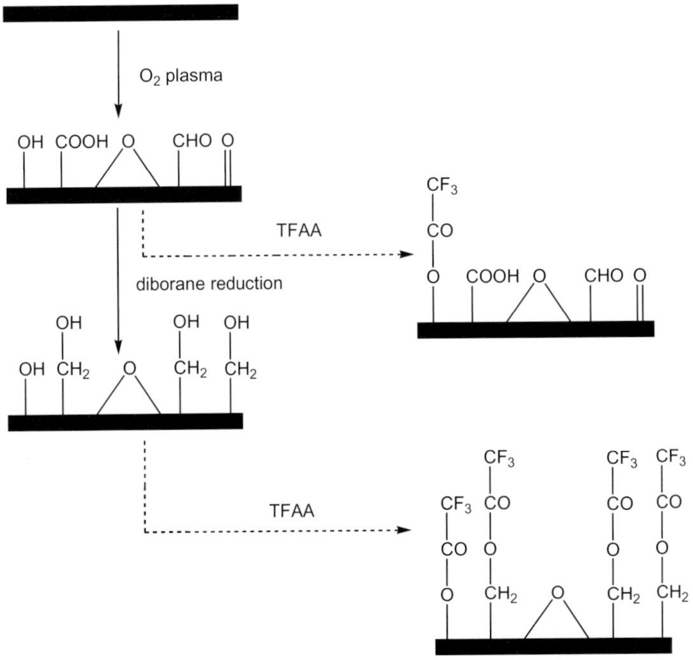

Scheme 9.3 Reduction of O-functional groups to OH groups and their labeling with TFAA.

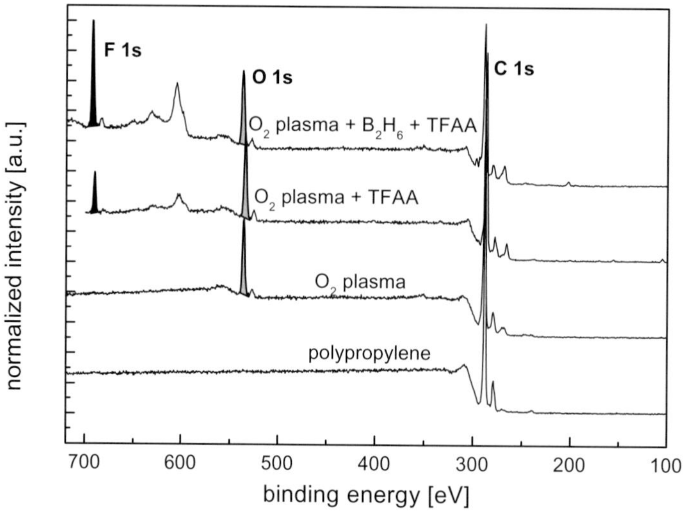

Figure 9.2 XPS survey scans of polypropylene surfaces, after exposure to oxygen plasma [continuous wave (cw)-plasma, 6 Pa, 2 s], after additional labeling with TFAA, and after additional reduction with diborane and TFAA derivatization.

Such produced hydrogen is now present in the plasma. Therefore, in the oxygen plasma OH species (radicals) can be formed, possibly by:

$$^{\bullet}O^{\bullet} + {}^{\bullet}H \rightarrow {}^{\bullet}OH$$

These ˙OH radicals can recombine with surface radicals sites at the polymer backbone and form hydroxy (alcohol) groups:

$$\text{-CH}_2\text{-CH}_2\text{-} + \textit{plasma} \rightarrow \text{-CH}_2\text{-C}^{\bullet}\text{H-} + {}^{\bullet}\text{H}\ (0.5\text{H}_2)$$

$$\text{-CH}_2\text{-C}^{\bullet}\text{H-} + {}^{\bullet}\text{OH} \rightarrow \text{-CH}_2\text{-CH(OH)-}$$

As demonstrated before, the auto-oxidation also produces alcohol groups at polymer surfaces. The thus-produced hydroperoxides may decay [12–14]:

$$\text{R-O-OH} \rightarrow \text{R-O}^{\bullet} + {}^{\bullet}\text{OH}$$

The OH radicals may recombine with C radical sites and the alkoxy radicals may tear hydrogen from neighboring alkyl chains and form alcohols:

$$\text{R-O}^{\bullet} + \text{R*-H} \rightarrow \text{R-OH} + {}^{\bullet}\text{R*}$$

Besides this introduction of OH groups, auto-oxidation also produces ketones, aldehydes, carboxylic groups, ethers, peroxides, peroxy acids, and so on [14]. Table 9.2 compares the efficiency and side-reactions of all methods.

Starting point for each reduction was an oxidation ratio adjusted to 27% O/C at a polyethylene surface [5]. As a recurring theme, the stronger the reduction

Table 9.2 Efficiency in conversion of all O-functional groups into OH groups.

Reduction	Efficiency	Side-reactions
Chemical reduction		
Na/NH_3	All O-functional groups reduced	Reduction C–F bonds to carbon [polytetrafluoroethylene (PTFE)]
Zn/HCl	All O-functional groups reduced	Partial reduction to hydrocarbons
$LiAlH_4$	All O-functional groups reduced	Ether groups scissioned, also ether groups in polymer backbones [polymer degradation, such as poly(ethylene glycol)]
$NaBH_4$	All O-functional groups reduced	Also ether groups in polymer backbones [polymer degradation, such as poly(ethylene glycol)]
B_2H_6	O-functional groups reduced without ether groups	
Vitride (Na-complex)	O-functional groups reduced without ether groups	Reduction is incomplete
Plasma-chemical reduction		
H_2-plasma	All O-functional groups reduced	Reduction to hydrocarbon, strong change of polymer structure, strong post-plasma oxidation
NH_3-plasma	All O-functional groups reduced and N-functional groups introduced	Reduction to hydrocarbon, strong change of polymer structure, strong post-plasma oxidation

power of the agent the higher the conversion percentage but also the loss in O-functional groups (O_{total}). This means that oxygen was removed from polyethylene surface and the remaining O_{total} was much lower than 27% O/C. Using $LiAlH_4$ the highest concentration of OH groups was achieved and 66% of all O-functional groups were converted into OH functionalities. However, only 75% of the original O-functionalities survived the reduction process. The missing 25% are reduced to hydrocarbons or may be dissolved within the process (Figure 9.3).

Using $LiAlH_4$, 11–14 OH/100 C-atoms could be produced in comparison to diborane reduction (8–11 OH) and Vitride (7–10 OH) (Figure 9.3). Application of Vitride (Na bis[2-methoxyethoxy]aluminium hydride) did not produce a higher yield in OH group formation.

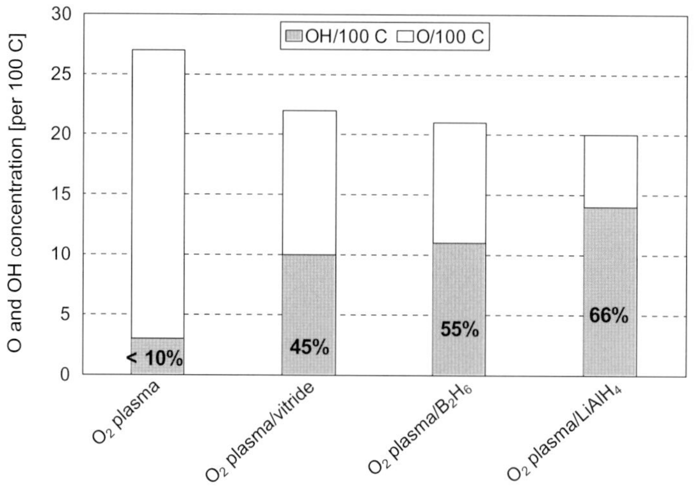

Figure 9.3 Yield in OH-group production using post-plasma chemical reduction (original O concentration ca. 27 O per 100 C).

9.3
Post-Plasma Chemical Grafting of Molecules, Oligomers, or Polymers

9.3.1
Grafting onto OH Groups

Hydroxy groups are consumed with silanes ($SiOCH_3$ or SiCl), isocyanates (CNO), or acids (COOH). A few examples are listed below [5]:

Nonylenesilane was grafted.

$$\blacksquare\text{-OH} + (CH_3O)_3\text{-Si-}(CH_2)_7\text{-CH=CH}_2$$

$$\rightarrow \blacksquare\text{-O-}(CH_3O)_2\text{-Si-}(CH_2)_7\text{-}CH{=}CH_2$$

The thus introduced C=C double bond was changed by electrophilic addition of bromine, ammonia, or hydrogen peroxide:

$$\rightarrow \blacksquare\text{-O-}(CH_3O)_2\text{-Si-}(CH_2)_7\text{-}CHBr\text{-}CH_2Br$$

Another example is the grafting of different amino acids onto polyolefin surfaces [15]. Here, alanine or cysteine is attached using the reaction pathway of consuming 3-aminopropyltriethyoxysilane with OH groups at polypropylene surfaces produced by oxygen plasma treatment and subsequent diborane reduction:

$$\blacksquare\text{-OH} + (CH_3O)_3\text{-Si-}(CH_2)_3\text{-NH}_2 \rightarrow \blacksquare\text{-O-}(OCH_3)_2\text{Si-}(CH_2)_3\text{-NH}_2 + CH_3\text{-OH}$$

The thus introduced amino end-groups were reacted with a chain-extender (glutaraldehyde) to produce new reactive end-groups (CHO):

$$\blacksquare\text{-O-OCH}_3)_2\text{Si-(CH}_2)_3\text{-NH}_2 + \text{OHC-(CH}_2)_3\text{-CHO} \rightarrow$$
$$\blacksquare\text{-O-(OCH}_3)_2\text{Si-(CH}_2)_3\text{-N=CH-(CH}_2)_3\text{-CHO}$$

The amino groups of amino acids reacted with the aldehyde end-groups:

$$\blacksquare\text{-O-OCH}_3)_2\text{Si-(CH}_2)_3\text{-N=CH-(CH}_2)_3\text{-CHO} + \text{NH}_2\text{-CH(CH}_3\text{)-COOH} \rightarrow$$
$$\blacksquare\text{-O-(OCH}_3)_2\text{Si-(CH}_2)_3\text{-N=CH-(CH}_2)_3\text{-CH=}\textit{N-CH(CH}_3\textit{)-COOH}$$

The OH groups are also easily consumed with isocyanates [16, 17]:

$$\blacksquare\text{-OH} + \text{OCN-R-NCO} \rightarrow \blacksquare\text{-O-CO-NH-R-}\textit{NCO}$$

The NCO end-group can be consumed with diols:

$$\blacksquare\text{-O-CO-NH-R-}\textit{NCO} + \text{OH-R}'\text{-OH} \rightarrow \blacksquare\text{-O-CO-NH-R-NH-CO-O-R}'\text{-OH}$$

or hydrolyzed:

$$\blacksquare\text{-O-CO-NH-R-}\textit{NCO} + \text{H}_2\text{O} \rightarrow \blacksquare\text{-O-CO-NH-R-}\textit{NH}_2 + \text{CO}_2$$

Another interesting example, using OH groups at polymer surfaces for chemical modification, is the bonding of radical-chain starters for subsequent initiation of photopolymerization [18]:

$$\blacksquare\text{-OH} + \text{ClCO-R-N=N-R-COCl} \rightarrow \blacksquare\text{-O-CO-R-N=N-R-COCl} + \text{HCl}$$

9.3.2
Grafting onto NH_2 Groups

Three methods have been described to produce monosort primary amino groups:

1) Ammonia plasma treatment with low selectivity, yield, and strong aging effects [19, 20];
2) plasma polymerization of allylamine [21–23];
3) bromination followed by reaction with ammonia or ethylenediamine [24]:

$$\blacksquare\text{-Br} + \text{NH}_3 \rightarrow \blacksquare\text{-NH}_2 + \text{HBr}$$

$$\blacksquare\text{-Br} + \text{H}_2\text{N-CH}_2\text{-CH}_2\text{-NH}_2 \rightarrow \blacksquare\text{-NH-CH}_2\text{-CH}_2\text{-NH}_2 + \text{HBr}$$

Amino groups react easily with aldehydes:

$$\blacksquare\text{-NH}_2 + \text{OHC-(CH}_2)_3\text{-CHO} \rightarrow \blacksquare\text{-N=CH-(CH}_2)_3\text{-}\textit{CHO}$$

Reactions of amino groups with isocyanates to form urea bonds have also been described [25]:

$$\blacksquare\text{-NH}_2 + \text{OCN-R-NCO} \rightarrow \blacksquare\text{-NH-CO-NH-R-}\textit{NCO}$$

$$\blacksquare\text{-NH}_2 + \text{OCN-R-NCO} \rightarrow \blacksquare\text{-NH-CO-NH-R-}\textit{NCO} + \text{H}_2\text{O} \rightarrow \blacksquare\text{-NH-CO-NH-R-}\textit{NH}_2$$

The obtained NCO-terminated polymer surfaces are useful materials for further grafting procedures. Subsequent reactions of these films with amino-terminated fluorescent labels like, for example, dansyl cadaverine (DNS-Ca-NH_2) and dansyl hydrazine (DNS-Hy-NH_2) result in fluorescent PP surfaces [26]:

$$▮\text{-NH-CO-NH-R-NCO} + \text{H}_2\text{N-Ca-DNS} \rightarrow ▮\text{-NH-CO-NH-R-NH-CO-NH-Ca-DNS}$$

$$▮\text{-NH-CO-NH-R-NCO} + \text{H}_2\text{N-Hy-DNS} \rightarrow ▮\text{-NH-CO-NH-R-NH-CO-NH-Hy-DNS}$$

9.3.3 Grafting onto COOH-Groups

Esterification of COOH groups is possible in the presence of carbodiimides that are added to bond the released water [16]:

$$▮\text{-COOH} + \text{HO-R (di-}\textit{tert}\text{-butylcarbodiimide and pyridine)} \rightarrow ▮\text{-}COOR + \text{H}_2\text{O}$$

Peptide (or amide) bond formation is also possible and can be used for grafting to the polymer surface:

$$▮\text{-COOH} + \text{NH}_2\text{-R-COOH} \rightarrow ▮\text{-CO-}NH\text{-}R\text{-}COOH + \text{H}_2\text{O}$$

9.4 Selective Plasma Bromination for Introduction of Monosort C–Br Bonds to Polyolefin Surfaces

9.4.1 General Remarks

This section introduces selective plasma processes for specific (monosort) polymer surface functionalization. The general lack of selective plasma processes is due to the available energy excess in low-pressure plasmas. The absence of any selectivity in a chemical reaction proceeding in plasma and at a polymer surface is a strong handicap in terms of introducing monotype functional groups of sufficient density onto polymer surfaces. This hinders the broader application of plasma processes.

Such monotype functional groups are the anchoring points for further chemical processing such as covalent grafting of any spacers, prepolymers, or functional biomolecules onto the polymer surface and, therefore, are badly needed. Selective monotype functionalizations should possess NH_2, OH, COOH, epoxy, carboxylic acid, sulfonic acid, double bonds, aldehyde, or other groups. At the beginning of this section a few ways are shown to produce them by plasma polymer and copolymer deposition. Selective plasma processes are characterized by the dominant production of one type of functional group. Only low concentrations of other types of functional groups are accepted and post-plasma oxidations should be absent [23, 26]. Selective plasma processes are rare with the plasma technique. The is due to the permanent excess of energy, the high degree of freedom for sticking plasma species from the gas phase onto the surface, and the unavoidable, permanent, and energy-rich plasma UV irradiation.

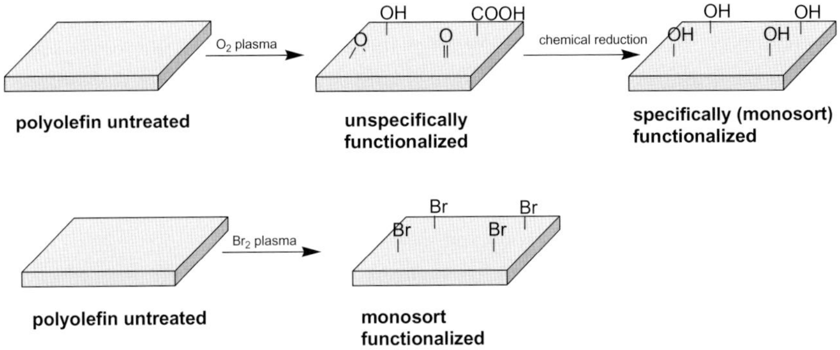

Scheme 9.4 Types of polymer surface functionalization.

Nearly all plasma-activated processes are radical reactions and additionally plasma UV irradiation further produces radicals at the surface, in surface-near layers, or even in the bulk of polymers [14]. However, not all C-radical sites are consumed during the plasma process. Thus, unsaturated and metastable trapped free radicals remain. They are the reason for post-plasma oxidation. This auto-oxidation is also not selective. It is responsible for the observed broad spectrum of O-functional groups. Thus, the need for a selective and specific surface (oxygen) functionalization was demonstrated (Scheme 9.4).

Unspecific (non-selective) surface functionalization is the normal situation. At this stage, the amateur begins to optimize the plasma parameters and hopes to achieve more selectivity in the formation of monosort functional groups. However, this does not make sense because the selectivity can only be marginally improved. There is the general conflict between the too-high energy in the plasma and the needed low energies for bond dissociation such as H-abstraction from the polymer backbone. In particular, the high-energy tail of the electron energy distribution function exceeds the binding energies in polymers significantly. The high ionization energies of atoms and molecules (>10 eV) and also the corresponding energies of other plasma-relevant features such as metastables, the short wavelength of the radiation ($\lambda < 200\,\text{nm}$), the permanent power input, and the need to sustain the discharge make it impossible to converge the energy level in the plasma to that of chemical bonds ($\Delta_R H \approx 3.0$–$4.5\,\text{eV}$). This fundamental discrepancy between plasma physics and chemical processing must be accepted. Pulsed plasma and remote plasma as well as lowering the wattage as much as possible may improve the situation marginally.

Nevertheless, surprisingly, one plasma-initiated functionalization reaction is more or less selective because of special circumstances. The reason for this lies in the thermodynamics of these reactions and in the absence of any possibility for side-reactions. Such a very rare case is plasma bromination using bromine (Br_2) or bromoform ($CHBr_3$) as precursors for the production of bromine atoms [27, 28]. This bromination produces monosort functionalization with C–Br groups. Bromine aims to fill the outer electron shell with electrons to adapt to the energetically favorable noble gas structure. Therefore, all halogen X (bromine) introductions have the same tendency to form exclusively C–X(Br) or X^- (Br^-). X^- can be removed easily from surface by any wash process. Only C–X groups remain. Other side-reactions, such as the trapping of radicals and their slow auto-oxidation, are present in the process of plasma fluorination (and chlorination) [29] but are absent in bromination with bromine or bromoform precursor [30]. Bromination is endothermic and does not produce additional enthalpy, which produces thermally initiated chain scissions and new radical sites.

Bromine-functionalized polyolefin surfaces have been used for wet-chemical graft reactions following nucleophilic substitution reactions (e.g., Williamson ether synthesis) to modify the polymer surfaces so as to improve adhesion to metals and to compatibilize them with biological systems.

9.4.2
History of the Plasma Bromination Process

Bromination was introduced by Friedrich *et al.* in 1989 (J. Friedrich, and Y. Novis, unpublished results, 1989) [31–34]. Later, the bromination process was investigated in more detail [35]. Then, the pulsed plasma technique was applied to brominate polyolefin surfaces using bromine and bromoform as plasma gas (vapor) and Br-containing monomers for depositing Br-carrying plasma polymer layers onto polyolefin surfaces [23].

9.4.3
Theoretical Considerations on the Plasma Bromination Process

The average electron energy and, therefore, position of the energy distribution function within the energy scale in the plasma depend on the ionization energy of the plasma gas molecules. The lower the ionization energy the lower is the average electron energy in the plasma. The ionization energies of the halogen-containing precursors decrease in succession from fluorine to iodine (Table 9.3) [36].

Table 9.3 Ionization potentials of halogen-containing plasma gases (and methane).

Precursor	Ionization potential (eV)
CF_4	17.8
F_2/F	15.8/17.4
CCl_4	11.5
$HCCl_3$	11.4
Cl_2/Cl	11.5/13.0
$CHBr_3$	10.5
Br_2/Br	10.5/11.8
CH_2I_2	9.3
CH_4	13.0

From Table 9.3, it can be expected that the average electron energy or electron temperature of bromine and iodine plasmas are more moderate than those of fluorine or noble gas plasmas such as from He, Ne, and Ar. It is assumed that an electron energy level nearer to the binding energies in polymers produces less degradation and crosslinking. Moreover, the radical formation and the associated auto-oxidation during the subsequent weeks or months may be lessened [20]. Bromoform ($CHBr_3$) and bromine (Br_2) have a sufficient vapor pressure for easy processing in plasma technology.

Under plasma conditions the bromoform molecule dissociates as proposed, thereby, on considering the different bond dissociation energies in the bromoform molecule:

$$CHBr_3 \rightarrow H^\bullet + {}^\bullet CBr_3 \; \Delta_D H^\theta = 389 \text{ kJ mol}^{-1}$$

$$CHBr_3 \rightarrow HCBr_2^\bullet + {}^\bullet Br \text{ and so on } \Delta_D H^\theta = 230 \text{ kJ mol}^{-1}$$

Bromine atoms (and also CBr_3 radicals) can add to olefinic double bonds:

$$2^\bullet Br \; (Br_2) + {>}C{=}C{<} \rightarrow {>}CBr{-}CBr{<}$$

and can attack aromatic rings or aliphatic chains by electrophilic or nucleophilic substitution as follows:

$$2^\bullet Br \; (Br_2) + C_6H_6 \rightarrow C_6H_5\text{-}Br + HBr$$

$$^\bullet Br + R\text{-}H \rightarrow R^\bullet + H\text{-}Br$$

$$^\bullet Br + R^\bullet \rightarrow CBr$$

or attack the C–H bond at allyl position:

$$2^\bullet Br \; (Br_2) + CH_2{=}CH\text{-}CH_3 \rightarrow CH_2{=}CH\text{-}CH_2Br + HBr$$

Using ethane as a model for polyethylene, the following thermodynamic calculation of the heat of reaction is possible considering Hess' rule and also entropy influences:

$$\Delta_R H^\theta = \Delta_D H^\theta{}_{CH3-CH2-H} - \Delta_D H^\theta{}_{H-Br} = 411 - 363 = +48 \text{ kJ mol}^{-1}$$

The positive reaction enthalpy shows that the reaction needs additional activation enthalpy (irradiation, plasma), in contrast to fluorination ($\Delta_R H^\theta = -155 \text{ kJ mol}^{-1}$) [37].

The electrophilic addition of Br to aromatic double bonds does not work:

$$Br_2 + C{=}C \neq CBr{-}CBr$$

unless additional activation energy is supplied:

$$2^\bullet Br + C{=}C \rightarrow CBr{-}CBr$$

and may also occur in graphite [38]:

$$2^\bullet Br + {>}C{=}C{<} \rightarrow {>}CBr{-}CBr{<}$$

It must be mentioned that chemically neutral bromine precursors are preferred for industrial processing because elemental halogens are chemically very aggressive. In particular, fluorination (of polyolefins) is thermodynamically preferred and therefore strongly exothermic so that it is difficult to control and not selective using gas-phase processing instead of liquid-phase processing [39]. Thermal control of fluorination is possible for processing in Freon® liquids under reflux conditions [40].

Otherwise, under self-heating conditions, fluorination with F_2 produces chain scissions and radical formation. For a given aromatic or aliphatic polymer the reactivity of radicals with it shows following succession [4]:

$$F^\bullet > HO^\bullet > Cl^\bullet > CH_3^\bullet > Br^\bullet > R{-}O{-}O^\bullet$$

To minimize the energy excess further and to avoid more efficiently any side-reactions, such as radical formation and auto-oxidation, the use of iodine instead bromine would seem evident because of the lower ionization energy and, therefore, because of the softest halogenation conditions. The C–I groups are also well-suited for further chemical reactions. Elemental iodine has a lower dissociation energy than bromoform and other neutral precursors. However, the energy profit from the formation of the H–I bond is too low:

$$I^\bullet\ R{-}H \rightarrow I{-}H + {}^\bullet R$$

$$\Delta_R H^\ominus{}_{298} = \Delta_R H^\ominus{}_{R\text{-}H} - \Delta_R H^\ominus{}_{I\text{-}H} = 411 - 295 = 116\ \text{kJ mol}^{-1}$$

Therefore, this reaction is endothermic and thermodynamically not possible. It may be that under plasma conditions this reaction occurs but it would be possess an insufficient yield. In contrast, iodine is well-known as a good inhibitor of radical reactions.

Thus, bromine is the best candidate for selective introduction of covalently bonded halogen atoms onto polymer surfaces. The bromine atom does not give energy profit; it is endothermic but only little additional plasma energy is needed. Therefore, the kinetic chain length of bromination is low. However, it shows the highest specificity (selectivity) in attacking different carbon C–H features. Bromine has an electrophilic character. Therefore, it attacks concentrations of high electron density. Tertiary C–H bonds reacted 1600 times faster and secondary C–H 32 times faster than primary C–H bonds (= 1), which is important if polypropylene is treated [4]. Under mild plasma conditions (low power input) the bromination of aliphatic polymers with bromoform is a selective reaction. Plasma bromination is dominated by the chemical reactions of free bromine atoms and only slightly superposed by reactions caused by the plasma energy excess. Among all other plasma processes for introduction of functional groups bromination is the most selective process and chemically useful and reactive C–Br groups are formed.

Bromoform as precursor is easy to process, bromine is difficult to dry (more side-reactions because of humidity traces) and it is chemically aggressive and corrodes the metal of the plasma reactor. Bromoform tends additionally to form Br-containing plasma polymer films. Therefore, the bromination process is dominated

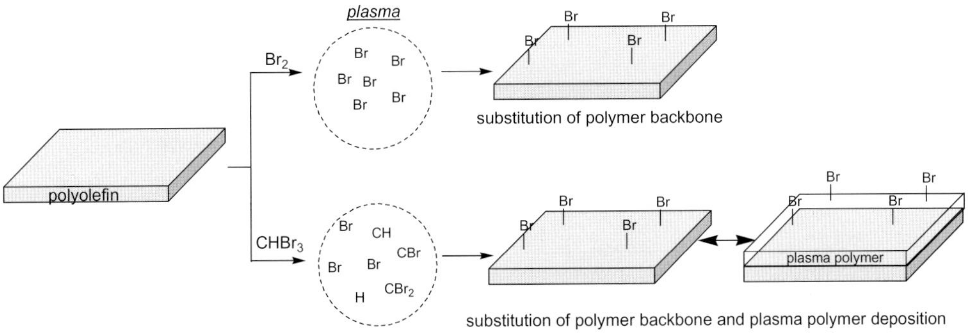

Scheme 9.5 Differences between bromoform and bromine plasma.

by Br-substitution at polymer substrate but is also accompanied with polymer deposition depending on plasma conditions and sample location in the plasma (Scheme 9.5) [33].

The plasma serves as source of free bromine atoms, as shown before, and film-forming CBr-fragments. Under specific conditions of plasma-induced bromoform dissociation the substitution process blocks the film formation process. Generally, elemental bromine is only suitable for substitution or addition and not for film formation [23].

Plasma-assisted bromination of polypropylene with elemental bromine yields 60 Br/100 C and with bromoform 100 Br/100 C:

$$Br_2 + plasma \rightarrow 2^{\bullet}Br$$

$$\text{-CH}_2\text{-CH(CH}_3\text{)-} + 2^{\bullet}Br \rightarrow \text{-CH}_2\text{-CBr(CH}_3\text{)-} + HBr$$

or:

$$\text{-CH}_2\text{-CH(CH}_3\text{)-} + plasma \rightarrow \text{-CH}_2\text{-C}^{\bullet}\text{(CH}_3\text{)-} + H^{\bullet}$$

$$\text{-CH}_2\text{-C}^{\bullet}\text{(CH}_3\text{)-} + {}^{\bullet}Br \rightarrow \text{-CH}_2\text{-CBr(CH}_3\text{)-}$$

Using bromoform alternative reactions are possible such as attachment of CBr_3:

$$\text{-CH}_2\text{-CH(CH}_3\text{)-} + {}^{\bullet}CBr_3 \rightarrow \text{-CH}_2\text{-C(CBr}_3\text{)(CH}_3\text{)-} + H^{\bullet}$$

or the formation of a thin plasma-polymer layer $C_nH_xBr_y$ by polyrecombination of bromoform fragments roughly as (cf. Scheme 9.5):

$${}^{\bullet}CBr_3 + n{}^{\bullet}CBr_2{}^{\bullet} + {}^{\bullet}CBr_3 \rightarrow Br_3C\text{-[}CBr_2\text{]}_n\text{-}CBr_3$$

The loss in Br concentration of plasma-brominated polymer samples during rinsing with tetrahydrofuran (THF) to remove low-molecular weight components confirms the above-assumed reaction to low-molecular weight linear species. Moreover, plasma-induced degradation to low-molecular weight components during bromination may also contribute to the soluble fraction (Scheme 9.6).

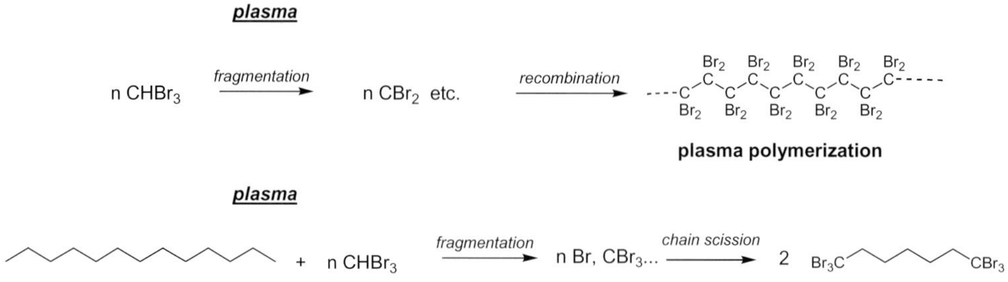

Scheme 9.6 Tentative chain bromination and plasma polymerization route.

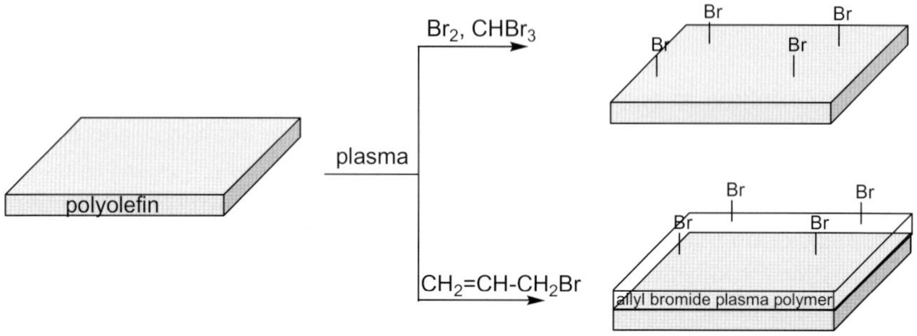

Scheme 9.7 Two routes for producing bromine-rich polyolefin surfaces; the use of allyl bromide (bottom route) also produces a bromine-rich thin polymer film.

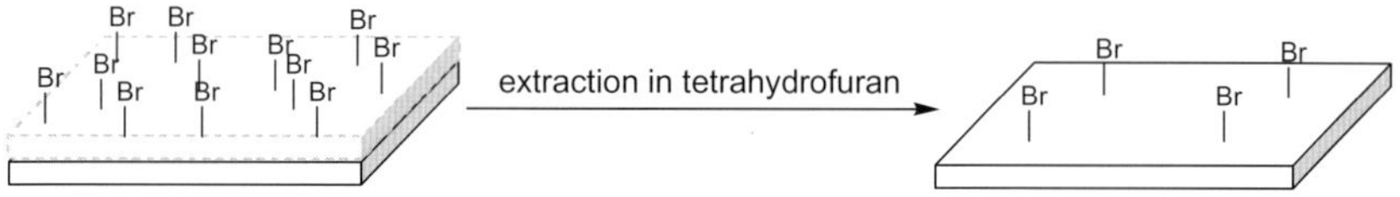

Scheme 9.8 Removal of loosely bonded material by solvent extraction.

Another possibility of polyolefin surface bromination is the deposition of bromine-containing plasma-polymers [23, 29]. Vinyl bromide and allyl bromide are suitable precursors of bromine-rich polymer thin film deposits (Scheme 9.7) [24].

For further wet-chemical processing it is necessary to remove loosely bonded bromine-containing fragments at the brominated polyolefin surfaces by an intense wash process using solvents (Scheme 9.8).

Some 50–75% of plasma-inserted C-Br groups could be removed from polyolefin surfaces by extraction with solvents [24].

Because the C–H and the C–C binding energies in aliphatic polymer backbones (PE, PP, PS . . .) are nearly identical (e.g., $-CH_2-H = 411$, $>CH-H = 396$, $=C-H = 389$, and $-CH_2-CH_2- = 370\,kJ\,mol^{-1}$) [4] and the entropy of the process should be have no influence (same number of educts and products), selective abstraction of H atoms from the polymer molecule is not possible (shown here for ethane):

$$H_3C\text{-}CH_2\text{-}H + Br_2 \rightarrow H_3C\text{-}CH_2\text{-}Br + HBr$$

and:

$$H_3C\text{-}CH_3 + Br_2 \rightarrow 2H_3C\text{-}Br$$

Thus, the C–C bond scissions and C–H substitution have similar probability or the C–C bond scission should dominate. As mentioned before, however, the C–C backbone is slightly shielded from too extensive chain scission [4]. In conclusion, the message is that the functionalization process is principally accompanied with the occurrence of a moderate number of chain scissions (polymer degradation). However, this fact can be generalized to all functionalizations and to all methods used for them – polyolefin surfaces cannot be functionalized without scissions of backbone. In contrast, poly(vinyl chloride) can be additionally chlorinated; however, here activated H–C bonds are available and chlorine atoms shield the backbone from the thermodynamically favored chain scissions.

Bromine-functionalized polyolefin surfaces can be used for several wet-chemical graft reactions.

9.4.4 Bromination Using Bromoform or Bromine Plasmas

The direct substitution of hydrogen from polymer backbones by bromine atoms was dominated by the use of bromine, bromoform, and *tert*-butyl bromide, selecting a special sample location in the reactor far from wall. This was roughly confirmed by co-exposure of Si-wafer surfaces, which did not show any or only marginal deposition of plasma polymers. The bromoform plasma was most efficient for introduction of bromine to a polypropylene surface in comparison to plasmas with elemental bromine or *tert*-butyl bromide as well as polymer-forming plasmas (Figure 9.4) [1].

The often-occurring post-plasma oxygen introduction amounts only 1–3% O/C, whereas more than 100 Br per 100 C could be introduced into the polypropylene surface. Obviously, from Figure 9.4, direct bromination of the polymer substrate has a higher yield in C–Br groups and less oxygen incorporation. The plasma polymer deposition can be made more efficient by addition of elemental bromine to the allyl bromide monomer, tentatively proposed as:

$$n CH_2{=}CH\text{-}CH_2Br + n Br_2 + \textit{plasma} \rightarrow \text{-}[CHBr\text{-}CBr(CH_2Br)]_n\text{-}$$

The high yield in Cl-introduction using chloroform may be attributed to the enhanced formation of a Cl-containing plasma polymer. Notably, the C–Cl bond is stronger than the C–Br bond and C–H contrawise. Therefore, the following

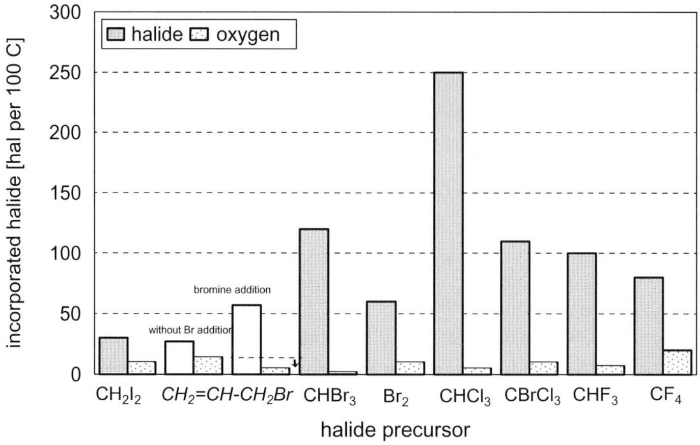

Figure 9.4 Efficiency of different bromination processes on polyethylene surfaces (each cw-rf plasma 100 W, 4 Pa, 180 s).

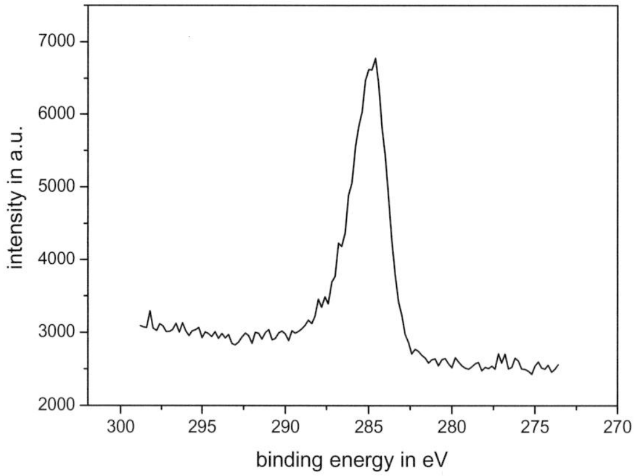

Figure 9.5 C1s peak of PP after bromination with bromoform using the pulsed rf plasma (100 W, 30 s, 6 Pa, duty cycle 0.1).

reaction also becomes probable, which may be responsible for high the yield in C-Cl by preferred attachment of CCl_3 instead of Cl:

$$CHCl_3 + plasma \rightarrow {}^{\bullet}CCl_3 + {}^{\bullet}H$$

$$CH^{\bullet} + {}^{\bullet}CCl_3 \rightarrow CH\text{-}CCl_3$$

XPS spectra show an unspecific broadening of the C1s signal (Figures 9.5) and the expected doublet of the Br2p peak (Figure 9.6).

The O1s peak only slightly rises above the noise. The existence of only each two bromine peaks ($Br3p_{1/2}$ and $Br3p_{3/2}$; $Br3d_{3/2}$ and $Br3d_{5/2}$) clearly indicates the

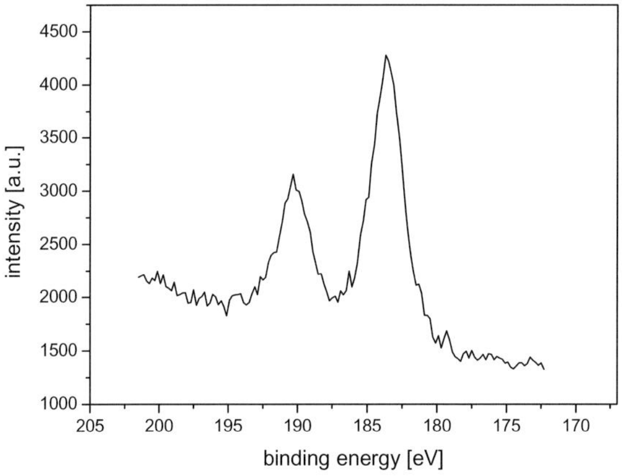

Figure 9.6 Br3p$_{1/2}$ and $_{3/2}$ peak of PP after bromination with bromoform using the pulsed rf plasma (100 W, 30 s, 6 Pa, duty cycle 0.1).

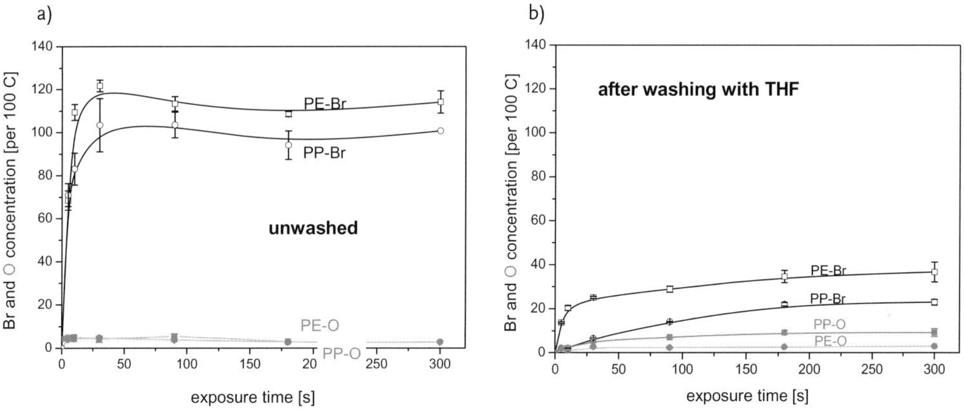

Figure 9.7 Br and O introduction onto polypropylene or polyethylene surfaces using the bromoform (tribromomethane) cw plasma process (8 Pa, 13.6 MHz, 100 W) (a) and (optionally) a post-plasma immersion of the sample in THF under ultrasonic exposure (b) (lines are guides for the eye only).

exclusive existence of only C-Br species at the surface of the brominated polypropylene.

The bromination depended on the plasma parameters. The introduction of bromine atoms onto the polypropylene surface increased linearly with time (Figure 9.7). The oxygen content did not exceed after intense extraction with tetrahydrofuran (THF) 4 O per 100 C atoms. This washed surface was the base for further chemical graft synthesis.

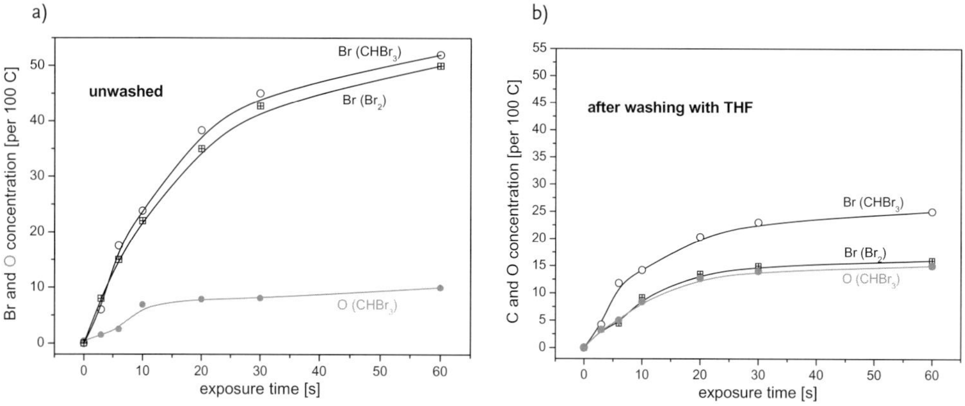

Figure 9.8 Br and O introduction onto polypropylene surfaces using the bromoform (tribromomethane) and bromine pulsed plasma process (8 Pa, 13.6 MHz, 100 W, duty cycle 0.1, 1000 Hz) (a) and post-plasma immersion of the sample in THF under ultrasonic exposure (b) (lines are guides for the eye only).

The maximal yield in Br introduction of the bromoform cw-plasma was 100 Br/100 C, after 180 s treatment, applied to polypropylene and 120 Br/100 C, after 180 s treatment, applied to polyethylene (cf. Figure 9.7), or in the case of elemental bromine cw-plasma the yield in both instances was 50 Br/100 C atoms (60 s). The post-plasma wash process in THF (30 °C) intensified the entrance of oxygen to the plasma-treated polymer foil, but only marginally.

Remember that the washing was used to remove plasma-produced mechanically unstable and therefore loosely bonded fragments, which may hinder a permanent chemical grafting of spacer molecules onto the polymer substrate. However, the wash process removed up to 50–75% of the polymer fragments with attached Br from the polyolefin surface (Figure 9.7). Notably, the remaining Br density (20–25 Br/100 C) is high enough to perform all grafting synthesis with high yield.

Using the pulsed plasma mode (duty cycle 0.1, 1000 Hz) the time-dependence of the Br introduction into polyolefin surfaces is similar to that of cw plasma, with, however more (post-plasma) oxygen introduction than with use of cw mode (Figure 9.8).

The bromination rates are comparable for the two modes. Pure bromine vapor plasma shows the same bromination rate as bromoform plasma.

As shown in Figure 9.9 the Br concentration at the washed polymer surface is strongly increased with growing power input to the plasma. It can be assumed that the dissociation rate of bromoform molecules increases with growing wattage and also that a stronger crosslinking appears and, therefore, the solvent resistivity of the brominated topmost surface layer should be improved. The very high Br concentrations measured after bromoform plasma exposure at polyethylene and

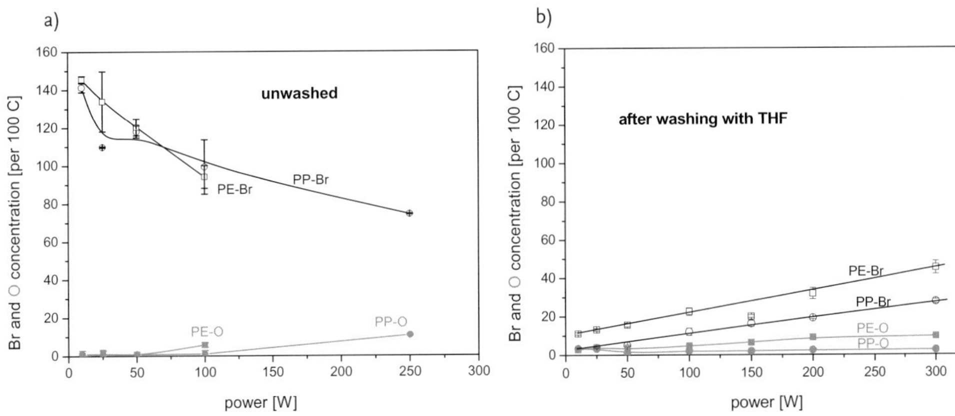

Figure 9.9 Comparison of the surface bromination of PE and PP using the cw mode of the rf bromoform (tribromomethane) plasma (13.56 MHz, 8 Pa, 180 s) unwashed (a) and after washing the samples in THF (b) according to the power input (lines are only guides for eyes).

polypropylene surfaces for low power input (<100 W) were attributed to the intensified deposition of a Br-rich plasma polymer under such special plasma conditions, as measured also using a quartz microbalance. This deposit was easily removed by washing with THF (Figure 9.9).

It was confirmed by Langmuir probe measurements of electron temperature in Br-containing plasma that a relatively low energy level is present in the bromoform plasma (3.1 eV average electron energy, $1.1 \times 10^{15}\,cm^{-3}$ electron density at 100 W). Besides the radical scavenging property of Br atoms ($C^{\bullet} + {}^{\bullet}Br \rightarrow C{-}Br$) this may be an additional reason for the observed absence of excessive auto-oxidation reactions.

9.4.5 Bromination Using Allyl Bromide Plasma

Another general possibility for producing bromine-rich surfaces is the deposition of thin bromine-containing plasma polymers. Vinyl bromide and allyl bromide are suitable candidates for thin film deposition. The highest yield in bromine concentration was achieved using allyl bromide (Figure 9.10).

This polymer deposition process is limited by the stoichiometry of the precursor molecule, for example, the resulting bromine concentration at the polymer surface correlates with the C/Br ratio in the precursor molecule. This threshold C/Br ratio for allyl bromide is 33 Br per 100 C. While passing through the plasma zone and depositing onto the polymer substrate about 20% of all bromine atoms are missed, so that a maximal 27 Br per 100 C remain (Figure 9.10). However, two disadvantages must be emphasized. First, the adhesion of allyl bromide plasma-polymer

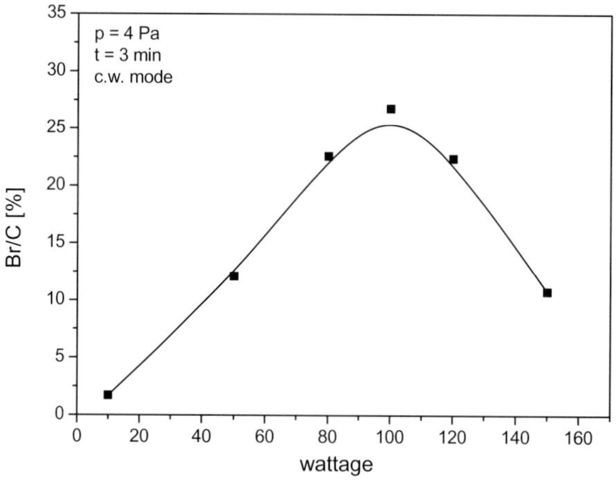

Figure 9.10 Bromination of polypropylene surfaces by deposition of an allyl bromide plasma polymer layer (cw-rf, 100 W, 4 Pa, 180 s).

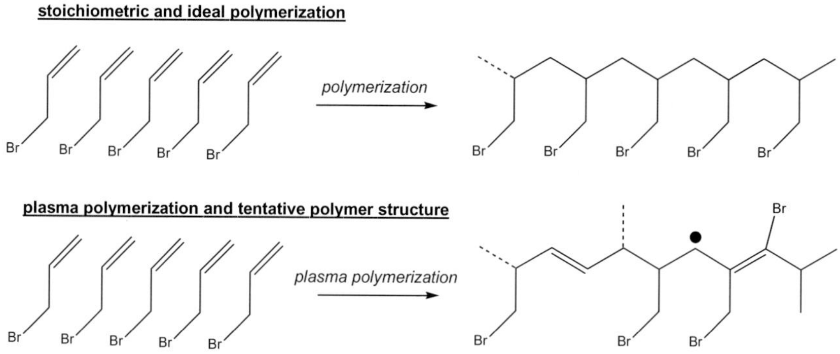

Scheme 9.9 Proposed model of plasma polymerized allyl bromide.

film to the polymer substrate was moderate and, second, the high concentration of trapped radicals within this layer. These radicals, responsible for the pick-up of oxygen from air, start the auto-oxidation. Some 5–7 undesired oxygen functionalities per 100 C were introduced, thus increasing the quantity of by-products (Scheme 9.9) [1].

A maximum in Br-introduction was estimated at about 100 W rf power input using the continuous-wave (cw) mode (Figure 9.10). However, this maximum of Br introduction was accompanied by the undesired maximum of about 7 co-introduced O species per 100 C. The optimum pressure ranges from 4 to 7 Pa under the given conditions of wattage (cw mode, 100 W) and reactor geometry.

9.4.6
Grafting onto Bromine Groups

The C–Br groups can be easily consumed by nucleophilic substitution reactions using diols and glycols (Williamson's ether synthesis) or diamines:

$$\blacksquare\text{-Br} + (\text{Na}^+)\text{O}^-\text{-R-OH} \rightarrow \blacksquare\text{-O-R-}\mathit{OH} + \text{NaBr}$$

$$\blacksquare\text{-Br} + (\text{Na}^+)\text{O-CH}_2\text{-CH}_2\text{-[O-CH}_2\text{-CH}_2]_n\text{-OH} \rightarrow$$
$$\blacksquare\text{-O-CH}_2\text{-CH}_2\text{-[O-CH}_2\text{-CH}_2]_n\text{-}\mathit{OH} + \text{NaBr}$$

$$\blacksquare\text{-Br} + (\text{Na}^+)\text{NH}^-\text{-R-NH}_2 \rightarrow \blacksquare\text{-NH-R-}\mathit{NH_2} + \text{NaBr}$$

Using the bromoform plasma and thus brominated polypropylene surfaces with 20 Br per 100 C different amounts of (poly(ethylene glycolate)s (PEGs or poly(ethylene oxide) = PEO) could be grafted depending on the length of grafted chains (3–8 glycol spacer per 100 C) and 5 diamines ($H_2N–[CH_2]_n–NH_2$) per 100 C (Figure 9.11). PEG 1000 was grafted as 1 PEG per 100 C atoms.

Because of the higher nucleophilicity these diamines also react without Na-assistance as shown for ethylenediamine:

$$\blacksquare\text{-Br} + \text{NH}_2\text{-CH}_2\text{-CH}_2\text{-NH}_2 \rightarrow \blacksquare\text{-NH-CH}_2\text{-CH}_2\text{-}\mathit{NH_2}$$

The yield in this reaction amounts to 20 ethylenediamines or using larger molecules 5 diaminohexane/100 C.

This sodium-free reaction allows the gas-phase reaction of ethylenediamine immediately after ending of plasma bromination by exposure of the brominated polyolefin surface to diamines. Using ethylenediamine, 22 ethylenediamine/100 C were permanently bound after a short extraction in THF [37].

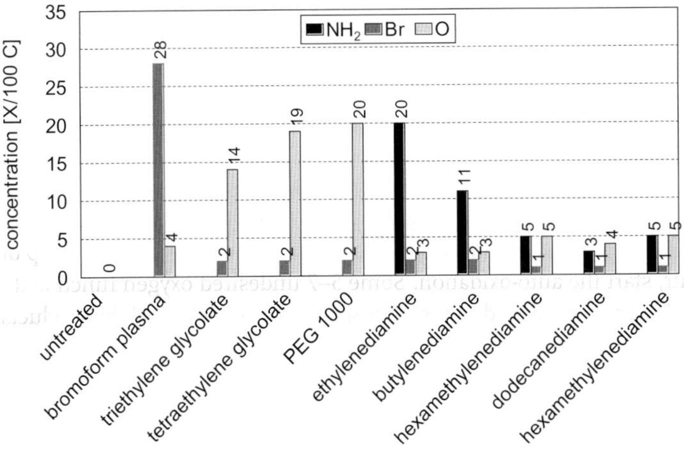

Figure 9.11 Maximal yield in grafting of glycols and amines onto bromoform-modified polypropylene surfaces.

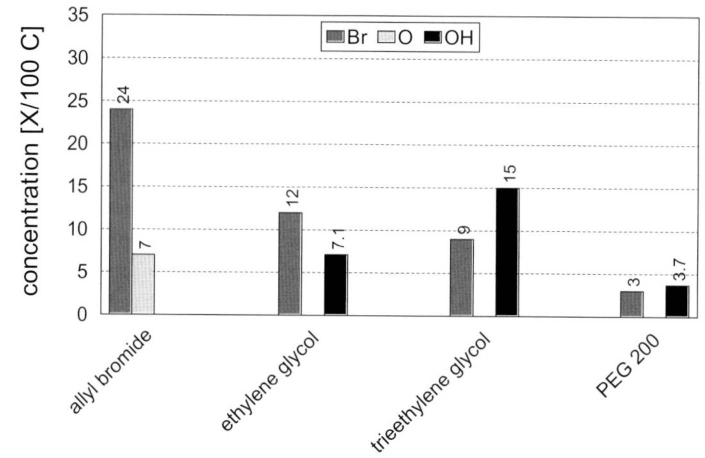

Figure 9.12 Yield in grafting of glycols starting from allyl bromide plasma-polymer.

Starting from plasma polymer layers made from allyl bromide with 24 Br/100 C and, however, characterized by the relatively high percentage of post-plasma introduced oxygen, 3.7–15 OH-terminated spacer molecules could be chemically grafted onto the brominated surface (Figure 9.12).

About 4 PEG200 chains were grafted onto C-Br, as also shown in Figure 9.12.

Similar yields were achieved using aliphatic diols (di-alcohols) of different chain lengths (Figure 9.13); 2–12 alcoholic spacers could be grafted, depending on chain length, using sodium assistance (Williamson synthesis).

Atomic transfer radical polymerization (ATRP) and click chemistry were also performed using brominated polyolefin substrates (Scheme 9.10) [37].

Other graft reactions of complex structure such as burst or comb are depicted in Figure 9.14. Applications in adhesion-promotion, new sensor structures, and new tribological surface properties of implants have been synthesized [37].

9.4.7
Yield in Density of Grafted Molecules at Polyolefin Surfaces

The grafting yield depended on the density of Br-anchoring groups at the polypropylene surface, their accessibility (depending on the size of the grafted molecule), on reactivity, on the general yield of the graft reaction, and so on. Thus, starting from about 10 C-Br groups per 100 C atoms about 4–9 molecules per 100 C atoms were grafted for small substituents, 1–4 for oligomeric substituents/100 C, and starting with 20–25 Br/100 C 15–18, and as record 22, small molecules (ethylenediamine) were grafted. Using voluminous molecules the graft yield was independent of the starting concentration of C-Br. The yield of C-Br consumption therefore depends on the volume of the grafted molecule and ranges between 10% and 90%. Table 9.4 compares grafting onto brominated and grafting onto oxygen plasma treated–diborane reduced polypropylene surfaces.

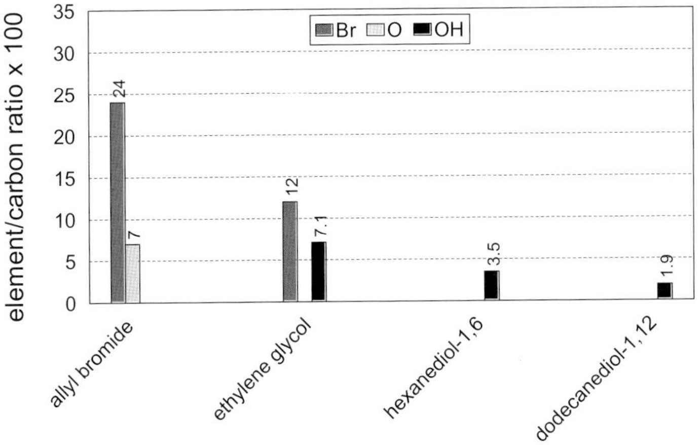

Figure 9.13 Yield in grafting of diols starting from allyl bromide plasmapolymer.

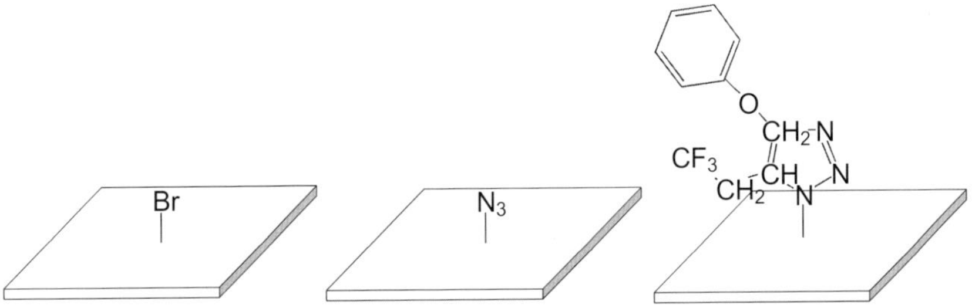

Scheme 9.10 Click reaction at brominated PE, then CBr converted into azide and consumed with substituted acetylene.

It is also possible to graft large cumbersome molecules with a dendrimer structure onto brominated polypropylene and polyethylene surfaces. Such molecules are needed to tailor metal–polymer interfaces. Such a spacer in metal–polymer interfaces should introduce covalent bonds, flexibilization, and water-repellent and barrier character. The spacers were bonded covalently to both the polymer and the metal and, thus, they are well-adherent, balancing mechanical stress along the interface, making it more durable against diffusion of humidity. To avoid redox reactions of evaporated electronegative metals with fluorine the use of hydrophobic siloxane-forming or containing molecules seems to be advantageous as water-repellent spacers (Chapter 7).

SiOSi-cage containing molecules such as POSS (polyhedral oligomers of silsesquioxanes) equipped with reactive functional groups positioned at side-groups offer the desired hydrophobic properties for interface tailoring. Here, as model, octa(amino-phenylene)-POSS ($C_{48}H_{48}N_8O_{12}Si_8$, molar mass: 1152 Da) was

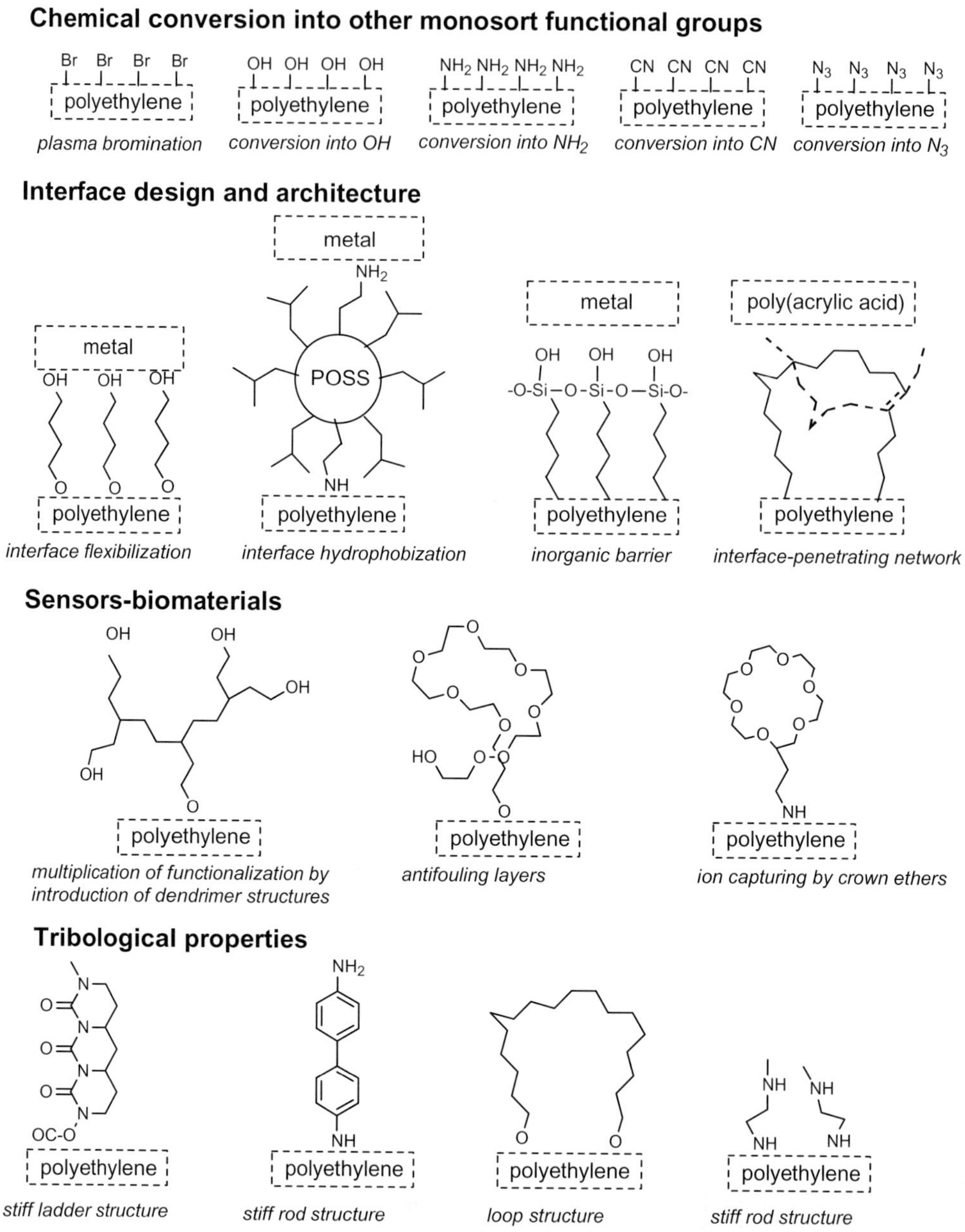

Figure 9.14 Molecules grafted onto brominated polyolefin surfaces.

Table 9.4 Spacer grafting onto OH and B groups at polypropylene surfaces.

Plasma treatment	Chemical processing	Number of OH per 100 C	O-total per 100 C	Grafted molecule	Number of grafted molecules per100 C	Number of OH as end-groups of the spacer per100 C
O_2 plasma	B_2H_6	10	20	Aminosilane	8	8 NH_2
O_2 plasma	B_2H_6	10	20	Alanine	5	–
O_2 plasma	B_2H_6	10	20	Cysteine	5	–
O_2 plasma	B_2H_6	10	20	Undecenylsilane	9	9 C=C
O_2 plasma	B_2H_6	10	20	Undecenylsilane	9	9
Plasma treatment	Chemical processing	Number of Br per 100 C	O-total per 100 C	Grafted molecule	Number of grafted molecules per100 C	Number of NH_2 as end-groups of the spacer per100 C
$CHBr_3$	–	20	9	Triethylene glycolate	6	6 OH
$CHBr_3$	–	20	7	Tetraethylene glycolate	5	5 OH
$CHBr_3$	–	20	7	PEG 1000	5	5 OH
$CHBr_3$	–	20	7	Hexamethylenediamine	5	5 NH_2

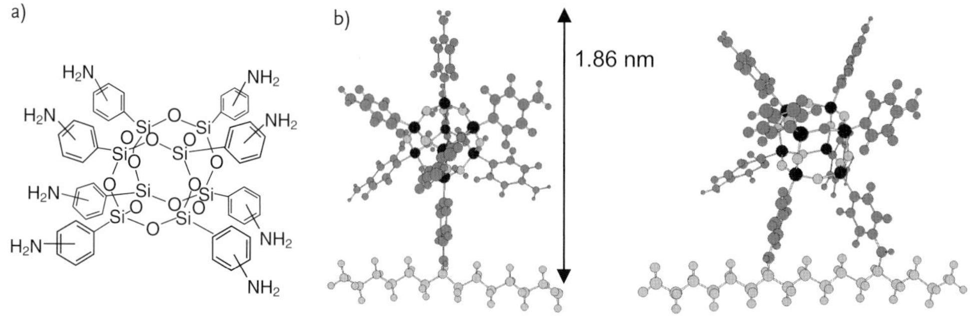

Figure 9.15 Octa(amino-phenylene)-POSS (a) and calculated configurations (b) after grafting onto partially brominated polyethylene surfaces by one or by two covalent bonds (black = Si, light shading = O).

grafted onto C-Br groups introduced onto polypropylene surfaces by exposure to bromoform plasma:

$$|\text{C-Br} + (\text{H}_2\text{N-phenylene})_8\text{-POSS} \rightarrow$$
$$|\text{C-NH-phenylene-POSS-(phenylene-NH}_2)_7 + \text{HBr}$$

XPS analysis confirmed the grafting of about 1.2 POSS molecules per 100 C atoms (Figure 9.15).

Considering the original number of about 20 Br per 100 C and the remaining number after reaction of 6–9 Br/100 C the linking of the amino-POSS to more than one C-Br group is probable. The single bonded and the multiple bonded cases are presented schematically in Figure 9.15b. Simple molecular dynamic calculations (MM2) generate these plots [37].

The diameter of this POSS molecule is about 1.9 nm. It occupies a volume of about 3 nm^2. Using this value and considering the graft density it can be roughly appreciated that ca. 80% of the polypropylene surface is covered by these POSS molecules. It must be remarked that this octamino-POSS was used only as model because unreacted NH_2 groups do not show the wanted hydrophobicity. To modifiy the interface only two functional groups are necessary, for bonding to polymer and to metal. The other side-chains of the siloxane cage may preferably consist of CH_3 or phenyl groups, thus producing more hydrophobicity.

The graft density of all diamines, glycols, and diols were determined by end-group derivatization with a fluorine-containing marker and measuring the introduced fluorine concentration using XPS as described before. Plotting the graft density versus molecule length, given by the number of CH_2 groups in the molecule, the correlation given in Figure 9.16 can be presented.

As mentioned before, extra-large molecules (molar mass 1000–5000 Da) can be grafted with 1 or 2 molecules per 100 C atoms; such grafted molecules are linked, for example, to about 4 nm^2 of the surface area. Smaller molecules can be grafted more densely (10–20 grafted molecules per 100 C atoms) to the same 4 nm^2, for example, 2–5 grafted molecules per 1 nm^2. It is assumed that nearly complete surface coverage is reached for small as well as for large molecules.

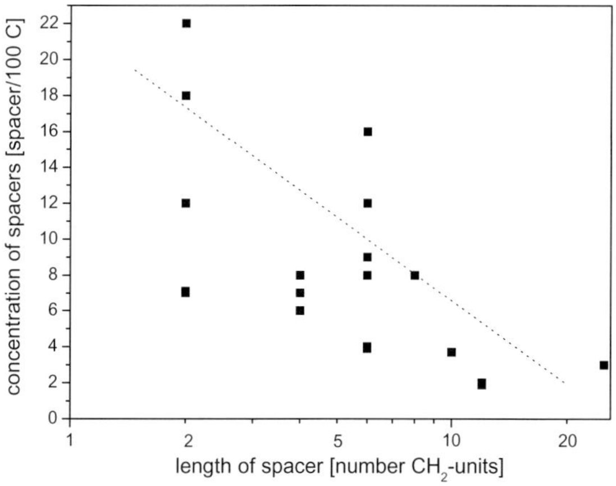

Figure 9.16 Dependence of the concentration of grafted spacer molecules on the length of spacer molecules (measured in terms of CH_2 groups).

9.4.8
Change of Surface Functionality

Because of the high density and long-time stability of plasma bromination the idea arose of re-functionalizing the C-Br group into the amino group in a simple chemical way, for example, to change the C-Br group into the NH_2 group. Notably, the plasma production of amino-group containing polymer surfaces is difficult, produces a low yield in NH_2 retention, shows several side-products, and is subject to undesired strong oxidation when exposed to air [41]. The substitution of Br in C-Br groups with NH_2 may be an alternative. Consumption with ammonia, though, was difficult. The best yield was 12 NH_2 per 100 C:

$$\blacksquare\text{-Br} + NH_3 \rightarrow \blacksquare\text{-}NH_2 + HBr$$

A higher yield in NH_2-groups was produced by grafting of ethylenediamine (20–22 NH_2/100 C):

$$\blacksquare\text{-Br} + NH_2\text{-}CH_2\text{-}CH_2\text{-}NH_2 \rightarrow \blacksquare\text{-}NH\text{-}CH_2\text{-}CH_2\text{-}NH_2 + HBr$$

The reaction with $NaNO_2$ followed by reduction to amino groups had a yield of about 4 NH_2 per 100 C:

$$\blacksquare\text{-Br} + NaNO_2 \rightarrow \blacksquare\text{-}NO_2$$

$$\blacksquare\text{-}NO_2 + NaBH_4 \rightarrow \blacksquare\text{-}NH_2$$

NaN_3 also generates 4 NH_2 per 100 C, by coupling the bromine with sodium azide:

$$\blacksquare\text{-Br} + NaN_3 \rightarrow \blacksquare\text{-}N_3$$

and exposing it to low pH:

$$\blacksquare\text{-}N_3\ (+\,H^+) \rightarrow \blacksquare\text{-}NH_2$$

Another route consists of consumption with potassium cyanide:

$$\blacksquare\text{-Br} + \text{KCN} \rightarrow \blacksquare\text{-}C{\equiv}N$$

and reduction of the formed nitrile group with $LiAlH_4$, leading to yields in NH_2 groups of about 3 NH_2 per 100 C:

$$\blacksquare\text{-C}{\equiv}\text{N} \rightarrow \blacksquare\text{-CH}_2\text{–}NH_2$$

It was found that a stable brominated polymer surface can easily be converted into a stable functionalized amino group. The chemical change from Br to NH_2 offers a new solution for stable amino-groups bearing polymer surfaces because amino group introduction via the ammonia plasma has no relevant yield [20] and the deposition of allylamine plasma-polymers produces only a yield of 8–18 NH_2 groups 100 C (e.g., 30–55% of the introduced amino groups remain in the respective allylamine plasma-polymer) connected with 10–20% post-plasma introduction of oxygen [21, 42–44]. After storage of the chemically converted NH_2 group containing polypropylene surface in air via bromination no relevant oxygen pick-up was measured (e.g., the surface functionalization was stable towards oxygen from ambient air).

Scheme 9.11 presents a schematic overview of chemical refunctionalizations starting from about 20 Br/100 C atoms at the polyolefin surface after plasma bromination.

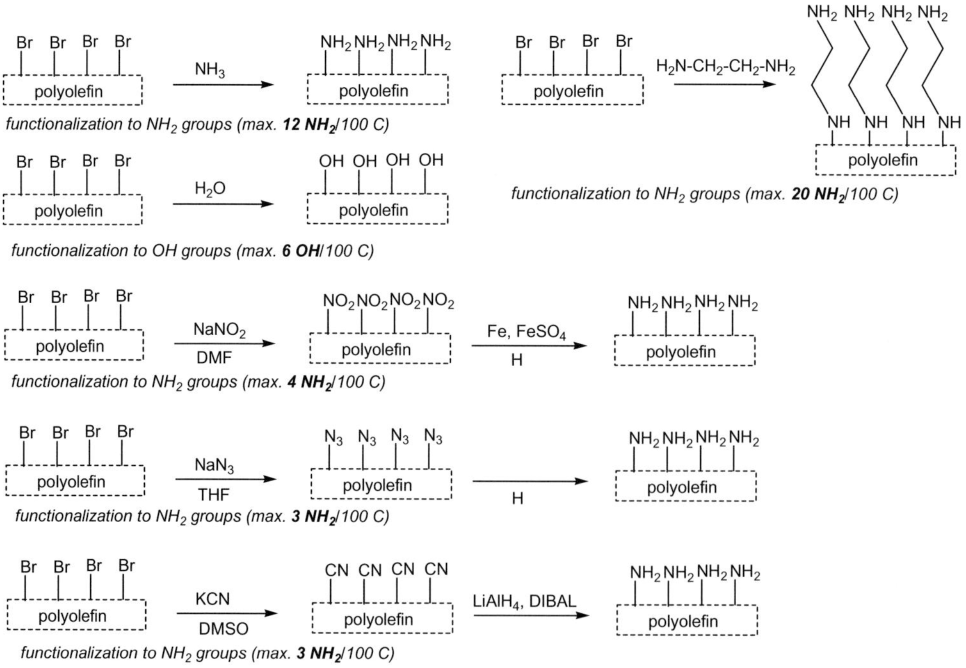

Scheme 9.11 Schematics of polyolefin refunctionalization after plasma bromination.

9.4.9
Surface Bromination of Polyolefins: Conclusions

It was possible to present the bromination as a selective plasma-chemical route to produce monotype functionalized polymer surfaces using the one-step bromoform (or bromine) plasma process. More than 50 bromines (Br_2) and >120 Br (bromoform) per 100 carbon atoms were introduced, of which about 20 Br per 100 C were permanently bonded to the polymer substrate. The low ionization potential of bromoform is responsible for the "soft" plasma conditions used during the bromination as well as the strong radical quenching property of bromine atoms. The alternative process, the deposition of thin plasma polymer top-coatings made of allyl or vinyl bromide, results also in 20–25 Br/100 C. Additional Br_2 increases the yield in Br to 60 Br/100 C but generally this process was accompanied by a very high post-plasma introduction of oxygen as undesired by-product.

Plasma bromination produces the highest yield in monosort functional groups and the highest selectivity in their production as compared with other types of polyolefin pretreatment for the formation of monosort functional groups (Figure 9.17).

Up to 80–90% of all bromines could be used for grafting reactions of small molecules of diols, glycols, and diamines independently of the production route if using bromoform, bromine, or allylamine. The larger the grafted molecule the lower was the consumption degree of bromine sites at the polymer surface. It was also possible to graft polymer chains onto the bromine sites in respectable yields.

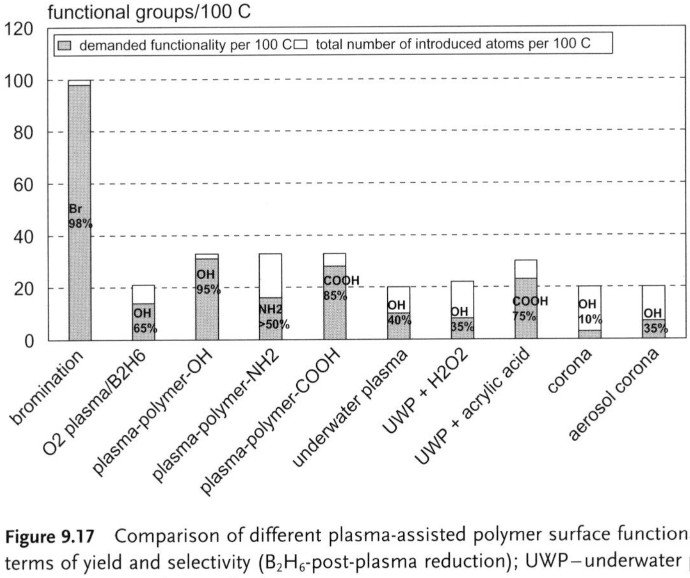

Figure 9.17 Comparison of different plasma-assisted polymer surface functionalizations in terms of yield and selectivity (B_2H_6-post-plasma reduction); UWP – underwater plasma; corona – dielectric barrier discharge.

Nearly all bromine atoms were removed from the surface by the graft process, thus only traces of Br remained. However, during chemical processing additional anchoring groups were lost. It is assumed that after washing with THF also during chemical processing additional Br-containing debris is dissolved and removed.

Another application of the highly selective plasma bromination may be the change from bromine functionalization to amino functionalization using ammonia gas, liquid ammonia, or diamines. Here, the way to a stable and dense amino-functionalized polymer surface is opened up, which promises more stable amino surfaces and a minimum of by-products.

The applications in focus are a new elastic type of metal–polymer bonding by introduction of flexible spacer via bromination and the use of amino-group-bearing polymer surfaces in biochips as well as biocompatible material surfaces.

9.4.10
Bromination of Poly(ethylene terephthalate)

XPS spectra show an unspecific broadening of the C1s and O1s signals of poly(ethylene terephthalate) (PET) and the disappearance of their fine structure (Figure 9.18). This PET had been exposed to the bromoform plasma for 180 s and shows the introduction of up to 23 Br/100 C-atoms. Detailed inspection of the structure of brominated PET was not possible using exclusively XPS. The appear-

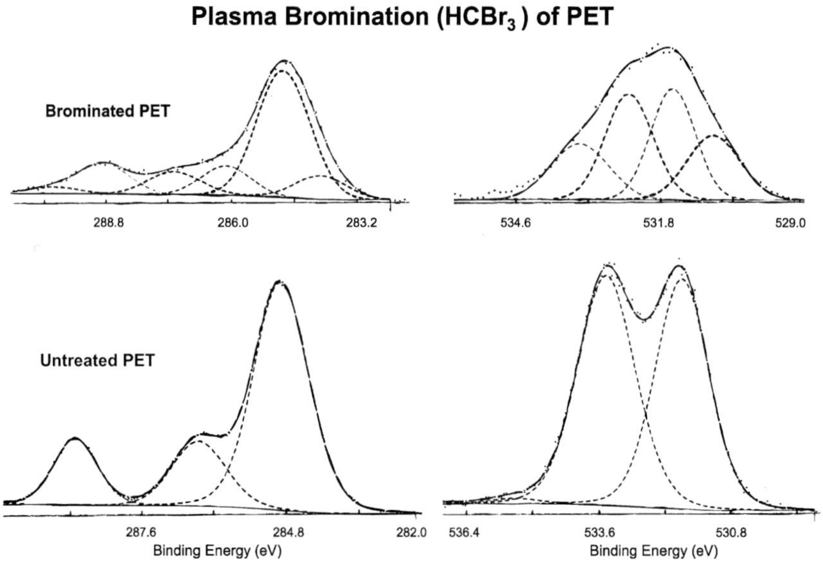

Figure 9.18 Bromination of poly(ethylene terephthalate) in a bromoform plasma (cw-rf plasma, 300 W, 10 Pa, 180 s) as shown by the XPS-measured C1s and O1s signals before and after plasma exposure.

ance of new oxidized (also brominated) C-species at the high binding side of the XPS-C1s spectrum and the decrease of the main signal at 285 eV indicates the bromination of the aromatic ring and its cracking.

9.5 Functionalization of Graphitic Surfaces

Graphitic materials play an important role in fiber-reinforced polymer composites containing carbon fibers, conducting graphite, fullerenes, or carbon nanotubes. Therefore, this class of materials is important to highlight. Often, it is necessary to couple these carbon materials by covalent bonds to polymer matrices or resins so as to improve the adhesion between them. First, carbon fibers were treated in ammonia plasma to introduce NH_2 groups and link them with epoxy groups ($CH_2{}^{O}CH$) of an epoxy resin matrix [45]:

$$\text{▌-}NH_2 + \text{epoxy-R} \rightarrow \text{▌-}NH\text{-}CH_2\text{-}CH(OH)\text{-}R$$

Plasma processing is a suitable method for overcoming the inertness of graphene structures and make them chemically interacting with polymers [38].

9.5.1 Bromination with Bromine Plasma

Graphite structures are formed from aromatic C=C double bonds substituted by further carbon atoms. Hydrogen atoms suited for substitution reactions are not present. Therefore, a completely different mechanism is observed, namely, the electrophilic addition to completely C-substituted aromatic double bonds. Generally, atoms or groups are easily added to olefinic double bonds. A prominent example is the bromination of residual C=C double bonds in polyolefins used for their identification. Electrophilic addition of bromine to aromatic double bonds does not work. Bromine attacks the aromatic ring by substitution of a hydrogen atom. It is introduced as a substituent without a change to the aromatic double bond. If the aromatic double bond does not possess hydrogen (is completely substituted by C residues) an electrophilic addition onto the aromatic double bond is possible using strong oxidizing agents such as fluorine at elevated temperature [46]. The reason for the inert behavior of this polyaromatic graphite system is its tendency to reconstruct the aromatic system by removing all introduced substituents. Addition to such aromatic double bonds requires a change of hybridization of participating carbon atoms from the sp^2 planar system to the tetrahedral sp^3 configuration. Consequently, sufficient reactivity is found at hydrogen-terminated graphene edges, isolated double bonds, defective sites, and amorphous carbon.

The introduction of two sp^3 tetrahedrons by the addition of two bromine atoms into the graphene layer was calculated as shown in Figure 9.19.

Addition of four bromines bends the graphene layer strongly and the addition of six or more bromines results in drastic deformations [38].

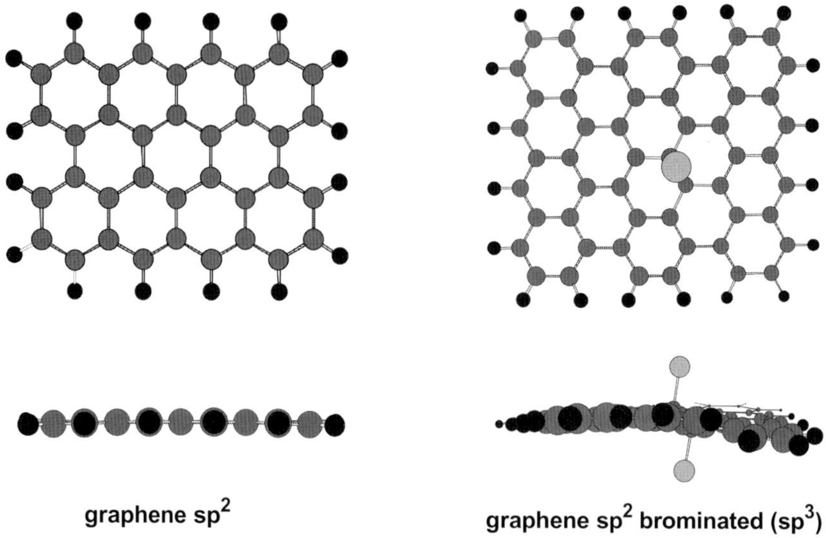

Figure 9.19 Schematic structure of graphene and brominated graphene calculated by MM2.

Plasma-chemical bromination was shown to exhibit outstanding bromination yield and functionalization selectivity on inert polyolefin surfaces. The thermodynamics of chemical polyolefin bromination is well known and shows a positive reaction enthalpy, that is, the process is endothermic and needs further (plasma or UV) activation energy to proceed. The radical substitution starts and ends in the sp^3 hybridization state of carbon.

The graphene structure and graphite sheets differ strongly from polyolefins. As shown in Figure 9.19 the transition of the sp^2 to sp^3 state and the inertia of the graphene structure are characteristic. Therefore, except at hydrogen-terminated graphene edges, isolated double bonds, and defective sites, bromination of the all-carbon domain of graphene proceeds via electrophilic addition to aromatic double bonds between carbon atoms, where substitution reactions are impossible since hydrogen atoms are absent. This addition of bromine transforms the graphitic sp^2-carbon atoms to sp^3 configuration and distorts the planar aromatic system due to a change from planar to tetrahedral bond geometries. The situation is similar to progressive hydration of graphene, where sp^2 to sp^3 transitions cause plane buckling until the – again planar – fully hydrated state of graphene, known as graphane, is reached [47, 48].

The large variety of existing carbon nanostructures (CNSs), including graphene, nanotubes, nanofibers, nanoribbons, nanocones, and fullerenes, has been in the focus of scientific interest throughout recent decades, resulting in many theoretical and experimental investigations, as well as practical applications, for example, References [49, 50]. For example, polymer nanocomposites (PNCs), which contain

CNSs dispersed in a polymer matrix, have found widespread academic and industrial interest since the 1990s [51]. Nowadays, they are widely adopted as excellent conductive and reinforced materials. Especially, carbon nanotubes (CNTs) in polymer composites have shown superior material properties of high relevance to automotive and aircraft industries, and promise to allow inclusion of beneficial effects like fire-resistance, anti-oxidative and electrically conductive properties, or sensor capabilities. However, the compatibility of a CNS with polymer matrices is often low. This impedes homogeneous dispersion by agglomeration of the CNS. There is strong practical and scientific interest in methods to enhance the compatibility of polymer and CNSs by covalent linking of organic residues, oligomers, or polymers, as well as to equip CNSs with additional functionalities, including anti-oxidative, flame-retardant, and conductive behavior.

Several chemical and physical processes are available to activate and to change the inert graphene structure, and will be discussed in the following. Most of them require harsh reaction conditions or are unspecific, that is, proceed with low selectivity.

Fullerenes react with liquid bromine preferentially to $C_{60}Br_{24}$, but species with lower bromination degree are also formed [52]. Brominated fullerenes split off all bromine atoms on heating due to the weakness of the C–Br bond. In fullerenes, the C–Br bond length of 0.203 nm exceeds the standard length of 0.196 nm [52]. Hydrolysis forms hydroxyl and epoxy groups easily. Moreover, the Bingel–Hirsch reaction uses bromine malonate for addition of organic residues by forming a cyclopropane structure [53, 54].

Carbon nanotubes are less reactive than fullerenes. Oxidations and reductions are possible together with subsequent grafting reactions on anchor groups. Fluorination with elemental fluorine proceeds easily and produces highly fluorinated CNTs [55].

Carbon materials are inert towards fluorine at room temperature. The reaction with fluorine requires about 300 °C. On fluorination of graphite, the F/C atomic ratio typically varies from 0.5 to 1 as the temperature is increased from 350 to 600 °C [55, 56]. At room temperature, the reactivity of fluorine is greatly improved by the presence of a volatile fluoride–anhydrous HF gaseous mixture. A degree of fluorination of $F/C = 1$ was achieved with iodine pentafluoride (IF_5) [57]. Substitution reactions with C–F bonds are possible using different Li alkyls. Plasma-chemical fluorination of carbonaceous materials has been studied in non-coating and in coating precursor gases [58, 59].

However, bromination, as another important type of halogenation, is reported to be not easily achieved. The accumulation of bromine on CNTs proceeds in an unusual order, perpendicularly to the nanotube surface. As a result, the area on which bromine accumulates is more oxidized than non-covered areas [60]. Intercalation of bromine into the graphene layers of graphite occurs readily at room temperature merely on immersion of graphite into liquid bromine or bromine vapor. It produces a quasi-stable compound of composition of up to C_8Br (stage one) [61]. When subjected to an atmosphere of bromine, graphite takes up bromine and swells by 55% [62]. Such a treatment process was applied for the selective

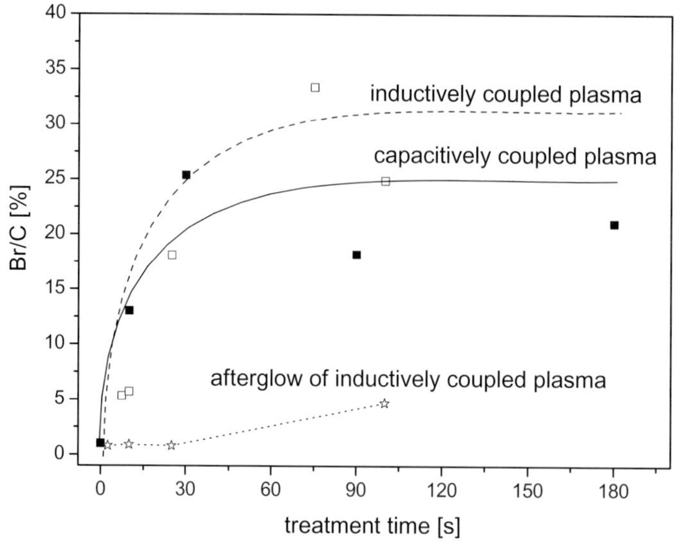

Figure 9.20 Introduction of Br and O versus exposure to cw-rf low-pressure Br_2 plasma using capacitively, inductively coupled or afterglow plasma.

intercalation of a carbon nanostructure, using liquid bromine as purification reagent [63]. It is further reported that electrochemically produced bromine gas introduces 2.8 at.% Br into carbon nanotubes [64].

The processing of HOPG (highly ordered pyrolytic graphite) surfaces with inductively coupled plasma in elemental bromine vapor showed a high bromination rate. The graphene surface structure was substantially brominated within 2 s plasma treatment at a power of 10 W (Figure 9.20).

This rate is similar to that observed on polyolefins for oxygen, bromine, or bromoform plasmas [24, 65]. Using polyolefins, the surface oxidation caused by low-pressure oxygen plasma treatment achieves a dynamic steady-state between oxygen group formation and oxygen functionality loss by formation of volatile oxygen-containing products within 2 s [15]. The plasma bromination dynamics observed here promotes on a similar time scale. At a plasma power of 50 W, bromine introduction to the HOPG surface increased to up to 37% Br. The bromination of natural graphite and carbon fibers was likewise found to increase with a rate similar to that of HOPG and polypropylene (Figure 9.21).

For intercalation studies, natural graphite (NG) and HOPG were exposed to a bromine atmosphere at the processing pressure but without plasma ignition. Less than 1% Br/C was found after exposure to bromine vapor at 8 Pa for 60 min. Therefore, intercalation of bromine into graphite layer interspaces is not significant on the time and pressure scale under study. In comparison to the other materials, a MWCNT (multiwall carbon nanotube) was brominated much more

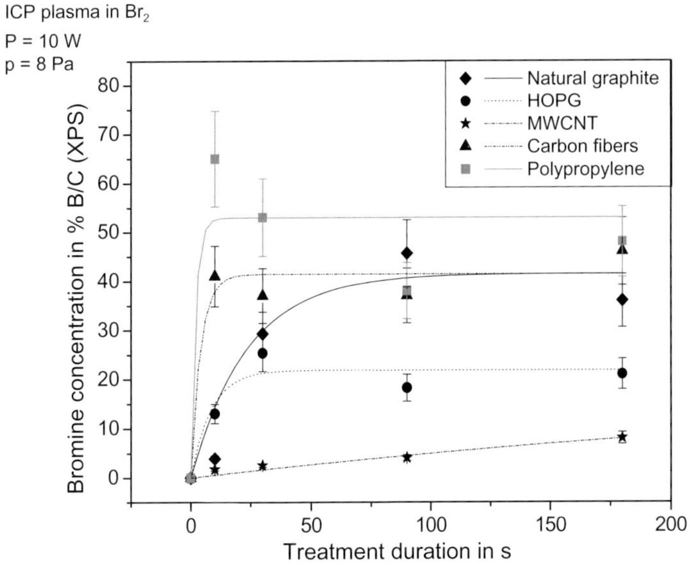

Figure 9.21 Introduction of Br and O versus exposure time to cw-rf low-pressure Br_2 plasma.

slowly. This can be attributed to the significantly higher specific surface area of about $250\,m^2\,g^{-1}$.

The oxygen content of HOPG, natural graphite, carbon fibers, polypropylene, and MWCNTs was higher than during earlier experiments on plasma bromination of polyethylene and polypropylene using the capacitively coupled bromoform plasma, where less than 3% O/C were found [1]. The present oxygen contamination may be caused by water residues in the bromine precursor, humidity at reactor walls, and vacuum leakage. The propensity of post-plasma reactions of C radical sites with oxygen from ambient air, which is strong for polymers, should be reduced for graphene-like surfaces due to a high surface mobility of radicals allowing for rapid radical recombination reactions [14]. None of the studied graphite materials showed significant degradation due to formation of low molecular weight oxidized material (LMWOM) [66, 67] during plasma treatment, as is typically observed on polymers, where the (highest brominated) top surface layers were lost on 15 min of washing in THF, resulting in a reduction of the bromine content to 50–75% of the initial value.

After plasma-chemical bromination, the XPS C1s peaks of HOPG and graphite exhibited a broadening at the high-energy side (Figure 9.22). The enhanced bromine concentration reduced the XPS C1s count rate. The Br3d peaks were symmetric and their structure required only a single sub-peak feature for interpretation.

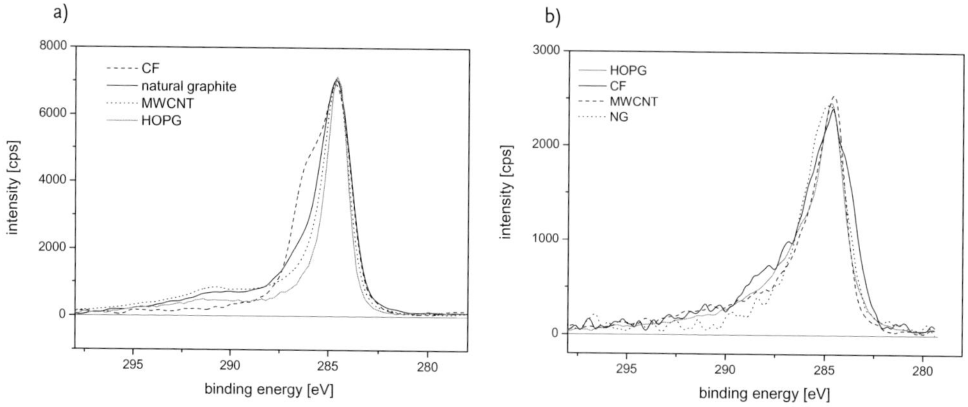

Figure 9.22 XPS C1s peaks of HOPG, NG (natural graphite), MWCNT, and CFs (carbon fibers) before (a) and after plasma-chemical bromination (b) (8 Pa pressure, 10 W input power, and 180 s ICP treatment duration in elemental bromine vapor). The XPS count rate after bromination (b) is reduced due to the reduced atomic carbon concentration of the surface after incorporation of bromine. To avoid overlapping, the individual spectra are displayed with a vertical offset of 1500 cps (a) and 1000 cps (b) each.

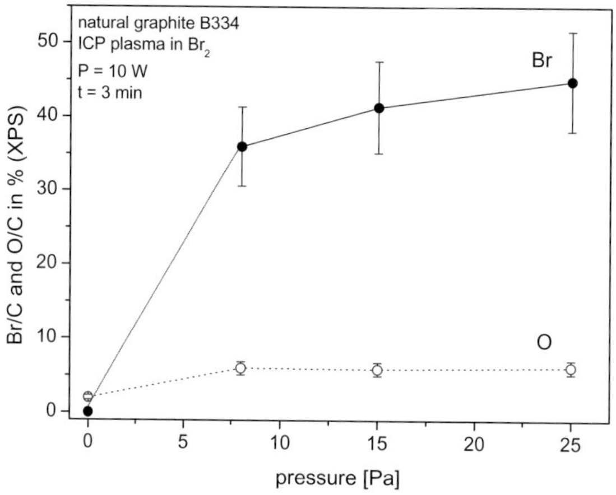

Figure 9.23 Pressure dependence of bromination efficiency and oxygen impurities on NG resulting from plasma processing in elemental bromine vapor for an exposure time of 180 s and 10 W plasma power.

9.5.2
Dependence of Bromination Rate on Plasma Parameters

The pressure dependence of plasma-chemical bromination was studied on natural graphite in detail (Figure 9.23). It was found that the bromine surface concentration on NG slightly increased with pressure.

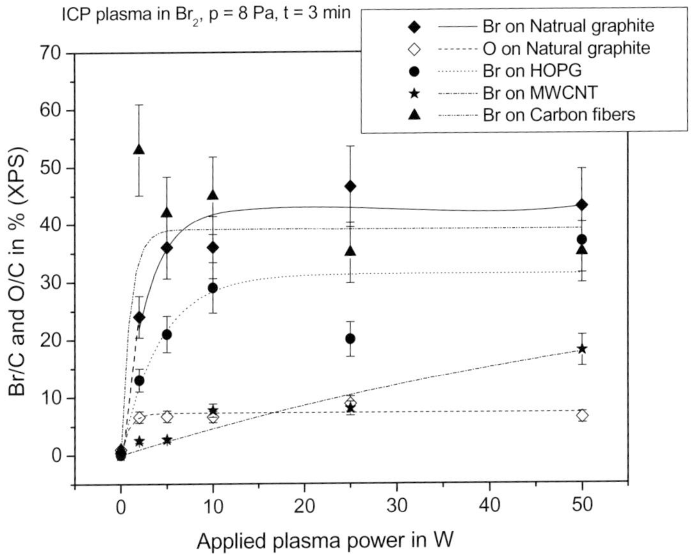

Figure 9.24 Plasma power dependence of the bromination of four graphite materials resulting from an exposure time of 180 s at 8 Pa pressure of elemental bromine vapor.

Figure 9.24 shows the plasma power dependency of the bromination process. The graphitic materials NG, HOPG, and CF, with small specific surface area, require significantly less plasma power than MWCNT (surface area of about $250\,m^2\,g^{-1}$). This indicates that a large surface area material requires a higher number of reactive bromination species, for example, bromine atoms, which are created at a higher rate at higher plasma power. For the other materials, the bromination efficiency levels off to a plateau already at low power values. They do not further profit from additionally produced reactive species, thus leading to optimal power efficiency at low plasma input power.

9.5.3
Alternative Plasma Bromination Precursors

Examples of less chemically aggressive alternative precursors for plasma-chemical bromination are bromoform ($CHBr_3$) and allyl bromide (CH_2=CH–CH_2Br). Plasma bromination experiments using bromoform resulted in up to 55% Br/C and may be accompanied by the deposition of a thin plasma polymer layer on the graphitic material. A further increase in bromination surface density is possible using mixtures of $CHBr_3$ and Br_2 as plasma precursor gas but is likewise accompanied by plasma polymer formation. Application to polypropylene yielded bromine concentrations of about 160% Br/C, which indicate the formation of C-Br_2 or C-Br_3 groups. Similar yields were achieved for plasma polymer-forming mixtures of Br_2 and allyl bromide.

9.5.4 Efficiency in Bromination of Carbon and Polymer Materials

Figure 9.25 compares the yield in bromination of different carbon and polymer materials.

All carbonaceous materials studied in the present work, except polypropylene, exhibit graphene-like surface structures. Some of them, especially MWCNTs and CFs, also contain defective graphite sites and amorphous carbon irregularities. For identical processing conditions, MWCNTs exhibited a significantly lower plasma bromination rate than the other, more compact materials. This was interpreted to be due to its higher specific surface area. HOPG, on the other hand, exhibits a very low surface area, of the order of 0.001 $m^2 g^{-1}$. This partially explains why its plasma bromination rate is higher than that on MWCNTs. However, since HOPG is highly regular in its structure it exhibits very few defects such as sp^3-hybridized carbon atoms and isolated double bonds. Compared to the likewise small surface area materials NG and CF (with about 40% Br/C), which exhibit more defective sites than HOPG, HOPG is less easily brominated (with 20–37% Br/C). The highest bromination yield was found on polypropylene, fully composed of sp^3-hybridized C, and on $CHBr_3$ or allyl bromide + Br_2 plasma-polymer coated surfaces. These findings suggest that carbon in a sp^3-hybridized state may be brominated more facilely by plasma-chemical processing than sp^2-hybridized carbon. It must be also considered that carbon fibers shadow one another much more than do plain materials, which are fully exposed to the plasma.

9.5.5 Grafting of Amines to Brominated Surfaces

HOPG graphene was brominated at low power input of the low-pressure inductively coupled plasma using elemental bromine. The bromine concentration amounted to 20 Br/100 C according to XPS. Subsequently, the C–Br was subjected

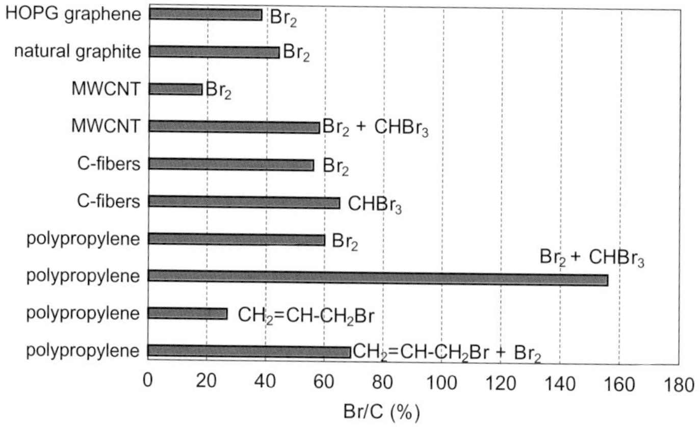

Figure 9.25 Comparison of bromination percentage at standard conditions and under optimal conditions.

to diaminohexane nucleophilic substitution. About 9 nitrogen atoms per 100 C were found by XPS, indicating grafting of 4–5 diaminohexane molecules per 100 C. This substitution efficiency corresponds to that measured earlier on polyolefin surfaces. During the reaction, the bromine was almost completely consumed. This suggests that soluble fragments were also brominated and then removed during the wash process. An example of a reactive C–Br species, which does not lead to a substitution product although lowering the Br content, is a tertiary bromide that will undergo elimination instead of fixing of reaction partners in substitution reactions. The fluorine concentration measured by XPS of 9 F/100 C after chemical derivatization of terminal amino groups with PFBA (pentafluorobenzaldehyde) is only about half the expected concentration. This difference led to the suggestion that a few diaminohexane molecules are bonded at both amino groups to graphite, thus forming loops.

For graft reactions of C–Br bonds on brominated HOPG with aminosilanes the observed substitution efficiency was much higher, even after intense washing with THF and water. Up to 60–70% of the C–Br groups were consumed by nucleophilic substitution:

$$\blacksquare\text{-Br} + H_2N\text{-}(CH_2)_3\text{-}Si(O\text{-}CH_2\text{-}CH_3)_3 \rightarrow \blacksquare\text{-NH-}(CH_2)_3\text{-}Si(O\text{-}CH_2\text{-}CH_3)_3 + HBr$$

Using controlled hydrolysis silanol groups are formed:

$$\blacksquare\text{-NH-}(CH_2)_3\text{-}Si(O\text{-}CH_2\text{-}CH_3)_3 + 3H_2O \rightarrow \blacksquare\text{-NH-}(CH_2)_3\text{-}Si(OH)_3 + 3CH_3\text{-}CH_2\text{-}OH$$

XPS showed the Si concentration to be about 40% higher than the N concentration. Therefore, the occurrence of a self-condensation of the silane compound was assumed. This should lead to free aminopropyl residues at the topmost surface of grafted HOPG, which was confirmed by PFBA derivatization and XPS.

Moreover, aminopropyl phosphonate was also grafted on polyolefins for adhesion improvement to metallic aluminium or as flame retardant agent:

$$\blacksquare\text{-Br} + H_2N\text{-}(CH_2)_3\text{-}PO(OH)_2 \rightarrow \blacksquare\text{-NH-}(CH_2)_3\text{-}PO(OH)_2 + HBr$$

9.5.6 Refunctionalization to OH Groups

As was shown for brominated polypropylene and polyethylene, the C–Br group can be replaced by –OH, $-NH_2$, $-N_3$, and so on. Here, the C–Br groups on HOPG surfaces were hydrolyzed under basic conditions to -OH groups. After chemical derivatization of hydrolyzed brominated surfaces with trifluoroacetic anhydride 3 -OH groups per 100 C were found by inspection of the resulting fluorine percentage using XPS.

9.5.7 NH_2 Introduction onto Carbon Surfaces

Early research on amino group introduction to carbon fibers for adhesion improvement in epoxy resin composites dates back to 1972 [20, 45]. It was continued in

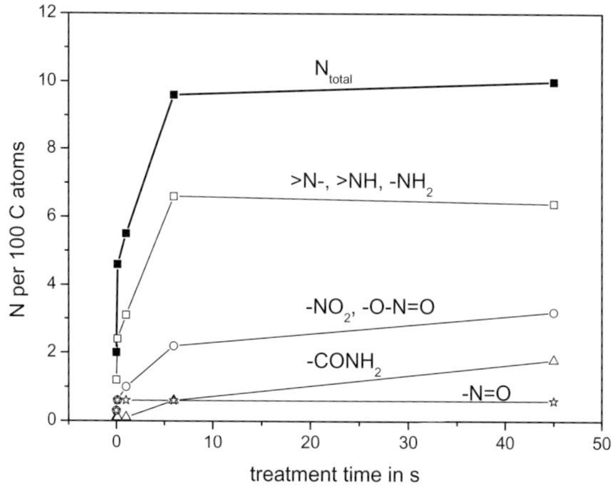

Figure 9.26 Amino group formation at a carbon surface during ammonia plasma exposure as measured by the fitted XPS N1s peak (cw-rf plasma, 300 W, 10 Pa).

the early 1980s [68, 69] and transferred to the graphene-like surface of HOPG (highly ordered pyrolytic graphite) as model surface in 1997 [70].

Ammonia plasma introduces some N-containing species to carbon fiber surfaces, among which a few primary amino groups can be found, but it acts also by hydrogenating as shown by secondary ion mass spectrometry (SIMS) [69, 71]. As already observed with ammonia plasma treated polyolefins nitrogen and (undesired) oxygen were incorporated into carbon fiber surfaces, as shown in Figure 9.26, depending on exposure time to the NH_3 plasma.

Independent of the type of introduced group or atoms, NH_2 groups or H, the hybridization of anchoring carbon atoms must be changed from sp^2 to sp^3. Hydrogenation from both sides of the graphene sheet results in the formation of graphane [72]. One-sided hydrogenation of graphene is expected to be disordered material in contrast to two-sided graphane [47, 73]. Annealing of graphane removes the hydrogen and the graphene is reconstructed.

Carbon fibers show a broadening of the C1s signal to the high binding energy side (Figure 9.27).

The double head of the C1s-peak at binding energies of 285.0 and 283.5 eV was interpreted as partial conversion of amorphous carbon and graphite structures into hydrocarbon by simultaneous hydrogenation of the ammonia plasma, as seen by the appearance of the new C1s signal at 285 eV, which is characteristic for hydrocarbons (Figure 9.27) [73]. Whereas the original graphite and carbon fibers show a peak at a binding energy of about 283.5 eV a new peak appears during the ammonia plasma treatment at 285.0 eV, which is assigned to the conversion of the graphite structure into a hydrocarbonaceous polymer layer. Along with the appearance of this peak a further not clearly resolved peak at higher binding energy indicates the oxidation of this sample on exposure to the air.

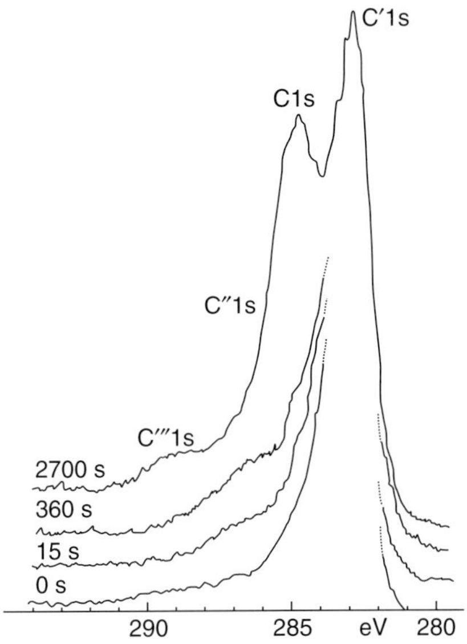

Figure 9.27 C1s signal at carbon fiber surfaces after exposure to ammonia gas plasma (30 s, cw-rf, 100 W, 10 Pa).

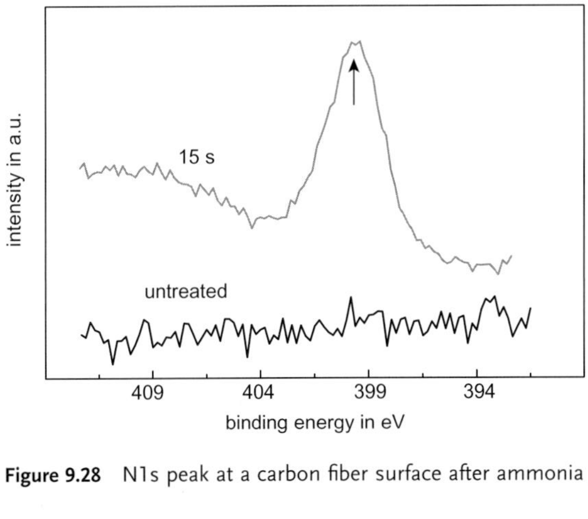

Figure 9.28 N1s peak at a carbon fiber surface after ammonia plasma exposure for 15 s.

The interaction between the ammonia plasma and the carbon fiber surfaces and the subsequent aging on exposure to ambient air results in the formation of amino groups, oxidized nitrogen species, and oxygen-containing functional groups as presented in Figure 9.26.

It must be added that the post-plasma introduction of oxygen functional groups exceeds the concentration of attached N-functional groups. The N1s signal shows a position near 399 eV in the XP-spectrum (Figure 9.28) [70].

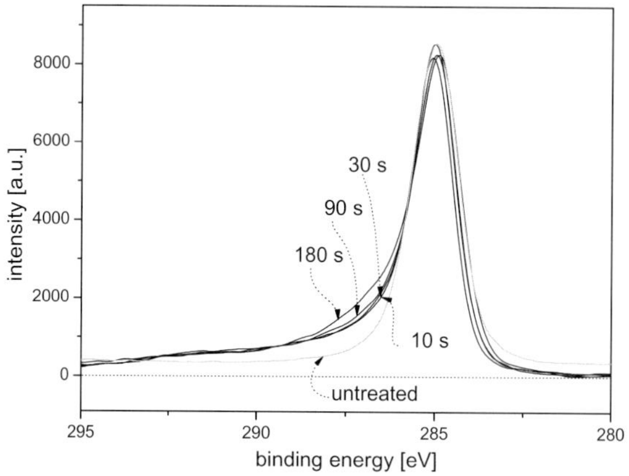

Figure 9.29 Changes in the C1s signal of HOPG exposed to a cw-rf low-pressure ammonia plasma.

A few different N-containing functional groups take this position, such as primary, secondary, tertiary, and quaternary amino/ammonium nitrogen, aniline, pyridine, nitrilo/cyano, azo and also amide groups.

The introduction of N-species into graphene structures of HOPG is much more difficult. Changes in the C1s signal are marginal (Figure 9.29). About 7% N/C could be introduced by exposure to ammonia plasma.

Several graft reactions have been performed onto brominated surfaces of graphene, nanotube, graphite, or carbon fiber for different applications such as adhesion-promotion, flame retardancy, or as stabilizer in polymer composites (Figure 9.30).

9.6 SiO_x Deposition

The deposition of glassy SiO_2 or sub-stoichiometric SiO_x layers is also thermodynamically preferred. SiO_2 is the stable end-product of any oxidation of elemental silicon. Therefore, the route of any oxidative plasma treatment of Si-containing substances or polymers delivers inevitably SiO_2 or more commonly SiO_x. Moreover, the deposition of any Si-containing organic precursor in the oxygen plasma produces SiO_x layers. Such popular precursor monomers are tetraethoxysilane (TEOS) or hexamethyldisiloxane (HMDSO) [74–78].

Such layers are applied in microelectronics, as corrosion-protective layers for metals or metalized polymers or as hard and abrasive coatings at polymer surfaces [79].

SiO_x layers are also used as barrier layers, which should hinder the permeation of oxygen gas (air), water, carbon dioxide, or flavor out of plastic bottles filled with drinking water, juice, or beer. These permeation barriers are deposited onto the inner surface of PET bottles to avoid mechanical scratching. The thickness of these

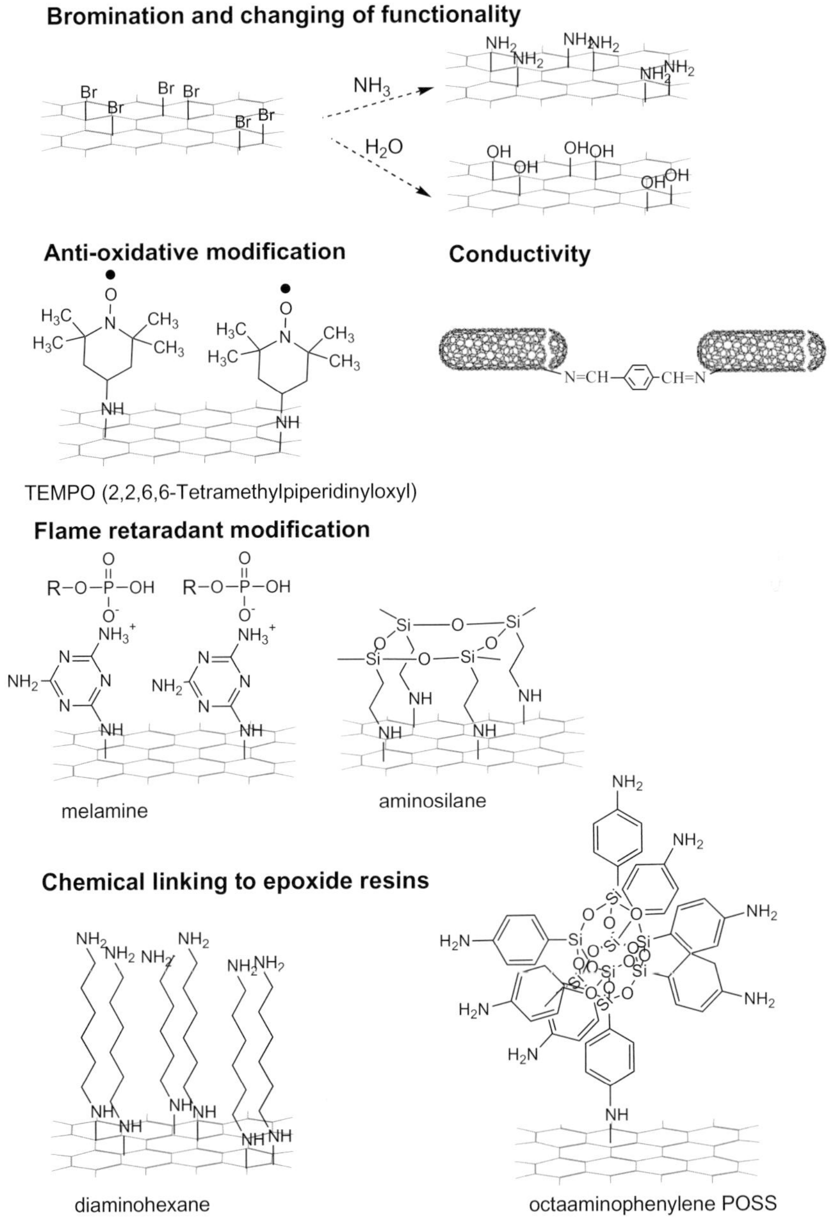

Figure 9.30 Graft reactions onto brominated carbon surfaces.

barriers does not exceed 20 nm for maximal barrier behavior. Fuel and solvents in polyethylene fuel tanks are also be coated by SiO_2 thin layers, thus forming a migration barrier.

The formation of SiO_2 or SiO_x layers is also very selective, as mentioned before, if it is in focus as end-product [80–86]. There is no other possibility than forming SiO_2. Formation of SiO_2 dominates if an excess of oxygen is added to the Si-precursor during deposition, as exemplified using HMDSO:

$$n(\mathrm{CH_3})_3\text{-Si-O-Si-}(\mathrm{CH_3})_3 + 26n\mathrm{O} \rightarrow \text{-[O-SiO-O-SiO-O]}_n\text{-} + 6n\,\mathrm{CO_2} + 9n\mathrm{H_2O}$$

SiO_x-layers were also deposited by atmospheric-pressure plasmas in air or using other oxidants [87, 88].

9.7 Grafting onto Radical Sites

The term radical was introduced by A. L. Lavoisier [89].

There are three possible ways to graft molecules or polymers onto plasma-modified polymer surfaces, as also mentioned above:

1) Consumption of functional groups on polymer surfaces with reactive groups or end-groups of molecules, oligomers, and polymers [23]. A prominent example is the reaction of primary amino groups with glutaraldehyde, resulting in a Schiff's base (azomethine):

$$\blacksquare\text{-NH}_2 + \text{OHC-(CH}_2)_3\text{-CHO} \rightarrow \blacksquare\text{-N=CH-CH}_2)_3\text{-CHO}$$

2) Grafting onto C-radical sites [90, 91]. Here, it is not clear yet what a defined graft reaction is. The radical–radical recombination is such a graft reaction. For example, NO–C-radical recombination is used to stabilize the polymer against post-plasma oxidation:

$$\blacksquare\text{-CH}_2^{\bullet} + {}^{\bullet}\text{NO} \rightarrow \blacksquare\text{-CH}_2\text{-NO}$$

or bromine:

$$\blacksquare\text{-CH}_2^{\bullet} + {}^{\bullet}\text{Br} \rightarrow \blacksquare\text{-CH}_2\text{-Br}$$

Another prominent example is the detection of C-radical sites by consumption with diphenylpicrylhydrazil (DPPH):

$$\text{DPPH [C}^{\bullet} + \text{(phenyl)}_2\text{-N-N}^{\bullet}\text{-arene(NO}_2)_3 \rightarrow \text{(phenyl)}_2\text{-N-N(C)-arene(NO}_2)_3]$$

Another variant is the starting of a graft polymerization, which may be stopped by recombination, disproportionation, or chain-transfer. Grafting of easily polymerizable vinyl or acrylic monomers is possible, such as styrene:

$$\blacksquare\text{-CH}_2^{\bullet} + \text{CH}_2\text{=CH-phenyl} \rightarrow \blacksquare\text{-CH}_2\text{-CH}_2\text{-C}^{\bullet}\text{H(phenyl), and so on}$$

This reaction needs an oxygen-free vacuum and no interruption between plasma finishing and monomer addition (oxygen-free).

3) Grafting onto post-plasma formed C–O–O$^{\bullet}$ (peroxy) radicals [92–94]. Here, the authors also produced C-radical sites by exposure to plasma but have considered the short lifetime of radicals and their reaction with traces of oxygen in the plasma, which cannot completely removed. Therefore, oxygen was introduced intentionally. In this way all C-radical sites were converted into peroxy radicals, which can be dissociated and can react with monomers:

$$\blacksquare\text{-CH}_3 + \textit{plasma} \rightarrow \blacksquare\text{-CH}_2^{\bullet} + {}^{\bullet}\text{H}$$

$$\blacksquare\text{-CH}_2^{\bullet} + {}^{\bullet}\text{O-O}^{\bullet} \rightarrow \blacksquare\text{-CH}_2\text{-O-O}^{\bullet}$$

$$\blacksquare\text{-CH}_2\text{-O-O}^{\bullet} + \text{R-H} \rightarrow \blacksquare\text{-CH}_2\text{-O-OH} + \text{R}^{\bullet}$$

$$\blacksquare\text{-CH}_2\text{-O-OH} + \textit{UV irradiation} \rightarrow \blacksquare\text{-CH}_2\text{-O}^{\bullet} + {}^{\bullet}\text{OH}$$

$$\blacksquare\text{-CH}_2\text{-O}^{\bullet} + \text{CH}_2\text{=CH-phenyl} \rightarrow \blacksquare\text{-CH}_2\text{-CH}_2\text{-C}^{\bullet}\text{H(phenyl), and so on}$$

9.7.1 Types of Produced Radicals

Kuzuya has extensively investigated the structure of plasma-produced carbon radicals in different polymers [95]. For example, in plasma-irradiated LDPE he found mid-chain alkyl radicals, mid-chain allyl radicals, and dangling bond sites as well as peroxy radicals. Tsuji has presented the mechanism of interaction of UV radiation with polymers and the formation of ESR detectable radicals [96].

9.7.2 Grafting onto C-Radical Sites

Plasma produces radicals in a surface-near layer by its particle bombardment [97] and much more probably by irradiation with its short-wavelength (20–200 nm) vacuum ultraviolet radiation [98]. For graft reactions of, most often, voluminous organic molecules with radicals must be situated at the topmost surface because large molecules cannot migrate into the polymer to bulk radicals [95]. However, oxygen molecules can diffuse into the polymer matrix and react with radicals trapped within the polymer bulk. At surface monomer, molecules and traces of oxygen in the plasma gas are in concurrence:

$$\blacksquare\text{-CH}_2^{\bullet} + \text{CH}_2\text{=CH-phenyl} \rightarrow \blacksquare\text{-CH}_2\text{-CH}_2\text{-C}^{\bullet}\text{H(phenyl)}$$

or:

$$\blacksquare\text{-CH}_2^{\bullet} + {}^{\bullet}\text{O-O}^{\bullet} \rightarrow \blacksquare\text{-CH}_2\text{-O-O}^{\bullet}$$

Oxygen is more reactive and wins this competition. Therefore, it must be carefully removed from plasma. Nevertheless, in the total absence of oxygen the graft polymerization is only initiated when the monomer can contact the radical within its lifetime. The typical lifetime of the hydroxyl radical ($^{\bullet}$OH) is 10^{-9} s, of alkoxy ($^{\bullet}$OR) 10^{-6}, of $^{\bullet}CH_3$ 10^{-4} s and peroxy radicals ($^{\bullet}$O-O-R) a few seconds [99]. If this precondition is fulfilled a polymer chain can grow from the C-radical site at the polymer surface. The open question is whether the monomer supply is sufficient when operated under low-pressure plasma conditions to avoid untimely chain termination. Nevertheless, fibrous microscopic objects have been observed, indicating that voluminous agglomerates were formed at a few places [91]. Monomers used were styrene or glycidyl methacrylate (GMA) [100]. The first announcement of styrene graft polymerization after noble gas plasma pre-activation was published in 1971 [101]. To avoid undesired oxygen traces in the reaction room temperature liquid monomer was poured into the plasma chamber over the polymer substrate for immediate graft polymerization in the presence of high monomer concentrations compared to gas-phase reaction under low-pressure [102, 103].

9.7.3 Post-Plasma Quenching of Radicals

The reaction of a radical quencher has been described before. Another method uses TEMPO [104]. It is prepared by oxidation of 2,2,6,6-tetramethylpiperidine ($R_1R_2N{-}O^{\bullet}$). This radical is widely used as a radical trap, as a structural probe for biological systems, and as a mediator in controlled free radical polymerization [105]. The reactions with it are reversible [106].

9.7.4 Grafting on Peroxide Radicals

To avoid the rapid deactivation of unstable C-radicals Suzuki consciously preferred the reaction with oxygen available after exposing the fresh plasma-treated polymer to ambient air [92]. As described before, C-radicals react immediately with the biradical O_2 in air to form peroxy radicals and subsequently hydroperoxides. After decay of the peroxy unit by irradiation alkoxy radicals can initiate the graft polymerization of vinyl or acrylic monomers such as acrylamide [107]. The acrylamide graft polymerization starts also from hydroperoxy sites, if their decomposition was incomplete:

$$\text{C-O-OH} \rightarrow \text{C-O}^{\bullet} + {}^{\bullet}\text{OH}$$

$$\text{C-O}^{\bullet} + \text{CH}_2\text{=CH-CONH}_2 \rightarrow \text{C-O–CH}_2\text{-CH}^{\bullet}(\text{CONH}_2) \text{ and so on}$$

The attachment via a peroxy group is fragile:

$$\text{C-O-O}^{\bullet} + \text{CH}_2\text{=CH-CONH}_2 \rightarrow \text{C-O-O-CH}_2\text{-CH}^{\bullet}(\text{CONH}_2) \text{ and so on}$$

The grafting yield depends on the concentration of plasma-produced radicals. Unfortunately, this radical production differs from plasma to plasma and from polymer to polymer strongly. Side-reactions with traces of oxygen and humidity are significant and plasma conditions also have an unknown influence on radical formation. Therefore, this process is not well reproducible and the number of ether/peroxy links cannot clearly be predicted. This graft technique has been reported for grafting on poly(tetrafluoroethylene) [107, 108] or on silicon rubbers [109].

9.7.5 Plasma Ashing

"Cold" plasma ashing was introduced for analysis (inductively coupled plasma mass spectrometry), archeological purposes [110], reduction of oxidized antiques, or excavation of polymer material for electron microscopy of polymer supermolecular structures [111–113].

Layer-like polymer structures or pores were identified using "plasma-etch gravimetry" [114]. Paper, hollow-fiber cellulose, prepregs, and resins were investigated microscopically using plasma ashing or "selective plasma-etching" techniques [115–118]. Interphases, stress-induced zones and surface-near layers were identified also by plasma-ashing or etching using electron microscopy [119, 120]. Single cellulose hollow-fibers, carbon fibers, and fibers after single fiber pull-out tests were characterized by electron microscopy using plasma ashing [121–123]. Trace elements in voluminous organic or polymer matrices can be excavated and in this way concentrated by ashing all organic material [124]. An important example was the plasma ashing of human lung tissue in the beginning of 1950s for excavation of asbestos fibers in the center of lung cancer [125]. Moreover, other micro-objects were ashed for microscopic analysis. The pioneering work has been presented by Thomas [126].

Initially, microelectronic production oxidative plasma-stripping (etching-ashing) of photo resists also formed oxides at the surface of chips, arising from catalyst, resist stabilizers, or other additives. This process was later eliminated. As the plasma is formed, many free radicals are created that could damage the wafer. The need to get rid of free radicals has increased, many machines now use a downstream plasma configuration, where plasma is formed remotely and the desired particles are channeled to the wafer.

Mineral-forming elements, such as silicon, incorporated in biological, organic or polymer matrices could be converted into oxides (SiO_2). Such oxide structures of plants, polymer composites, or tissues give information on the internal structure of a sample and the distribution of the mineral-forming elements in the substrate [127]. This process was also called petrification. The opposite route is also possible, namely, to reduce oxides and sulfides to metals, as applied for the purification and cleaning of antique coins.

References

1 Friedrich, J.F., Mix, R., Schulze, R.-D., Meyer-Plath, A., Joshi, R., and Wettmarshausen, S. (2008) *Plasma Proc. Polym.*, **5**, 407–423.

2 Lee, S.D., Sarmadi, M., Denes, F., and Shohet, J.L. (1997) *Plasma Polym.*, **2**, 177.

3 Denes, F., Young, A., and Sarmadi, M. (1197) *J. Photopolym. Sci. Technol.*, **10**, 91.

4 Beckert, R., Fanghänel, E., Habicher, W.D., Metz, P., Pavel, D., and Schwetlick, K. (2004) *Organikum*, Wiley-VCH Verlag GmbH, Weinheim.

5 Kühn, G., Weidner, St., Decker, R., Ghode, A., and Friedrich, J. (1999) *Surf. Coat. Technol.*, **116–119**, 796–801.

6 Nuzzo, R.G., and Smolinsky, G. (1984) *Macromolecules*, **17**, 1013–1019.

7 Friedrich, J.F., Unger, W.E.S., Lippitz, A., Koprinarov, I., Weidner, S., Kühn, G., and Vogel, L. (1998) in *Metallized Plastics 5&6: Fundamental and Applied Aspects* (ed. K.L. Mittal), VSP, Utrecht, pp. 271–293.

8 Everhart, D.S. and Reilley, C.N. (1981) *Anal. Chem.*, **53**, 665–676.

9 Brown, H.C., Schlesinger, H.I., and Burg, A.B. (1939) *J. Am. Chem. Soc.*, **61**, 671.

10 Kühn, G., Weidner, St., Decker, R., Ghode, A., and Friedrich, J. (1999) *Surf. Coat. Technol.*, **116–119**, 748.

11 Geng, Sh., Friedrich, J., Gähde, J., and Guo, L. (1999) *J. Appl. Polym. Sci.*, **71**, 1231–1237.

12 Gerenser, L.J., Elman, J.F., Mason, M.G., and Pochan, J. (1985) *Polymer*, **26**, 1162.

13 Nakayama, Y., Takahashi, K., and Sasamoto, T. (1996) *Surf. Interface Anal.*, **24**, 711.

14 Friedrich, J., Kühn, G., and Gähde, J. (1979) *Acta Polym.*, **30**, 470–477.

15 Friedrich, J., Unger, W., Lippitz, A., Koprinarov, I., Ghode, A., Geng, Sh., and Kühn, G. (2003) *Composite Interface*, **10**, 139–172.

16 Friedrich, J., Kühn, G., Mix, R., Retzko, I., Gerstung, V., Weidner, St., Schulze, R.-D., and Unger, W. (2003) in *Polyimides and Other High Temperature Polymers: Synthesis, Characterization and Applications* (ed. K.L. Mittal), VSP, Utrecht, p. 359.

17 Hoffmann, K., Mix, R., Hoffmann, K., Buschmann, H.-J., Friedrich, J.F., and Resch-Genger, U. (2009) *J. Fluoresc.*, **19**, 229–237.

18 Samal, R.K., Iwata, H., and Ikada, Y. (1983) in *Physicochemical Aspects of Polymer Surfaces*, vol. 2 (ed. K.L. Mittal), Plenum Press, New York, p. 801.

19 Hollahan, J.R., Stafford, B.B., Falb, R.D., and Payne, S.T. (1969) *J. Appl. Polym. Sci.*, **13**, 807.

20 Friedrich, J., Gähde, J., Frommelt, H., and Wittrich, H. (1976) *Faserforsch. Textiltechn./Z. Polymerenforsch.*, **27**, 604–608.

21 Gombotz, W.R. and Hoffman, A.S. (1988) *J. Appl. Polym. Sci., Appl. Polym. Symp.*, **42**, 285–304.

22 Ameen, A.P., Short, R.D., and Ward, R.J. (1994) *Polymer*, **35**, 4382.

23 Kühn, G., Ghode, A., Weidner, St., Retzko, I., Unger, W.E.S., and Friedrich, J.F. (2000) in *Polymer Surface Modification: Relevance to Adhesion*, vol. 2 (ed. K.L. Mittal), VSP, Utrecht, pp. 45–64.

24 Wettmarshausen, S., Mittmann, H.-U., Kühn, G., Hidde, G., and Friedrich, J.F. (2007) *Plasma Proc. Polym.*, **4**, 832–839.

25 Resch-Genger, U., Hoffmann, K., Mix, R., and Friedrich, J.F. (2007) *Langmuir*, **23**, 8411–8416.

26 Hoffmann, K., Resch-Genger, U., Mix, R., and Friedrich, J.F. (2006) *J. Fluoresc.*, **16**, 441–448.

27 Wann, J.-H., Chen, X., Chen, J.-J., Calderon, J.G., and Timmons, R.B. (1997) *Plasmas Polym.*, **24**, 245.

28 Friedrich, J.F., Kühn, G., and Gähde, J. (1979) *Acta Polym.*, **30**, 47.

29 Friedrich, J., Kühn, G., Schulz, U., Jansen, K., and Möller, B. (2002) *Vakuum Forschung Praxis*, **14**, 285.

30 Wettmarshausen, S., Mix, R., Meyer-Plath, A., Mittmann, H.-U., and Friedrich, J. (2009) in *Polymer Surface Modification*, vol. V (ed. K.L. Mittal), Brill, Leiden, pp. 3–18.

31 Friedrich, J., Loeschcke, I., and Lutgen, P. (April 1990) in *Proceeding in Adhesion and Surface Analysis, Loughborough April 1990* (ed. D.M. Brewis), The Adhesion Society, Loughborough, pp. 125–127.

32 Friedrich, J. (1991) Plasma modification of polymers, in *Polymer-Solid Interfaces* (eds J.J. Pireaux, P. Bertrand, and J.L. Bredas), Institute of Physics Publishing, Bristol, pp. 443–454.

33 Wettmarshausen, S., Mittmann, H.-U., Kühn, G., Hidde, G., and Friedrich, J.F. (2007) *Plasma Proc. Polym.*, **4**, 832.

34 Friedrich, J.F., Mix, R., Schulze, R.-D., Meyer-Plath, A., Joshi, R., and Wettmarshausen, S. (2008) *Plasma Proc. Polym.*, **5**, 407.

35 Kiss, E., Samu, J., Toth, A., and Bertoti, I. (1996) *Langmuir*, **12**, 1651.

36 Wedenejew, W.J., Gurwitsch, L.W., Kondratjew, W.H., Medwedew, W.A., and Frankewitsch, E.L. (1971) *Energie Chemischer Bindungen*, VEB Deutscher Verlag der Grundstoffindustrie, Leipzig.

37 Friedrich, J., Wettmarshausen, S., Hidde, G., and Hennecke, M. (2009) *Surf. Coat. Technol.*, **203**, 3647–3655.

38 Friedrich, J.F., Wettmarshausen, S., Hanelt, S., Mach, R., Mix, R., Zeynalov, E., and Meyer-Plath, A. (2010) *Carbon*, **48**, 3884–3894.

39 Friedrich, J.F., Kühn, G., Schulz, U., Jansen, K., Bertus, A., Fischer, S., and Möller, B. (2003) *J. Adhesion Sci. Technol.*, **17**, 1127.

40 Lagow, R. and Margrave, J.L. (1979) *Prog. Inorg. Chem.*, **26**, 161.

41 Friedrich, J., Mix, R., Kühn, G., Retzko, I., Schönhals, A., and Unger, W. (2003) *Composite Interface*, **10**, 173–223.

42 Müller, M. and Oehr, C. (1999) *Surf. Coat. Technol.*, **116–119**, 802.

43 Choukourov, A., Biederman, H., Slavinska, D., Trchova, M., and Holländer, A. (2003) *Surf. Coat. Technol.*, **174–175**, 86.

44 Inagaki, N., Narushima, K., Kuwabara, K., and Tamura, K. (2005) *J. Adhesion Sci. Technol.*, **19**, 1189.

45 Gähde, J., Wittrich, H., Schlosser, E., Friedrich, J., and Kaiser, G. DD-patent 106 052, Nov. 08, 1972.

46 Rüdorff, W. and Rüdorff, G. (1947) *Z. Anorg. Allg. Chem.*, **253**, 253.

47 Elias, D.C., Nair, R.R., Mohiuddin, T.M.G., Morozov, S.V., Blake, P., and Halsall, M.P. (2009) *Science*, **323** (**5914**), 610–613.

48 Sofo, J.O., Chaudhari, A.S., and Barber, G.D. (2007) *Phys. Rev. B*, **75** (**15**), 153401.

49 Luzzi, D.E. and Smith, B.W. (2000) *Carbon*, **38**, 1751–1756.

50 Baughman, R.H., Zakhidov, A.A., and de Heer, W.A. (2002) *Science*, **297** (**5582**), 787–792.

51 Wypych, F. and Satyanarayana, K.G. (2005) *J. Colloid Interface Sci.*, **285**, 532–543.

52 Krüger, A. (2007) *Neue Kohlenstoffmaterialien*, Vieweg+Teubner.

53 Bingel, C. (1993) *Chem. Ber.*, **126**, 1957–1959.

54 Hirsch, A. and Brettreich, M. (2005) *Chemistry and Reactions*, Wiley-VCH Verlag GmbH, Weinheim.

55 Hamwi, A., Alvergnat, H., Bonnamy, S., and Beguin, F. (1997) *Carbon*, **35**, 723–728.

56 Rüdorf, W. and Rüdorff, G. (1947) *Z. Anorg. Chem.*, **253**, 281–296.

57 Hamwi, A., Daoud, M., and Cousseins, J. (1988) *Synth. Met.*, **26**, 89–98.

58 Cicala, G., Milella, A., Palumbo, F., Rossini, P., Favia, P., and d'Agostino, R. (2002) *Macromolecules*, **35**, 8920–8922.

59 Barlow, A., Birch, A., Deslandes, A., and Quinton, J. (2006) Proceedings ICONN '06 – International Conference on Nanoscience and Nanotechnology, Brisbane, 2006.

60 Jin, Z.X., Xu, G.Q., and Goh, S.H. (2000) *Carbon*, **38**, 1135–1139.

61 Rüdorff, W. (1941) *Z. Anorg. Allg. Chem.*, **245**, 383–390.

62 Saunders, G.A., Ubbelohde, A.R., and Young, D.A. (1963) *Proc. R. Soc. London, Ser. A*, **271**, 499–511.

63 Chen, Y.K., Green, M.L.H., Griffin, J.L., Hammer, J., Lago, R.M., and Tsang, S.C. (1996) *Adv. Mater.*, **8**, 1012–1015.

64 Unger, E., Graham, A., Kreupl, F., Liebau, M., and Hoenlein, W. (2002) *Curr. Appl. Phys.*, **2**, 107–111.

65 Friedrich, J.F., Unger, W., Lippitz, A., Giebler, I., Koprinarov, I., and Weidner, S. (2000) in *Polymer Surface Modification: Relevance to Adhesion* (ed. K.L. Mittal), Brill Academic Publishers, pp. 137–172.

66 Bikerman, J. (1968) *Science of Adhesive Joints*, 2nd edn, Academic Press.

67 Strobel, M., Corn, S., Lyons, C.S., and Korba, G.A. (1987) *J. Polym. Sci., A*, **25**, 1295–1307.

68 Ivanova, V.P., Andreevskaja, G.D., Friedrich, J.F., and Gähde, J. (1980) *Acta Polym.*, **31**, 752–756.
69 Friedrich, J., Ivanova-Mumjeva, V.G., Andreevskaja, G.D., Gähde, J., Loeschcke, I., and Throl, U. (1983) *Acta Polym.*, **34**, 171–177.
70 Friedrich, J.F., Schulz, E., Weidner, S., and Kühn, G. (1997) Proceedings Techtextil-Symposium 97, 5.23, Frankfurt a. M., 1997.
71 Min, H., Wettmarshausen, S., Friedrich, J., and Unger, W.E.S. (2011) *J. Anal. At. Spectrom.* **26**, 1157–1165.
72 Sofo, J.O. (2007) *Phys. Rev. B*, **75**, 153401–153404.
73 Friedrich, J., Gähde, J., Ivanova-Mumeva, V.G., Andreevskaya, G.D., Reiner, H.-D., Ebert, I., and Richter, Kh. (1981) *Acta Polym.*, **32**, 36–40.
74 Vasile, M.J. and Smolinsky, G. (1972) *J. Electrochem. Soc.*, **119**, 451.
75 Akovali, G. and Bölük, M.Y. (1981) *Polym. Eng.*, **21**, 658.
76 Wrobel, A.M., Wertheimer, M.R., Dib, J., and Schreiber, H.P. (1979) *Polymer Prepr.*, **20**, 723.
77 Wrobel, A.M., Kryszewski, M., and Gazicki, M. (1983) *Macromol. Sci., Part A: Chem.*, **20**, 583.
78 Inagaki, N. (1996) *Plasma Surface Modification and Plasma Polymerization*, Technomic, Lancaster.
79 Benítez, F., Martínez, E., and Esteve, J. (2000) *Thin Solid Films*, **377–388**, 109–114.
80 Creatore, M., Palumbo, F., and d'Agostino, R. (2002) *Plasma Polym.*, **7**, 291–310.
81 Sarmadi, A.M., Ying, T.H., and Denes, F. (1995) *Eur. Polym. J.*, **31**, 847–857.
82 Li, K., and Meichsner, J. (1999) *Surf. Coat. Technol.*, **116–119**, 841–847.
83 Korzec, D., Theirich, D., Werner, F., Traub, K., and Engemann, J. (1995) *Surf. Coat. Technol.*, **74**, 67–74.
84 Alexander, M.R., Jones, F.R., and Short, R.D. (1997) *Plasma Polym.*, **2**, 277–300.
85 Hegemann, D., Vohrer, U., Oehr, C., and Riedel, R. (1999) *Surf. Coat. Technol.*, **116–119**, 1033–1036.
86 Alexander, M.R., Short, R.D., Jones, F.R., Michaeli, W., and Blomfield, C.J. (1999) *Appl. Surf. Sci.*, **137**, 179–183.
87 Foest, R., Adler, F., Sigeneger, F., and Schmidt, M. (2003) *Surf. Coat. Technol.*, **163–164**, 323–330.
88 Massines, F., Gherardi, N., Fornelli, A., and Martin, S. (2005) *Surf. Coat. Technol.*, **200**, 1855–1861.
89 Rüchardt, C. (1992) Radikale–eine chemische theorie in historischer sicht, *Sitzungsberichte Heidelberger Akad. Wissenschaften*, 319–345, Mathematischnaturwissenschaftliche Klasse.
90 Yasuda, H. (1992) *J. Macromol. Sci., A Chem.*, **10**, 383–420.
91 Geckeler, K.E., Gebhardt, R., and Grünwald, H. (1997) *Naturwissenschaften*, **84**, 150–151.
92 Suzuki, M., Kishida, A., Iwata, H., and Ikada, Y. (1986) *Macromolecules*, **19**, 1804–1808.
93 Tan, K.L., Woon, L.L., Wong, H.K., Kang, E.T., and Neoh, K.G. (1993) *Macromolecules*, **26**, 2832–2839.
94 König, U., Nitschke, M., Menning, A., Eberth, G., Pilz, M., Arnhold, C., Simon, F., Adam, G., and Werner, C. (2002) *Colloids Surf., B*, **24**, 63–71.
95 Kuzuya, M., Noguchi, A., Ishikawa, M., Koide, A., Sawada, K., Ito, A., and Noda, N. (1991) *J. Phys. Chem.*, **95**, 2398.
96 Tsuji, K. (1973) *Adv. Polym. Sci.*, **12**, 131–190. Springer-Verlag, Berlin.
97 Hansen, R.H. and Schonhorn, H. (1966) *J. Polym. Sci., B*, **4**, 203.
98 Hudis, M.V. (1972) *J. Appl. Polym. Sci.*, **16**, 2397.
99 Pearsons, A.F. (2003) *An Introduction to Radical Chemistry*, Blackwell Science, Oxford.
100 Inagaki, N., Tasaka, S., and Horikawa, Y. (1991) *Polym. Bull.*, **26**, 283.
101 McCallum, J.R. and Rankin, C.T. (1971) *J. Polym. Sci., B*, **9**, 751.
102 Yamaguchi, T., Nakao, S.-I., and Kimura, S. (1991) *Macromolecules*, **24**, 5522.
103 Yamaguchi, T., Nakao, S.-I., and Kimura, S. (1996) *J. Polym. Sci., A*, **34**, 1203.
104 Lebedev, O.L. and Kazarnovskii, S.N. (1960) *Zh. Obshch. Khim.*, **30**, 1631–1635.
105 Montanari, F., Quici, S., Henry-Riyad, H., and Tidwell, T.T. (2005) *Encyclopedia*

of Reagents for Organic Synthesis, John Wiley & Sons, Inc., Hoboken.

106 Veregin, R.P.N., Georges, M.K., Kazmaier, P.M., and Hamer, G.K. (1993) *Macromolecules*, **26**, 5316.

107 Kang, E.T., Zhang, J., Cui, C.Q., Lim, T.B., and Neoh, K.G. (1998) *J. Adhesion Sci. Technol.*, **12**, 1205.

108 Zhang, Y., Huan, A.C.H., Tan, L., and Kang, E.T. (2000) *Nucl. Instrum. Methods Phys. Res., Sect. B*, **168**, 29.

109 Völcker, N., Klee, D., Höcker, H., and Langfeld, S. (2001) *J. Mater. Sci. Mater. Med.*, **12**, 111.

110 Jakes, K. and Mitchell, J. (1996) *J. Archeol. Sci.*, **23**, 149–156.

111 Thomas, R.S. (1976) *J. Macromol. Sci.*, **A10**, 255–346.

112 Grasenick, F. (1956) *Radex-Rundschau*, **4/5**, 226246.

113 Spit, B.J. (1967) *Faaserforsch. Textiltechnik – Z. Polymerenforsch.*, **18**, 161–168.

114 Friedrich, J. and Gähde, J. (1981) *Plaste Kautsch.*, **28**, 620–626.

115 Friedrich, J., Throl, U., Gähde, J., and Schierhorn, E. (1982) *Acta Polym.*, **33**, 405–410.

116 Throl, U., Gähde, J., Friedrich, J., and Schierhorn, E. (1982) *Acta Polym.*, **33**, 561–566.

117 Throl, U., Gähde, J., and Friedrich, J. (1982) *Acta Polym.*, **33**, 667–673.

118 Renekker, D.H. and Bolz, L.H. (1976) *J. Macromol. Sci., A*, **10**, 599–609.

119 Kaempf, G. and Orth, H. (1975) *J. Macromol. Sci.-Phys. B*, **11**, 151–164.

120 Possart, W. and Friedrich, J. (1986) *Plaste Kautsch.*, **33**, 273–279.

121 Friedrich, J., Lehmann, M., Throl, U., and Raubach, H. (1986) *Acta Polym.*, **37**, 655–661.

122 Friedrich, J., Deutsch, K., and Throl, U. (1988) *Acta Polym.*, **39**, 406–411.

123 Friedrich, J., Lehmann, M., Raubach, H., and Throl, U. (1989) *Acta Polym.*, **40**, 19–24.

124 Tsuji, O., Wydeven, T., and Hozumi, K. (1977) *Microchem. J.*, **22**, 229–235.

125 Höper, W.E. (2008) *Asbest in Der Moderne. Industrielle Produktion, Verarbeitung, Verbot, Substitution Und Entsorgung*, Waxmann Verlag, Münster/New York.

126 Thomas, R.S. (1962) Demonstration of structure-bound mineral constituents in thin sectioned bacterial spores by ultramicroincineration, in *Fifth International Congress for Electron Microscopy*, vol. 2 (ed. S.S. Brease, Jr.), Academic Press, New York, p. PR-11.

127 Friedrich, J., Throl, U., Deutsch, K., Seibt, H., and Engelbrecht, L. (1989) DD-AS 294 089 A5, August 15.

10 Atmospheric-Pressure Plasmas

10.1 General

More than 99% of matter in the universe is in the plasma state [1]. Low-pressure (dark room of the universe, ionosphere, etc.) and high-pressure (son, stars, etc.) plasmas are present. On Earth, different natural plasmas are known such as lightning, Earth aurora or flames, which can be assigned to the plasma state because of its (low) ionization. Lightning in the primeval atmosphere of Earth is responsible for the genesis of life [2–3].

Different kinds of technically used atmospheric-pressure plasmas exist, such as arc plasma used for welding, cutting, coating, powder formation, light emission (UV/visible), and so on, (low-temperature) atmospheric pressure glow discharge (APGD), corona and dielectric barrier discharge, thermal or non-thermal plasma jets, spark, surface/sliding, and micro discharges. Industrial applications to polymers are welding, heating, melting, radiation emission, surface modification, metal and polymer coating, etching, cleaning, and so on [4].

Polyolefins such as polyethylene and polypropylene are often subjected to atmospheric-pressure plasmas [5]. They do not possess functional groups in their chemical structure and, therefore, such groups are not present at their surface. Thus, any physical or chemical interactions and, therefore, sufficient adhesion to other polymer, inorganic, or metal coatings cannot be achieved. To introduce functional groups onto aliphatic chains only oxidative processes are chemically possible. Such an oxidation is known as the Bashkirov reaction and is used in industry. Using air [6–7] and catalyst, alcohols (organic borates) [8] or fatty acids (permanganate) [9] were produced at elevated temperatures. Even at room temperature polyolefins could be extensively oxidized using fluorine, the strongest oxidation agent [10–11]. Fluorine oxidation proceeds as hydrogen substitution, is nearly complete and, therefore, produces perfluorinated paraffins and partially fluorinated polyolefins [12]. However, the equivalence of C–C (370 kJ mol^{-1}, polyethylene) and C–H (397 kJ mol^{-1}, polyethylene) bond dissociation energies produces also C–C chain scissions, which means polymer degradation. Thus, low-molecular weight oxidized material (LMWOM) is always present at the polyolefin surface after any oxidation, for example, fluorination, bromination, or

The Plasma Chemistry of Polymer Surfaces: Advanced Techniques for Surface Design, First Edition.
Jörg Friedrich.

oxidation in air [13]. Such LMWOM formation, associated with the formation of a weak boundary layer [14], has been extensively investigated by Strobel and coworkers [15].

The polyolefin surface oxidation is directed to form polar oxygen-containing groups at the surface and is carried out industrially by flame treatment, which was introduced by Kreidl [16]. Fluorination or oxyfluorination [17, 18], exposure to low-pressure glow discharges [5, 19], and atmospheric-pressure dielectric barrier discharge (DBD, often called "corona") are other treatments for oxidizing the polyolefin surface [20–22]. Irradiation with short-wavelength UV radiation produced by excimer lamps is a more modern process [23–25].

10.2
Dielectric Barrier Discharge (DBD) Treatment

Polyolefin surface modification by atmospheric-pressure dielectric barrier (corona) discharges in air is very popular and extensively applied in industrial processing [21–22, 26–42].

The effect of introducing polar groups onto the polyolefin surface is often temporary, in particular when using the dielectric barrier discharge (DBD) treatment [43]. In the presence of oxygen a broad variety of O-functional groups are produced [44].

Strobel *et al.* determined the penetration depth of the plasma-emitted radiation of the atmospheric-pressure plasma into polypropylene foils as 10–1000 nm, characterized by an oxidized layer at the polymer surface [45]. The observed globular structure [46–47] was attributed to LMWOM produced by "corona" discharge at a threshold energy of $<4\,kJ\,m^{-2}$ for the polymer surface or $>4\,kJ\,m^{-2}$ for the bulk polymer [48–49].

Sun *et al.* investigated the influence of typical antioxidants, the erucamide slip agent, and fatty acid amines on the treatability of PE (polyethylene) and PP (polypropylene) [50]. The presence of antioxidants required longer treatment times to achieve a desired level of oxidation and, therefore, surface energy. The hydrophobic recovery, the time-dependent disappearance of surface polar groups, was also caused by additive migration, especially that of slip agents. Re-treatment was proposed as one way for effective surface modification of additive-containing polyolefins.

Novák *et al.* studied the corona treatment in air of isotactic PP (i-PP) and low-density polyethylene (LDPE) with the aim of enhancing the adhesion properties towards poly(vinyl acetate) [51]. They found that the hydrophobic recovery of i-PP was influenced by the polymer crystallinity. Higher fractions of amorphous regions gave rise to higher concentration of hydroperoxides and consequentially to higher surface energy. Additives in the polymer, originating from processing of the foils, influenced the diffusion ability of the modified chains (plasticizing). The diffusion of additives to the surface also produced lower surface energy (sweating out). This fact is significant for the long-time stability of corona-modified surfaces.

Guimond and Wertheimer compared the polymer degradation of BOPP (biaxially orientated polypropylene) and LDPE using two different dielectric barrier discharge treatment types – an air corona and a nitrogen atmospheric pressure glow discharge (APGD) [47]. While N_2 APGD treatments were more efficient in increasing the surface energy, producing a high degree of functionalization, and giving negligible polymer degradation the air corona treatment only slightly improved the surface energy. The hydrophobic recovery was estimated for both treatment methods using contact angle measurements. A strong fall in surface energy was observed within one week for both types of plasma treatment. The mechanism of aging was discussed in terms of thermodynamically driven re-orientations of functional groups and diffusion away from the surface into the bulk to balance their concentration both at the surface and in the bulk.

The surface topography of corona-treated BOPP films is determined by the existence of agglomerates of low molecular weight products, as confirmed by Jones *et al.* [52]. They found that at a low corona energy of about $0.1\,\mathrm{J\,cm^{-2}}$ the produced agglomerates were soluble in water, even after dipping into water for 2 min. Lynch *et al.* studied the influence of atmospheric-pressure plasma on the surface properties (contact angle, composition, and topography) of PE using various gas compositions [53]. They measured a higher long-term wettability using corona treatment in oxygen and nitrogen atmosphere than in air.

The rate of oxygen introduction is nearly the same for DBD, low-pressure cw-rf oxygen plasma, and a plasma jet (blown-out spark discharge in air) (Figure 10.1).

The oxygen introduction by DBD treatment is reflected in OH- and C=O group formation on polypropylene surfaces as evident in a series of ATR-IR-spectra (Figure 10.2).

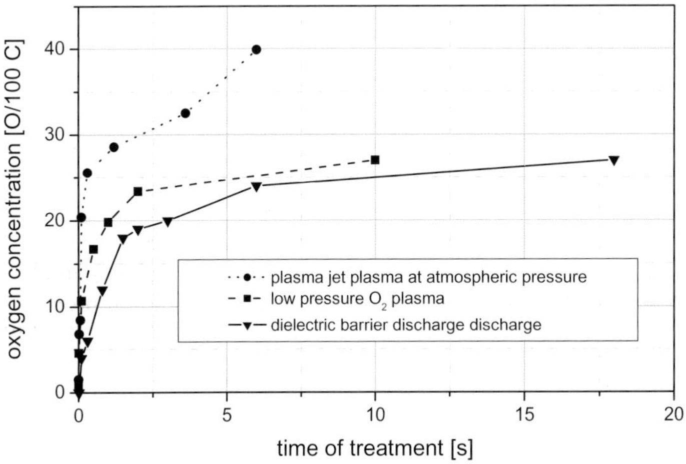

Figure 10.1 Comparison of O-introduction into polypropylene by atmospheric-pressure plasma in air and by low-pressure plasma in oxygen.

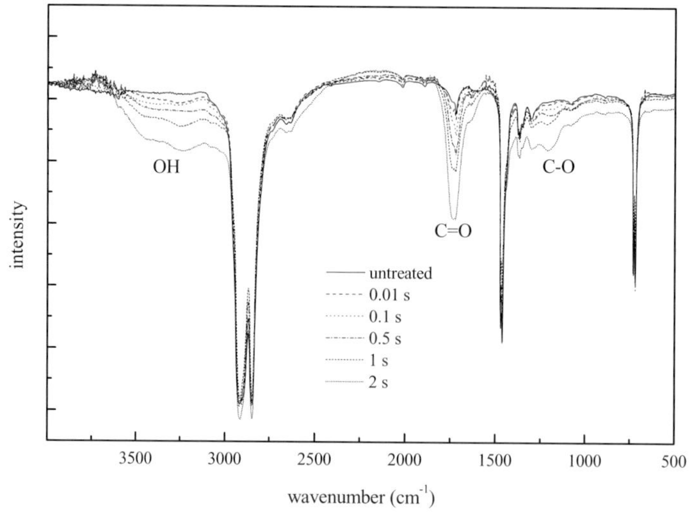

Figure 10.2 ATR-FTIR spectra of polypropylene exposed for different times to the plasma of dielectric barrier discharge (DBD) in air.

The comparison of DBD, plasma jet treatment (blown-out spark discharge in air), and low-pressure oxygen in terms of the ability to incorporate oxygen and therefore in increasing surface energy manifests also the significant role of degradation products (LMWOM) (Figure 10.3).

Evidently, DBD treatment introduced oxygen did not increase the surface energy as much as low-pressure oxygen plasma. The plasma jet produces the highest oxygen introduction, however, by undesired metal oxides from electrodes deposited onto polypropylene surface, but not the highest surface energy.

The C1s peak fitting into CH (aliphatic, C–C, C=C), C–O (OH, epoxy, ether), C=O (ketone, aldehyde), and O–C=O (acid, ester) components for DBD, plasma jet, and low-pressure oxygen plasma treatment shows slight differences (Figure 10.4).

The higher oxidized species (C=O and O–C=O) are present in higher percentage after exposure to atmospheric-pressure plasmas than to the low-pressure oxygen plasma. Moreover, nitrogen introduction was also observed [11].

The surface energy of polypropylene was increased by its exposure to the DBD in air, in particular by an increase in the polar contribution (Figure 10.5). The polar contribution corresponds well with the introduction of O-functional groups (cf. Figure 10.3).

The molar masses of polymers exposed to the DBD in air decreases, as demonstrated by size exclusion chromatography (SEC; also known as GPC–gel permeation chromatography), and is also evident for exposure to the low-pressure oxygen plasma, as shown in Figure 10.6. Noticeably, monomers, dimers, and trimers dominate.

“Super-oxidation” by exposure to the jet plasma (air, oxygen, air + oxygen) also occurs, as shown by the differences in O-concentration (XPS data) on comparing

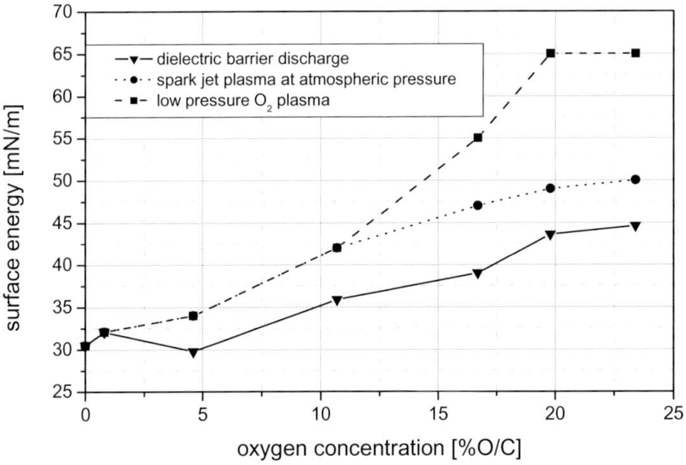

Figure 10.3 Dependence of surface energy on oxygen concentration at polypropylene surfaces for three kinds of plasma pretreatment.

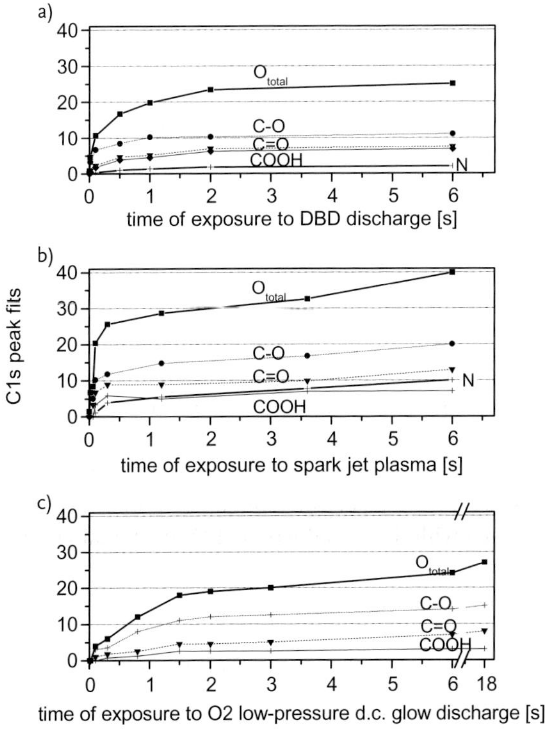

Figure 10.4 C1s peak fits and O as well as N concentration of polypropylene surfaces versus exposure time to DBD, plasma jet, and low-pressure oxygen plasma.

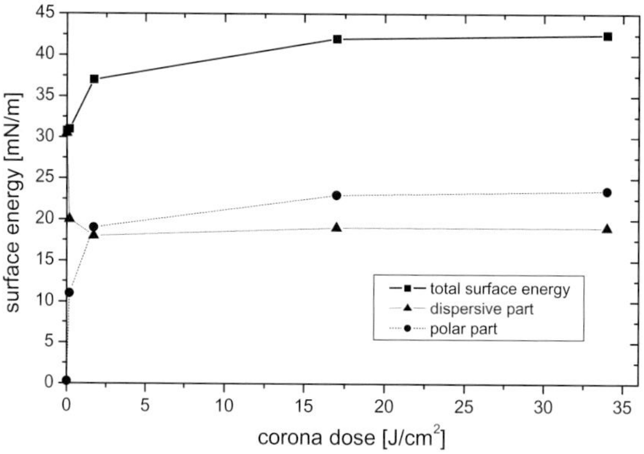

Figure 10.5 Increase of surface energy of polypropylene during exposure to dielectric barrier discharge in air.

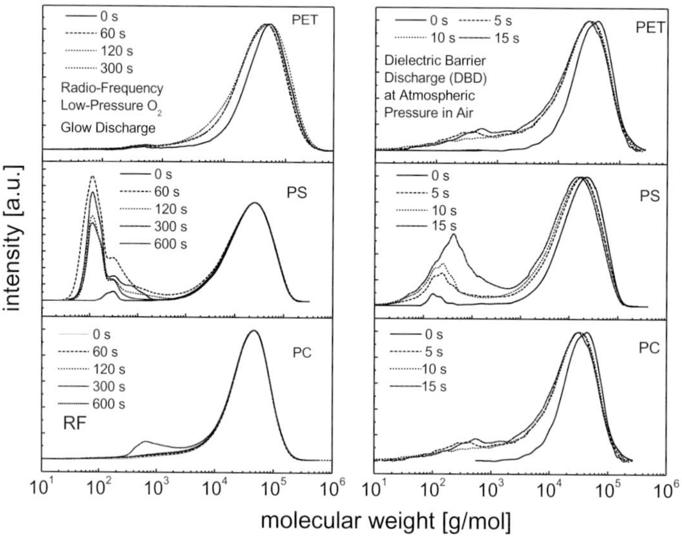

Figure 10.6 SEC results of polymer degradation on exposure to low-pressure oxygen rf plasma or to the atmospheric barrier discharge (DBD) in air. PET = poly(ethylene terephthalate); PS = polystyrene; PC = polycarbonate.

its time-dependence with that of exposing PET to low-pressure oxygen plasma (Figure 10.7).

Also in this case, electrode material was sputtered from the electrode and deposited as oxygen-rich metal oxide onto poly(ethylene terephthalate), thus simulating strong oxidation of PET. In reality, plasma exposure decreased O-concentration by decarbonylation or decarboxylation (Norrish-I rearrangement of ester groups), from theoretically 40 O/100 C and 37 O/100 C in the untreated commercial PET sample to about 30 O/100 C. Given that the ester group together with the ethylene

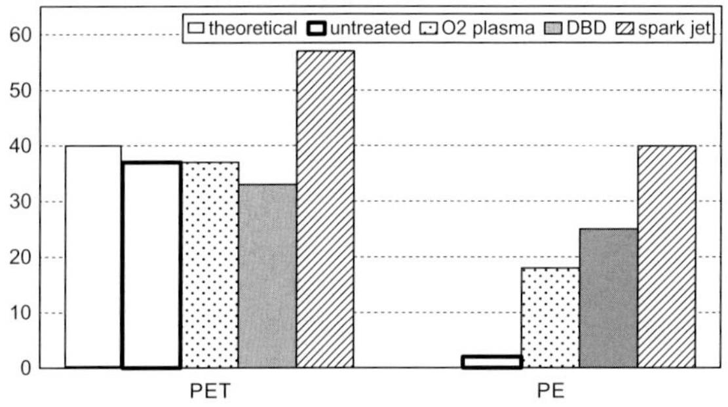

Figure 10.7 Oxygen concentration of PET and PE after 18 s exposure to atmospheric plasma jet and dielectric barrier discharge (DBD) in air as well as to low-pressure rf oxygen discharge for each 18 s.

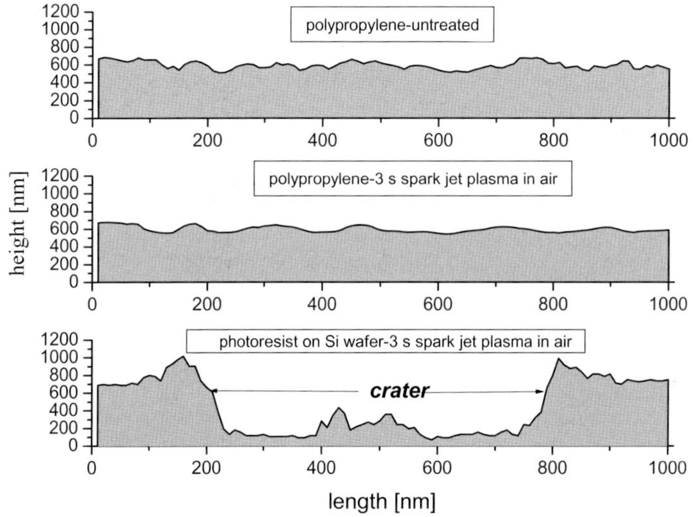

Figure 10.8 Impact of a streamer produced by a spark jet plasma in air on a polypropylene surface, as detected by atomic force microscopy (AFM).

glycol unit forms the PET backbone, it is obvious that any loss in oxygen destroys the polymer structure dramatically and decreases strongly the molar mass.

Spark, corona, and streamer of dielectric barrier discharges leave many marks of micro-plasmas at polypropylene surfaces (Figure 10.8).

Moreover, the molar mass of polymers (PC, PS, and PET) becomes decreased strongly during exposure to barrier discharge (or plasma jet), as shown in Figure 10.6. Details of the mass loss of PET exposed to the DBD or to oxygen low-pressure plasma are depicted in Figure 10.9. Increasing polypropylene temperature from 20 °C to 80 °C produces higher surface energy and polar contribution exposing this polymer to the spark jet plasma (Figure 10.10). Temperature also influences

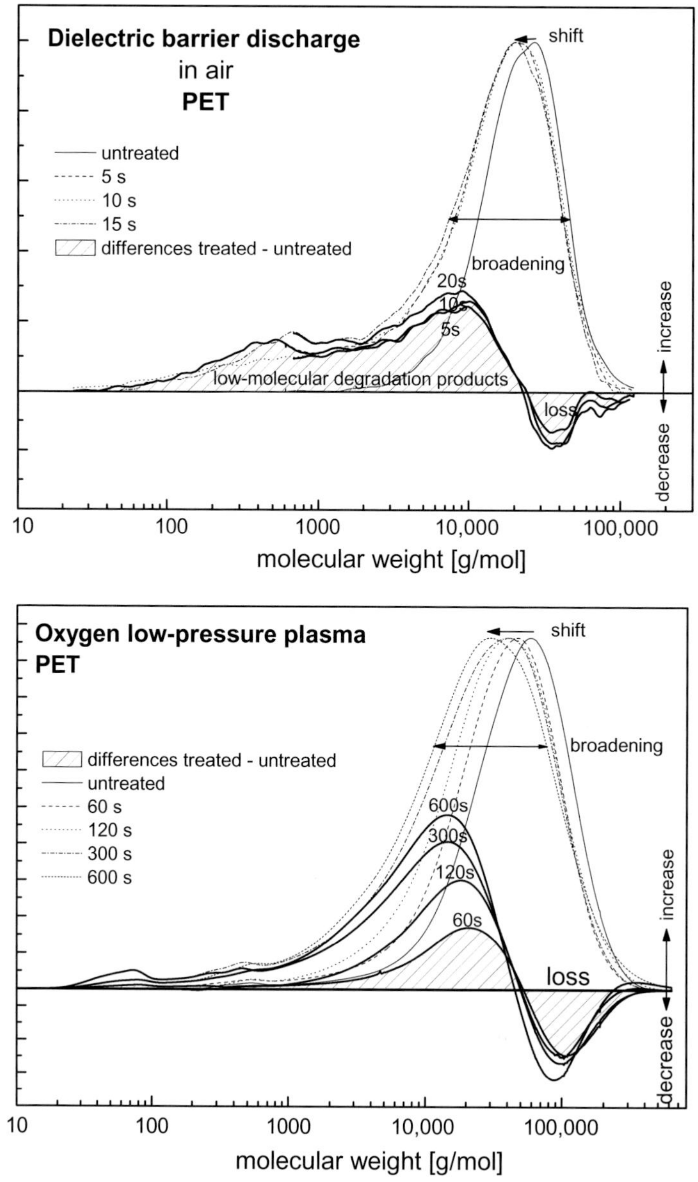

Figure 10.9 Details of the change in the molar mass of PET with exposure to dielectric barrier discharge in air and to low-pressure oxygen plasma, measured by means of size exclusion chromatography (SEC).

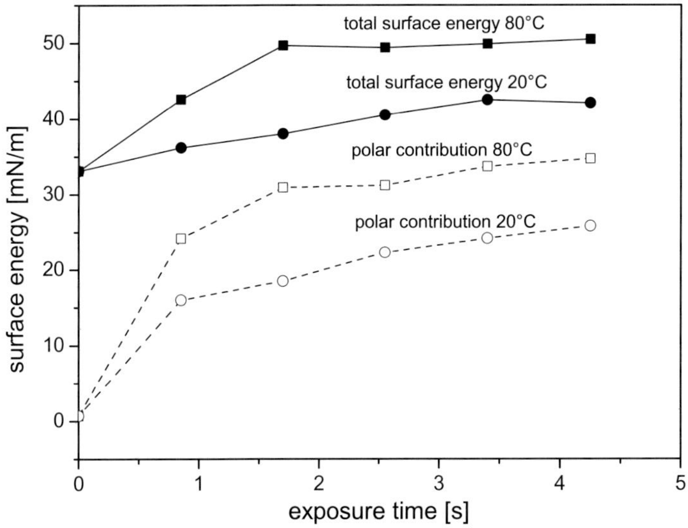

Figure 10.10 Surface energy and polar contribution of polypropylene in dependence on exposure time to the dielectric barrier discharge in air and for two temperatures.

introduction of O-functional groups into polyolefin surfaces as shown for polypropylene exposed to the dielectric barrier discharge in air at two temperatures (Figure 10.10). The higher increase in surface energy at 80 °C may also be interpreted as indicator of more extensive polymer degradation.

10.3
Polymerization by Introduction of Gases, Vapors, or Aerosols into a DBD

In 2002, Dow Corning Plasma Solutions developed an atmospheric-pressure plasma liquid deposition using an ultrasonic atomizer [54]. Ward *et al.* reported the fast atmospheric-pressure glow discharge deposition of cyclic siloxane monomers (octamethylcyclotetrasiloxane, tetramethylcyclotetrasiloxane) in the presence of He or He/O_2 (99 : 1) as process gases [55]. They sprayed monomers into the plasma, started a plasma polymerization, and deposited 300 nm thick polymer films on PE. Depending on the used process gas SiO_2-rich or hydrophobic coatings were deposited with an oxygen barrier of polyethylene improved by a factor of 4–6. In a further paper the authors describe the generation of acrylic acid polymer films deposited by the same method using He as plasma process gas [56]. The deposited poly (acrylic acid) films reduced the oxygen permeation of the PE by a factor of about seven and also improved the single-lap shear strength of nylon strips.

O'Hare *et al.* were one of the first to apply nebulized liquid precursors and injected the produced aerosols directly into the atmospheric-pressure plasma for the production of anti-microbiological coatings [57]. They applied aqueous solutions of acrylic acid or poly(ethylene glycol methacrylate) or diacrylate monomers in combination with quaternary ammonium salts of commercial broad-spectrum anti-microbiological substances/disinfectants. The deposited plasma polymer contained the entrapped cetalkonium or benzalkonium salts, thus retaining their biological and chemical properties

Recently, Twomey *et al.* investigated the effect of plasma parameters on the chemistry and morphology of aerosol-assisted plasma deposits of hexamethyldisiloxane (HMDSO), polydimethylsiloxane (PDMS), and tetramethylsiloxane (TMSO) [58].

Wu employed different types of atmospheric- and low-pressure plasma pretreatments with regard to improving the adhesion properties of polymers [59] as did Drost, Friedrich, and Garbassi [60–62].

Deposition of plasma polymers by corona or dielectric barrier discharges was first tested in the 1960s [63]. Exposure of benzene vapor in a corona discharge produced not only polymers but also phenyl radicals, thus forming biphenyl, terphenyl, acetylene, fulvene, cyclohexadiene, and so on. In 1979, Donohoe and Wydeven [64] described plasma polymerization of a gas mixture of ethylene (about 4 vol.%) and helium, delivering soft coatings of a typical composition of $C_2H_{3.26}O_{0.23}$, which were well adherent to glass substrates and deposited at rates of 0.1–0.2 nm s^{-1}.

More recently, interest was revitalized to produce plasma polymers using atmospheric-pressure plasmas [41, 65–67]. An alternative way is the nebulizing of monomer solutions and their introduction into an atmospheric-pressure glow discharge for plasma polymer formation and deposition [66, 68, 69]. Using benzene or octamethylcyclotetrasiloxane as monomers with nitrogen in the DBD plasma, Janca and Pavelka had deposited clear, soft, and well adhering films at deposition rates of 0.2–0.5 nm s^{-1} [70]. Silicon- and fluorine-containing coatings were also produced [71, 72] as well as carboxylic-group containing polymer layers [73]. Technical details and developments of plasma polymerization under atmospheric-pressure plasma conditions have been presented by Salge [74].

By injecting monomers into the plasma, the probability of complete monomer fragmentation and random recombination is also very high in the atmospheric plasma and, thus, irregularly structured polymer layers were often produced. To avoid this fragmentation–recombination process the alternative is to introduce complete macromolecules equipped with the desired functional groups into the plasma and deposit them under activation of the layer-forming macromolecules and the substrate surface by aerosol-DBD plasma [75], which was first claimed by Hoechst [76].

10.4
Introduction of Polymer Molecules into the Atmospheric-Pressure Plasma and Their Deposition as Thin Polymer Films (Aerosol-DBD)

The atmospheric-pressure aerosol plasma is produced within the electrode gap by injection of the aerosol in the presence of the substrate, which is modified or coated with an ultrathin polymer film (Figure 10.11).

Figure 10.11 Aerosol dielectric barrier discharge (DBD) with injection of the nebulized liquid or polymer solution into the electrode gap (plasma).

The different nature and different reactivity of atmospheric-pressure plasma introduced O-functional groups are a strong hindrance for all subsequent chemical reactions with these groups. For this purpose, monotype functional groups in high density at the polyolefin surface are needed, so that only one type of covalent bond to the coating (or other material) atoms or molecules is formed. However, such selectivity in monotype surface functionalization in sufficient density is not possible to achieve using the common barrier or corona discharges in air or by plasma polymerization. As a pragmatic alternative, polyolefin surfaces were coated with very thin layers of functional-group carrying polymers acting as adhesion-promoting interlayers in polymer composites (Figure 10.12).

Obviously, the atmospheric-pressure plasma in air produces an unspecific surface functionalization, which can be laboriously unified by wet-chemical reduction of carbonyl features using $LiAlH_4$ [77]. For deposition of such a primer layer onto the polyolefin surface, a polymer solution was aerosolized and introduced into the DBD plasma [78]. The working hypothesis was that the macromolecules in the aerosol droplets are plasma-activated, for example, radicals are formed. Simultaneously, radicals are also formed at the polyolefin surface on exposure to the plasma. Thus, a covalent bonding of the coating molecules to the substrate molecules by radical–radical recombination was expected (Figure 10.13) [79].

Each macromolecule of the coating polymer possesses the desired monotype functional groups in high concentration. Thus, OH, NH_2, or COOH group-rich polymer and copolymer layers were deposited onto polyethylene or polypropylene surfaces. These different functional groups were used to test the adhesion of thin thermally (Al, Cu) or sputtered (Ti) metal layers.

To produce OH-rich surfaces poly(ethylene glycol)–poly(vinyl alcohol) copolymer, for NH_2-rich surfaces poly(vinylamine), and for COOH-rich surfaces

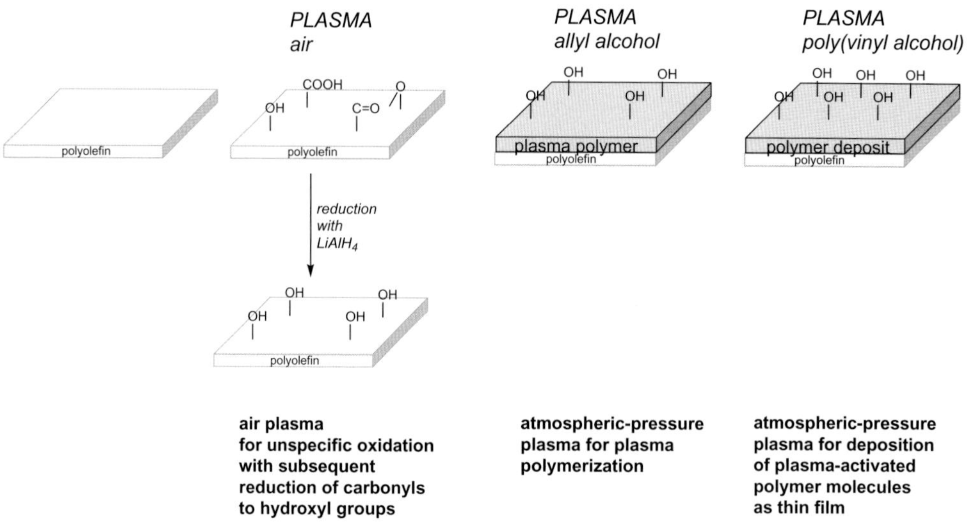

Figure 10.12 Schematics of polymer surface functionalization by atmospheric-pressure plasmas.

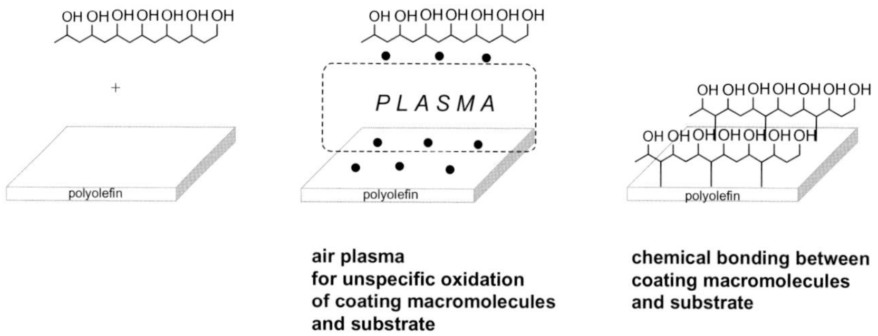

Figure 10.13 Schematics of expected link formed between aerosol coating and substrate.

poly(acrylic acid) layers were deposited (20 nm). These polymer solutions were also used for deposition of these polymers by employing the electrospray ionization (ESI) technique [78]. Notably, aerosol-DBD and ESI techniques need similar technical equipment, as demonstrated later. However, the deposition mechanism of polymer films is different. DBD consists of a series of streamer discharge channels between two ceramic coated electrodes. The polymer foil, whose surface is treated, passes continuously through the gap between the two electrodes filled with the plasma zone. This type of discharge under atmospheric pressure in presence of air produces O- (and N-) functional groups, degradation, and roughens the surface of the polyolefin substrate. However, aerosol-DBD modifies not only the polyolefin

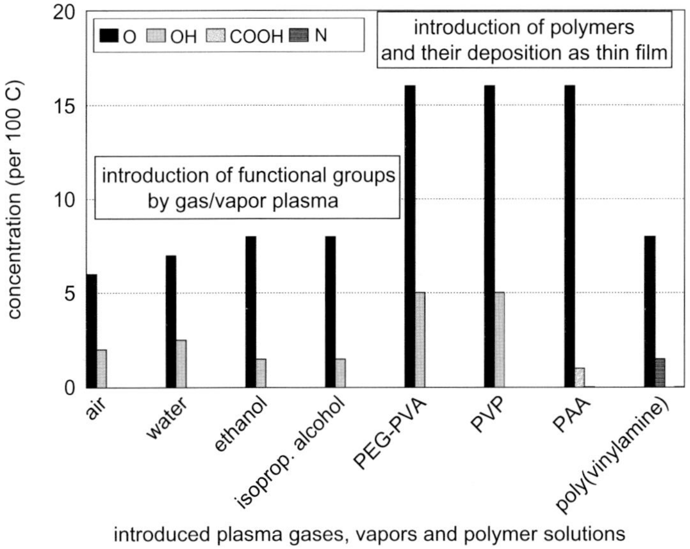

Figure 10.14 Concentrations of functional groups introduced by different DBD treatments ($t = 0.01$ s); PEG-PVA = graft copolymer of poly(ethylene glycol) (PEG) and poly(vinyl alcohol) (PVA), PVP = poly(vinylpyrrolidone), and PAA = poly(acrylic acid).

substrate but also the (polymer) molecules in the aerosol droplets, as described before (cf. Figure 10.13).

Using the DBD technique in air aerosols of water, ethanol, and isopropanol as well as copolymers and homopolymers with functional groups were introduced into the DBD plasma. The measured yield in oxidation was each 6–8% O/C and the yield in hydroxyl groups amounts was 1–2 OH/100 C after derivatization with trifluoroacetic anhydride (TFAA) and measuring the fluorine percentage (Figure 10.14).

The deposition of functional groups bearing polymers by the aerosol-DBD technique produces a much higher oxygen percentage as well as 1–5 functional groups per 100 C (cf. Figure 10.13).

The energy density (wattage × residence time) did not influence significantly the yield in oxygen introduction (Figure 10.15); however, it was strongly dependent on the admixed gas, vapor, or polymer. DBD treatment in air and with water aerosol produced the highest O introduction but not the best adhesion properties, as demonstrated later.

Figure 10.16 compares the C1s and O1s signals of aerosol-DBD deposited polymer films (Kollicoat-PEG-PVA graft copolymer) with those of electrospray-deposited (ESI) films (see next section) and cast deposited films.

Evidently, DBD deposited films show much loss in functional group as well as structure of copolymer considering the displacement of sub-peaks of the C1s

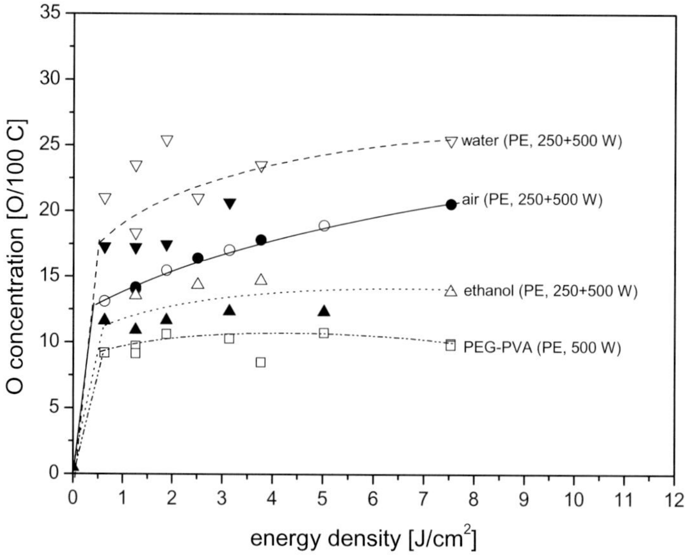

Figure 10.15 Dependence of oxygen introduction on energy density for simple DBD treatment, aerosol DBD treatment with water and with ethanol, and layer deposition of PEG-PVA copolymer (Kollicoat® IR), note hollow symbols are assigned to 250 W and full symbols to 500 W.

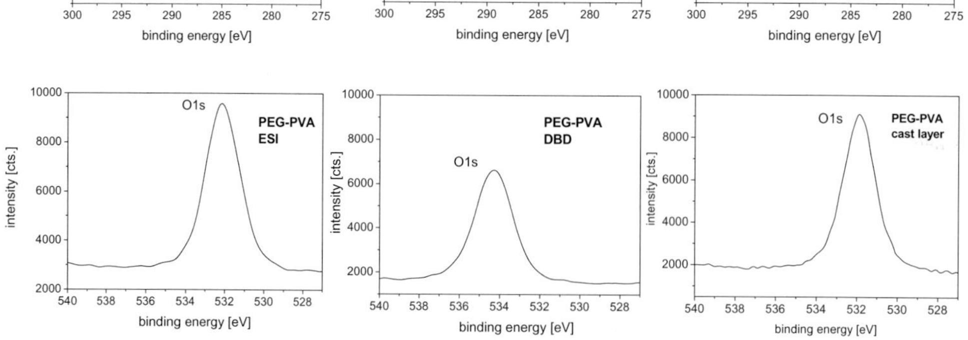

Figure 10.16 C1s and O1s peaks of PEG-PVA copolymer films deposited (a) by ESI, (b) by aerosol-DBD ("corona"), and (c) as cast films.

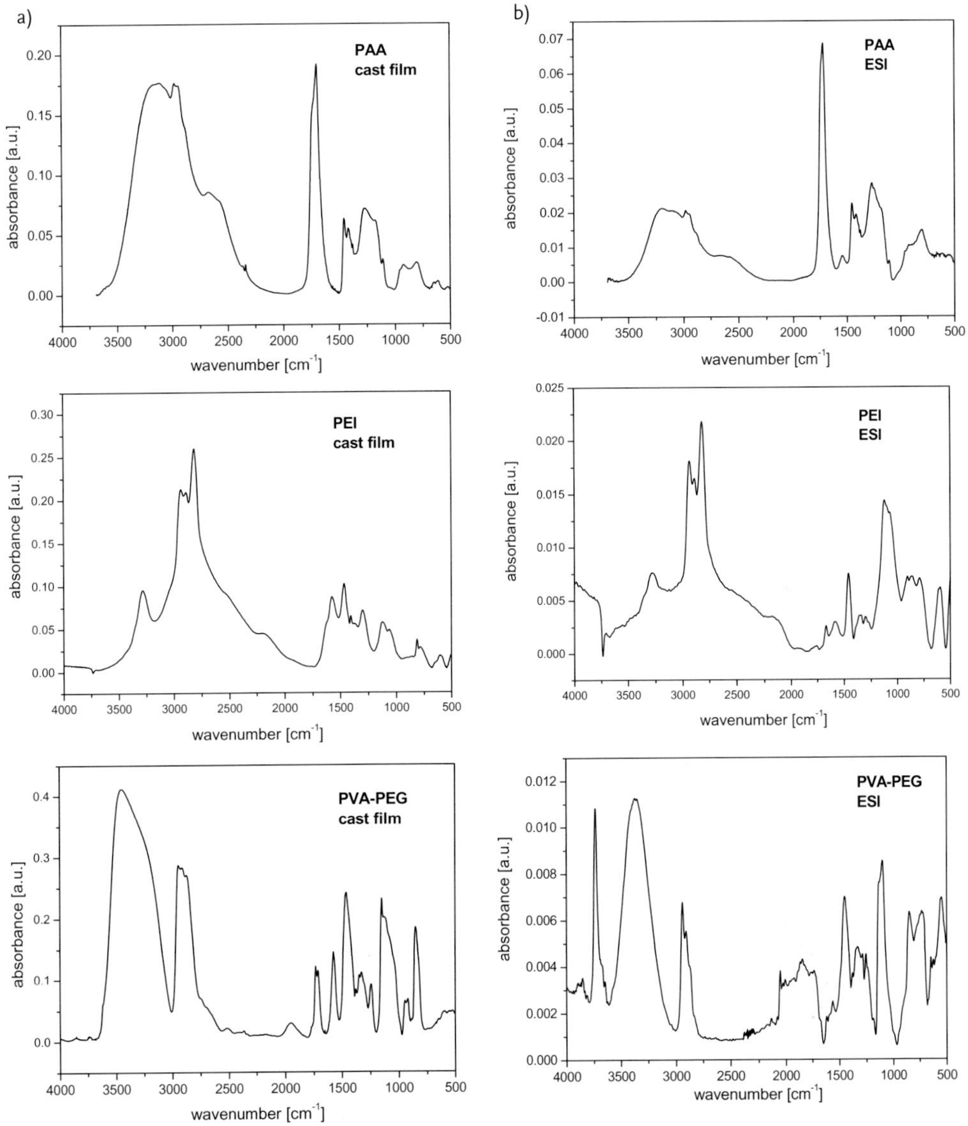

Figure 10.17 IR grazing incidence reflection spectra (72° to the surface normal) of poly(acrylic acid) (PAA), poly(ethyleneimine) (PEI), and PEG-PVA copolymer layers on gold surfaces produced by (a) casting and (b) ESI deposition.

signal and the broadening and decreasing in intensity of the O1s signal. In contrast, ESI-films show nearly total accordance to the reference films.

Infrared reflectance absorption spectroscopy (IRRAS) under grazing incidence reflectance (GIR) conditions confirm the similarity of ESI and reference films and the much changed structure of aerosol-DBD films (Figure 10.17). The "cast" or spin-coating films were prepared as reference by using commercial polymers.

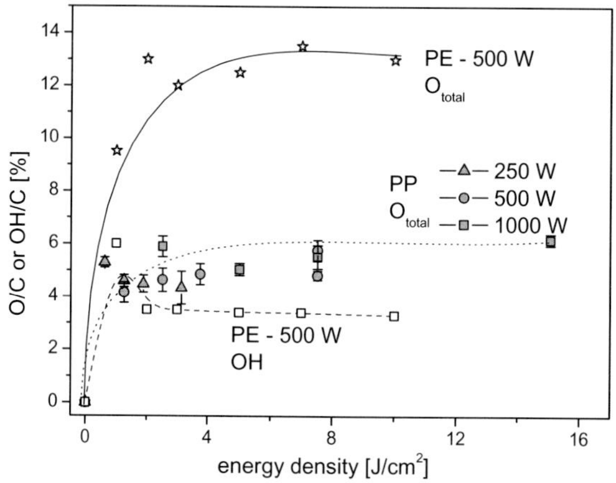

Figure 10.18 O and OH incorporation into PP and PE surfaces by g-PEG-PVA copolymer deposition versus energy density (with varying wattage).

The survival of g-PEG-PVA (Kollicoat IR) and its OH groups depended on the energy dose or, as shown in Figure 10.18, density; 1–2 J cm^{-2} were sufficient to achieve the maximal O and OH concentration. Higher energy densities did not increase further the O and OH concentration at PP surfaces.

Copolymer (1%), dissolved in water, was nebulized and deposited as polymer layers in the DBD discharge region. To remove weakly or unbound fractions of deposited copolymer layers the foils were washed in water or alcohol.

The graft-poly(ethylene glycol)–poly(vinyl alcohol) copolymer (g-PEG-PVA, Kollicoat IR) layer showed about 5 O per 100 C after deposition on PP surfaces and nearly 9 O per 100 C on PE after washing with water (Figure 10.14). Using the derivatization of OH groups with TFAA of water-washed samples it could be shown that 3–4 OH groups per 100 C had survived at the polyethylene surface and about 2 OH at polypropylene surfaces by employing low energy densities. At higher energy densities the concentration of remaining OH groups decreased, as did their fractions among all oxygen species.

A pinhole-free PEG-PVA layer at the surface of polyethylene should show theoretically more than 50 O/100 C, which would include about 25 OH/100 C. Thus, it can be concluded that the coating with copolymer was incomplete again or the copolymer was removed by the wash process and/or the copolymer was strongly degraded. The low yields in O introduction by incomplete copolymer coating of polypropylene and polyethylene were also reflected in relatively high contact angles. These angles were in the range 60–70° for polyethylene and 85–90° for polypropylene after washing the coated samples with water.

Aerosol-DBD deposited poly(vinylpyrrolidone) did not adhere well to polyolefin surfaces and did not survive the wash process with water. Thus, 3 N and 20 O/100 C were measured as maximum at the highest energy density on PE and PP surfaces. A complete coating with non-degraded poly(vinylpyrrolidone) would produce

16 N and 16 O per 100 C atoms (cf. Figure 10.14). Thus, regarding the N-percentage only 20% coverage of surface can be assumed; moreover, oxidation of the deposited layer was also found due to the presence of air in this aerosol-spray process.

Amino groups at the polymer surface were produced by deposition of Lupramine, a high molecular weight poly(vinylamine) (PVAm) dissolved in water. In the XPS spectrum of PVAm aerosol modified PE surfaces about 2 N/100 C remained after washing with water, of which 1 NH_2 group per 100 C was found, independent of the applied energy density. The oxygen content amounted to 9 O/100 C after washing with water and up to 12 O/100 C were found after applying an ethanol washing (cf. Figure 10.14). These results can be explained again by incomplete coverage of the surface or by strong degradation of poly(vinylamine) and oxidation by the plasma in predominant air. Additionally, only half of the remaining N was found as primary amino groups.

The surface modification of PE by deposition of a nebulized poly(acrylic acid) (PAA) solution in the DBD zone produced increasing oxygen content on PE and PP with increasing energy density; however, the COOH group concentration remained constant as well as the water contact angle of water-washed samples. Ethanol washing reduced the incorporated oxygen by more than 50% and also the number of COOH groups was reduced. However, the wettability was found to be unchanged as measured by the water contact angle, which was in the range 68–72°.

Poly(acrylic acid) (PAA) possesses an elemental composition of 67 O or 33 COOH groups per 100 C. A maximum of 25 O/100 C were measured immediately after its deposition as a PAA layer but only 1 COOH/100 C after washing with water. The COOH groups were derivatized with trifluoroethanol and the fluorine introduction was then measured by XPS. Water solubility of the deposited film and the decomposition of COOH groups on exposure to the DBD plasma are the reasons for the low COOH concentration.

To sum up, all aerosol DBD processes using low molecular weight polar substances (water, ethanol, isopropanol) showed nearly unchanged water contact angles of polyolefin surfaces in comparison to the well-known and often applied simple DBD treatment in air. The air and water DBD produced the highest yield in oxygen introduced on PP and PE surfaces. In contrast, surface modification by polymer layer deposition improved the wetting of polyolefin surfaces significantly. However, the contact angles were only moderately lowered rather than strongly lowered as expected. Moreover, XPS analysis of the deposited layer showed a much lower concentration of functional groups than that found in the original polymer material. The reason for this deficit may be incomplete coverage of the polyolefin surface by the adhesion-promoting polymer layer or its removal during the wash process. However, a series of atomic force microscopy (AFM) micrographs of the polyolefin surface after coating with the functional-group carrying polymer layers did not show islands or holes; only a homogeneous ribbed topography was observed. Therefore, a more probable explanation may be that polymer layer degradation had occurred during plasma exposure, but the depicted AFM topography does not exclude the incompleteness of polymer coating layer. The number of functional groups that survived in the deposited layer did not exceed 10% of the original concentration.

10.5
DBD Treatment of Polyolefin Surfaces for Improving Adhesion in Metal–Polymer Composites

Exposure of polyolefins to the dielectric barrier discharge (DBD or "corona") in air or after deposition of polymer layers using the aerosol-DBD or the ESI process is an appropriate treatment for improving metal–polymer adhesion. Oxygen-containing groups are well-suited for interaction with metal atoms of metal coatings. Metal–polymer interactions include also the formation of M–O–C covalent bonds with extra-high bond energy. A precondition is the existence of OH groups at the polymer surface for the formation of M–O–C bonds. Generally, COOH groups are also able to form strong bonds with metals, such as bidentate bonds with aluminium atoms (see Chapter 7) [80]. Each 150 nm thick Al layer was deposited onto DBD-treated polyolefin surfaces by thermal evaporation. The thus produced metal–polyolefin system consists of the polyolefin foil, the DBD-modified polyolefin surface (optional), the adhesion-promoting polymer layer (aerosol-DBD or ESI), and an aluminium layer (Scheme 10.1).

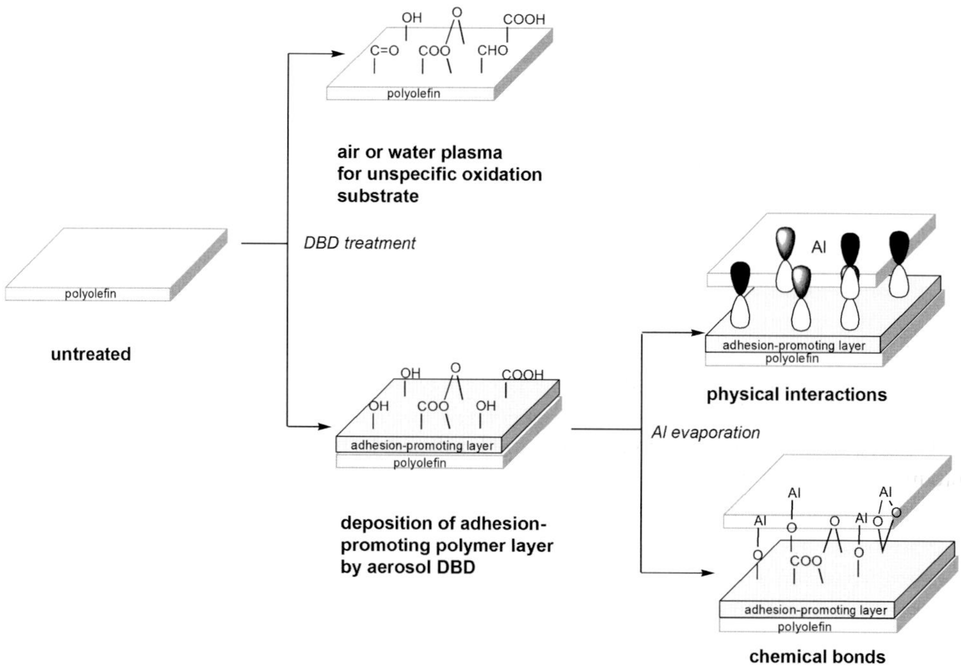

Scheme 10.1 Schematics of DBD pretreatments and resulting adhesion to aluminium by physical and chemical interactions.

Already, the simple exposure to DBD in air produced well-adhered Al–PP systems (Figure 10.19a). The energy density introduced into the DBD plasma did not influence the metal–PP peel strength significantly. The peeling propagates between the plasma-modified polypropylene and the unmodified bulk polypropylene, as measured by XPS of peeled surfaces, thus indicating the formation of a "weak boundary layer" by exposure to the DBD plasma. Water, ethanol, or isopropanol DBD treated Al–PP composites showed similar peel strengths, the same peel front propagation, and the same failure mechanism (Figure 10.19b).

Using an energy input of 250 W for the DBD, a maximum air-peel strength of 360 $N\,m^{-1}$ was measured and after addition of a water aerosol the peel strength was 280 $N\,m^{-1}$. In contrast, the deposition of the PEG-PVA slightly improved the peel strength to 400 $N\,m^{-1}$ at 250 W energy input. The locus of peeling in the Al-(PEG-PVA)-PP composite was found along the PP-(PEG-PVA) interface, showing that the interactions between the adhesion-promoting layer and the polypropylene were weak. Using low-pressure glow discharge deposited acrylic acid plasma polymer layer as adhesion promoter in the same Al–PP the peel front propagated within the polypropylene (cohesive failure) [69].

The degradation products at polyolefin surfaces produced by exposure to DBD may be counterproductive. This is evidenced by the strong loss of material observed during washing (Figure 10.20).

Note that storage of plasma-treated polyolefin surfaces lowers strongly the surface energy (Figure 10.21) as well as the adhesion of composites (Figure 10.22).

Poly(acrylic acid) layers deposited in the aerosol-DBD at atmospheric pressure strongly improved the peel strength of Al to the PP substrate. Even at low energy density, the peel strength measured was >1000 $N\,m^{-1}$ (Figure 10.19c). The Al–polymer composites with PAA could not be peeled using the peel test procedure, evidencing strong interactions between PAA-PP and PAA-Al, that is, at both interfaces of this composite [79]. Poly(vinylamine) showed a poor peel strength when used as adhesion-promoting layer.

10.6 Electrospray Ionization (ESI) Technique

Nebulizing of a polymer solution within a high-voltage electrical field (3–12 kV) produces a totally new type of spray. The charged droplets of spray become much smaller by solvent evaporation and sustain charge repulsion (Coulomb explosion) during passage from the capillary electrode to the counter electrode or target. Thus, the droplets decay in the electrical field into smaller ones stepwise and simultaneously the corona discharge ionizes the molecules directly or by charge transfer from solvent ions (APCI – atmospheric-pressure chemical ionization). In the absence of corona discharge polymer degradation within the droplets in the gas phase is avoided. In addition to the MALDI (matrix-assisted laser desorption ionization) process [81], the ESI process [82–84] allows the transfer of complete

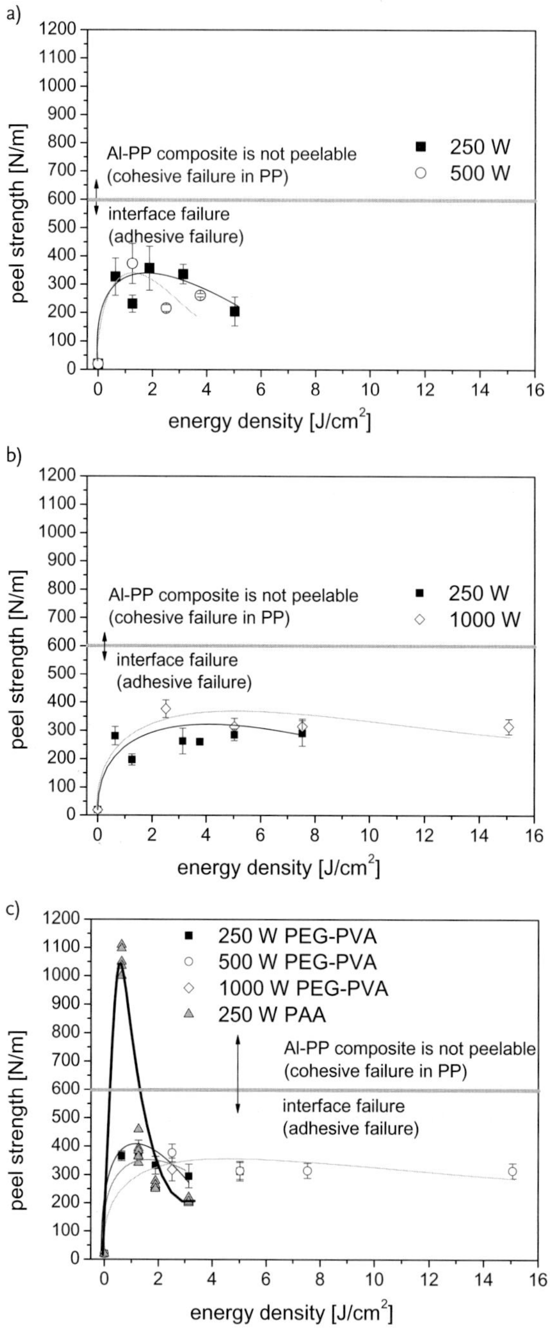

Figure 10.19 Peel strength of Al–PP composites. PP was pretreated by atmospheric-pressure DBD: (a) air, (b) water, and (c) PEG-PVA and PAA.

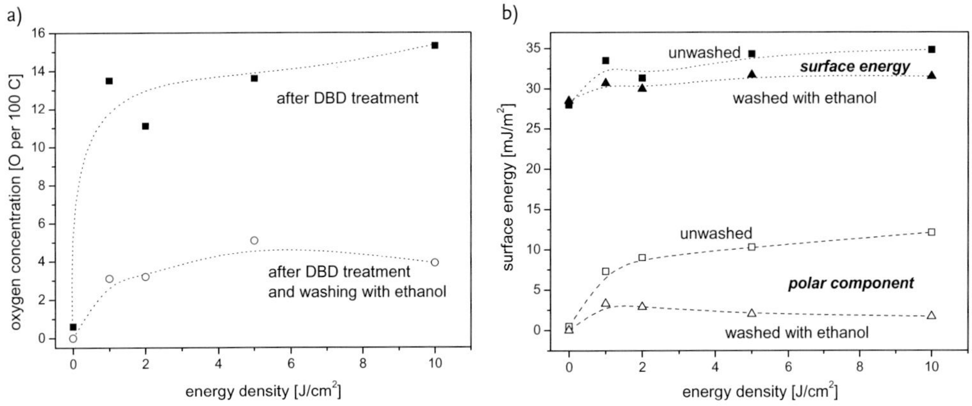

Figure 10.20 Oxygen uptake (a) and changes in surface energy (b) on exposure of polypropylene foil to a barrier discharge at atmospheric pressure as a function of applied energy density.

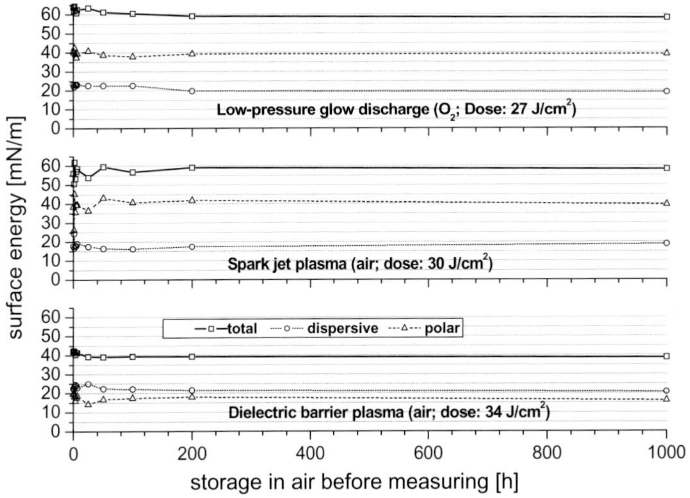

Figure 10.21 Loss in surface energy during exposure to air (storage) after low-pressure oxygen rf discharge treatment and atmospheric DBD and spark jet modification.

and intact single macromolecules into the gas phase under soft ionization conditions.

Thus, the droplet diameters in the aerosol-DBD and ESI processes differ by about three orders of magnitude. The ESI process ends in the separation of multiple-charged macromolecules in the gas phase (Figure 10.23).

The regularity of chemical composition and structure of deposited thin films converges to that of classic polymers in the following sequence: DBD (totally

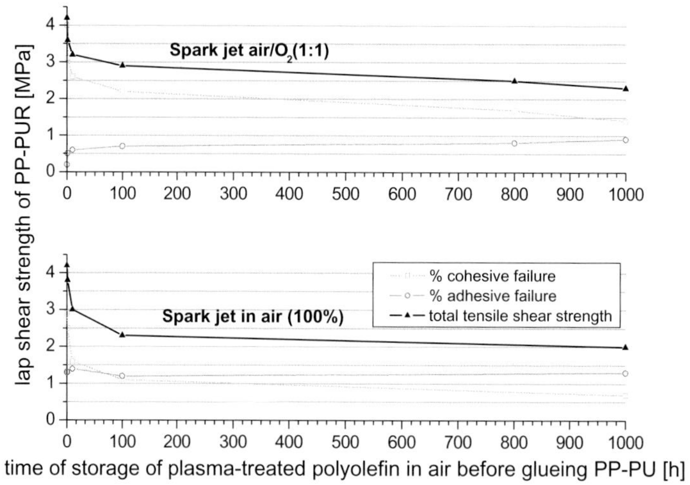

Figure 10.22 Loss in tensile shear strength of polyurethane–polypropylene composites on exposure of plasma-treated polymers to air (storage) before gluing (forming the composite).

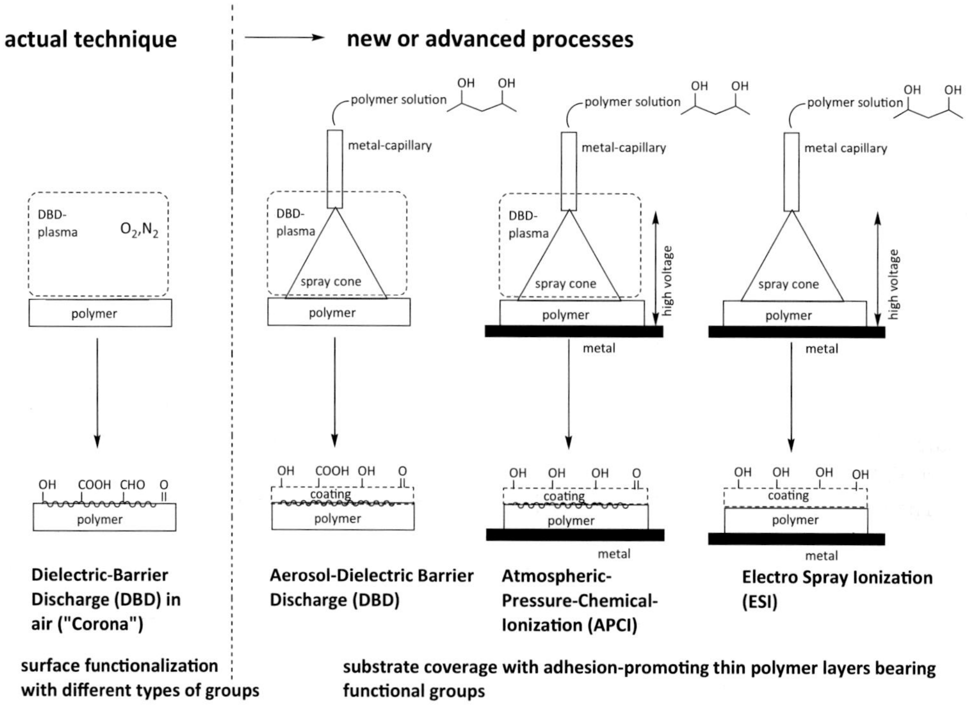

Figure 10.23 Schematics of air DBD, aerosol DBD, atmospheric-pressure chemical ionization (APCI), and electrospray ionization (ESI).

irregular structure at surface of polymer substrate) → aerosol-DBD (degraded substrate and film structure) → APCI (partially degraded film structure) → ESI (regular, non-degraded film structure) (Figure 10.23).

The electrospray is a technique for producing isolated macromolecules in the gas phase. Highly diluted polymer solutions are sprayed through a (heated) stainless steel capillary as aerosol. Between the capillary and the counter electrode, which is covered with the polymer sample, a high voltage is applied. The droplets of the spray form a cone (Taylor cone). The solvent evaporates from the sprayed aerosol droplets during transport to the counter electrode. Therefore, the droplets become smaller and smaller. Existing ions in the droplets and newly formed ions in the presence of the high electrical field strength come closer and closer together. Thus, the distance between charges in the droplets becomes too small (Rayleigh limit), and the same charges produce strong repulsion and the droplets decay into smaller ones (known as Coulomb explosion). At the end of this explosion cascade, all macromolecules are separated in the gas phase (Figure 10.24). The charged residue model (CRM) suggests that electrospray droplets undergo solvent evaporation and droplet fission in succession, eventually leading to progeny droplets that contain, on average, one macromolecular ion [84]. The ions observed by mass spectrometry were single macromolecular (*M*) ions and were formed by the addition of H^+ ($[M + H]^+$) or by addition of another type of cation such as Na^+ ($[M + Na]^+$), or by the abstraction of H^+ ($[M - H]^-$). Multiply-charged ions such as $[M + nH]^{n+}$ are often observed. Many charged states are observed for large macromolecules. These macromolecular ions form, after de-charging, a very thin polymer layer at the counter electrode. The charge flow from the capillary electrode to the counter electrode, transported by the charged droplets and macromolecules, has the character of a dark discharge (without any glow), and thus no degradation of macromolecules is possible. Positive or negative charged ions may be produced depending on the current polarity. Heating of the solution accelerates the solvent evaporation process.

Two problems exist: the poor adhesion of unmodified (virgin) macromolecules to unmodified polyolefin substrates and the charging of isolated polymer substrates, which repel the macromolecular ions. Corona pretreatment of the polymer substrate should increase the polymer layer adhesion. Continuous changing of current polarity (pulsing) or the use of an electron beam can neutralize any surface charging. Technologically, the electrospray is an advancement of the aerosol-DBD.

The electrospray technique is also used in mass spectrometry to analyze high-molecular weight macromolecules, and is known as ESI-ToF MS (electrospray ionization time-of-flight mass spectrometry) [82–84]. Sakata and other researchers used electrospray as an alternative method for evaporation of pyrolectric poly(vinylidene fluoride) (PVDF) most often within a corona field [85]. In 2006, this technique was used by Friedrich and coworkers to spray polymer solutions and deposit thin polymer layers [86, 87]. Arefi-Khonsari's group also used such equipment, similar to APCI, for plasma polymerization of monomers under atmospheric pressure conditions as mentioned before [66].

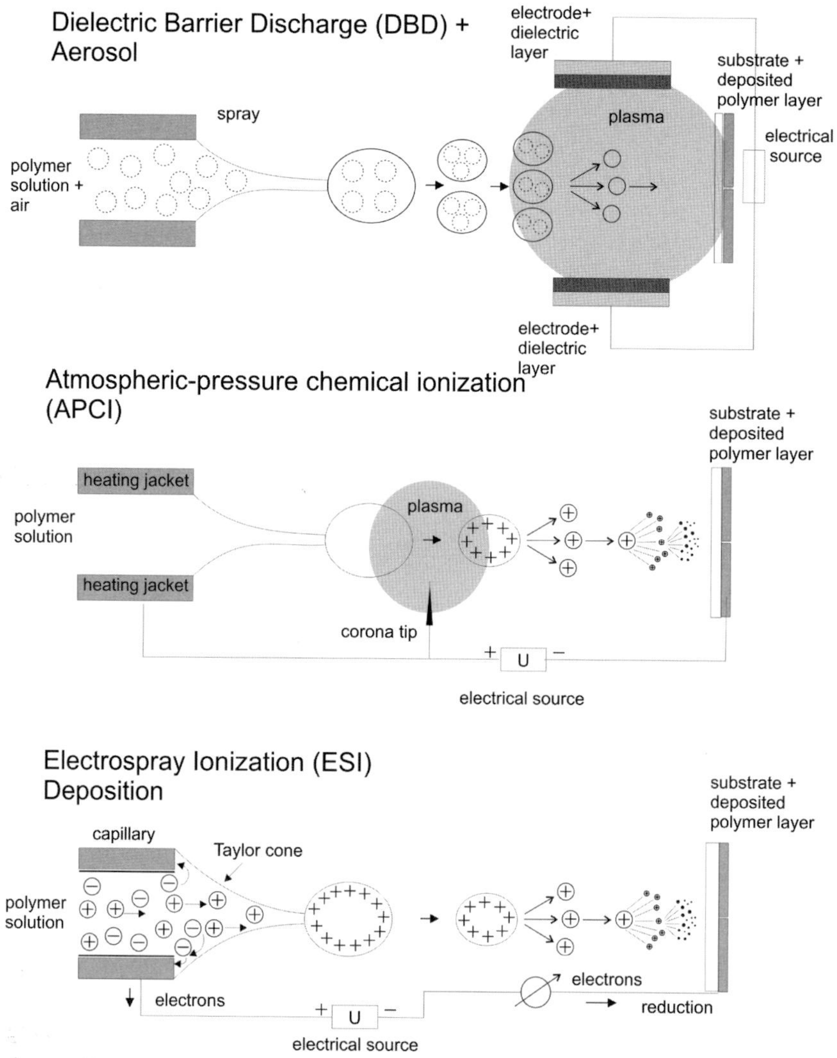

Figure 10.24 Schematics of the aerosol-DBD, APCI, and ESI processes.

At that time it was unknown that high-molecular weight polymers could be transferred to the gas phase without any fragmentation. The only way out was to depolymerize polymers to monomers and their subsequent polymerization at the substrate surface. Under low-pressure conditions Poll tested such thermal depolymerization followed by plasma polymerization using poly(tetrafluoroethylene) (PTFE), depolymerizing it to tetrafluoroethylene and then producing a PTFE-like plasma polymer [88]. Such plasma polymer layers possess a more or less regular composition if the dominant polymerization process is chain growth. Thus, Biederman and Slavinska subjected several polymers to a sputter process characterized by fragmentation or depolymerization of the polymer and film formation in the adsorption (condensate) layer by recombination of fragments or by monomer chain-growth polymerization [89].

In contrast to such low-pressure processes with recombination of fragments and chain growth polymerization, the ESI process brings intact and complete macromolecules into the gas phase without any fragmentation or depolymerization during their deposition [83, 84]. It should be added again that matrix-assisted laser desorption/ionization also permits the transfer of macromolecules to the gas phase [79].

10.6.1
ESI + Plasma

In the presence of a plasma glow (very high voltage, APCI mode) weak degradation of poly(methyl methacrylate) was found using MALDI mass spectrometry to characterize the deposited layer (Figure 10.25).

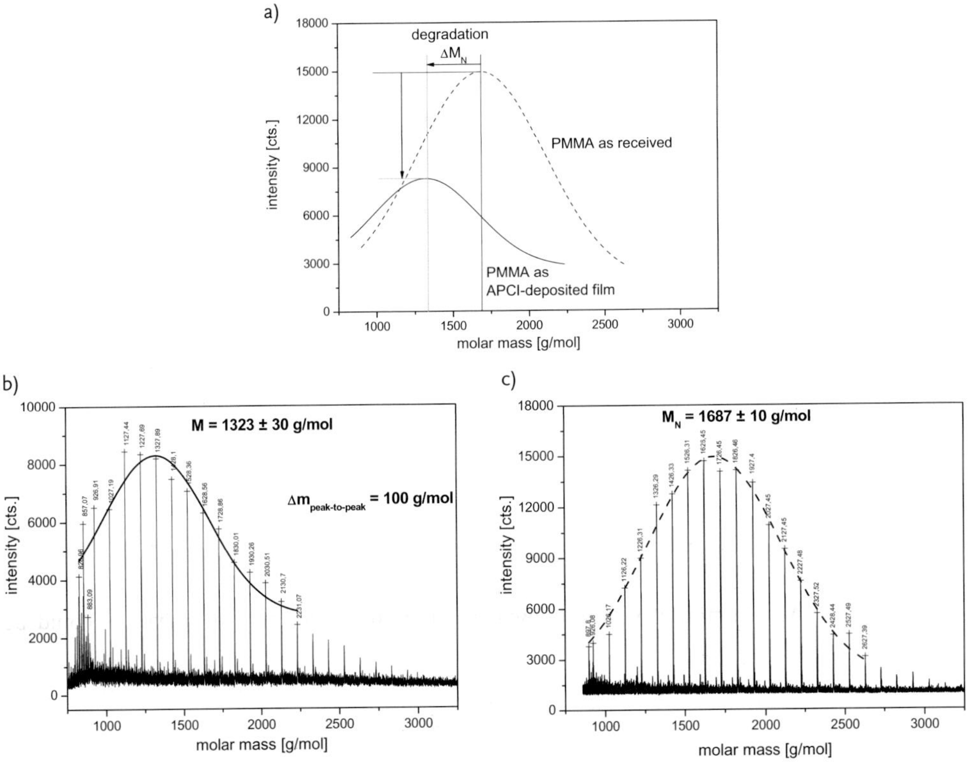

Figure 10.25 Electrospray deposition of poly(methyl methacrylate) (PMMA) in the presence of soft plasma glow (a) and analysis by matrix-assisted laser desorption/ionization time-of-flight (MALDI-ToF) mass spectrometry of the deposited surface layer (b) and of the original polymer (c).

The peak-to-peak distance was unchanged and amounted to 100 g mol^{-1} (Da), which corresponds to the repeat unit of PMMA. Therefore, the basic structure of the polymer was unchanged and new structures, caused by new degradation products, were not found.

To achieve sufficient adhesion of the ESI-deposited PMMA layer, the polypropylene substrate was modified by short exposure to the atmospheric barrier discharge, thus introducing polar groups onto the polymer surface and increasing its surface energy. These polar groups at the PP surface would interact with the PMMA molecules of the deposited layer.

10.6.2
ESI without Plasma

Polymer layer deposition is also possible without using additional ionization sources. To accomplish this, the field strength must be strong enough so that field ionization occurs or solvent molecules with polar (water, alcohols, acetonitrile, etc.) or ionic structure (acids) transfer their charge (protons or alkali ions) to the polymer molecule. Under such spray conditions, no glow is visible. The absence of any discharge avoids fragmentation of macromolecules. The deposited polymer layer shows (nearly) the same structure and stoichiometry as the original polymer when comparing the theoretical oxygen percentage with that of the deposited PMMA layer (Figure 10.26). After a 1 min deposition nearly complete coverage of the Au surface is seen, as indicated by the 36 O/100 C elemental composition.

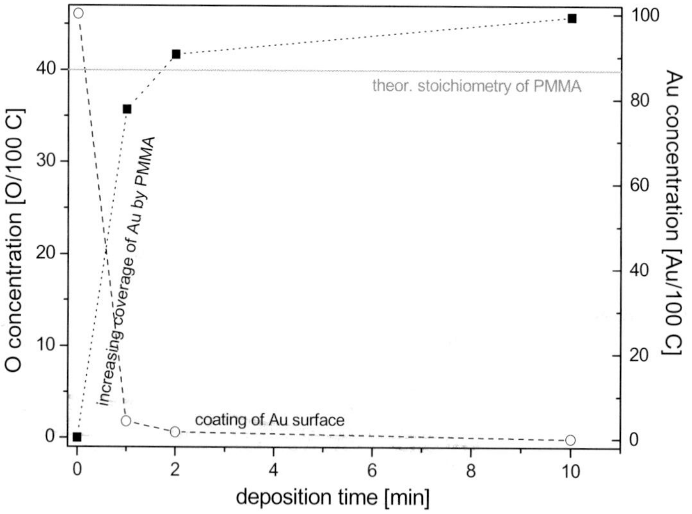

Figure 10.26 Variation of O concentration in a deposited PMMA layer on a Au-coated Si-wafer with the deposition time of the ESI-spray process (contamination of gold surface by ambient air was eliminated and a pure Au surface was assumed).

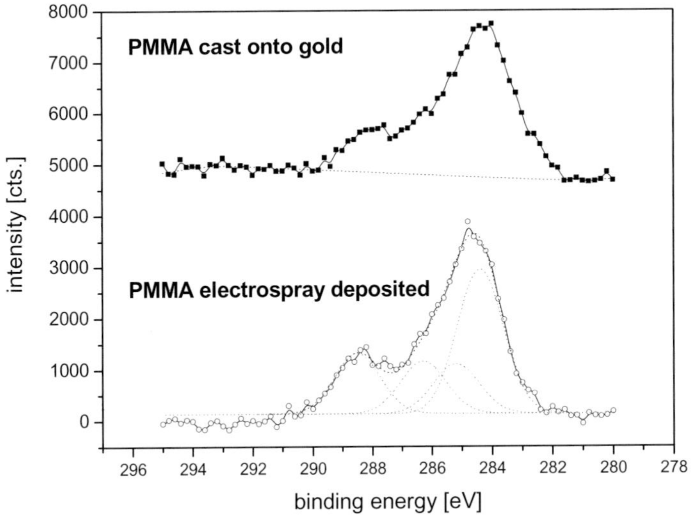

Figure 10.27 C1s peaks of PMMA layers deposited onto Au-coated Si-wafer by casting of PMMA solution or by ESI spraying.

The PMMA was not significantly altered in structure or composition during the ESI process, as shown by the XPS-measured O concentration in the original and in the ESI-sprayed layer. The C1s signals of the cast film and the ESI-deposited film are nearly identical (Figure 10.27).

10.6.3
Comparison of Aerosol-DBD and Electrospray

Reference coatings made of PEG-PVA copolymer show ca. 36 O per 100 C, the ESI coating has 39 O/100 C, and the aerosol-DBD coating possesses a maximum of 19 O/100 C. Similar results were obtained using poly(acrylic acid) and poly(ethyleneimine) coatings. Thus, obviously, the ESI process did not significantly alter the structure or stoichiometry of the deposited coatings, in contrast to the aerosol-DBD process.

The aerosol-DBD process ("corona") broadens strongly the O1s peak of the PEG-PVA copolymer and produces carbon species in the C1s peak, which have three bonds to oxygen atoms (carboxylic and ester groups). These groups are not present in the original copolymer (Figure 10.16). In contrast, the ESI sprayed PEG-PVA copolymer shows only weak changes in the C1s peak compared to the untreated reference copolymer. The O1s peak remains unchanged.

In Figure 10.17 the IR spectra of reference materials and ESI-deposited layers of PEG-PVA copolymer, poly(ethyleneimine), and poly(acrylic acid) were compared and found to be nearly identical. However, it must be considered that some

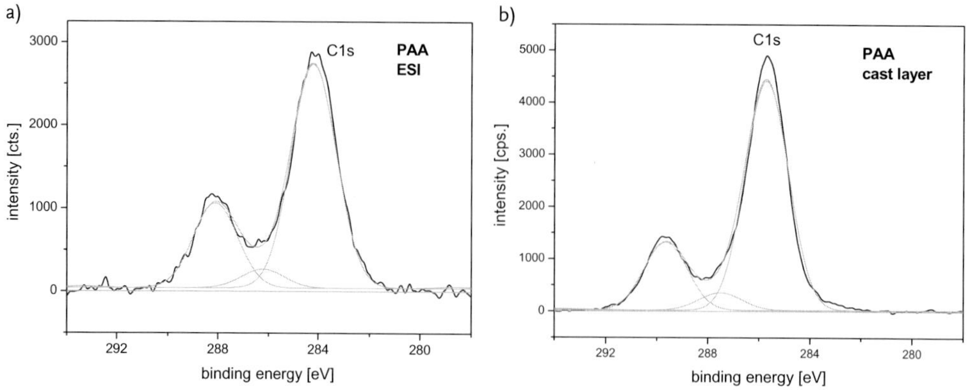

Figure 10.28 C1s peaks of poly(acrylic acid) (PAA) deposited onto Au surfaces by (a) ESI process and (b) casting.

thin ESI layers grow in an island-like manner and thus the gold substrate showed through the polymer coating in cases of insufficient thickness.

Poly(acrylic acid) is sensitive to interaction with any plasma. This was found when acrylic acid was polymerized in a low-pressure plasma and the deposited plasma polymer layer analyzed using IR and XPS [78, 79]. The carboxylic group may be considered as a precursor for the plasma-initiated elimination of CO and CO_2, in the polymer as well as in the monomer [66]. In addition, the ester group tends to decompose [90]; thus, deficiencies of oxygen or carboxylic or ester groups are indicators of plasma-initiated degradation.

The C1s peaks of the cast and ESI deposited poly(acrylic acid) layers did not show any significant differences (Figure 10.28).

PMMA could also be deposited without structural degradation by applying the ESI spray method, as shown by FTIR-GIR spectra of cast and ESI deposited layers (Figure 10.29). All bands of PMMA were found in both spectra.

10.6.4
Topography

The topography of ESI-deposited PMMA is shown in AFM pictures in Figure 10.30. The substrate was a Si wafer and was coated with 10 and 50 nm PMMA layers. The growth mechanism of the polymer layer (Figure 10.30, from left to right) is more or less island-like (corresponding to the Volmer–Weber growth mechanism). XPS measurements had shown that PMMA layers thicker than 10 nm became pin-hole free, as presented in Figure 10.27. The ESI spray blows away a large fraction of solvent molecules and small droplets, especially if charging of the substrate surface is also possible.

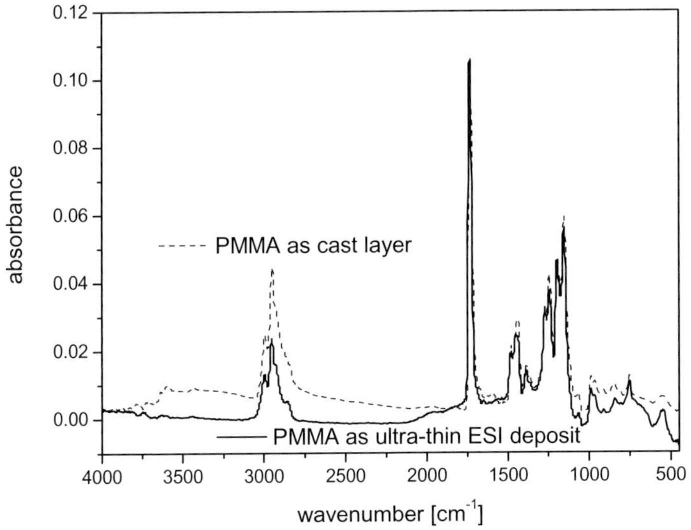

Figure 10.29 Comparison of IR spectra of PMMA layers produced by casting and by ESI spraying as recorded using the grazing incidence reflection mode, normalized to the C=O band (1700 cm^{-1}).

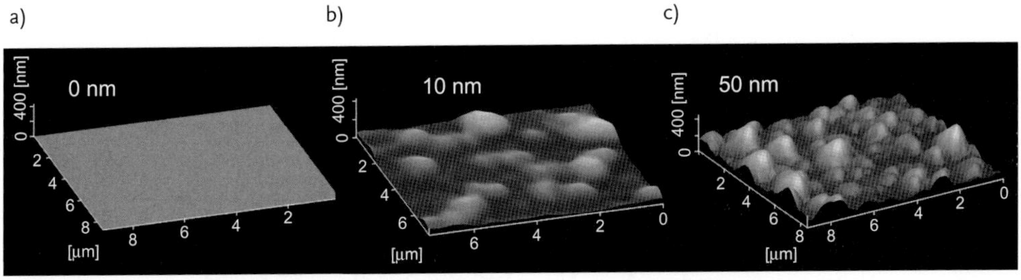

Figure 10.30 AFM images of Au-coated Si wafer before (a) and after deposition of a 10 nm PMMA layer (b) and a 50 nm PMMA layer (c) using the ESI deposition technique.

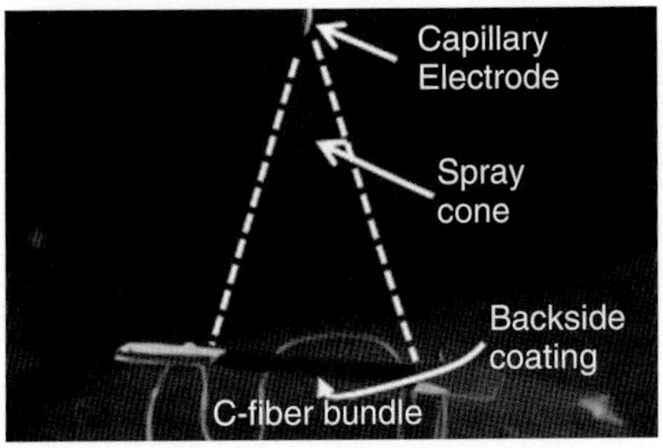

Figure 10.31 ESI photograph with capillary, spray cone (accented by white hatched lines), and C-fiber bundle contacted and grounded by glue and Al foil.

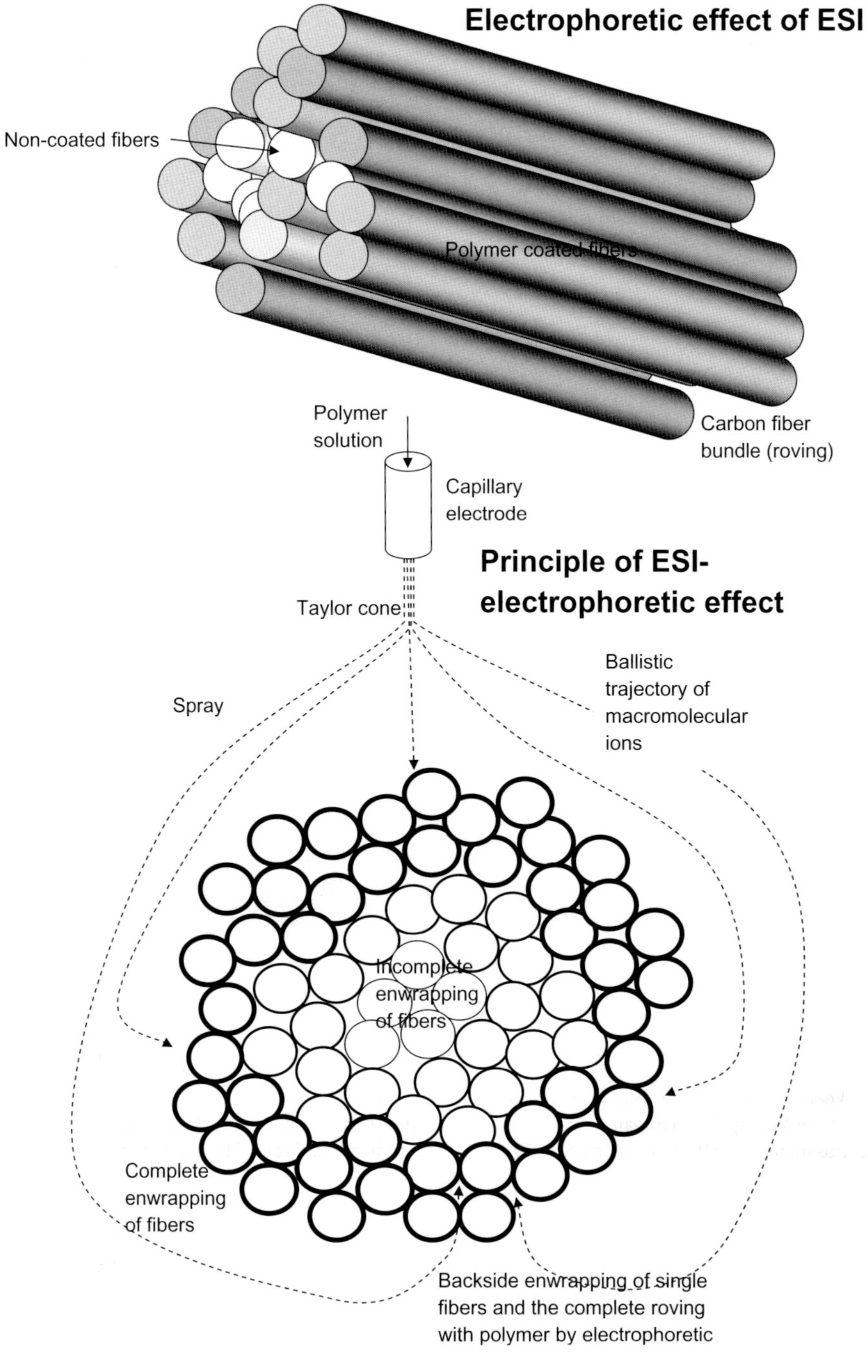

Figure 10.32 Scheme of electrophoretic effect of ESI demonstrated for a carbon fiber bundle (darkened fibers or thick circles indicate thick ESI coating) and its backside coating (no direct spraying is possible because of geometry).

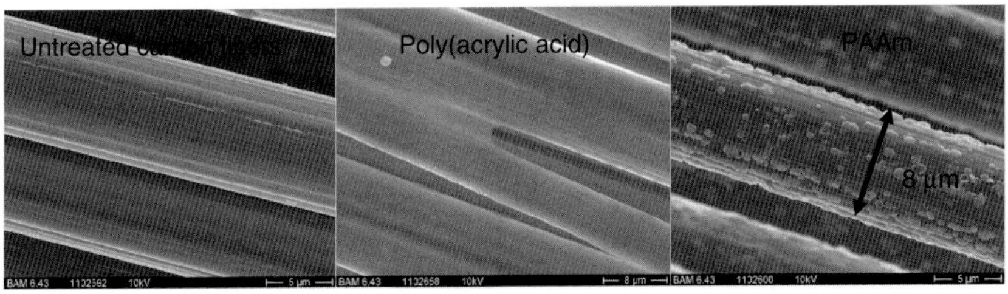

Figure 10.33 Carbon fibers before treatment, after deposition of a ca. 50-nm-thick layer poly(acrylic acid), and after deposition of a ca. 200-nm-thick layer poly(allylamine).

10.6.5 Electrophoretic Effect of ESI

Pin-hole free polymer layers <10 nm were produced by the electrophoretic effect of ESI in case of conductive substrates; that is, holes show higher current flow and were automatically closed. Thus, the backside and inner surfaces of carbon fiber rovings were completely enwrapped with ultrathin polymer layers. These layers can be used as adhesion promoter in polymer-fiber composites (Figures 10.31 and 10.32).

C-fibers were homogeneously coated with poly(acrylic acid) and island-like with poly(allylamine) (Figure 10.33).

References

1 Wikipedia (2011) "Plasma," de.wikipedia.org (accessed on 15 November 2011)
2 Oparin, A.I. (1924) *Proiskhozhdenye Zhiznyi*, Moskowski rabozhi, Moscow.
3 Miller, S.L. (1953) *Science*, **117**, 3046.
4 Blasek, G. and Bräuer, G. (2010) *Vakuum-Plasma-Technologien*, Eugen G. Leuze Verlag, Bad Saulgau.
5 Rossmann, K. (1956) *J. Polym. Sci.*, **19**, 141.
6 Fischer, F. and Tropsch, H. (1926) *Brennstoff-Chem.*, **7**, 97.
7 Weissermel, K. and Arpe, H.-J. (2003) *Industrial Organic Chemistry*, 4th edn, Wiley-VCH Verlag GmbH, Weinheim.
8 Bashkirov, A.N. and Chertkov, I.B. (1947) *Dokl. Akad. Nauk*, 817–824.
9 Fanghänel, E., Beckert, R., Habicher, W.D., Metz, P., Pavel, D., and Schwetlick, K. (2004) *Organikum*, 22nd edn, Wiley-VCH Verlag GmbH, Weinheim.
10 Habenicht, G. (2009) *Kleben*, 6th edn, Springer, Berlin.
11 (a) Friedrich, J., Wigant, L., Unger, W., Lippitz, A., Erdmann, J., Gorsler, H.-V., Prescher, D., and Wittrich, H. (1995) *Surf. Coat. Technol.*, **74–75**, 910; (b) Friedrich, J., Gross, Th., Lippitz, A., Rohrer, P., Saur, W., and Unger, W. (1993) *Surf. Coat. Technol.*, **23**, 267–278.
12 Friedrich, J., Kühn, G., Schulz, U., Jansen, K., Bertus, A., Fischer, S., and Möller, B. (2003) *J. Adhesion Sci. Technol.*, **17**, 1127.
13 Friedrich, J., Wettmarshausen, S., and Hennecke, M. (2009) *Surf. Coat. Technol.*, **203**, 3647–3655.

14 Bikerman, J.J. (1968) *The Science of Adhesive Joints*, 2nd edn, Academic Press, New York.
15 Strobel, M., Corn, S., Lyons, C.S., and Korba, G.A. (1987) *J. Polym. Sci., Part A: Polym. Chem.*, **25**, 129.
16 Kreidl, W.H. (1959) *Kunststoffe*, **49**, 71.
17 Adcock, J.H. and Lagow, R.J. (1974) *J. Am. Chem. Soc.*, **96**, 7588.
18 Schonhorn, H. and Hansen, R.H. (1967) *J. Appl. Polym. Sci.*, **12**, 1231.
19 Lange, J. and Wyser, Y. (2003) *Packag. Sci. Technol.*, **16**, 149.
20 Kim, C.Y., Evans, U., and Goring, D.A.I. (1971) *J. Appl. Polym. Sci.*, **15**, 1357.
21 Owens, D.K. (1975) *J. Appl. Polym. Sci.*, **19**, 265.
22 Owens, D.K. (1975) *J. Appl. Polym. Sci.*, **19**, 3315.
23 Holländer, A., Kröpke, S., and Ehrentreich-Förster, E. (2007) *Plasma Proc. Polym.*, **4**, S1052–S1056.
24 Kogelschatz, U., Esrom, H., Zhang, J.-Y., and Boyd, I.W. (2000) *Appl. Surf. Sci.*, **168**, 29–36.
25 Truica-Marasescu, F.-E. and Wertheimer, M.R. (2005) *Macromol. Chem. Phys.*, **206**, 744–757.
26 Novak, I., Pollak, V., and Chodak, I. (1998) *Angew. Makromol Chem.*, **260**, 47.
27 Park, S.-J. and Jin, J.-S. (2001) *J. Colloid Interface Sci.*, **236**, 155.
28 Friedrich, J., Wigant, L., Unger, W., Lippitz, A., and Wittrich, H. (1998) *Surf. Coat. Technol.*, **98**, 879.
29 Kim, C.Y., Evans, U., and Goring, D.A.I. (1971) *J. Appl. Polym. Sci.*, **15**, 1365.
30 Courval, G.J., Gray, D.G., and Goring, D.A.I. (1976) *J. Polym. Sci., Polym. Lett. Ed.*, **14**, 231.
31 Kim, C.Y., Evans, U., and Goring, D.A.I. (1972) *J. Appl. Polym. Sci.*, **15**, 1965.
32 Carley, J.F. and Kitze, P.T. (1978) *J. Polym. Eng. Sci.*, **18**, 326.
33 Briggs, D., Rance, D.G., Kendall, C.R., and Blythe, A.R. (1980) *Polymer*, **21**, 895.
34 Strobel, M., Dunatov, C., Strobel, J.M., Lyons, C.S., Perron, S.J., and Morgan, M.C. (1989) *J. Adhesion Sci. Technol.*, **3**, 321.
35 Potente, H. and Krüger, R. (1979) *Adhäsion*, **23**, 381.
36 Stradal, M. and Goring, D.A.I. (1977) *Polym. Eng. Sci.*, **1**, 38.
37 van der Linden, R.V. (1979) *Kunststoffe*, **69**, 71.
38 Rabel, W. (1971) *Farbe Lack*, **77**, 997.
39 Hansmann, J. (1979) *Adhäsion*, **23**, 136.
40 Amouroux, J., Goldman, M., and Revoil, M.F. (1982) *J. Polym. Sci., Polym. Chem. Ed.*, **20**, 1373.
41 Sawada, Y. (2005) in *Polymer Surface Modification and Polymer Coating by Dry Process Technologies* (ed. S. Iwamori), Research Signpost, Trivandrum, p. 19.
42 Novák, I. and Florian, S. (2001) *Polym. Int.*, **50**, 49.
43 Friedrich, J., Unger, W., Lippitz, A., Gross, Th., Rohrer, P., Saur, W., Erdmann, J., and Gorsler, H.-V. (1996) in *Polymer Surface Modification: Relevance to Adhesion* (ed. K.L. Mittal), VSP, Utrecht, pp. 49–72.
44 Clark, D.T., and Dilks, A. (1979) *J. Polym. Sci., Polym. Chem. Ed.*, **17**, 957.
45 Strobel, M., Walzak, M.J., Hill, J.M., Lin, A., Karbashewski, E., and Lyons, C.S. (1995) *J. Adhesion Sci. Technol.*, **9**, 365.
46 Overney, R.M., Lüthi, R., Haefke, H., Frommer, J., Meyer, E., Güntherodt, H.-J., Hild, S., and Fuhrmann, J. (1993) *J. Appl. Surf. Sci.*, **64**, 197.
47 Guimond, S. and Wertheimer, M.R. (2004) *J. Appl. Polym. Sci.*, **94**, 1291–1303.
48 O'Hare, L.-A., Smith, J.A., Leadley, S.R., Parbhoo, B., Goodwin, A.J., and Watts, J.F. (2002) *Surf. Interface Anal.*, **33**, 617.
49 O'Hare, L.-A., Leadley, S., and Parbhoo, B. (2002) *Surf. Interface Anal.*, **33**, 335.
50 Sun, C., Zhang, D., and Wadsworth, L.C. (1999) *Adv. Polym. Technol.*, **18**, 171.
51 Novák, I., Pollák, V., and Chodák, I. (2006) *Plasma Proc. Polym.*, **3**, 355.
52 Jones, V., Strobel, M., and Prokosch, M.J. (2005) *Plasma Proc. Polym.*, **2**, 547.
53 Lynch, J.B., Spence, P.D., Baker, D.E., and Postlethwaite, T.A. (1999) *J. Appl. Polym. Sci.*, **71**, 319.
54 Goodwin, A.J., Merlin, P.J., Badyal, J.P.S., and Ward, L. (2002) PCT Patent WO 0225548.
55 Ward, L.J., Schofield, W.C.E., and Badyal, J.P.S. (2003) *Langmuir*, **19**, 2110.
56 Ward, L.J., Schofield, W.C.E., and Badyal, J.P.S. (2003) *Chem. Mater*, **15**, 1466.

57 O'Hare, L., O'Neill, L., and Goodwin, A.J. (2006) *Surf. Interface Anal.*, **38**, 1519.

58 Twomey, B., Rahman, M., Byrne, G., Hynes, A., O'Hare, L.-A., and Dowling, D. (2008) *Plasma Proc. Polym.*, **5**, 737.

59 Wu, S. (1982) *Polymer Interface and Adhesion*, Marcel Dekker, New York.

60 Drost, H. (1978) *Plasmachemistry*, Akademie-Verlag, Berlin.

61 Friedrich, J. (1980) *Chem. Tech.*, **32**, 393.

62 Garbassi, F., Morra, M., and Occhiello, E. (1998) *Polymer Surfaces-from Physics to Technology*, John Wiley & Sons, Ltd, Chichester.

63 Ranney, M.W. and Connor, W.F. (1969) *Adv. Chem. Ser.*, **80**, 287.

64 Donohoe, K. and Wydeven, T. (1979) *J. Appl. Polym. Sci.* **32**, 2591–2601.

65 Klages, C.-P., Höpfner, K., Kläke, N., and Thyen, R. (2000) *Plasmas Polym.*, **5**, 79.

66 Tatoulian, M., Arefi-Khonsari, F., and Borra, J.-P. (2007) *Plasma Proc. Polym.*, **4**, 360.

67 Thyen, R., Weber, A., and Klages, C.P. (1997) *Surf. Coat. Technol.*, **97**, 426.

68 Ward, L.J., Scofield, W.C.E., Badyal, J.P.S., Goodwin, A.J., and Merlin, P.J. (2003) *Chem. Mater.*, **15**, 1466.

69 Ward, L.J., Scofield, W.C.E., Badyal, J.P.S., Goodwin, A.J., and Merlin, P.J. (2003) *Langmuir*, **19**, 2110.

70 Janca, J. and Pavelka, P. (1984) *Scripta Fac. Sci. Nat. Univ. Purk. Brun.*, **14**, 21.

71 Segers, M. and Dhali, S.K. (1991) *J. Electrochem. Soc.*, **138**, 2741.

72 Yokoyama, T., Kogoma, M., Kanazawa, S., Moriwaki, T., and Okazaki, S. (1990) *J. Phys. D Appl. Phys.*, **23**, 374.

73 Thomas, M., von Hausen, M., Klages, C.-P., and Baumhof, P. (2007) *Plasma Proc. Polym.*, **4**, S475–S481.

74 Salge, J. (1996) *Surf. Coat. Technol.*, **80**, 1.

75 Friedrich, J.F., Mix, R., Schulze, R.-D., Meyer-Plath, A., Joshi, R., and Wettmarshausen, S. (2008) *Plasma Proc. Polym.*, **5**, 407–423.

76 Dinter, P., Bothe, L., and Gribbin, J.D. (1987) Hoechst patent DE 3705482.

77 Kühn, G., Weidner, St., Decker, R., Ghode, A., and Friedrich, J. (1999) *Surf. Coat. Technol.*, **116–119**, 796–801.

78 Mix, R., Friedrich, J., and Rau, A. (2009) *Plasma Proc. Polym.*, **9**, 566–574.

79 Friedrich, J., Mix, R., Schulze, R.-D., and Rau, A. (2010) *J. Adhesion Sci. Technol.*, **24**, 1329–1350.

80 Alexander, M.R., Beamson, G., Blomfield, C.J., Leggett, G., and Duc, T.M. (2001) *J. Electron. Spectrosc. Relat. Phenom.*, **121**, 19–32.

81 Karas, M., Bachmann, D., Bahr, U., and Hillenkamp, F. (1987) *Int. J. Mass Spectrom. Ion Process.*, **78**, 53.

82 Fenn, J.B. (2003) *Angew. Chem. Int. Ed.*, **42**, 3871.

83 Dole, M., Mack, L.L., Hines, R.L., Mobley, R.C., Ferguson, L.D., and Alice, M.B. (1986) *J. Chem. Phys.*, **49**, 2240.

84 Fenn, J.B., Mann, M., Meng, C.K., Wong, S.F., and Whitehouse, C.M. (1986) *Science*, **246**, 64.

85 (a) Sakata, J. and Mochizuki, M. (1991) *Thin Solid Films*, **195**, 175–184; (b) Choy, K.-L. and Bai, W. (2000) *Thin Solid Films*, **372**, 6–9.

86 Friedrich, J. (2006) Proceeding 10th International Conference on Plasma Surface Engineering (PSE), Garmisch-Partenkirchen, Germany, 10–15 September 2006.

87 Friedrich, J., Mix, R., Meyer-Plath, A., Schulze, R.-D., and Wettmarshausen, S. (2007) Proceedings of the 19th International Symposium on Plasma Chemistry (ISPC-19), Kyoto, Japan, 26–31August 2007.

88 Poll, H.-U., Arzt, M., and Wickleder, K.-H. (1976) *Eur. Polym. J.*, **12**, 505.

89 Biederman, H. and Slavinska, D. (2000) *Surf. Coat. Technol.*, **125**, 371.

90 Friedrich, J., Loeschcke, I., Frommelt, H., Reiner, H.-D., Zimmermann, H., and Lutgen, P. (1991) *J. Polym. Deegrad. Stabil.*, **31**, 97–119.

11
Plasma Polymerization

11.1
Historical

The general intention of plasma polymerization is to produce ultrathin and pinhole-free polymer layers with defined and regular structure but with variable composition; the polymer layers also need to be durable and resistant towards aging, oxidation, shrinking, and so on, comparable with classic polymers. A second intention is the polymerization of monomers possessing functional groups to produce polymer films with monosort functional groups. These monotype functional groups in high concentration may serve as anchoring points for chemical graft synthesis or may improve adhesive interactions to other solids in polymer composites. However, it will be shown that these desired regular structures cannot be realized principally; nevertheless, many technical applications are known.

The first report on the formation of plasma polymers in the vapor of oil exposed to an atmospheric-pressure electrical discharge was presented by Dutch researchers in 1796 [1]. More detailed research was presented by well-known chemists in the nineteenth century such as Berthelot, who developed the arc synthesis of acetylene [2] or P. de Wilde [3] and also P. and A. Thenard [4]. For the conversion of methane in the plasma Linder and Davis observed the formation of hydrogen and acetylene, which then forms plasma polymer. They also found the formation of aromatics such as benzene, styrene, biphenyl, and naphthalene as well as plasma polymers as side-products if aliphatic precursors were exposed to plasma [5]. New results of extensive research on chemical synthesis and plasma polymerization as well as copolymerization to variable products were published by German scientists in 1950s [6].

In the 1960s Jesch *et al.* analyzed the structure of plasma polymers and found the formation of unsaturations (C=C, C≡C), the occurrence of insolubility (crosslinking), and oxygen attachment to radicals on exposure to air [7]. The most important finding was the general absence of indications for methylene chains ($\rho[CH_2]_{\geq 4}$) at 730 cm^{-1} or 720 + 731 cm^{-1}, as was also confirmed by Wightman [8].

Neiswender also proposed the intermediate formation of acetylene and found also numerous radicals in the polymer layer [9]. He proposed first an energy-related dose factor to better compare the different plasma conditions. Subsequently, this

The Plasma Chemistry of Polymer Surfaces: Advanced Techniques for Surface Design, First Edition.
Jörg Friedrich.

idea was perfected and made popular by Yasuda (Yasuda factor – YF = W/MF, where W = wattage in J s^{-1}, F = monomer flow rate in mol s^{-1}, and M = molecular weight of monomer in kg mol^{-1}) [10]. Westwood adapted the G-value known from radiation chemistry and calculated it for several plasma processes [11].

The preferred locus of plasma polymerization was discussed, controversially, as either in the adsorption layer or in the gas phase. An ionic chain mechanism was assumed [11–13] and most often a radical mechanism [14–17]. The most mentioned indications for a radical mechanism were the high concentration of trapped radicals in plasma polymers and the dominance of neutral energy-rich plasma species in the gas phase. Kobayashi combined the radical chain growth mechanism with that of acetylene as key intermediate in one model [17]. Tibbitt proposed a speculative model of the structure of the ethylene plasma polymer, including the formation of aromatic phenyl rings (Figure 11.1) [18, 19].

Yasuda proposed a new mechanistic concept in the 1970s and 1980s, the so-called "atomic polymerization" for high W/MF (wattage/molar mass × monomer flow) factors [20]. All monomer molecules become extensively fragmented in powerful plasmas, most often into single atoms. These atoms and small fragments recombine randomly in a first-order process, rearrange and become newly activated, newly fragmented, and so on. Therefore, the monomer composition and structure does not characterize greatly the resulting plasma polymer structure and composition (Figure 11.1).

These irregular structures are also produced by the permanent bombardment of the growing polymer layer with plasma particles and energy-rich UV photons [21, 22]. Neiswender [9], Dinan [23], Suhr/Rosskamp [24, 25], and Friedrich [26] also attributed the plasma polymerization in the main to a random polyrecombination of radicals but additionally to a speculative sticking of C_2 species, thus forming their multiples (C_4, C_6, C_8) such as acetylene, cyclobutadiene, diacetylene, phenylacetylene, [2 + 2] adduct of cyclobutadiene, octatetraene, and styrene as well as plasma polymers (Figure 11.1).

Stille elucidated the reactions of benzene in plasma, such as dehydrogenation to phenyl ring formation, phenyl dimerization, ring cracking to hexatriene biradicals, and polymer formation (Figure 11.1) [27, 28]. Therefore, benzene, toluene, and xylene are well-suited "monomers" for continuous-wave (cw) plasma polymerization with extensive fragmentation and polyrecombination demonstrating the exotic character of plasma polymerization caused by excess energy [23, 28–30], as was already discussed in the 1930s [5, 31].

In the 1960–1970s extensive and also controversial discussion on the mechanism of polymerization was continued, such as radical versus ionic versus chemical chain growth, ion–molecule, and fragmentation–recombination together with kinetics [11, 13, 14, 20, 32–35]. Recently, this debate was revitalized by comments of Short, Hegemann, and d'Agostino [36, 37].

Neiswender [9], Bradley [38], Westwood [11], Yasuda [10, 20], and Friedrich [26] found that the structure of plasma polymers is more regular and defined in the sense of classic polymers the lower the power input (per monomer molecule = W/MF) to the plasma. On the other hand, the low plasma power input

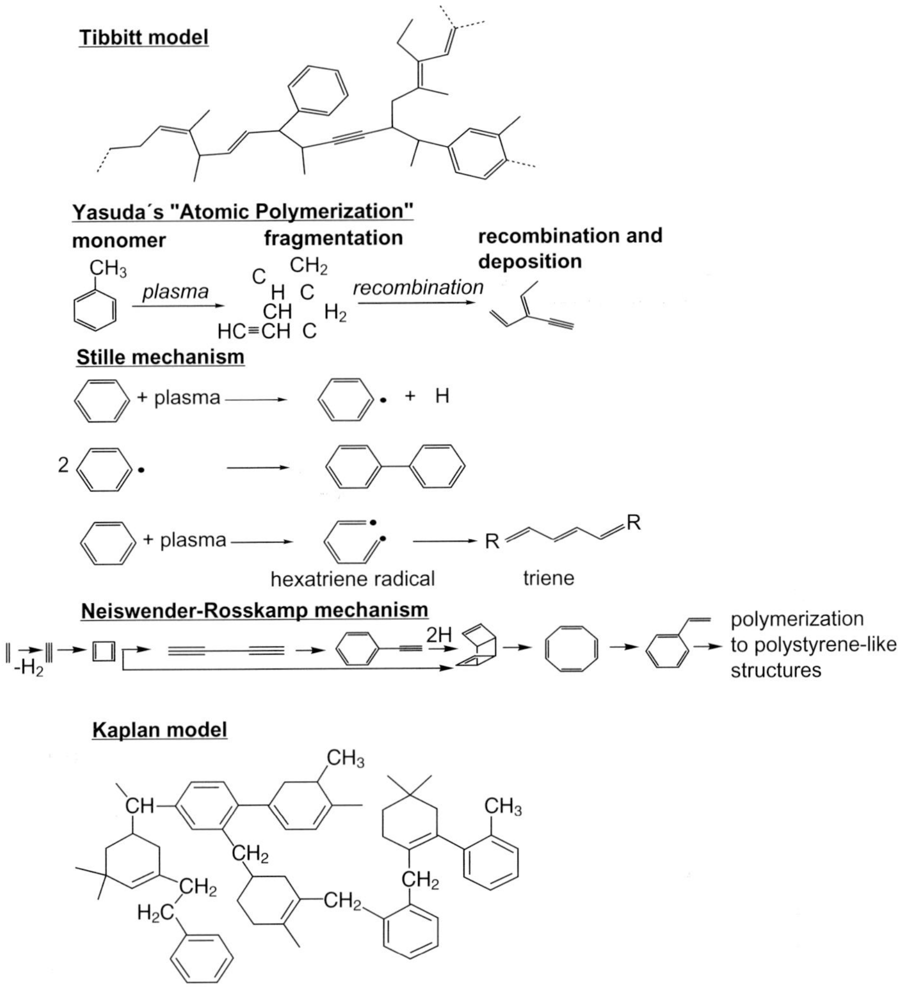

Figure 11.1 Proposed model of plasma polymer produced from ethylene with branches, crosslinks, aromatic rings, and double bonds, "atomic polymerization," benzene ring cracking and acetylene–cyclobutadiene–benzene–styrene mechanism, and polystyrene formation.

induced the inclusion of monomer and oligomer molecules into the polymer structure [17, 39]. Thus, the thermal stability of polymer layers is lowered because of the evaporation of these components at temperatures lower than 100 °C. Tibbitt distinguished between low-molecular weight (soluble) and crosslinked ethylene plasma polymers [39], as is also found for plasma polymerized toluene (Figure 11.1) [40, 41].

A new innovation of plasma polymerization was the introduction of the plasma pulsing technique by Tiller [42], Yasuda [43, 44], and continued by Shen and Bell

[45, 46] and intensely used by Timmons [47–49] and perfected by development of the pressure-pulse technique by Friedrich [50–54].

11.2 General Intention and Applications

The goal of plasma polymerization is to deposit ultrathin, well-adherent, and regularly structured polymer layers (Figure 11.2).

Monosort functional-group containing plasma polymers should retain all functional groups during plasma polymerization. In reality the plasma polymers lose the monomer structure and partially their functional groups during the deposition and also during post-plasma exposure to air.

As mentioned before, ultrathin and pinhole-free polymer layers of variable composition can be deposited as plasma polymer layers but generally with far from defined and regular structure, which makes them durable and resistant towards aging, oxidation, shrinking, and so on. The retention of functional groups of monomer molecules during plasma polymerization is much higher than that of the complete monomer structure in the deposited polymer. Thus, layers with high concentrations of a monosort functional group were an important application for adhesion-promoting layers or as anchoring points for graft reactions. Since the 1970s, adhesion-promoting plasma polymer layers have been of increasing technical interest [38, 54–57]. The degree of retained functional groups in the deposited polymer (φ) is given by $\varphi = [c_{fg}M]_n / nc_{fg}M$, where M = monomer, n = number of monomers, and c_{fg} = concentration of functional groups per monomer or per monomer unit; the resulting φ is most often 50–90% (Figure 11.3) [51].

Plasma polymers are well-suited for coating of solids such as membranes, semiconductors, metals, textiles, or polymers, with a minimal thickness of 10 nm and a maximal thickness of a few 100 nm. A clever choice of "monomer" allows us to

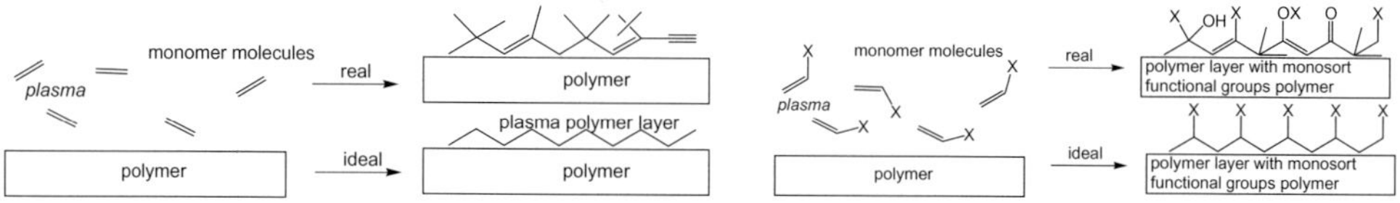

Figure 11.2 Schematics of ideally structured and composed plasma polymers with and without monosort functional groups.

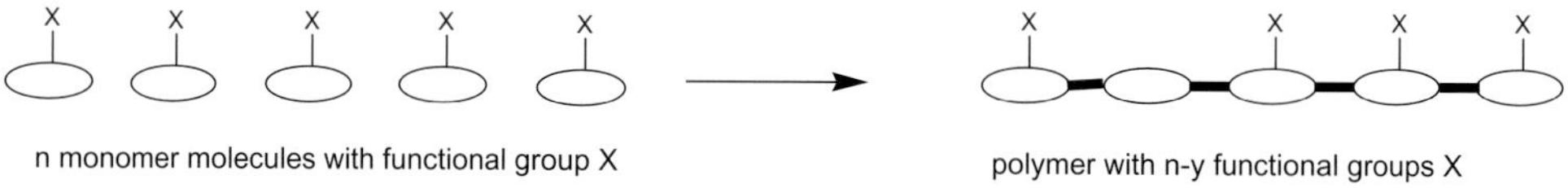

Figure 11.3 Retention (80%) of functional groups.

produce corrosion inhibition, anti-fogging, chemical, scratch and abrasion resistance, adherence, lubrification, flame-resistance, permeability, antistatic, barrier, and optical properties. Notably, the substrate properties are not influenced strongly by this deposition process.

The first announced application of plasma polymers was their use as separator membranes in nuclear battery devices introduced by Goodman in 1960 [58]. Further attempts were the coating of textiles for water-repellence and better washability, such as coating of PET textiles with plasma polymerized poly(acrylic acid), capping of microelectronic devices, and coating of PET tire cord [59–62].

Thin plasma polymer layers were applied as barrier layers [63–66] or as corrosion-inhibiting coatings on aluminum components in the car industry and so on [67, 68]. Thin polymer films were produced either by exposure to the low-pressure glow discharge plasma [58] or by plasma-enhanced chemical vapor deposition (PECVD) [69] or more recently by atmospheric-pressure glow discharges (APGD) [70] as well as by atmospheric-pressure dielectric barrier discharges (DBDs) [71].

11.3 Mechanism of Plasma Polymerization

Polymerization of classic monomers (vinyl, acrylic, etc.) and low-molecular weight substances ("monomers" or, better, precursor) is initiated by any gas plasma at atmospheric or low pressure. In a few cases inorganic gases and compounds may also form organic polymers, as evidenced by the formation of amino acids in the primeval atmosphere of Earth [72, 73]. In the low-pressure (non-isothermal) glow discharge plasma the following deposition modes may be distinguished:

- *(Chemical) radical chain growth polymerization* starting from plasma-produced radical fragment to linear or branched but regularly structured homopolymers;
- *ionic chain growth polymerization* (cationic, anionic) to linear/branched regularly formed products;
- *ion–molecule reactions*;
- *monomer fragmentation–polyrecombination* to randomly and irregularly structured polymers;
- *monomer conversion into polymer-forming intermediates* with partially defined structural elements;
- *co-monomer fragmentation–recombination of all fragments* to randomly and irregularly structured "copolymers;"
- *radical chain growth copolymerization* to linear/branched regularly formed copolymers with alternating, block, or graft structure;
- *chemical grafting onto radical sites or functional groups of plasma polymers or plasma-exposed polymer surfaces* ("graft copolymers") [74, 75].

11.3.1 Plasma-Induced Radical Chain-Growth Polymerization Mechanism

Plasma-produced radicals, radical fragments, radical-sites at solid surfaces or gaseous monomers, and so on initiate a classic chain-growth polymerization to polymer (P) by continuous addition of monomer molecules (M), such as acrylic or vinyl monomers:

$$M + \textit{plasma} \rightarrow \text{fragments}^{\bullet}$$

$$\text{fragment}^{\bullet} + M \rightarrow \text{fragment-M}^{\bullet} \ldots \rightarrow \text{fragment-M}^{\bullet} + M \rightarrow P_n^{\bullet}$$

Free radical recombination dominates the ending of chain growth:

$$2P_n^{\bullet} \rightarrow P_n\text{-}P_n$$

or disproportionation: $P_1^{\bullet} + P_1^{\bullet} \rightarrow P_1 + P_2$, or radical transfer: $P_1^{\bullet} + RQ \rightarrow P_1Q + R^{\bullet}$, or termination by initiator radicals: $P_n^{\bullet} + {}^{\bullet}\text{fragment} \rightarrow P_n\text{-fragment}$ [76].

An ionic chain growth mechanism similar to that of radicals may also occur in the plasma: $\text{Fragment}^+ + M \rightarrow P_i^+$ and $P_i^+ + M \rightarrow P^+_{i+1.}$

Polymerization reactions are equilibrium reactions, which are controlled by thermodynamics. They proceed if the standard Gibbs energy is negative (exothermic):

$$\Delta G^0{}_p = \Delta H^0{}_p - T\Delta S^0{}_p = -RT \log_e K_n \tag{11.1}$$

where

$\Delta H^0{}_p$ is the standard polymerization enthalpy,
$\Delta S^0{}_p$ is the standard polymerization entropy,
T is temperature,
K_n is the equilibrium constant.

Plasma-induced chain growth polymerization is a chemical process only started by radicals, which are produced by the plasma exposure, similar to radiation-initiated polymerization, which is started by photons, electrons, or alpha particles [77]. The activation energy needed to start the chain growth polymerization for vinyl monomers is $\Delta G \approx +120\,\text{kJ mol}^{-1}$ (ca. 1 eV) [78, 79]. Since 5000–20 000 additions per second occur under atmospheric pressure, each with ca. $\Delta G \approx -20\,\text{kJ mol}^{-1}$, this complete reaction is strongly exothermic and needs ca. 0.1–2 s for chemical polymerizations (= time between start and termination of the growing chain) [76]. Classic initiator radicals have low dissociation energy, decompose slowly (thermally), and have therefore a relatively long life-time and are inserted in the growing polymer chain [80, 81]. Plasma-produced C-radicals are generated by random dissociation of strong C–C or C–H covalent σ bonds (fragmentation) of monomer molecules and have high energy, react rapidly, and have a short life-time of $<10^{-10}$ s [76, 82]. C-radical sites at polymer chains are formed on exposure to the plasma by hydrogen abstraction or C–C bond scission:

$$\sim\text{-CH}_2\text{-CH}_2\text{-}\sim + \textit{plasma} \rightarrow \sim\text{-CH}_2\text{-CH}^{\bullet}\text{-}\sim + 0.5H_2$$

Table 11.1 Standard dissociation energy (SDE) of aliphatic compounds.

Bond	$CH_3–CH_3$	$(CH_3)_3C–H$	$(CH_3)_2CH–H$	$CH_3–CH_2–H$	$CH_3–H$	$C_6H_5–H$
SDE ($kJ\,mol^{-1}$)	370	385	396	411	435	458

$$\sim\text{-}CH_2\text{-}CH_2\text{-}\sim + \textit{plasma} \rightarrow \sim\text{-}CH_2^{\bullet} + {}^{\bullet}CH\text{-}\sim$$

Table 11.1 lists the standard dissociation energies of such bonds.

Disregarding the entropy term at first, for a raw estimation of the priority in bond scission, C–C bond scission is favored in comparison to that of C–H bonds. Thus, biradicals are produced by π-bond scission if easily polymerizable acrylic, vinyl-, or diene monomers are used. However, simple π-bond activation to a biradical leads often to rapid recombination because of the low collision rates and low-pressure conditions and therefore the lack of possibilities to react with new partners.

Thermodynamically, the polymerization produces, according to Eq. (3.1), an energy gain, which can be explained by transferring the π-bond into a (second) σ-bond and by loss of degrees of freedom (many monomers to one polymer), thus causing negative ΔS. Most often the enthalpy is much greater than the entropy term. Only at high temperature does compensation of the two terms occur (ceiling temperature) and the polymerization stops. At room temperature the ΔG of styrene polymerization is $-20\,kJ\,mol^{-1}$, that is, the polymerization is exothermic and proceeds spontaneously under release of enthalpy in the presence of an initiator radical. For an average polymerization degree of $X = 1000$ polystyrene has a molar mass of about $100\,000\,g\,mol^{-1}$, and $110\,kJ\,mol^{-1}$ were needed for initiation. The free reaction enthalpy amounts approximately to $\Delta G \approx X\Delta H_{Polym} + \Delta H_{Start} \approx -1000 \times 20 + 110\,kJ\,mol^{-1} \approx -20\,000\,kJ\,mol^{-1}$.

The kinetic chain length (ν) characterizes the number of monomer molecules that were added by one initiation step before termination occurs. It is the ratio of chain propagation (R_p) and termination rates (R_t): $\nu = R_p/\Sigma R_t$. It grows with increasing monomer concentration and decreases with increasing initiator concentration. If the termination is dominated by recombination of two polymer radicals then $\nu \approx [M][I]^{-1/2}$. Unfortunately, in the low-pressure plasma only a very low concentration of monomer molecules is present but there is a relatively high initiator concentration; therefore, short kinetic chain lengths occur. Recombination, disproportionation, and chain transfer are dependent on collisions of two partners and thus are hindered by low pressure.

An important co-occurring reaction is the radical reaction with traces of molecular oxygen in the plasma phase or that after finishing the plasma treatment and exposing the polymer to oxygen from air: $P^{\bullet} + {}^{\bullet}O\text{-}O^{\bullet} \rightarrow P\text{-}O\text{-}O^{\bullet} \rightarrow +RH \rightarrow P\text{-}O\text{-}OH + R^{\bullet} \rightarrow$ decay $\rightarrow$ broad variety of oxidized products [83].

Another start mechanism (homogeneous catalysis) uses $CBrCl_3$ as plasma gas and supposes its dissociation [84]. A bromine radical starts the chemical chain growing reaction:

$$Br^{\bullet} + CH_2{=}CH\text{-}CH_3 \rightarrow CH_2{=}CH\text{-}CH_2^{\bullet} + HBr$$

and:

$$CH_2{=}CH\text{-}CH_2^{\bullet} + BrCCl_3 \rightarrow CH_2{=}CH\text{-}CH_2Br + {}^{\bullet}CCl_3$$

which adds to:

$${}^{\bullet}CCl_3 + CH_2{=}CH\text{-}CH_3 \rightarrow Cl_3\text{-}CH_2{=}CH\text{-}CH_2^{\bullet} \text{ and so on}$$

A particular case is the free radical polymerization of allyl monomers. Macro-radicals are generated similar to free radical polymerization of vinyl monomers. However, termination of a kinetic chain occurs by degradative chain transfer to monomers, leading to self-inhibition:

$$\sim CH_2\text{-}CH^{\bullet}(CH_2\text{-}R) + CH_2{=}CH(CH_2\text{-}R) \rightarrow \sim CH_2\text{-}CH_2(CH_2\text{-}R) + CH_2{=}CH\text{-}C^{\bullet}H\text{-}R$$

As the propagation rate is not much greater than the transfer rate the polymerization degree is usually not greater than 10–20. However, chain transfer is also associated with collisions and may therefore be insignificant under low-pressure conditions [85].

11.3.2 Ion–Molecule Reactions

Ordinary ion–molecule reactions are not simple monomer additions as with radical or ionic polymerization. The simplest reaction is charge transfer without a change of molecular weight from A to A: $A^+ + A \rightarrow A + A^+$ or for different atoms A and B: $A^+ + B \rightarrow A + B^+ + \Delta E$ (energy surplus).

More probable is dissociative charge transfer, for example:

$$He^+ + CH_4 \rightarrow C(H_{4-x})^+ + 4xH + He^0$$

An ion–molecule reaction can also produce an increase in molecular weight of the main product. The reaction enthalpy is contained in the side-product, such as in methane ion–molecule reactions [86]:

$$CH_3^+ + CH_4 \rightarrow C_2H_5^+ + H_2$$

$$C_2H_5^+ + CH_4 \rightarrow C_3H_7^+ + H_2 \text{ and so on}$$

11.3.3 Fragmentation–(Poly)recombination ("Plasma Polymerization")

Furthermost from chemical chain-growth polymerization is the polymerization mechanism that occurs in continuous-wave (cw) plasma. Complete fragmentation of monomer molecules to atoms or fragments occurs:

$$ABCDEF + \textit{plasma} \rightarrow A + B + CD + E + F$$

The permanent shower of high-energy particles and radiation to monomer molecules and polymer surface produces fragmentation and atomization largely independent of the nature of monomer. Such fragmented reactive (classic) monomers, saturated precursors, aromatic substances, or inorganic gases and vapors form (easily) randomly composed plasma polymers of high irregularity by (poly)recombination of fragments and atoms: $n(A + B + CD + E + F) \rightarrow [FCABDE]_n$.

Moreover, hydrogen abstraction, CO, CO_2, and H_2O elimination, cracking of aromatic rings, and so on occur. A special reaction pathway is the formation of unsaturated film-forming intermediates, such as ethylene, acetylene, cyclobutadiene, and so on, which also contribute to the plasma polymer formation. These intermediates produce a few defined structural elements within the polymer, similar to polystyrene. They are typical products of endothermic processes and need for their formation the energy excess present in the plasma.

The higher the energy doses per monomer molecule the more fragmentation proceeds [10, 87]. A lower energy dose benefits the survival of monomer structures in the plasma polymer. Calculations performed by Friedrich based on experimental data show that classic vinyl monomers consume about 10–100 eV per molecule during their residence time in the plasma for 100% conversion of monomers into polymer layer [26]. Low-molecular weight alkanes (aliphatic compounds) such as *n*-hexane need about 1000 eV for a 60% conversion into plasma polymer [40]. Such high energy flow to one monomer molecule produces nearly complete dissociation/fragmentation into atoms and small fragments. A simple calculation shows that the dissociation of all bonds of an *n*-hexane molecule ($5 \times$ C–C and $14 \times$ C–H) needs about 70 eV, whereas 1000 eV were transferred from plasma. The impossibility of any selectivity is obvious. End-products of such plasma polymerizations are plasma polymers and hydrogen. Yasuda denoted this process as "atomic polymerization" [87], while Friedrich called it "quasi-hydrogen plasma" [40].

11.4 Plasma Polymerization in Adsorption Layer or Gas Phase

Polymer growth in the adsorption layer is favored because of low-pressure conditions with high mean free paths and low collision rates (Figure 11.4) [30, 35, 88–92]. However, it was also found that with growing pressure (>133 Pa) powder formation occurs and the polymer formation was changed to the gas phase (Figure 11.4) [16, 29, 93–96].

During deposition of plasma polymer permanent vacuum UV (VUV) irradiation ($\lambda \approx$ 100–200 nm ~ 6–12 eV, C–C dissociation enthalpy $\approx$ 3.7 eV) from the plasma produces H-abstraction, double bond formation, radical formation, crosslinking or degradation, loss in supermolecular structure, and oxidation via peroxide formation (Figure 11.5) [97].

Another assumption is the formation of a more defined "skin" on plasma polymers. This skin is formed after switching-off the plasma by reaction of (classic) monomer molecules with radicals at the layer surface. Thus, this topmost layer is post-plasma chemically produced and should present a more regular structure and composition (Figure 11.5) [98].

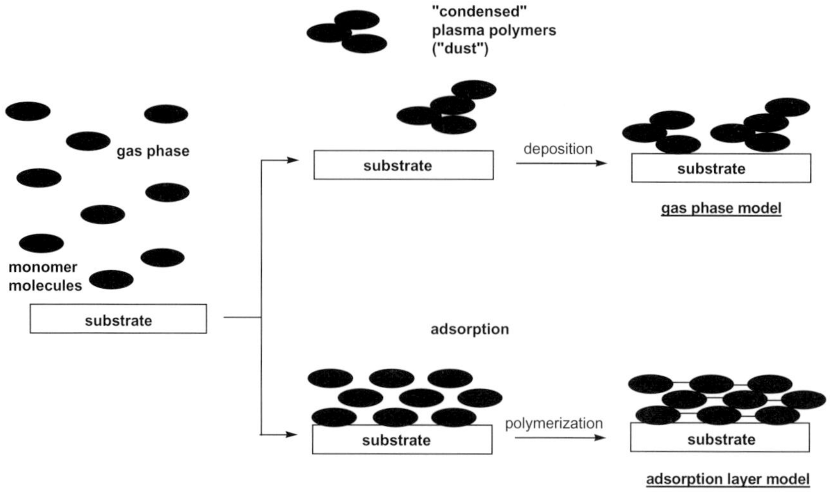

Figure 11.4 Gas phase and adsorption layer model for location of plasma polymerization.

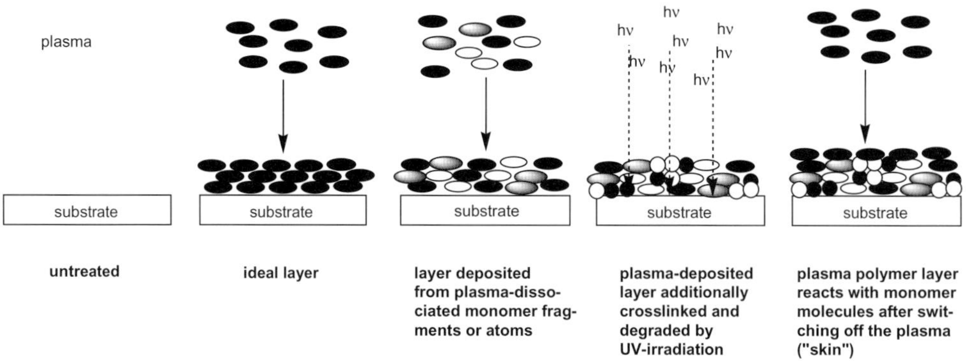

Figure 11.5 Effect, shown schematically, of plasma-UV irradiation and monomer grafting after switching-off the discharge on plasma polymer structure.

11.5
Side-Reactions

The actual process of polymer formation can be characterized as polyrecombination of (radical) fragments and atoms. Such a multi-recombination process is strongly exothermic and produces much excess enthalpy, which must be dissipated and transferred. Therefore, it can be speculated that local thermochemical

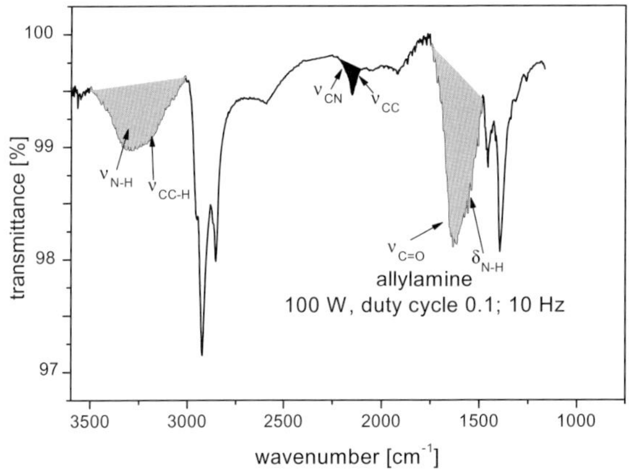

Figure 11.6 FTIR-ATR spectra of plasma-polymerized allylamine with NH-/NH_2-specific bands (in gray) and a significant CN band.

processes also contribute to the analyzed irregular structure of plasma polymers. Endothermic dehydrogenation of monomers and plasma polymers is characteristic of such a high-energy reaction. The formation of acetylene C≡C bonds and the dehydrogenation of amino groups to nitrile groups in homo- and copolymers of allylamine are examples [98]. The IR spectrum of the allylamine plasma polymer shows a moderate band of the $\nu_{C\equiv N}$ stretching vibration at approximately 2150 cm^{-1} (Figure 11.6). The absorbance (extinction) of this band correlates with an approximately 10% transfer of NH_2 groups into C≡N groups as measured with nitrile-containing reference polymers:

$$\blacksquare\text{-CH}_2\text{-NH}_2 + \textit{plasma} \rightarrow \blacksquare\text{-C}\equiv\text{N} + 2\text{H}_2$$

The relative broadness of the $\nu_{C\equiv N}$ stretching vibration (large FWHM) may be due to the superposition with $\nu_{C\equiv C}$ also appearing as $\nu_{\equiv C\text{-}H}$ near 3300 cm^{-1}:

$$\text{R-CH}_2\text{-CH}_2\text{-R}' + \textit{plasma} \rightarrow \text{R-C}\equiv\text{C-R}' + 2\text{H}_2$$

A similar abnormal reaction was also found for the formation of aromatic rings in plasma polymers produced from ethylene [39]:

$$\text{R-CH}_2\text{-CH}_2\text{-CH}_2\text{-CH}_2\text{-CH}_2\text{-CH}_2\text{-R}' + \textit{plasma} \rightarrow \text{R-aryl-R}' + 3\text{H}_2$$

This formation and also the cracking of aromatic rings in benzene, toluene, and xylene and their plasma polymerization are also examples of exotic reactions in plasmas due to the excess energy [23, 27–30, 92].

For probability and steric reasons the polyrecombination is incomplete. A high concentration of radicals remains trapped in the deposited plasma polymer layer or is secondarily produced by UV irradiation from plasma [7, 14]. These radicals react with oxygen, which is either dissolved in the polymer or diffuses into the polymer when it is exposed to ambient air. This peroxide formation is the starting point for the previously mentioned auto-oxidation. The reaction of plasma polymers with oxygen from air may proceed vigorously. After milling plasma polymers at low temperature (−196 °C) the numerous C-radicals in the powdered plasma polymer come into contact with oxygen from air and react under self-ignition of flaming, then blackening, and release of smoke (J. Friedrich, private communication).

Crystalline and supermolecular structures (spherulites, rod-like, lamellae, etc.) were generally absent in plasma polymers [14, 35, 99]. However, oligomers with rod-like molecular structure such as biphenyl, diphenyl ether, or diacetylene were frequently formed and may produce orientation (mesogen) [100]. More recently, it was discovered that fluorine-containing plasma polymers also show separate fibrous structures of perfluoroalkyl chains with high crystallinity [101, 102].

11.6 Quasi-hydrogen Plasma

The release of hydrogen from monomer molecules and plasma polymers and enriching it in the gas (plasma) phase has been called "quasi-hydrogen plasma." It may be remembered as the production of amorphous carbon layers (a-CH) with maximal hydrogen release. An example of such a loss of hydrogen during plasma polymerization is the pulsed-plasma deposition of allyl alcohol–styrene copolymers using different duty cycles (Figure 11.7).

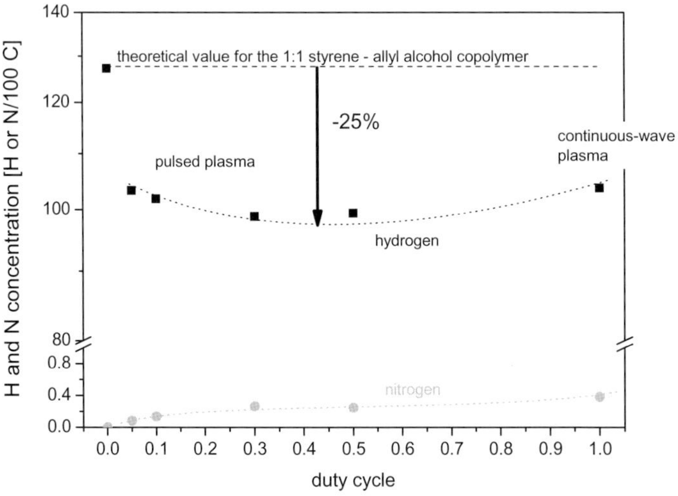

Figure 11.7 Theoretical and CHN analysis-measured hydrogen ratio in a plasma-deposited styrene–allyl alcohol (1 : 1) copolymer.

The dominance of hydrogen in the plasma increases the electron temperature because of low cross-sections and low collision rates in plasma consisting of hydrogen molecules, atoms, and protons. The higher electron energy in such hydrogen-dominated plasma produces a higher fragmentation rate and degree of monomer molecules (cf. "atomic polymerization") [40].

Toluene and d_8-toluene with and without additional hydrogen or deuterium were plasma polymerized in a DC glow discharge under reduced pressure using a tube-type reactor made from glass [40]. Electron temperatures were measured using a Langmuir probe at different ports along the tube-type reactor. Moreover, at these ports deposited plasma polymer layers were analyzed by CH analysis and IR spectroscopy. The gas-phase composition was also measured by gas chromatography (thermal conductivity detector). Enrichment of hydrogen (and somewhat methane) in the argon plasma and a complementary strong hydrogen deficiency in plasma polymers (up to 75%) were measured.

After injection of toluene into the plasma tube (argon direct current low-pressure glow discharge) these molecules decrease electron density and temperature because of their large electron capture cross-section [98]. Afterwards, in the second half of the tube, the enrichment with hydrogen dominated and the electron temperature exceeds that of argon gas significantly.

Contrariwise to the electron temperature the portion of aromatic rings and CH_3 groups in the deposited plasma polymer decreased from injection point to the end of tube. The layers near the injection point were colorless while those deposited in the zone behind the injection point were yellow to dark brown. The brown layers have a low hydrogen percentage (H/C = 0.3; Figure 11.8).

Only addition of the threefold molar ratio of hydrogen gas to the toluene–argon plasma could balance the loss in hydrogen in the deposited layers (Figure 11.8).

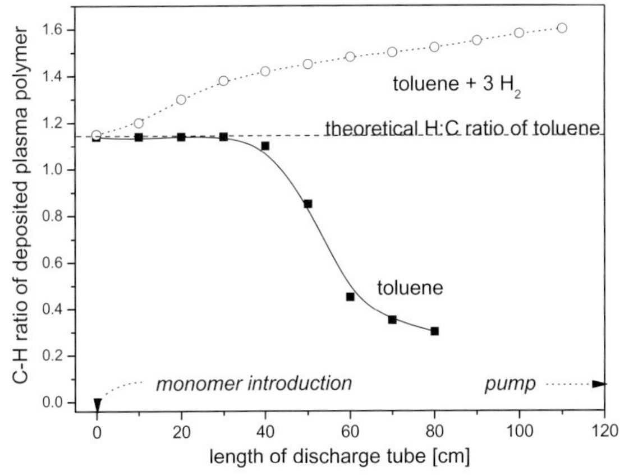

Figure 11.8 Variation of C:H ratio in DC-plasma polymerized toluene layers with the length the monomer passes through a tube-type reactor.

Using d_8-toluene and a threefold amount of hydrogen gas (H_2) a significant D→H-exchange was measured on analyzing the deposited plasma polymer by IR spectroscopy and CH analyses (Figure 11.9). Strong C–H stretching and deformation signals were found and, vice versa, strong C–D signals on using D_2 and h_8-toluene (Figure 11.9).

The plasma polymer layer of the second part of the tube contained no aromatics (Figure 11.10).

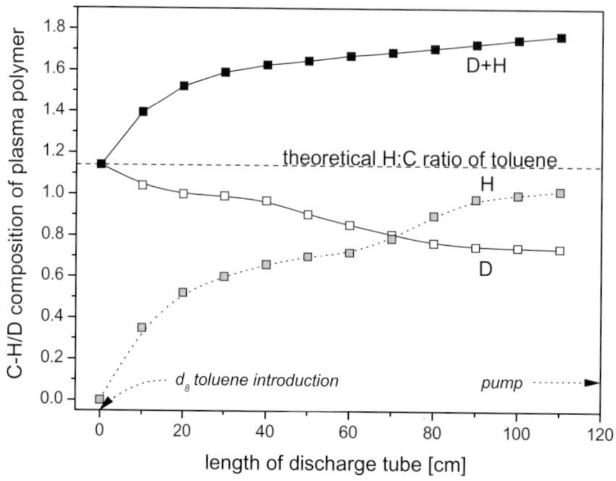

Figure 11.9 Dependence on tube length of D↔H exchange in a plasma polymer deposited in a d_8-toluene–3H_2 DC plasma atmosphere.

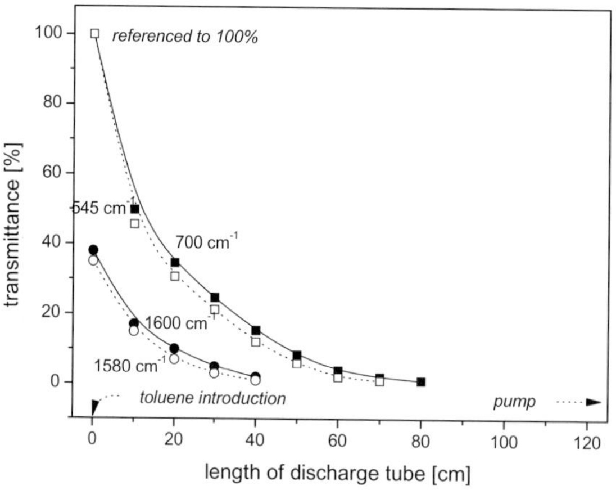

Figure 11.10 Variation of the loss of aromatic rings in h_8-toluene and d_8-toluene plasma polymer layers with the locus of deposition.

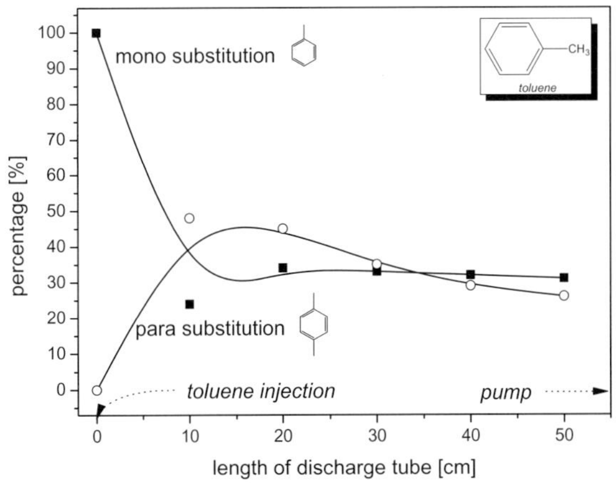

Figure 11.11 Toluene plasma polymerization in a low-pressure DC discharge and the types of substitution of aromatic ring structures observed.

It could also be shown that the mono-substitution, which may have a polystyrene-like structure, is changed successively, beginning from the toluene injection point downstream to di-substitution (preferentially para substitution) as measured by IR (Figure 11.11) [40].

Similar D–H exchange was found using d_{74}-hexatriacontane and *d*-polyethylene exposed to NH_3 plasma and, vice versa, h_{74}-hexatriacontane and *h*-PE exposed to ND_3 plasma [103].

11.7 Kinetic Models Based on Ionic Mechanism

The first model for plasma polymerization, based on an ionic mechanism of polymer formation within the adsorption layer, was presented by Williams and Hayes, in 1966 [34]. The polymerization rate was defined as $m = \alpha It$, where m is the deposited mass of polymer per area unit, α is a constant that depends on adsorption, I is the current density at electrodes (AC, kHz), and t is the deposition time.

An ionic mechanism and polymerization in the adsorption layer were also assumptions in the model published by Poll, in 1970. He considered transport rates of ions to the surface, and postulated ambipolar diffusion and control of the plasma boundary layer [104], which was also supplemented by Carchano [105].

The model proposed an ionic initiation step:

$$P + e^- \rightarrow P^+ + e^{-\prime}$$

followed by neutralization of positively charged ions and formation of radicals [86]:

$$P^+ + e^- \rightarrow P^* \rightarrow 2P^\bullet$$

It follows that:

$$dP^\bullet/dP = k_{1j}^2$$

where j is the current density.

The polymer growth is the product of radical and monomer concentration. Adapting the Langmuir adsorption isotherm as model he proposed: $dn/dt = \gamma\Phi(1 - n/n_0) - n/\tau$, where γ is the sticking coefficient, Φ is the average collision number of thermal fluctuation, n_0 is the coverage at saturation, τ is the average residence time (in plasma), $\tau_0 = 10^{-13}$ s, and p is the gas pressure. Considering the cross-section of monomer particles for polymerization the following polymer growing rule for P was derived:

$$P = \gamma\Phi(1 + Z_e\gamma\Phi/J_0 2qn_\infty)^{-1}$$

with $1/n_\infty = 1/n_0 + 1/\gamma\Phi Z$.

Westwood considered exclusively an ionic (cationic) chain propagation mechanism for the plasma polymerization of vinyl chloride, as published in 1971, because of preferred polymer deposition onto the cathode in direct current glow discharges [12]. He interpreted radical reactions observed by Denaro [14] as concurrent reactions. The conversion into plasma polymer was expressed in terms of the G-value. The measured G-values were <1, evidencing the absence of an ionic chain-growth polymerization and the dominance of termination reactions.

Thompson and Mayhan also presented, in 1972, a cationic mechanism, assigning it as the most probable mechanism because the addition of radical scavengers to the plasma did not change the polymer deposition rate [13]. Critically, all radical scavengers take part in the plasma processing, are also fragmented, or re-arranged and, thus, they are not able to quench radicals definitely. The excitation step was proposed to be $M + e^- \rightarrow M^{**} + e^{-\prime}$ followed by formation of M^+ and also M^-, $M^{\bullet+}$, $M^{\bullet-}$, or ion pair formation P^+ and Q^- (which needs much energy, >10 eV). These species were also formed directly by inelastic collisions with electrons, such as:

$$M + e^- \rightarrow M^+ + e^{-\prime} + e^{-\prime\prime}$$

Chain propagation occurs that follows a cationic polymerization mechanism:

$$M^+ + M \rightarrow MM^+$$

$$MM^+ + nM \rightarrow M[M_n]M^+$$

Branching and crosslinking were explained by formation of additional positive charges along the formed polymer chain due to impact with energetic plasma particles. These mid-chain radicals are able to add (graft) monomer molecules. Polycation formation is proposed, which is plausible for polar and ionic polymers, as is known from ESI (electrospray ionization) [106]: $MMM[M_n]M^+ + \Delta E \rightarrow MM^+M[M_n]M^+$. The newly formed charge at the growing chain starts the growth of a side-chain:

$$MM^{+}M[M_{n}]M^{+} + M \rightarrow MM(M^{+})M[M_{n}]M^{+}$$

Crosslinking may also occur on this way. Termination of all processes occurs by ion–electron recombination or charge transfer to the monomer. Monomer fragmentation was not considered. The life-time of the cations is very short for continuous chain propagation [107]. Moreover, a cationic mechanism needs nucleophilic monomers such as styrene. The deposition rate was derived using an Arrhenius plot [13].

Vasile and Smolinsky investigated the plasma phase for charged and neutral reactive intermediates. Their conclusion was that unsaturated charged intermediates play the most important role during plasma polymerization [13, 16].

11.8
Kinetic Models of Plasma-Polymer Layer Deposition Based on a Radical Mechanism

Most scientists in the field have been fascinated by the possibility of "polymerizing" all kind of gaseous or evaporable organic substances. Fragmentation to short residues and atoms, dehydrogenation, recombination of radicals, and formation of unsaturated polymer forming intermediates and their polymerization were held responsible for the generation of this new class of polymers. Non-polymerizable intermediates should be removed with the exhaust gas.

The fragmentation and polyrecombination mechanism (atomic-polymerization) is concurrent with the (plasma-initiated) chemical (radical) polymerization. The chemical chain growth polymerization may be exemplified for methyl methacrylate:

$$\mathrm{nCH_2{=}C(CH_3)\text{-}COO\text{-}CH_3} + \Delta E \rightarrow \mathrm{\text{-}[CH_2\text{-}C(CH_3\text{-}COO\text{-}CH_3]}_n + \Delta E'$$

for chemical polymerization E(= energy) may be produced by plasma, and the fragmentation–recombination ("atomic polymerization"):

$$n\mathrm{CH_2{=}C(CH_3)\text{-}COO\text{-}CH_3} + x\Delta E \rightarrow n\mathrm{C} + n\mathrm{CH} + n\mathrm{CH_2} + n\mathrm{CH_3} + \mathrm{nCO} + \mathrm{nCO_2},$$
$$n\mathrm{H_2}\ \text{and so on} \rightarrow \mathrm{C{=}CH\text{-}C(OH)(CH_3)\text{-}CH{=}CH(CH_2\text{-}CH{=}CH_2)\text{-}}$$

Most kinetic models with a radical mechanism consider these two different processes.

Kassel proposed the plasma-electrical conversion of methane into acetylene before polymerization because methane is not able to polymerize in a direct way [108]:

$$\mathrm{CH_4}\ (k_1) \rightarrow \mathrm{C_2H_6},\ \mathrm{C_2H_6}\ (k_2) \rightarrow \mathrm{C_2H_4},\ \mathrm{C_2H_4}\ (k_3) \rightarrow \mathrm{C_2H_2},\ \mathrm{C_2H_4}\ (k_4)$$
$$\rightarrow \text{end-products},\ \mathrm{CH_4}\ (k_5) \rightarrow \mathrm{C_2H_2},\ \mathrm{C_2H_6}\ (k_6) \rightarrow \mathrm{C_2H_2},\ \mathrm{CH_4}\ (k_7) \rightarrow \mathrm{C_2H_4},$$
$$\mathrm{C_2H_4}\ (k_8) \rightarrow \text{end-products}.$$

Intermediates are able to form layer-depositing gaseous products (cf. *Neiswender mechanism*) [9]:

$$C_2H_2 \rightarrow C_4H_4 \rightarrow C_6H_6 \rightarrow C_8H_8$$

Neiswender referenced the energy introduced into the plasma (W) to the number of monomer molecules (B) during the transition in the plasma zone (T). Thus, the energy dose per mole of monomer (E, J mol^{-1} or Ws mol^{-1}) is given by $E = PT/B$ [9].

Atomic-polymerization, designated by Drost as "plasma electrical polymerization" [86], is characterized by complete fragmentation of precursor molecules to reactive atoms and fragments, which pool together to give a random structured conglomerate of low-molecular weight oligomers and polymers. High molar mass products and crosslinking may be produced by random radical recombination and irradiation with plasma VUV radiation [39, 107]. Therefore, the structure is statistically determined and consists of an irregular sequence of fragments. Nevertheless, besides completely irregular structures a few chemically well-defined structures also often occur in plasma polymerization, as mentioned before [19, 52, 98].

Thus, Bell confirmed the favored formation of acetylene bonds in the ethylene plasma because of its low positive reaction enthalpy:

$$C_2H_4 + e^- \rightarrow C_2H_2 + H_2 + e^{-\prime} \; E = 1.8 \text{ eV}$$

$$C_2H_4 + e^- \rightarrow 2CH_2 + e^{-\prime} \; E = 6.3 \text{ eV}$$

$$C_2H_4 + e^- \rightarrow C_2H_4^+ + 2e^- \; E = 10.5 \text{ eV}$$

$$C_2H_4 + e^- \rightarrow C2H_2^+ + H_2 + e^{-\prime} \; E = 13.1 \text{ eV}$$

$$C_2H_4 + e^- \rightarrow C_2H_3^+ + H + 2e^- \; E = 13.3 \text{ eV}$$

Acetylene was considered to be a key intermediate in nearly all plasma polymerizations; thus, under retention of the triple bond a substitution reaction is also possible: $HC{\equiv}CH + X_2 \rightarrow HC{\equiv}CX + HX$, hydrogen can be added: $HC{\equiv}CH + H_2 \rightarrow H_2C{=}CH_2$, it can form benzene structures: $3HC{\equiv}CH \rightarrow C_6H_6$, or polymerize with participation of the triple bond: $n HC{\equiv}CH \rightarrow$ -(HC=CH)$_n$- [9, 109].

Only a few reaction pathways to plasma polymers are unambiguous, and are determined from chemical thermodynamics, such as the formation of SiO_2-like layers by Si-containing precursors in the presence of oxygen in the plasma, such as hexamethyldisiloxane + O_2:

$$(CH_3)_3\text{-Si-O-Si-}(CH_3)_3 + 12O_2 \rightarrow 2SiO_2 + 6CO_2 + 9H_2O$$

The same is true for the deposition of bromine-containing precursors as plasma polymers [110, 111]. The octet rule allows only formation of C–Br or Br^- if oxygen is absent. Since Br^- formation is unlikely because of the radical mechanism and the absence of counter ions let assume that all Br^- is bonded in HBr, which is removed by the vacuum pump; thus bromoform produces oxygen-free plasma polymers with a Br/C ratio of about 1 : 1:

$$CHBr_3 + \textit{plasma} \rightarrow CHBr_2^{\bullet} + {}^{\bullet}Br$$

followed by:

$$CHBr_2^{\bullet} + \textit{plasma} \rightarrow {}^{\bullet}CHBr^{\bullet} + {}^{\bullet}Br$$

and finally:

$$n^{\bullet}\mathrm{CHBr}^{\bullet} \rightarrow \text{-}[\mathrm{CHBr}]_n$$

Neglecting these two or three exceptions the following reaction pathways are in the foreground:

1. chemical chain-growth polymerization with very short kinetic chain length;
2. fragmentation and random (poly-)recombination under high excess energy conditions to oligomers and irregularly structured polymers with high concentrations of radicals, unsaturations, undesired oxygen, and so on;
3. monomer fragmentation and re-arrangement to give defined intermediates that can polymerize and form structures with a few similarities to classic ones;
4. crosslinking by random radical recombination and UV irradiation from plasma.

Plasma polymerization via fragmentation–polyrecombination is the most energy-consuming route and therefore is maximally inefficient. It was demonstrated earlier by the example of hexane plasma polymerization that the introduced plasma energy is high enough to dissociate all bonds in the molecule several times.

Denaro proposed three steps within a radical mechanism:

1. chain growing: $R_n^{\bullet} + M \rightarrow R_{n+1}^{\bullet}$,
2. radical recombination: $R_n^{\bullet} + R_m^{\bullet} \rightarrow P_{n+m}$,
3. trapping of radicals (loss of radicals): $R_n^{\bullet} \rightarrow R_n^{\bullet}$ (trapped).

Using the BET equation for multilayer adsorption and other assumptions the following rate equation was derived: $R = rp(p + A + [A^2 + Br]^{-1/2})^{-1}$, where r is the radical formation rate, p is the monomer pressure, and A and B are constants [14].

The idea of polymerizable intermediates was found also in a paper from Tiller [112]. He proposed the transformation of the precursor into an intermediate, which can form a polymer (layer). The transformation from precursor into the intermediate determines the plasma polymer deposition rate [112].

In 1971, Yasuda created a widely accepted (pseudo-)kinetic model based on electronic excitation or activation of monomer molecules (M) or gas atoms (X). He considered different reaction rate constants (k_x) for each sub-reaction [113]:

$\mathrm{M} \rightarrow \mathrm{M}^* \ (k_1)$ and $\mathrm{X} \rightarrow \mathrm{X}^* \ (k_2)$,

energy transfer $\mathrm{X}^* + \mathrm{M} \rightarrow \mathrm{X} + \mathrm{M}^* \ (k_3)$

or by Penning ionization, initiation step $\mathrm{M}^* \rightarrow \mathrm{M}^{\bullet} \ (k_i)$

and chain-growth $\mathrm{M}^{\bullet} + \mathrm{M} \rightarrow \mathrm{MM}^{\bullet} \ (k_p)$ or $\mathrm{M}_n^{\bullet} + \mathrm{M} \rightarrow \mathrm{M}_{n+1}^{\bullet}$

The polymer growth rate (R) was then written as:

$$R = kik_pk_1[\mathrm{M}]^2(1 + k_2k_3/k_1[\mathrm{X}])$$

where

k_1, k_2, k_3, k_p = rate constants for each sub-reaction,
[M] = monomer concentration,
[X] = carrier gas concentration.

Summarizing and simplifying all sub-reactions, the overall growth rate was proposed as:

$$R = a[p_M]^2[1 + bp_{gas}]$$

where

p_M = monomer pressure,
p_{gas} = total gas pressure,
a and b = constants with dimensions.

Tiller and Friedrich have considered the classic radical chain growth mechanism (A) and the polyrecombination of plasma-produced radical fragments (B) [114, 115]. Moreover, the formation of plasma polymer clusters in the gas phase was also included (B_2); deposition of a plasma polymer layer, either directly via the gas phase (with monomer concentration, M_0), or in the so-called adsorption layer (monomer concentration, M_{ads}), formation of plasma polymer in the latter being symbolized by B_1. Then, the layer deposition rate P_r is: $P_r = P_{rA} + P_{rB1} + P_{rB2}$.

Tiller proposed the following elementary processes:

- Radiation and plasma particle activation of monomer molecules, chemical chain growth processes, and recombination of radicals within the adsorption layer:

$$M_{ads} \rightarrow R^\bullet\ (k_1);\ R^\bullet + M_{ads} \rightarrow R_2^\bullet\ (k2);\ R_n^\bullet + R_m^\bullet \rightarrow P_{n+m}(k_3)$$

- (B_1) irradiative and plasma particle activation, fragmentation, and recombination in the adsorption layer:

$$M_{ads} \rightarrow R^\bullet\ (k_4);\ R^\bullet + R'^\bullet \rightarrow P\ (k_5)$$

- (B_2) irradiative and plasma particle activation, fragmentation, and recombination in the gas phase layer:

$$A_{ads} \rightarrow R^\bullet\ (k_4);\ M_0 \rightarrow R_0^\bullet;\ R^\bullet + R_0^\bullet \rightarrow P$$

For P_{rA} and chemical chain growth:

$$P_{rA} = k_2[R^\bullet][M_{ads}] = k_2(k_1/k_3\ [h\nu])^{1/2}[M]^{3/2}$$

and P_{rB1}, the layer growth by recombination in the adsorption layer, is: $P_{rB1} = k_5[R^\bullet]^2 = k_4k_5[I_{h\nu}][M_{ads}]$ and in the gas phase: $P_{rB2} = k_7[I_{0h\nu}][M_0]$. The total polymerization rate is:

$$P_r = k_2(k_1/k_3)^{1/2}[M_{ads}]^{3/2}[I_{h\nu}]^{1/2} + k_4k_5[I_{h\nu}][M_{ads}] + k_7[I_{0h\nu}][M_0]$$

Morita postulated a radical mechanism also in radiofrequency (rf) plasmas. Under low-frequency conditions charge carriers are oscillating and then an ionic mechanism should dominate [116].

Shen and Bell considered bond dissociation energies and proposed a radical mechanism based on the 10^4 higher concentrations of radicals in the plasma than that of ions [117].

Lam and Baddour assumed that activation of molecules takes place in the gas phase and chain growth in the adsorption layer [118].

Tibbitt has also created a kinetic model for a mechanism that explains the structure of the two proposed ethylene plasma polymer species, a low-molecular weight oligomer and a crosslinked polymer [39]. He also considered a radical polymerization mechanism with a plasma initiation step and chain growth in terms of adsorption coefficients and plasma parameters:

$$e + M_g + k_i \rightarrow 2R_g^{\bullet} + e' \text{ initiation}$$

$$S + R_g^{\bullet} + k_{pg} \rightarrow R_s^{\bullet} \text{ radical adsorption}$$

$$R_g^{\bullet}\,Mg + k_{pg} \rightarrow 2R_{g+1}^{\bullet} \text{ homogeneous chain growth}$$

$$R_s^{\bullet}\,Mg + k_{ps} \rightarrow 2R_{s+1}^{\bullet} \text{ heterogeneous chain growth}$$

where: e represents an electron, M the monomer, R the radical, s the substrate, g the gas, and k is the rate constant.

The polymerization rate was then:

$$r_p = (0.5dk_{pg} + K_R k_{ps})[M_g][R_g]$$

where

K_R = adsorption coefficient of radicals at the surface of electrodes,

$[R_s]$ = $K_R[R_g]$, where $[R_s]$ = radical concentration at surface and $[R_g]$ = radical concentration in gas phase

d = distance between electrodes.

The equation can be solved if monomer and radical concentrations are known:

$$r_p = [M_g]_0^2\, 2ac[1 - \exp - (a - b)\tau]/[(a - b)\exp(a + b)\tau]$$

where

$a = k_i[e]$,

$b = (2/d)k_a[S]$,

$c = (0.5d)k_i + kp_s$,

$\tau = V/Q$ with Q = monomer flow and V = volume of plasma zone.

It was found again that introduction of high power into the plasma enhances the irregularity of polymer structure but increases the molecular weight to unlimited values (crosslinked). On the other hand, introduction of low power improves the regularity of plasma polymers but limits the molecular weights to a few hundred or thousand $g\,mol^{-1}$ [7].

As mentioned before, Yasuda developed also the idea of "atomic polymerization" [87, 119]. This concept assumes that the monomer molecules were fragmented in the plasma to single atoms. These atoms and fragments recombine (polyrecombination) [26, 77]. New insight into the mechanism of plasma polymerization is also offered by Yasuda's CAP-mechanism (competitive ablation and polymerization), in which there is the simultaneous occurrence of etching and deposition. This mechanism distinguishes also between "atomic (physical)" or "plasma-state" and "molecular (chemical)" or "plasma-induced" polymerization [43, 44]. Yasuda showed that the CAP mechanism is a competition between polymer deposition and polymer etching of deposited layer, with a surplus for deposition (*"echternacher springprozession"*): $R_{total} = R_{depos} - R_{etch}$.

Systematic work on deposition rates of different monomers was performed by Yasuda and Inagaki [94, 95, 113, 120]. The measured deposition rates were strongly dependent on plasma conditions, type of plasma and monomer, and plasma chamber geometries. Its validation is relative and can only be used for raw orientation. It is very important to note that the plasma volume depends on plasma parameters. It can expand or shrink. Therefore, the deposition rate at a fixed point in the reactor is also influenced by simple shrinking or expanding of plasma. Results were achieved with different types of reactor, plasma and electrode geometry, substrate temperatures, flow conditions, and other plasma parameters, which may influence the kinetics and therefore the proposed theory and mechanism. Therefore, no universal and widely accepted kinetic model can be recommended. Moreover, it must be remembered the goal of thin polymer film deposition is the production of defined and regular structures without defects. The state-of-the art plasma polymerization is far from this goal. New ideas, new principal techniques must be developed.

11.9
Dependence on Plasma Parameter

Polymer deposition in fragmentation-dominated monomer plasma is determined by complex collision and radiation processes (Figure 11.12).

A considerable fraction of electron energies, especially those of the high energy tail of the electron energy distribution (Figure 11.13) exceeds significantly the energy needed for inelastic collisions such as dissociation, fragmentation, or cracking of aromatic rings (Figure 11.12) [121]. Moreover, continuous exposure of all plasma-deposited products to the energy-rich plasma-UV-irradiation additionally introduces defects and irregularities *ex post*.

The pressure (p) dependence of the deposition rate in an electrode-less capacitive coupled rf discharge has a p^2 dependence, which becomes saturated at higher pressure (Figure 11.14) [94, 113].

In contrast to Yasuda's interpretation it can be speculated that the deposition rate was only dependent on the characteristics of the vacuum pump, that is, pump speed and the related monomer flow. With sufficient power provided (nearly) 100% monomer can be converted into plasma polymer; if the power is low only a fraction of monomer molecules is converted into polymers, in particular at high pressure.

The absorbed energy per monomer molecule during the transition through the plasma zone determines the deposition rate. Yasuda proposed a linear dependence

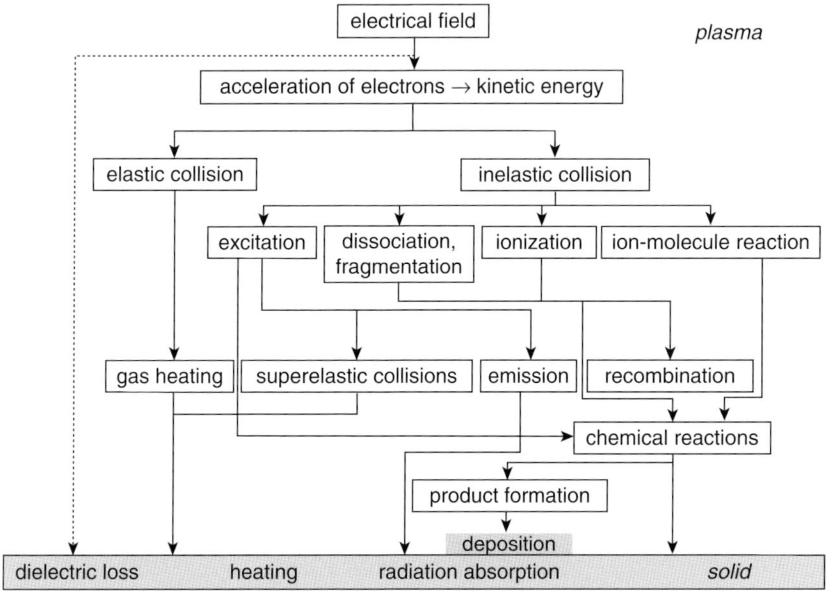

Figure 11.12 Energy distribution to collision and radiation processes during plasma polymerization [122].

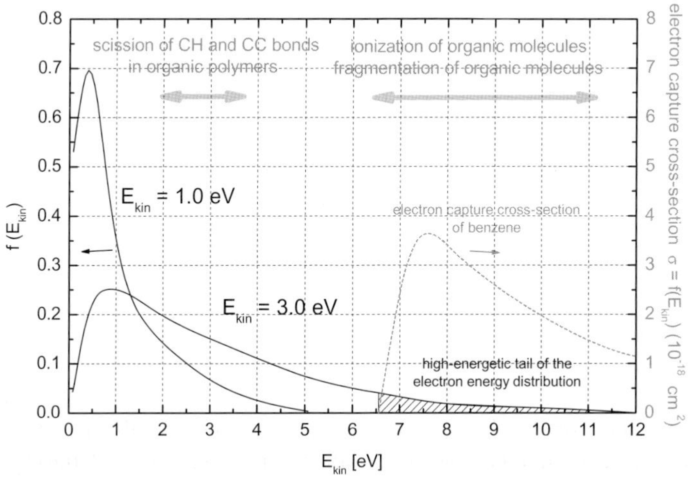

Figure 11.13 Electron energy distribution function and electron capture cross-section of benzene as well as the dissociation and ionization energies.

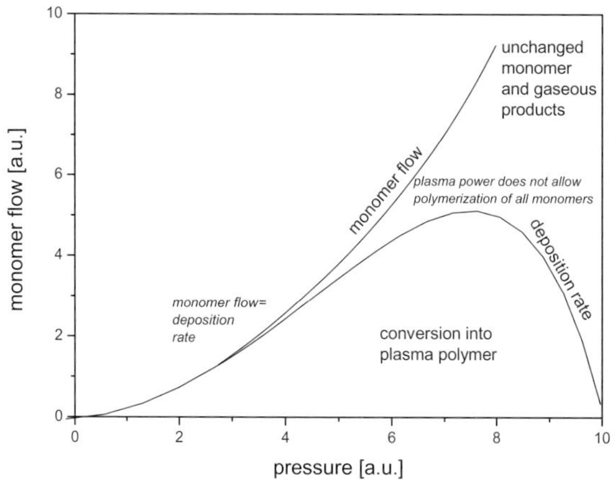

Figure 11.14 Pressure-dependence of deposition rates of organic monomers in low-pressure glow discharges.

of deposition rate (R_0) on monomer flow (F_w): $R_0 = kF_w$, where k is a monomer-specific rate constant [94]. Using low wattages, characterized by a low Yasuda factor ($Y = W/FM$), the linearity between deposition rate and flow rate is not realized [10]. Yasuda has presented the example of tetrafluoroethylene (TFE) polymerization and its dependence on plasma dose (Yasuda factor) [10]. Using a low dose, a low deposition rate and formation of Teflon-like material are also observed; using a high dose, carbon-like material and high deposition rates are found.

The type of monomer also greatly influences the deposition rate. In contrast to saturated olefin monomers, low wattage and high monomer flow (low Yasuda factor) also produce relatively high deposition rates if using chemically polymerizable vinyl, acrylic, diene, or aromatic monomers [26]. As mentioned before, Yasuda, Denaro, Bradley, Inagaki, and other authors have systematically tested many organic substances as monomers for plasma polymer deposition [14, 38, 94, 95, 113, 120].

Friedrich has investigated the dependence of plasma polymer deposition rates of a homologous series of monomers on molar mass of monomer, pressure, and wattage (Figure 11.15) [123].

A linear dependence of deposition rate on molar mass, pressure (over a large range), and degree of unsaturation (H/C ratio) was found. The effect of power input on the deposition rate was different for saturated and unsaturated monomers. Unsaturated precursors (classic monomers with low H/C ratio) show deposition rates independent of power or even decreasing with increasing power. Saturated (aliphatic) precursors need much more energy for deposition, which increases linearly with power.

Yasuda has introduced families of monomers and therefrom deposited plasma polymers following the involved concentration of trapped radical sites detected by

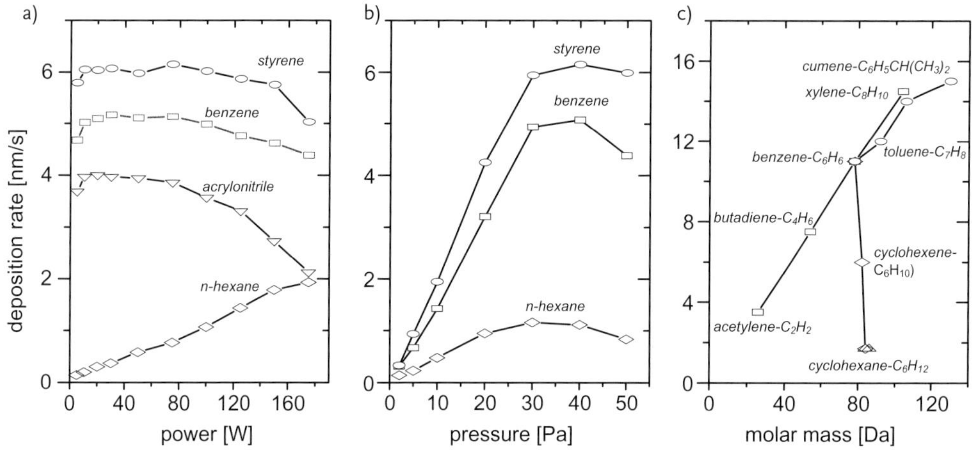

Figure 11.15 Variation of deposition rates of a homologous series of monomers with (a) wattage, (b) pressure, and (c) molar mass in the cw-rf low-pressure plasma.

electron spin resonance (ESR) [124]. The hydrogen release during plasma polymerization in the closed system (not pumping down of any products) was also considered as characteristic (measured by the increase of pressure during plasma exposure). Polymerization of monomers lowers the pressure while, on the other hand hydrogen, is removed from monomer or polymer by formation of C=C double bonds, radical formation, and crosslinking and increases the pressure [119, 122, 124]. On the basis of this hydrogen release Yasuda found some structural categories for the common behavior of monomer series [10, 124]. The significance of hydrogen release during plasma polymerization has been mentioned before ("quasi-hydrogen plasma").

11.10 Structure of Plasma Polymers

The composition and structure of plasma polymers were investigated by bulk and surface analytical methods. Bulk methods used were IR transmission spectroscopy [7, 40], dielectric relaxation spectroscopy [39, 97, 98, 125, 126], elemental composition [122, 127–132], electron spin resonance (ESR) [5, 133–136], thermoluminescence [137, 138], thermogravimetry [13, 139, 140], ^{1}H and broad-line NMR [41, 135]. All the results of bulk analysis reflect the total irregularity in structure and composition of plasma polymers independent of the mode of production, cw-plasma or pulsed plasma. The most revealing criterion for irregularity of plasma polymer structure and composition shows the non-existence of the ρ_{CH2} IR signal for ≤4CH_2 units of plasma-polymerized polyethylene at the wavenumber 730 cm^{-1}

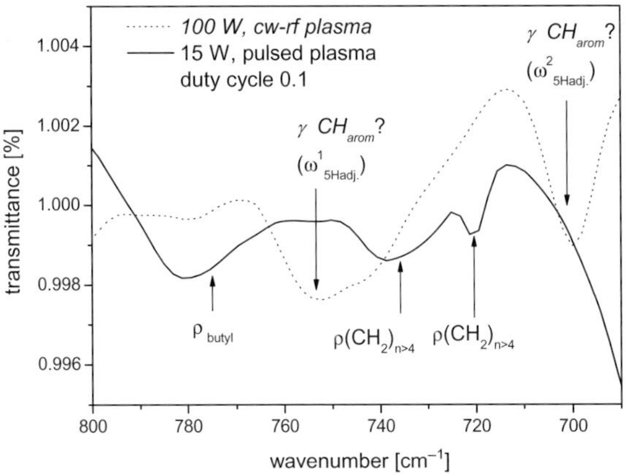

Figure 11.16 Indications of chemically polymerized (regularly structured) sequences in the $700\,cm^{-1}$ region for plasma polymerized ethylene given by the rocking vibration characteristic for $(CH_2)_n$ with $n \leq 4$.

(or a doublet at 720 and $731\,cm^{-1}$) [17]. This signal cannot be clearly identified (too weak or absent) for the 100 W cw plasma produced polyethylene (Figure 11.16); however, it may exist as a weak signal using the pulsed-plasma mode [128].

Aromatic rings should also be formed from ethylene, as observed by Stille, Hay, and Tibbitt [29, 39, 141] in contrast to Kronick *et al.*, who did not observe aromatic rings derived from ethylene plasma polymerization [7]. A similar problem is the retention of aromatic rings and their substitution, as also discussed above. Thus, Kolotyrkin and Friedrich have investigated the cw-plasma polymerization of toluene in detail [28, 40]. In the case of styrene, aromatic rings are also under-represented in plasma polymers in relation to the monomer composition. It can be speculated that the phenyl group is split off, as is known from oxygen plasma exposure of commercial polystyrene [142]. A different explanation is ring cracking. Moreover, the mono-substitution is weakened, as shown before and in the respective IR spectra (Figure 11.17).

A very high polydispersity (broadness of molar mass distribution) is characteristic for the (small) soluble fraction of plasma polymers, which can only be separated by chromatographic methods [22, 143]. As mentioned before, the high-energy UV irradiation of plasma on already deposited plasma polymer layers produces extraordinary random and irregular structures with crosslinking, many defects, and radicals [22]. Another source of aromatic rings within plasma polymer layers is embedding of benzene (styrene), biphenyl, terphenyl, and oligomeric phenylenes or phenylacetylenes [27, 144]. Other specifics of plasma polymer layers were discussed before, the content in non-aromatic double and triple bonds as well as the high concentration in CH_3 groups [7, 51–53– 92, 97]. The CH_3 (methyl) groups were produced by hydrogenation of broken chains:

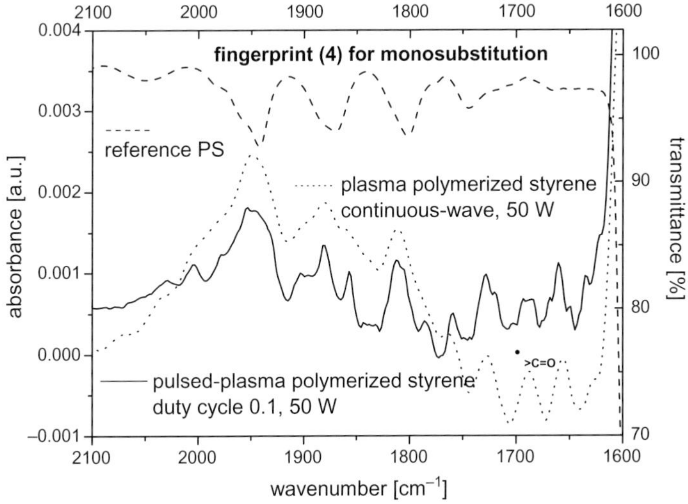

Figure 11.17 Comparison of commercial polystyrene with plasma polymerized in the IR region 1700–2000 cm^{-1}.

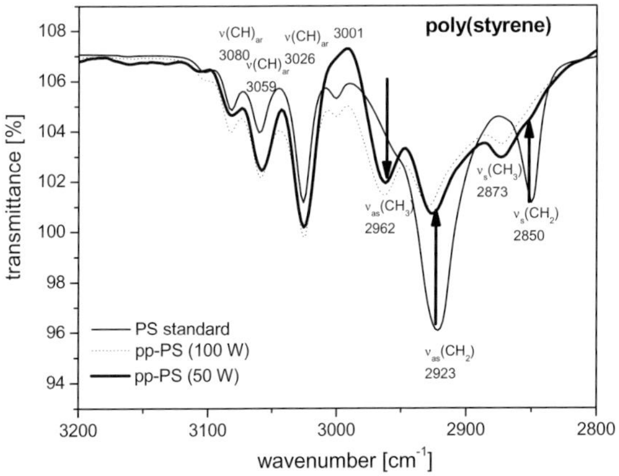

Figure 11.18 Comparison of IR spectra of pulsed-plasma PS (polystyrene) (cw-rf plasma, 50 W; 0.015 ms plasma on/0.085 ms plasma off; 4.5 Pa; 80 sccm) and commercial PS standard (150 000 g mol^{-1}) using grazing-incidence-reflectance FTIR ($\alpha = 70°$).

$$\sim\text{-CH}_2\text{-CH}_2\text{-}\sim + H_2 + \textit{plasma} \rightarrow \sim\text{-CH}_3 + H_3C\text{-}\sim$$

which can easily be demonstrated by the appearance of the $\nu_{as}CH_3$ vibration at 2962 cm^{-1} (Figure 11.18) [51].

Plasma polymers did not show a sharp melting point but, conversely, a broad weakening beginning at low temperatures (<100 °C) [13]. Sometimes, they become black during heating [92] or begin to smoke on exposure to air (J. Friedrich,

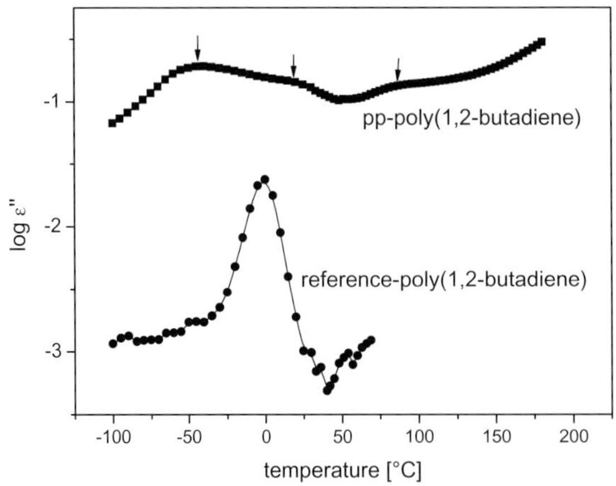

Figure 11.19 Dielectric relaxation (log ε''–dielectric loss) of pulsed-plasma polymerized and chemically polymerized poly(1,2-butadiene) versus temperature.

unpublished observation, 1976). The density differs from that of classic polymers [145]. Dielectric relaxation spectroscopy (DRS) detects all kinds of dipoles in the plasma polymer (functional groups, radicals, defects) and is an excellent indicator for the regularity of polymer structure and the occurrence of defects [146]. At first glance commercial reference-polybutadiene and pulsed-plasma polybutadiene show totally different behavior. The level of signals for the plasma polymer is about two orders of magnitude higher. It is must be conceded that the plasma deposits are not polymers (Figure 11.19).

11.11 Afterglow (Remote or Downstream) Plasmas

After switching-off the discharge a short afterglow is observable [147]. Long-living excited plasma species leave the plasma zone, convectively driven, and emit fluorescence and (forbidden) phosphorescence radiation also far from the plasma origin such as observed in nitrogen discharges [148]. Such metastable states are able to transport their energy far from the discharge and transfer it by inelastic collisions (Penning ionization), such as:

$$Ar^m \rightarrow Ar + h\nu$$

$$Ar^m + B \rightarrow Ar + B^+ + e^-$$

or:

$$Ar^m + C \rightarrow Ar + C^{\bullet +} + e^-$$

Table 11.2 Energies of lowest metastable excited states [149].

Species	Lowest metastable state (X^m) (eV)	Ionization potential (eV)
H_2	10.2	15.4
N_2	9.2	15.8
O_2	12.5	8.5
He	19.8	24.5
Ne	16.6	21.5
Ar	11.8	15.8

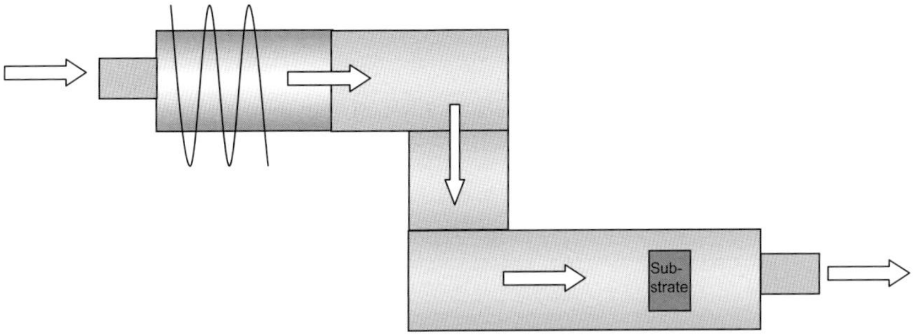

Figure 11.20 Schematics of afterglow arrangement.

The energies for the lowest triplet state given in Table 11.2 demonstrate the capability for initiating bond scissions in polymers and its activation [excitation–ionization energies of molecules (atoms)]. Therefore, the metastables are also able energetically to initiate degradation or fragmentation reactions [120].

The life-time of metastable intermediates is much higher than that of excited states (10^{-8} s) and amounts to 10^{-3}–1 s. Therefore, a convective transport to field- and energy-rich UV-free (resonance or fluorescence radiation) zones of the reactor is possible. This drift distance separates short-lived excited states, ions, and electrons from metastables. Moreover, optical decoupling is also possible by insertion of a kink in the tube-type reactor, thus avoiding exposure of the monomer and polymer to energy-rich UV irradiation in the afterglow zone. Figure 11.20 shows such apparatus schematically.

The removal of highly energetic species leads to the exclusive survival of metastable species. Their concentration decreases during their movement along the gas drift and, therefore, the reactivity in the afterglow zone also decreases with increasing distance from the plasma source. Superelastic collisions may sustain the plasma in the afterglow zone for awhile by releasing the energy stored in rovibronic degrees of freedom of the atoms and molecules that drift from the plasma. Rovibronic coupling denotes simultaneous interactions between

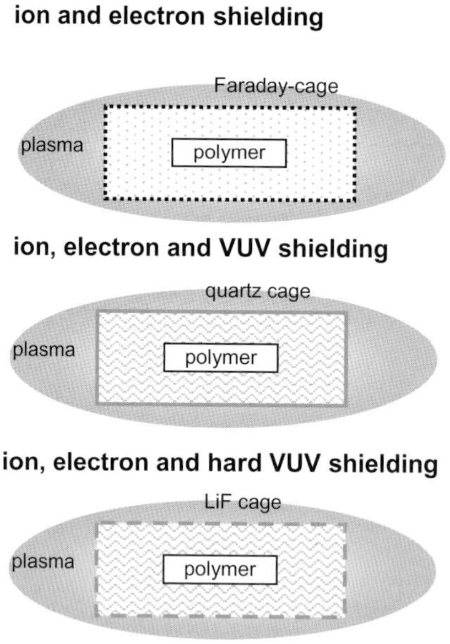

Figure 11.21 Variants of remote plasma: ion and electron shielding (Faraday cage), VUV shielding (quartz cage), and VUV ($\lambda < 120$ nm) shielding (LiF cage).

rotational, vibrational, and electronic degrees of freedom in a molecule. Here, the Franck–Condon principle is also to be regarded because of the coupled electronic and vibration excitation. Especially in molecular gases, the plasma chemistry in the afterglow is significantly different from that in the plasma glow. The profit in selectivity is low compared to the exposure of polymers within the plasma because of the high energy of metastables (>10 eV) compared to the much lower dissociation energies of C–C and C–H bonds in polymers (ca. 3.7 eV). Therefore, many endothermal reactions can also be initiated by metastables, leading also to irregularly structured reaction products. A plasma afterglow can be temporal, with an interrupted (pulsed) plasma source (cf. Section 12, pulsed plasma).

It is also possible to hold off charge carriers by using a Faraday cage (metal mesh). Plasma irradiation can be also limited and adjusted by use of quartz, MgF_2, or LiF-windows (Figure 11.21).

A few authors have used the remote plasma for deposition of plasma polymer layers, such as from hexamethyldisiloxane [150] or from C_4F_8 [151] or from triethylene glycol for application to the adsorption of proteins [152].

11.12 Powder Formation

Under certain plasma conditions polymer powder is formed in the plasma. Fragments and molecules condense to particles in the gas phase, grow, and are depos-

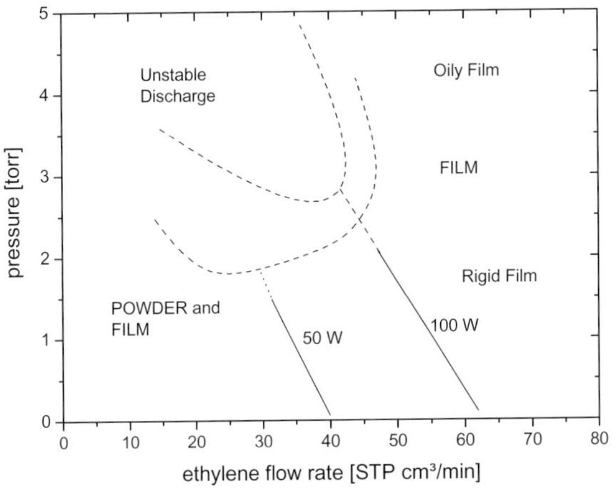

Figure 11.22 Influence of monomer pressure and flow on type of products (see Reference [84]).

ited as flaked plasma-polymer layer [11, 17, 84, 93, 153]. For applications of plasma polymers as barrier layer or in microelectronics such powder deposition is undesired. Kobayashi has created a "phase diagram" for the dependence of film and powder formation on pressure and monomer flow (Figure 11.22).

Low pressure and low monomer flow assist powder formation in the plasma, as does the distance between the electrodes. These conditions also enhance the formation of radicals and, thus, the chance of monomer grafting onto such seeds increases [93]. The powders were strongly crosslinked. Other authors also found powder formation at high pressure, which was more probable than low pressure [152, 154, 155]. Negative ions should also initiate polymerization to powders in the gas phase because negative ions can be fixed in the plasma boundary layer and thus they are able to form powder particles by ion–molecule reactions [156]. Aromatic rings containing monomers tend most strongly to powder formation [153, 157]. Such powder or dust particles in the plasma can also be used as plasma probe for characterizing plasma and plasma boundary layers [158].

11.13 Plasma Catalysis

Drost has distinguished between inner and exterior catalysis in plasma chemistry [86]. He considered Penning ionization by superelastic collisions as inner catalysis. Metastable excited noble gas atoms such as Ar* were produced by collisions with electrons: $Ar + e^- \rightarrow Ar^* + e^{-\prime}$, then Ar* impinges on monomer molecules and produces monomer ions: $Ar^* + M \rightarrow Ar + M^+ + e^-$ [86]. It is disputable as to whether such elementary processes may be assigned as inner catalysis.

Numerous activities are known to remove air pollutants using plasma and catalysis [159, 160].

Plasma catalysis is a process that delivers more products at same energy and material input. It is widely known in plasma physics that often undesired pollutants or additions (ca. 10^{-4}) can dramatically decrease ignition and burning voltages [161]. For example, mercury is successfully added for plasma chemical cracking of methane or for ammonia synthesis as sensitizer [162–164].

Such sensitizing or catalysis effects were also observed for plasma polymerization. Bromotrichloromethane addition to the propene plasma increased significantly the polymer deposition rate [84]. Bromotrichloromethane is a radical scavenger and potential chain transfer agent. The deposition rate of propylene was increased by multiples using such a sensitizer. The following reactions were held responsible for this effect: $BrCCl_3 + \textit{plasma} \rightarrow Br^{\bullet} + {}^{\bullet}CCl_3$ as the most probable dissociation pathway of the $BrCCl_3$ molecule. Atomic bromine removes a hydrogen atom from a polypropylene molecule: $Br^{\bullet} + CH_2{=}CH\text{-}CH_3 \rightarrow C_3H_5^{\bullet} + HBr$, and also the remaining CCl_3 radical: ${}^{\bullet}CCl_3 + CH_2{=}CH\text{-}CH_3 \rightarrow C_3H_5^{\bullet} + HCCl_3$. A few other reactions may also be of relevance: $C^{\bullet} + {}^{\bullet}Br \rightarrow C\text{-}Br$ and $C{=}C + {}^{\bullet}CCl_3 \rightarrow {}^{\bullet}C\text{-}C\text{-}CCl_3$ as well as $BrCCl_3 + Cl_3C\text{-}C\text{-}C^{\bullet} \rightarrow Cl_3C\text{-}C\text{-}C\text{-}Br + {}^{\bullet}CCl_3$.

Similar effects provoked by addition of halogen-containing substances were also reported, such as by Kobayashi. He used dichlorodifluoromethane as sensitizer [165].

11.14 Copolymerization in Continuous-Wave Plasma Mode

Copolymerization of two organic or one organic + one inorganic or two inorganic substances in the continuous-wave plasma mode can be easily performed if the substances or “co-monomers” or precursors are strongly fragmented and put together again by (random) polyrecombination to give a motley collection of molecules. Such process has no similarity to chemical (radical) copolymerization. The thus produced “copolymers” or “pseudo-copolymers” are a random conglomerate of all types of fragments and atoms in close accordance to cw plasma polymerized (homo)polymers. Nevertheless, the cw plasma technique allows is to combine all types of fragments to form a new copolymer-like network.

The most exotic example, mentioned in Chapter 5, concerns the production of amino acids from inorganic gases in the imitated primitive Earth atmosphere using a spark plasma [72, 73].

In this way non-polymerizable gases, such as acetylene + CO + H_2O, acetylene + H_2 + N_2, and acetylene + H_2O, can be mixed in any ratio and can be form a random structured “pseudo-copolymer” [166, 167]. “Pseudo” was introduced by the author to distinguish the product from chemically copolymerized copolymers. The structure and composition of these deposited pseudo-copolymer layers have no similarity to those of classic copolymers. The elemental composition reflects

Table 11.3 Elemental composition (number of atoms per 100 C) of plasma polymers and copolymers produced under cw-plasma conditions.

Monomer	Polymer composition				Monomer composition	
	C	H	O	N	C	H
Styrene (low wattage)	100	90–100	1–3		100	100
Styrene (high wattage)	100	10	1–6		100	100
Benzene	100	90–105	10–15		100	100
Ethylbenzene	100	100–125	10–15		100	125
Cyclohexane	100	110–125	7–10		100	200
Cyclopentane	100	100–138	7		100	200
Benzene + H_2 (1 : 3)	100	130–170	14–20		100	200
Benzene + N_2 (1 : 3)	100	108	18	10	100	100

the exotic composition and, therefore, anarchic structure of deposited layers (Table 11.3) [114].

A few more similarities with classic copolymerizations were found for the copolymerization of acrylic acid and octadiene [168]. However, a classic copolymerization using octadiene and the easily copolymerizable acrylic acid may occur in the polymer-chemical sense also under cw-plasma and low-pressure conditions. The chemical character of copolymerization and the type of resulting copolymer or homopolymer depend on the "copolymerization parameters" as explained in more detail in the next section. It can be argued that most plasma-chemical copolymerizations are only continuations of fragmentation–polyrecombination polymerization and not copolymerizations in the chemical sense. Chemical copolymer formation may play a role in plasma copolymerization of vinyl acetate and styrene [169]; however, it plays a more important role under the conditions of pulsed-plasma copolymerization [170, 171].

Plasma copolymerization in the bulk was also presented; however, since this polymerization method leads to products similar to conventional polymers it is only noted here [172]. It does not include the typical "atomic" or fragmentation–polyrecombination polymerization normally encountered in gas-phase, radiofrequency or microwave, glow discharge equipment.

Several irregular copolymerizations in the chemical sense are presented in the literature. Hexamethyldisiloxane (HMDSO) and methyl methacrylate [169] as well as tetrafluoroethylene were plasma "copolymerized" [173]. Another attempt was made by Inagaki, who has "copolymerized" acrylic acid and CO_2 to maximize the number of resulting COOH groups for modifying deposits of membranes [174]. In addition, Munro *et al.* have copolymerized perfluorobenzene and tetramethyltin [175].

The plasma copolymerization of ethylene and tetrafluoroethylene (TFE) was performed by Kobayashi [165] and investigated in more detail by Golub *et al.* [176].

Plasma-produced copolymers with acetylene for optical applications have been described by Yasuda *et al.* [177, 178] and also by Chinese researchers [179].

Fixation of carbon dioxide by plasma copolymerization with ethylene was also reported [180]. CO_2 as "co-monomer" in plasma copolymer formation was also used to enhance the production of COOH groups in the deposited polymer layer of acrylic acid [174].

References

1 Bondt, N., Deimann, J.R., Paets van Trostwijk, A., and Lauwerenburg, A. cited in (1796) *J. Fourcroy Ann. Chem.*, **21**, 58.

2 (a) Berthelot, M. (1863) *Ann. Chim. Phys.*, **6/7**, 53; (b) Berthelot, M. (1866) *Ann. Chim. Phys.*, **9**, 413; (c) Berthelot, M. (1867) *Ann. Chim. Phys.*, **12**, 5; (d) Berthelot, M. (1869) *Compt. Red.*, **67**, 1141; (e) Berthelot, M. (1876) *Compt. Red.*, **62**, 1283.

3 de Wilde, P. (1874) *Ber. D Chem. Ges.*, **7**, 352.

4 Thenard, P. and Thenard, A. (1874) *Compt. Rend.*, **78**, 219.

5 Linder, E. and Davis, A. (1931) *J. Phys. Chem.*, **35**, 3649.

6 (a) Schüler, H. and Reinebeck, L. (1951) *Z. Naturforsch., Teil A*, **6**, 271; (b) Schüler, H. and Reinebeck, L. (1952) *Z. Naturforsch.*, **7a**, 285; (c) Schüler, H. and Reinebeck, L. (1954) *Z. Naturforsch.*, **9a**, 350; (d) Schüler, H. and Stockburger, M. (1959) *Z. Naturforsch.*, **14a**, 981; (e) Schüler, H., Prchal, K., and Kloppenburg, E., (1960) *Z. Naturforsch.*, **15a**, 308; (f) König, H. and Hellwig, G., *Physik, Z.* (1951) **129**, 491.

7 (a) Jesch, K., Bloor, J.E., and Kronick, P.L. (1966) *J. Polym. Sci. A*, **1**, 1487; (b) Kronick, P.L., Jesch, K., and Bloor, J.E. (1969) *J. Polym. Sci., A*, **1** (**7**), 767.

8 Vastola, F.J. and Wightman, J.P. (1964) *J. Appl. Chem.*, **14**, 69.

9 Neiswender, D.D. (1969) *Adv. Chem. Ser.*, **80**, 338.

10 Yasuda, H. and Hirotsu, T. (1973) *J. Polym. Sci., Polym. Chem. Ed.*, **16**, 743.

11 Westwood, A.R. (1971) *Eur. Polym. J.*, **7**, 363.

12 Poll, H.-U. (1970) *Z. Angew. Physik*, **4**, 260.

13 Thompson, L.F. and Mayhan, K.G. (1972) *J. Appl. Polym. Sci.*, **16**, 2291 and 2317.

14 (a) Denaro, A.R., Owens, P.A., and Crawshaw, A. (1968) *Eur. Polym. J.*, **4**, 93; (b) Denaro, A.R., Owens, P.A., and Crawshaw, A. (1969) *Eur. Polym. J.*, **5**, 471.

15 Baddour, R.F. and Timmins, R.S. (1967) *The Application of Plasmas to Chemical Processing*, Pergamon, Oxford.

16 Vasile, M.J. and Smolinsky, G. (1972) *J. Electrochem. Soc.*, **119**, 451.

17 (a) Kobayashi, H., Bell, A.T., and Shen, M. (1973) *J. Appl. Polym. Sci.*, **17**, 885; (b) Kobayashi, H., Bell, A.T., and Shen, M. (1973) *Macromolecules*, **7**, 227.

18 Tibbitt, J.M., Jensen, R., Bell, A.T., and Shen, M. (1977) *Macromolecules*, **10**, 647.

19 Tibbitt, J.M., Shen, M., and Bell, A.T. (1976) *J. Macromol. Sci. Chem., A*, **10**, 1623.

20 (a) Yasuda, H. (1981) *J. Polym. Sci. Macromol. Rev.*, **16**, 199; (b) Yasuda, H. (1985) *Plasma Polymerization*, Academic Press, Orlando.

21 Hansen, R.H. and Schonhorn, H. (1966) *J. Polym. Sci., B*, **4**, 203.

22 Hudis, M. (1972) *J. Appl. Polym. Sci.*, **16**, 2397.

23 Dinan, F.J., Fridman, S., and Schirrmann, P.J. (1969) *Adv. Chem. Ser.*, **80**, 289.

24 Suhr, H. (1972) *Angew. Chem. Int. Ed. Engl.*, **11**, 781.

25 Rosskamp, G. (1972) Untersuchungen über Umlagerungen, Eliminierungen und Polymerisationen im Plasma einer Glimmentladung – ein beitrag zum Cyclobutadienproblem. PhD thesis, University of Tübingen, Germany.
26 Friedrich, J., Gähde, J., Frommelt, H., and Wittrich, H. (1976) *Faserforsch. Textiltechn./Z. Polymerenforsch.*, **27**, 517.
27 Stille, J.K., Sung, R.L., and van der Kooi, J. (1965) *J. Org. Chem.*, **30**, 3116.
28 Tuzov, L.S., Gilman, A.B., Schtschurorov, A.N., and Kolotyrkin, V.M. (1967) *Vysokomol. Soedin.*, **11**, 2414.
29 Hay, P.M. (1969) *Adv. Chem. Ser.*, **30**, 350.
30 Gilman, A.B., Kolotyrkin, V.M., and Tunizki, N.N. (1970) *Kinet. Katal.*, **10**, 1267.
31 Harkin, W.D. and Gans, D.M. (1930) *J. Am. Chem. Soc.*, **52**, 5156.
32 Möllenstedt, G. and Speidel, R. (1961) *Z. Angew. Phys.*, **13**, 231.
33 Bradley, A. and Hammes, J.P. (1963) *J. Electrochem. Soc.*, **110**, 15.
34 Williams, D.T. and Hayes, M.W. (1966) *Nature*, **209**, 769.
35 Kolotyrkin, V.M., Gilman, A.B., and Tsapuk, A.K. (1967) *Uspechii Chim.*, **8**, 1380.
36 d'Agostino, R., Favia, P., Förch, R., Oehr, C., and Wertheimer, M.R. (2010) *Plasma Proc. Polym.*, **7**, 363.
37 Hegemann, D., Steele, D.A., and Short, R.D. (2010) *Plasma Proc. Polym.*, **7**, 365.
38 Bradley, A. (1979) *Ind. Eng. Chem., Prod. Res. Dev.*, **9**, 101.
39 Tibbitt, J.M., Bell, A.T., and Shen, M. (1976) *J. Macromol. Sci. Chem., A*, **10**, 519–533.
40 Friedrich, J., Wittrich, H., and Gähde, J. (1978) *Faserforsch. Textiltechn./Z. Polymerenforsch.*, **29**, 481.
41 Kaplan, S. and Dilks, A. (1983) *J. Polym. Sci., Polym. Chem. Ed.*, **21**, 1919.
42 Meisel, J. and Tiller, H.-J. (1972) *Z. Chem.*, **7**, 275.
43 Yasuda, H.K. and Hsu, T. (1977) *J. Polym. Sci., Polym. Chem. Ed.*, **15**, 81.
44 Yasuda, H.K. and Hsu, T. (1976) *J. Appl. Polym. Sci.*, **20**, 1769.
45 Nakajima, K., Bell, A.T., Shen, M., and Millard, M.M. (1979) *J. Appl. Polym. Sci.*, **23**, 2627.
46 Vinzant, J.W., Shen, M., and Bell, A.T. (1978) *ACS Polym. Prepr.*, **19**, 453.
47 Savage, C.R. and Timmons, R.B. (1991) *Polym. Mater. Sci. Eng.*, **64**, 95.
48 Savage, C.R., Timmons, R.B., and Lin, J.W. (1991) *Chem. Mater.*, **3**, 575.
49 Calderon, J.G. and Timmons, R.B. (1998) *Macromolecules*, **31**, 3216.
50 Kühn, G., Weidner, St., Decker, R., Ghode, A., and Friedrich, J. (1999) *Surf. Coat. Technol.*, **116–119**, 748.
51 Friedrich, J., Retzko, I., Kühn, G., Unger, W., and Lippitz, A. (2001) in *Metallized Plastics 7: Fundamentals and Applied Aspects* (ed. K.L. Mittal), VSP, Utrecht, p. 117.
52 Retzko, I., Friedrich, J.F., Lippitz, A., and Unger, W.E.S. (2001) *J. Electron. Spectrosc. Relat. Phenom.*, **121**, 111.
53 Kühn, G., Retzko, I., Lippitz, A., Unger, W., and Friedrich, J. (2001) *Surf. Coat. Technol.*, **142–144**, 494.
54 Friedrich, J., Gähde, J., Frommelt, H., and Wittrich, H. (1976) *Faserforsch. Textiltechn./Z. Polymerenforsch.*, **27**, 604–608.
55 Friedrich, J. (1980) *Chem. Technol.*, **32**, 393–403.
56 Friedrich, J., Loeschcke, I., and Gähde, J. (1986) *Acta Polym.*, **37**, 687–695.
57 Friedrich, J.F., Mix, R., Schulze, R.-D., Meyer-Plath, A., Joshi, R., and Wettmarshausen, S. (2008) *Plasma Proc. Polym.*, **5**, 407–423.
58 Goodman, J. (1960) *J. Polym. Sci.*, **44**, 551.
59 Bradley, A. and Fales, J.D. (1971) *Chem. Technol.*, **4**, 232.
60 Licari, J.J. (1970) *Plastic Coatings for Electronics*, McGraw-Hill Book Co., New York.
61 Gregor, L.V. (1968) *IBM J. Res. Dev.*, **12**, 140.
62 Lawton, E.L. (1974) *J. Appl. Polym. Sci.*, **18**, 1557.
63 Yasuda, H.K. (1984) *J. Membr. Sci.*, **18**, 273–284.
64 Inagaki, N., Kobayashi, N., and Matsushima, M. (1988) *J. Membr. Sci.*, **38**, 85–95.
65 Agres, L., Segui, Y., del Sol, R., and Raynaud, P. (1996) *J. Appl. Polym. Sci.*, **61**, 2015–2022.

66 Fracassi, F., d'Agostino, R., Favia, P., and van Sambeck, M. (1993) *Plasma Sources Sci. Technol.*, **2**, 106.
67 Stein, H., Zehender, E., and Blaich, B. DE pat. 2 625 448 (1976).
68 (a)Dittmer, G. (1978) *Ind. Res. Devel. Sept.*, 169–183; (b) Grünwald, H., Jung, M., Kukla, R., Adam, R., and Krempel-Hesse, J. (1998) in *Metallized Plastics 5-6: Fundamental and Applied Aspects* (ed. K.L. Mittal), VSP, Utrecht.
69 Hess, D.W. and Graves, D.B. (1993) in *Chemical Vapour Deposition: Principles and Applications* (eds M.L. Hitchmann and K.E. Jensen), Academic Press, London, pp. 385–431.
70 Massines, F., Rabehi, A., Decomps, P., Gadri, R.B., Segui, P., and Mayoux, C. (1998) *J. Appl. Phys.*, **83**, 2950–2957.
71 Mishra, K.K., Khardekar, R.K., Singh, R., and Pant, H.C. (2002) *Rev. Sci. Instrum.*, **73**, 3251–3257.
72 Miller, S.L. (1953) A production of amino acids under possible primitive Earth conditions. *Science*, **117**, 528.
73 Miller, S.L. and Urey, H.C. (1959) Organic compound synthesis on the primitive Earth. *Science*, **130**, 245.
74 (a) Korschak, V.V. (1963) *Dokl. Akad. Nauk*, **151**, 1332; (b) Tuzov, L.S., Gilman, A.B., Schtschurorov, A.N., and Kolotyrkin, V.M. (1967) *Vysokomol. Soedin.* **11**, 2414.
75 Kuprianov, S.E. (1965) *J. Eksp. Teor. Fisikii*, **48**, 468.
76 Elias, H.-G. (1997) *An Introduction to Polymer Science*, Wiley-VCH Verlag GmbH, Weinheim.
77 Stiller, W. and Friedrich, J. (1981) *Z. Chem.*, **21**, 91–118.
78 Houwink, R. and Staverman, A.J. (1962) *Chemie und Technologie der Kunststoffe*, Akademische Verlagsgesellschaft Geest&Portig, Leipzig.
79 Stein, D.J. and Mosthaf, H. (1968) *Angew. Makromol. Chem.*, **2**, 39–50.
80 Wedenejew, W.J., Gurwitsch, L.W., Kondratjew, W.H., Medwedew, W.A., and Frankewitsch, E.L. (1971) *Energien Chemischer Bindungen, Ionisationspotentiale Und Elektronenaffinitäten*, VEB Deutscher Verlag für Grundstoffindustrie, Leipzig.
81 Fanghänel, E. (ed.) (2004) *Organikum*, Wiley-VCH Verlag GmbH, Weinheim.
82 Brandrup, J. and Immergut, H.E. (2003) *Polymer Handbook*, 4th edn, S. II, New York: Wiley.
83 Friedrich, J., Kühn, G., and Gähde, J. (1979) *Acta Polym.*, **30**, 470–477.
84 Sharma, A.K., Millich, F., and Hellmuth, E.W. (1979) in *Plasma Polymerization* (eds M. Shen and A.T. Bell), ACS Symposium Series, 108, American Chemical Society, Washington, p. 53.
85 Friedrich, J., Kühn, G., and Mix, R. (2006) *Prog. Colloid Polym. Sci.*, **132**, 62–71.
86 Drost, H. (1978) *Plasmachemie*, Akademie-Verlag, Berlin.
87 Yasuda, H.K. (1981) *J. Polym. Sci. Macromol. Rev.*, **16**, 119.
88 Williams, T. and Hayes, M.W. (1966) *Nature*, **209**, 769.
89 Haller, I. and White, P. (1967) *J. Phys. Chem.*, **9**, 1784.
90 Mearns, A.M. (1969) *Thin Solid Films*, **3**, 201.
91 Hollahan, J.R. and McKeever, R.P. (1969) *Adv. Chem. Ser.*, **80**, 272.
92 Hollahan, J.R. (1972) *Makromol. Chem.*, **154**, 303.
93 Tkachuk, B.V., Bushin, V.V., Kolotyrkin, V.M., and Smetankina, N.P. (1967) *Vysokomol. Soedin. A*, **9**, 2018.
94 Yasuda, H. and Lamaze, C.E. (1973) *J. Appl. Polym. Sci.*, **17**, 1519.
95 Yasuda, H. and Lamaze, C.E. (1973) *J. Appl. Polym. Sci.*, **17**, 1533.
96 Simionescu, Cr., Asandei, N., Dénes, F., Sandulovici, M., and Popa, Gh. (1969) *Eur. Polym. J.*, **5**, 427.
97 Friedrich, J., Kühn, G., Mix, R., Fritz, A., and Schönhals, A. (2003) *J. Adhesion Sci. Technol.*, **17**, 1591.
98 Friedrich, J., Kühn, G., Mix, R., Retzko, I., Gerstung, V., Weidner, St., Schulze, R.-D., and Unger, W. (2003) in *Polyimides and Other High Temperature Polymers: Synthesis, Characterization and Applications* (ed. K.L. Mittal), VSP, Utrecht, pp. 359–388.
99 Hinze, D. and Poll, H.-U. (1974) *Plaste Kautsch.*, **3**, 194.
100 Kryszewski, M. (1973) *Plaste Kautsch.*, **20**, 885.

101 d'Agostino, R., Cramarossa, F., Fracassi, F., and Iluzzi, F. (1990) in *Plasma Deposition, Treatment and Etching of Polymers* (ed. R. d'Agostino), Academic Press, New York.
102 Limb, S.J., Lau, K.K.S., Edell, D.J., Gleason, E.F., and Gleason, K.K. (1999) *Plasma Polym.*, **4**, 21.
103 Wettmarshausen, S., Min, H., Unger, W., Jäger, C., Hidde, G., and Friedrich, J. (2011) *Plasma Chem. Plasma Proc.*, 551–572.
104 Poll, H.-U. (1970) *Z. Angew. Phys.*, 4, 260.
105 Carchano, H. (1974) *J. Chem. Phys.*, **61**, 3634.
106 Fenn, J.B. (2003) *Angew. Chem. Int. Ed.*, **42**, 3871.
107 Friedrich, J., Mix, R., and Kühn, G. (2005) in *Plasma Processing and Polymers* (eds R. d'Agostino, P. Favia, C. Oehr, and M.R. Wertheimer), Wiley-VCH Verlag GmbH, Weinheim, pp. 3–21.
108 Kassel, L.S. (1932) *J. Am. Chem. Soc.*, **54**, 3949.
109 Bell, A.T. (1976) *J. Macromol. Sci.-Chem., A*, **10**, 369.
110 Wettmarshausen, S., Mittmann, H.-U., Kühn, G., Hidde, G., and Friedrich, J.F. (2007) *Plasma Proc. Polym.*, **4**, 832–839.
111 Friedrich, J., Wettmarshausen, S., and Hennecke, M. (2009) *Surf. Coat. Technol.*, **203**, 3647–3655.
112 Tiller, H.-J., Pelzl, G., and Dumke, K. (1972) in *Physik und Technik Des Plasmas*, Bd. III, Deutsche Physikalische Gesellschaft, Vortragsband Suhl, 176 pp.
113 Yasuda, H. and Lamaze, C.E. (1971) *J. Appl. Polym. Sci.*, **15**, 2277.
114 Friedrich, J., Gähde, J., Frommelt, H., and Wittrich, H. (1976) *Faserforsch. Textiltechn./Z. Polymer.*, **27**, 599–603.
115 Tiller, H.-J., Wagner, U., Fink, P., and Meyer, K. (1977) *Plaste Kautsch.*, **24**, 619.
116 Morita, S., Sawa, G., and Ieda, M. (1976) *J. Macromol. Sci.-Chem.*, **10**, 501.
117 Shen, M. and Bell, A.T. (1979) *Plasma Polymer. ACS Symp. Ser.*, **108**, 1.
118 Lam, D.K., Baddour, R.F., and Stancell, A.F. (1976) *J. Macromol. Sci.-Chem.*, **10**, 421.
119 Friedrich, J. (1981) *Contr. Plasma Phys.*, **21**, 261.
120 Inagaki, N. (1996) *Plasma Surface Modification and Plasma Polymerization*, Technomic, Lancaster, PA.
121 Hertz, G. and Rompe, R. (1973) *Plasmaphysik*, Akademie-Verlag, Berlin.
122 Friedrich, J. (1974) Untersuchungen zur Modifizierung von Feststoffoberflächen im Nichtisothermen Plasma einer Hochfrequenzentladung. PhD thesis, Berlin.
123 Friedrich, J., Gähde, J., Frommelt, H., and Wittrich, H. (1976) *Faserforsch. Textiltechn./Z. Polymer.*, **27**, 517–522.
124 Yasuda, H. (1976) *J. Macromol. Sci. Chem., A*, **10**, 383.
125 Chowdhury, F.-U.-Z. and Bhuiyan, A.H. (2000) *Thin Solid Films*, **370**, 78.
126 Stundzia, V., Biederman, H., Slavınska, D., Nedbal, J., Hlidek, P., Poskus, A., Mackus, P.K., and Howson, R.P. (2000) *J. Phys. D Appl. Phys.*, **33**, 719–724.
127 Bradley, A. and Hammes, J. (1963) *J. Electrochem. Soc.*, **110**, 59.
128 Friedrich, J., Mix, R., Kühn, G., Retzko, I., Schönhals, A., and Unger, W. (2003) *Composite Interface*, **10**, 173–223.
129 Yasuda, H.K. (1979) in *Plasma Polymerization* (eds M. Shen and A.T. Bell), ACS Symposium Series, vol. 108. American Chemical Society, Washington D.C., p. 37.
130 Yasuda, H., Lamaze, C.E., and Sakaokou, K. (1973) *J. Appl. Polym. Sci.*, **17**, 137.
131 Yasuda, H. and Lamaze, C.E. (1973) *J. Appl. Polym. Sci.*, **17**, 201.
132 Stuart, M. (1963) *Nature*, **6**, 59.
133 Kuzuya, M., Noguchi, A., Ito, H., Kondo, S., and Noda, N. (1991) *J. Polym. Sci., Part A: Polym. Chem.*, **29**, 1.
134 Kuzuya, M., Kondo, S., Sugito, M., and Yamashiro, T. (1998) *Macromolecules*, **31**, 3230.
135 Dilks, A., Kaplan, S., and van Laeken, A. (1981) *J. Polym. Sci., Polym. Chem. Ed.*, **19**, 2987.
136 Haupt, M., Barz, J., and Oehr, C. (2008) *Plasma Proc. Polym.*, **5**, 33.
137 Krüger, S., Schulze, R.-D., Brademann-Jock, K., and Friedrich, J. (2006) *Surf. Coat. Technol.*, **201**, 543–552.

138 Krüger, S., Schulze, R.-D., Brademann-Jock, K., Swaraj, S., and Friedrich, J. (2006) *Vakuum Forschung Praxis*, **18**, 32–37.

139 Wrobel, A.M., Kowalski, J., Grebowicz, J., and Kryszewski, M. (1982) *J. Macromol. Sci. A*, **17**, 433–452.

140 Li, Y., Liu, L., and Fang, Y. (2003) *Surf. Coat.*, **52**, 285–290.

141 Swift, F., Sung, R.L., Doyle, J., and Stille, J.K. (1965) *J. Org. Chem.*, **30**, 3114.

142 Friedrich, J.F., Unger, W.E.S., Lippitz, A., Giebler, R., Koprinarov, I., Weidner, St., and Kühn, G. (2000) in *Polymer Surface Modification: Relevance to Adhesion*, vol. 2 (ed. K.L. Mittal), VSP, Utrecht, pp. 137–172.

143 Mix, R., Gerstung, V., Falkenhagen, J., and Friedrich, J. (2007) *J. Adhesion Sci. Technol.*, **21**, 487–507.

144 Streitwieser, A. and Ward, H.R. (1963) *J. Am. Chem. Soc.*, **85**, 539.

145 Knickmeyer, W.W., Peace, B.W., and Mayhan, K.G. (1974) *J. Appl. Polym. Sci.*, **18**, 301.

146 Kremers, F. and Schönhals, A. (2007) *Broadband Dielectric Spectroscopy*, Springer, Berlin.

147 Herzberg, G. (1928) *Z. Physik A*, **46**, 878–895.

148 Lewis, P. (1900) *Annal. Phys.*, **307**, 459–468.

149 Brockhaus Encyclopedia (1968) *Atom-Struktur Der Materie*, VEB Bibliographisches Institut, Leipzig.

150 Lee, S.H. and Lee, D.C. (1998) *Thin Solid Films*, **325**, 838.

151 Ningel, K.P., Theirich, D., and Engemann, J. (1998) *Surf. Coat. Technol.*, **98**, 1142.

152 Beyer, D., Knoll, W., Ringsdorf, H., Wang, J.H., Timmons, R.-B., and Sluka, P. (1998) *J. Biomed. Mater. Res.*, **36**, 181.

153 Liepins, R. and Sakaoku, K. (1972) *J. Appl. Polym. Sci.*, **16**, 2633.

154 Sandrina, L., Silverstein, M.S., and Sacher, E. (2001) *Polymer*, **42**, 3761.

155 Schultrich, B. (2007) *Vakuum*, **19**, 32.

156 Hollenstein, Ch., Dorier, J.-L., Dutta, J., Sansonnens, L., and Howling, A.A. (1994) *Plasma Sources Sci. Technol.*, **3**, 278.

157 (a) Brown, K.C. (1971) *Eur. Polym. J.*, **7**, 363; (b) Hays, P.M. (1969) *Adv. Chem. Ser.* **80**, 350.

158 Kersten, H., Deutsch, H., Otte, M., Swinkels, G.H.P.M., and Kroesen, G.M.W. (2000) *Thin Solid Films*, **377–378**, 530.

159 Kizling, M.B. and Jaras, S.G. (1996) *Appl. Catal. A Gen.*, **147**, 1.

160 Dahiya, R.P., Mishra, S.K., and Veefkind, A. (1993) *IEEE Trans. Plasma Sci.*, **21**, 346.

161 Rutscher, A. and Deutsch, H. (1983) *Wissensspeicher Plasmaphysik*, VEB Fachbuchverlag, Leipzig.

162 Kondratjew, W.N. and Nikitin, J.J. (1970), *Chemie Und Physik Des Niedertemperaturplasmas*, Verlag der Moskauer Universität, 41 pp. S. 41.

163 Filipov, J.W., Lebedev, W.P., Salaman, W.W., and Kobosev, N.I. (1950) *Z. Phys. Chem., UdSSR*, **24**, 1009.

164 Polak, L.S. (1966) *Pure Appl. Chem.*, **13**, 345.

165 Kobayashi, H., Shen, M., and Bell, A.T. (1974) *J., Macromol. Sci.-Chem.*, **A8**, 1345.

166 Yasuda, H. and Marsh, H.C. (1975) *J. Appl. Polym. Sci.*, **19**, 2881.

167 Yasuda, H., Marsh, H.C., Bumgarner, M.O., and Morosoff, N. (1975) *J. Appl. Polym. Sci.*, **19**, 2845.

168 Alexander, M.R. and Duc, T.M. (1999) *Polymer*, **40**, 5479–5488.

169 Urrutia, M.S., Schreiber, H.P., and Wertheimer, M.R. (1988) *J. Appl. Polym. Sci., Appl. Polym. Symp.*, **42**, 305.

170 Friedrich, J., Mix, R., and Kühn, G. (2005) in *Plasma Processing and Polymers* (eds R. d'Agostino, P. Favia, C. Oehr, and M.R. Wertheimer), Wiley-VCH Verlag GmbH, Weinheim, pp. 3–12.

171 Friedrich, J., Mix, R., Schulze, R.-D., and Kühn, G. (2005) in *Adhesion* (ed. W. Possart), Wiley-VCH Verlag GmbH, Weinheim, pp. 265–288.

172 Simionescu, C.I. and Simionescu, B.C. (1991) *Polym. Prepr.*, **32**, 434.

173 Sakata, J., Yamamoto, M., and Tajima, I. (1988) *J. Polym. Sci., Part A: Polym. Chem.*, **26**, 1721.

174 Inagaki, N. and Matsunga, M. (1985) *Polym. Bull.*, **13**, 349.

175 Munro, H.S. and Till, C. (1985) *Thin Solid Films*, **131**, 255–260.

176 Golub, M.A., Wydeven, T., and Cormia, R.D. (1992) *J. Polym. Sci., Part A: Polym. Chem.*, **30**, 2683–2692.

177 Yasuda, H., Bumgarner, H.O., Marsh, H.C., Devito, D.P., Wolbarsht, M.L., Reed, J.W., Bessler, M., Landers, M.B., Hercules, D.M., and Carver, J. (1975) *J. Biomed. Mater. Res*, **9**, 629–643.

178 Yasuda, H., Marsh, H.C., and Tsai, J. (1978) *J. Appl. Polym. Sci.*, **19**, 456.

179 Yeping, L., Yue-E, F., Rong, F., and Jinyun, X. (2001) *Radiat. Phys. Chem.*, **60**, 637–642.

180 Terajima, T. and Koinuma, H. (2004) *Macromol. Rapid Commun.*, **25**, 312–314.

12
Pulsed-Plasma Polymerization

12.1
Introduction

As mentioned several times before, plasma polymerization under continuous-wave conditions produces the nearly complete fragmentation of monomer molecules during exposure to the plasma. The random polyrecombination of fragments and atoms renders the chemical composition and structure of plasma polymers completely irregular. It was noted that the structure of monomers is for the most part far from that of classic polymers but the retention of functional groups, introduced with the monomer molecules, is much higher. To overcome this general disadvantage of plasma polymers attention was directed towards chemical processes to improve substantially the retention of functional groups independent of the overall structure of the plasma polymer. The polymer should be formed in a predominantly chemical way, which promises more defined polymerization, regularity in structure and composition, and a high degree of functional group retention. The principal route to chemically defined structures consisted also of minimizing all monomer (and polymer) exposure to the plasma.

12.2
Basics

Short plasma pulses (a few μs) activate vinyl or acrylic monomer molecules, produce radicals, and start the plasma polymerization. After the pulse has ended the radicals initiate a purely chemical radical chain reaction in the following (long) plasma-off period (μs to ms). Thus, the plasma polymer should consist of more chemically regular products than those of the continuous-wave (cw) mode with predominantly random radical recombination. Ideally, the composition of the plasma polymer depends on the pulse-on/pulse-off ratio. Then, the shorter the plasma pulse and the longer the off-time the more the chemical product should dominate. The irregular structured fragmentation–recombination product has a rate of formation given by $R_{on} = k_1 R_{fragm} + k_2 R_{chem}$. In addition to fragmentation, chemical chain propagation may also occur under plasma exposure during the

The Plasma Chemistry of Polymer Surfaces: Advanced Techniques for Surface Design, First Edition.
Jörg Friedrich.

pulses, characterized by continuous radical production and re-initiation. The deposition rate in the plasma-less period is produced exclusively by chain growth polymerization without plasma exposure and re-initiation: $R_{off} = k_3 R_{chain}$. Introducing the duty cycle, d_c ($d_c = t_{on}/t_{on} + t_{off}$), the deposition rate for pulsed-plasma R_{pp} is now:

$$R_{pp} = R_{on} + R_{off} = d_c(k_1 R_{fragm} + k_2 R_{chem}) + (1 - d_c)(k_3 R_{chain})$$

Chain propagation of chemical polymerization is limited by the (low) probability of attaching the next monomer to the radical at the growing chain end under the conditions of a plasma sustained at a pressure of about 10 Pa. Deactivation of chain propagation (short kinetic chain length) under vacuum conditions is expected by recombination of neighboring radicals in the polymer. Another important concurrence is the reaction of C-radical sites with traces of (molecular) oxygen in the plasma gas, leading to the formation of peroxy radicals and hydroperoxides. Migration (delocalization) of radical sites into the deposited polymer layer may also be considered. Recombination in the gas phase, disproportionation, and radical transfer to monomer molecules need partners. To compensate for the loss of active radical sites during the plasma-off period, the radical formation and the radical chain growth start again after every new pulse. Theoretically, a single plasma pulse should be sufficient to start an endless radical chain growth polymerization to high molar masses, as is characteristic for polymerization in liquid phase or under high-pressure conditions. However, under low pressure plasma conditions fresh starter radicals must be produced repeatedly in pulsed plasmas with long plasma-off periods and short plasma pulses. Such pulsed plasma, with long plasma-off periods and short plasma pulses, offers a good compromise for efficient production of polymer structures with a minimum of irregularities.

It was proposed that the resulting plasma polymer has a structure and composition close to those found in radical polymerized polymers. Only a few irregularities should be involved in such plasma polymers incorporated by each plasma pulse (Figure 12.1).

During the plasma pulse (t_{on}) monomer fragmentation and substrate activation occur. Radicals are produced, which initiate a gas-phase radical polymerization during the plasma-off periods (t_{off}). During chain growth no plasma can disturb the polymerization and introduce defects. However, for ignition of plasma a higher start voltage is needed than for continuously burning the plasma during the pulse duration ($U_{ign} > U_{burn}$). If the plasma-off periods are prolonged so that no remaining charge carriers are in the gas phase the re-ignition voltage increases. At the end of a plasma pulse a short afterglow exists (Figure 12.2). During the plasma-off period only chemical reactions of radicals or fragments are possible.

The electrical pulse does have not a right-angular profile but, instead, a ramp; nevertheless, the plasma ignites from only a minimum ignition voltage, as can be seen by the SEERS (self-exciting electron resonance spectroscopy) sensor signal. The sensor in the wall of a diode-like reactor detects the ignition of plasma after

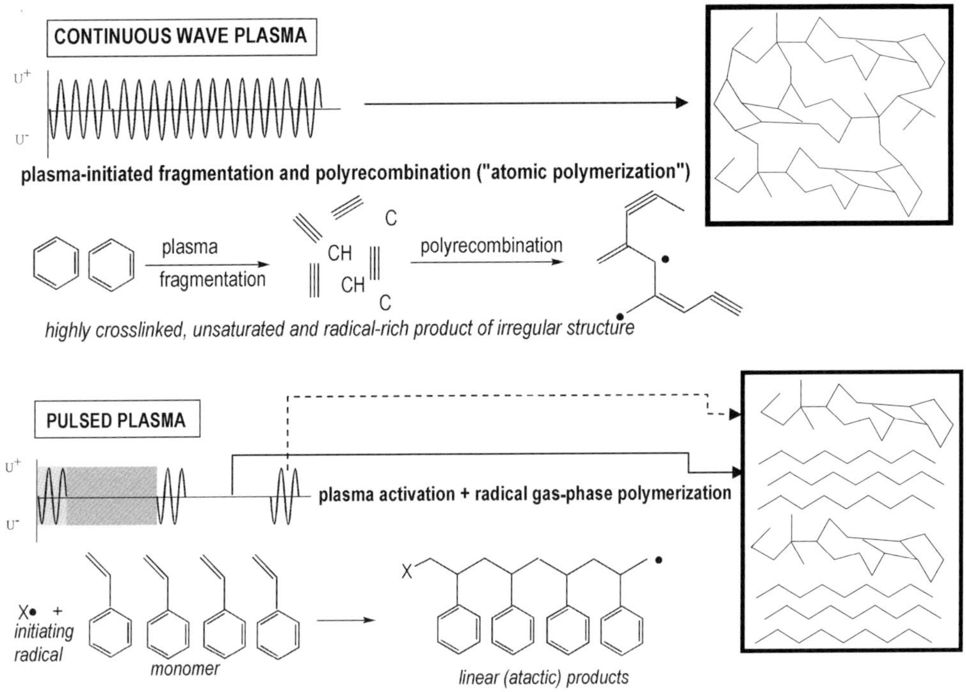

Figure 12.1 Schematics of continuous-wave (cw) and pulsed-plasma (pp) produced polymerization and their products.

achieving nearly the maximum of voltage during the pulse. It also shows a short afterglow (Figure 12.2).

It could be shown that all monomers were also deposited in the plasma-off periods [1]. However, obviously, the post-plasma activity rapidly decreases and high kinetic chain lengths can not be expected because the monomer concentration is too low (low pressure). Nevertheless, the deposition rates of monomers in the pulsed-plasma mode are much higher than in the cw mode, especially if all rates were referenced to the same plasma-on exposure time (Table 12.1).

The deposition rate (R_{pp}) was formulated as the sum of the chemical contribution during the plasma-off period (R_{off}) and fragmentation–polyrecombination product during plasma pulses (R_{on}) and by neglecting chemical contributions during the pulse duration: $R_{pp} = R_{on} + R_{off}$. In the case of exclusively fragmentation–polyrecombination is $R_{pp} = R_{on}$. Therefore, the difference between R_{on} (deposition rate in continuous-wave plasma) and R_{pp} may be a measure of the chemical contribution to the total deposition rate and, hence, the deposition rate in the continuous-wave plasma (R_{cw}) is given be $R_{cw} = R_{on}$. When $R_{on} << R_{off}$ the chemical polymerization should dominate.

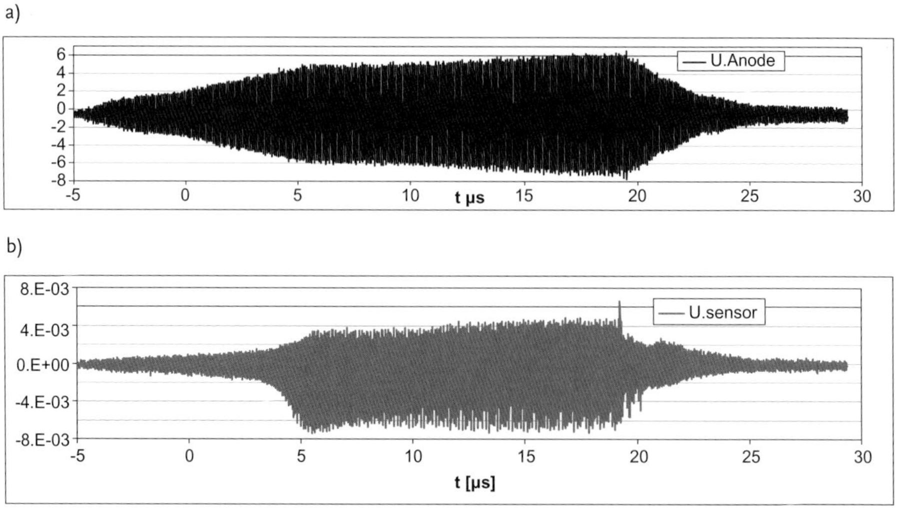

Figure 12.2 Single plasma pulse (24 ms, 10 kHz) (a) and plasma potential (b).

Table 12.1 Deposition rates of pulsed-plasma polymerization (duty cycle-DC = 0.1).

Monomer	Continuous-wave plasma	Pulsed plasma (duty cycle 0.1)	Referenced to 100% plasma-on	Pulsed plasma (duty cycle 0.1); Yasuda [2, 3]
Acetylene	100	90	900	77
Ethylene	100	100	1000	102
Styrene	100	220	2200	84
Acrylic acid	100	105	1050	220
Allyl alcohol	100	65	650	–
Allylamine	100	65	650	–
n-Hexane	100	–	–	10

As mentioned before, the products formed in the plasma-off period should be regularly structured in contrast to those formed in the plasma-on period. Thus, a mixture of regular and irregular structured product should be produced. The percentage of regularly composed polymer depends on the reaction rate of chain growth without plasma-assistance and of course on the duty cycle and pulse frequency. More or less regularly structured pulsed-plasma polymers are unknown. It must be concluded that the regularly composed layer fraction is also bombarded by the particle shower from the plasma and more probably is irradiated by far UV radiation and, thereby, is decomposed and crosslinked. Thermal effects may also contribute to the dominance of irregularly structured products. In addition, defects and radical formation also occur in the same way as in cw-plasma produced plasma polymers. Alternatively, shorter pulse durations and longer plasma-off periods may hinder the formation of additional defects and radicals in the plasma polymer, but the deposition rates decreases strongly. This effect evidences that pure chemical layer deposition is low under the conditions of low pressure. One reason for this may be the higher re-ignition voltage needed for plasma pulse succession with long plasma-off periods. This may produce "hard" plasma conditions at the beginning of pulses and, thereby, produce more defects. Using a low duty cycle, $d_c = t_{on}/(t_{on} + t_{off})$, that is, $d_c \leq 0.1$ the regularity of structure and composition decreases again in contrast to expectation.

Besides the duty cycle the repetition frequency of pulses plays an important role. A typical frequency is 1 kHz but also 100 or 10 Hz have been used. This pulse repetition frequency differs from the rf-frequency, which is most often 13.6 MHz.

The following mechanism during pulsed and cw plasma polymerization is probable (cf. Figure 12.1):

Initiation (plasma pulse): $M + \text{plasma} \rightarrow (^\bullet)M^\bullet$ formation of initiator radical

Initiation (cw plasma): $M + \text{plasma} \rightarrow A^\bullet + B^\bullet + C^\bullet + D^\bullet$ fragmentation

Chain growth (plasma-off): $M^\bullet + M \rightarrow P^\bullet$ radical chain growth

Chain-growth (cw plasma): $A^\bullet + B^\bullet + C^\bullet + D^\bullet \rightarrow \text{BDCA}$ polyrecombination

Post-deposition reactions continuation of chain growth polymerization, auto-oxidation.

12.3 Presented Work on Pulsed-Plasma Polymerization

Several monomers, such as acetylene, ethylene, butadiene, and styrene, have been deposited as thin polymer films by pulsed plasmas of low power [4]. Acetylene can react by opening the triple bond. A substitution of H at the triple bond by any other fragment was also observed [5]. Ethylene does not form a plasma polymer in the continuous wave plasma that includes any structure that is comparable to chemically polymerized polyethylene [6–18]. Here, using the pulsed-plasma mode a very weak vibration was assigned to the missing $\rho(CH_2)_{\geq 4}$ vibration characteristic for $[(CH_2)_n, n \geq 4]$ in IR spectra [19].

By opening the vinyl bond of the styrene molecule by plasma activation a radical chain polymerization can be started. The *G*-value of this reaction (polymerized monomers per 100 eV) is >20 [1]. Therefore, it is likely that radical chain growths polymer contributes essentially to the formation of the pulsed-plasma produced polystyrene (pp-PS).

This opens up a way of enhancing the pure chemical radical polymerization in the gas phase. Pulsed-plasma polymerization was introduced first by Tiller in 1972 [19], later continued by Yasuda [2, 20] and Shen and Bell [21, 22], and then further developed by Timmons [5, 6] and Friedrich [23–25].

12.4 Role of Monomers in Pulsed-Plasma Polymerization

To initiate a radical chain growth the monomer must be easily polymerizable (by any plasma-produced radical). For this purpose, preferentially acrylic and vinyl monomers are suited. However, acrylic monomers tend to decompose at the ester group by CO and CO_2 release. Vinyl monomers lose a few of their functional groups. Dienes form crosslinked structures, as probed using butadiene [25]. As mentioned before, allyl monomers could be polymerized in the pulsed plasma with partially high retention of their functional groups. The ability of radical polymerization also depends on the reactivity or stabilization of the radical intermediate. Especially, resonance stabilization can hinder the chain propagation. However, in contrast, plasma polymerization of allyl monomers is easy. Olefins, such as ethylene, propylene, tetrafluoroethylene, or hexafluoropropene are principally suited for radical polymerization; however, on using the pulsed plasma most often crosslinked and irregularly structured plasma polymers are produced.

Ring-opening polymerization is also possible in the plasma. Diaminocyclohexane (DACH), the above-discussed cyclobutadiene, and benzene are characteristic examples of successful plasma polymerization. Nevertheless, special processes based on ring opening and exclusive use of pulsed-plasma mode are unknown. Condensation reactions initiated by either plasma or pulsed plasma have not been described.

Surprisingly, acetylene shows a high deposition rate in pulsed plasma but it is not known as a classic monomer. On exposure to plasma it may polymerize either by opening the triple bond to give double bond containing polyacetylene or by H-substitution to form polyacetylenes with triple bonds.

The regularity of structure is (slightly) improved by using the pulse plasma technique but at the same time the number of applicable monomers becomes strongly limited and is focused on vinyl and acrylic monomers.

The retention degree, which is a measure of inserted regular structures and chemical quality of produced polymer layers, may be shown by the following the reaction:

$$n\mathrm{CH_2{=}CH\text{-}X_0} + \textit{pulsed plasma} \rightarrow \text{-}[\mathrm{CH_2\text{-}CHX}]_n\text{-}$$

Retention of functional groups as shown by percentages of functional groups in the deposited layers ranges from 50% to 95% (often 80–90%) as described later in detail [1]. Retention of complete structure elements introduced by monomer molecules into the produced plasma polymer is difficult to attain. Backbone, side-chains, and functional groups were often completely degraded [23, 26, 27]. There is also no universal indicator for structural retention. Therefore, it is impossible to present such a “structural” retention degree. It is approximately more or less 0–10% on utilizing the results of bulk analytical investigations such as thermogravimetry, thermoluminescence, dielectric relaxation spectroscopy, and so on [1, 4, 27]. In contrast, surface analytical methods often show more similarities to classic polymer structures, as will be shown below [28]. A "skin" effect was made possible for these surprising relationships to classic polymers.

High post-plasma activity (auto-oxidation) is also characteristic for pulsed-plasma polymers because of the high concentrations of radicals also found in this type of plasma polymer. In particular, if non-classic monomers were used the only possible plasma polymerization mechanism is that of fragmentation–recombination and, thus, inevitably destruction of monomer structure, radicals, and other defects occurs ($R_{pp} \approx R_{on}$) [18]. During exposure to ambient air the trapped radicals can add oxygen from air – up to 20% O/C ratio for ethylene, acetylene, or butadiene [24].

As mentioned before, the low-pressure condition of pulsed plasma (1–100 Pa) hinders any chain growth reactions in the plasma-off period. Higher deposition rates are not possible because of a strong lack in monomer molecules. Therefore, chain termination dominates and leads to short kinetic chain lengths. Chemical polymerizations are performed in liquid phase or under high-pressure conditions and, thus, the chain growth is not hindered by monomer supply. The difference in pressure between such classic polymerizations and pulsed-plasma polymerization amounts approximately to a factor of 10^8. Therefore, under low-pressure conditions the life-time of initiating radicals must be very long, otherwise the chain reaction is terminated by recombination, disproportionation, chain transfer or by the above-mentioned reaction with oxygen.

To avoid the ending of plasma pulse initiated polymerization by termination after completion of a pulse, the chain-growing reaction is re-started by the next pulse. Because of the low monomer density in the gas phase or at the surface of growing plasma polymer, chain transfer to gas phase monomer molecule is hindered. This might explain the high yield in allyl monomer plasma polymerization. Allyl monomers are chemically very difficult to polymerize because of "degradative chain transfer" [29]. Hence the retention degree of functional groups is high for allyl pulsed-plasma polymerization; for example, for allyl alcohol about 90% of all OH groups were retained, for allyl bromide this is 75%, and for allylamine >50% – in all cases the radical chain growth mechanism should be present [30, 31]. However, the fragmentation–polyrecombination process is also present, as shown by spectroscopic identification of irregular structural elements [25, 32–41].

The dominance or continuance of the fragmentation–polyrecombination reaction is obvious on comparison of IR spectra of cw-plasma and pulsed-plasma polymerized ethylene in the range 700–800 cm^{-1}, as shown before. The rocking vibration at ca. 720/730 cm^{-1} characteristic for classic polyethylene is completely absent in cw-plasma polymerized polyethylene [41–44] and only suggestively appears when using the pulsed-plasma polymerization mode [23]. It means that at least two ethylene units are linked together.

The deposition rates of different monomers under cw- and pulsed- ($d_c = 0.1$; $f = 1$ kHz) plasma conditions were measured by Yasuda [2]. Under pulsed-plasma conditions (duty cycle of 0.1) cyclohexane shows only 10% of the deposition rate in comparison to that under cw-plasma conditions, which is characteristic for

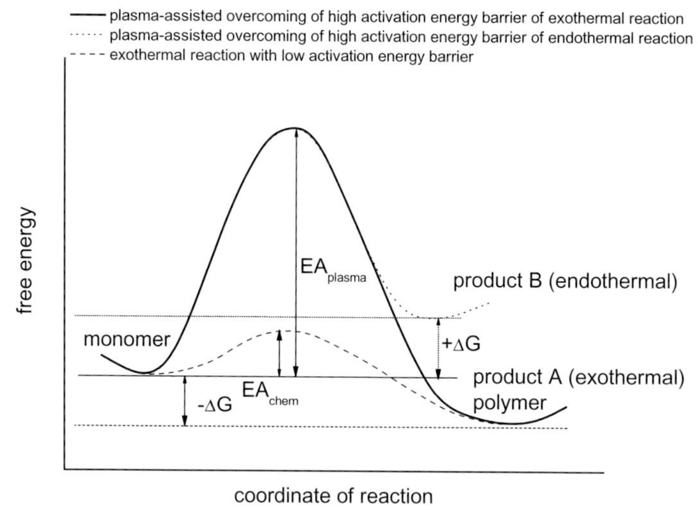

Figure 12.3 Scheme of activation energy for initiation of polymerization.

fragmentation–polyrecombination, as expected for aliphatic "monomers" (Table 12.1). In contrast, acrylic acid showed a doubling of deposition rate using the pulsed mode Table 12.1). The effective plasma-on time under pulsed-plasma conditions was about 10% of the total deposition time but the deposition rate was 200%, that is, chemical polymerization can be expected in comparison to the cw-plasma polymerization. Styrene is the prototype for a spontaneously polymerizing monomer under exposure to small doses of light, heat, or particle collision for initiation of a radical chain-growth polymerization as exothermal reaction (Figure 12.3) [1].

Figure 12.3 depicts the two principal courses of reaction energy. The aliphatic precursor (black line) needs much more activation energy for its fragmentation at the activation barrier, while the recombination to polymer product lowers the energy but to a level higher than the level of energy at beginning. The vinyl monomer needs much lower activation energy for activation of the π-bond and forms the polymer under energy profit characteristic of exothermal polymerization.

The minimal energy transferred to one monomer molecule during passage through the plasma zone to start the polymerization was calculated as approximately 0.4 eV for the styrene monomer [45]. A polymerization degree (X) of approximately $X = 25$ was calculated and confirmed by size-exclusion chromatography (SEC) measurements [46].

12.5 Dark Reactions

It was of interest to determine when this chemical contribution to the styrene plasma polymerization proceeds, during plasma pulses or during plasma-off time

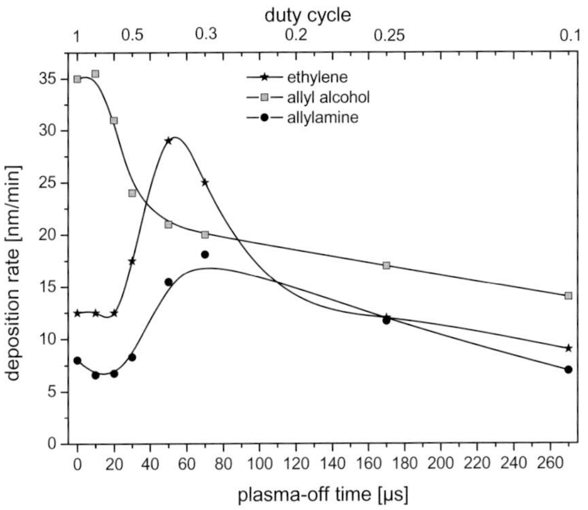

Figure 12.4 Deposition rate in pulsed-plasma mode during plasma-off versus its duration or duty cycle. The plasma pulse duration was 30 ms and wattage 100 W.

(dark phase). Figure 12.4 confirms that all monomers show post-plasma polymer deposition during plasma-off periods. However, obviously, the post-plasma activity rapidly decreases.

Allyl alcohol starts, in the plasma-less intermediate period, from a high level of deposition rate after ending of the plasma pulse. The deposition rate falls to a common level such as those of allylamine and ethylene (Figure 12.4). These other monomers show at the beginning of the plasma-off period an intermediate maximum. Beyond these maxima the deposition rate of all three monomers is similar and decreases slightly as assumed.

12.6 Pressure-Pulsed Plasma

The idea behind pressure-pulsed plasma was to increase the monomer pressure only during plasma-off periods. Therefore, the number of sticking monomer molecules at the surface can be increased considerably and inhibition of chain growth by disproportionation, recombination, chain transfer, or diffusion may be avoided (Figure 12.5).

To initiate a new plasma pulse the system was pumped down from high pressure at the end of the plasma-off time (possibly 0.1 s) to low pressure before starting the next pulse. Thus, the plasma could be easily (normally) initiated. At the end of the plasma pulse and beginning of the plasma-less period vinyl monomer was pressed into the reactor. An increasing monomer supply should prolong the kinetic chain length of the radical polymerization. The realized pressure increase amounts to a factor of ca. 10^2. The pressure-pulsed plasma was generated in a

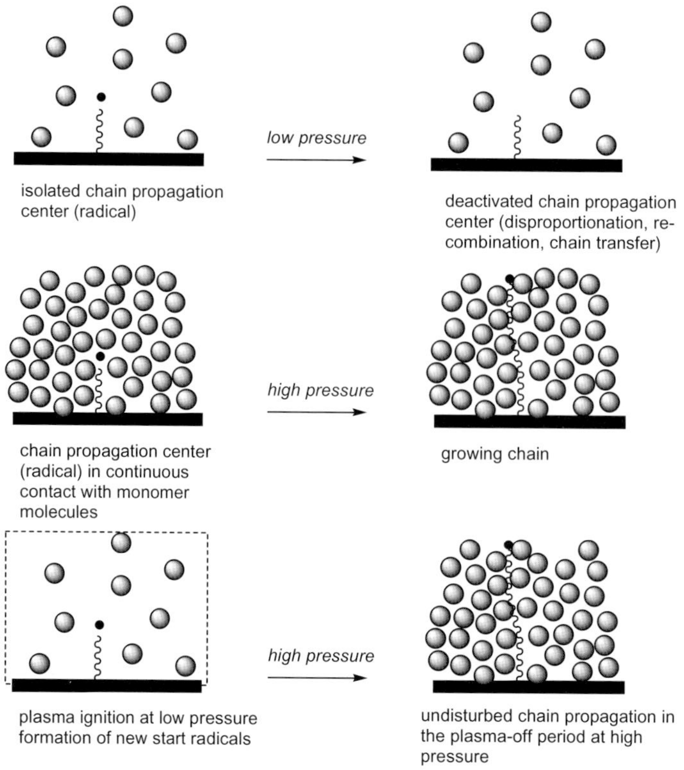

Figure 12.5 Schematic view of the correlation between high pressure, high sticking rate, and chain propagation for chemical gas-phase polymerization. The right-hand column depicts the principal particle densities for low-pressure plasma ignition and high-pressure chemical chain propagation.

special reactor with a parallel plate electrode system. The reactor volume was small ($200\,cm^3$). The monomer gas (or monomer vapor) was injected by piezo-crystal valves. To realize the necessary rapid pumping down of the high pressure in the reactor at each pulse (frequency 1–10 Hz) a rapidly working valve with a large cross-section to the vacuum pump was used. The valve was constructed as an oscillating plate valve (or rotary slide) and was electronically piloted. To increase the pumping rate during the end of each cycle a large vacuum buffer volume between valve and pump was mounted. Synchronization between pulsing of the rf current for ignition the plasma and pulsing the pressure was electronically managed (Figure 12.6). The deposition rates were measured using a quartz microbalance.

Indeed, the deposition rates could be improved by the intermediate pressure pulse (Figure 12.7). As seen for the simple pulsed plasma, allyl alcohol and buta-

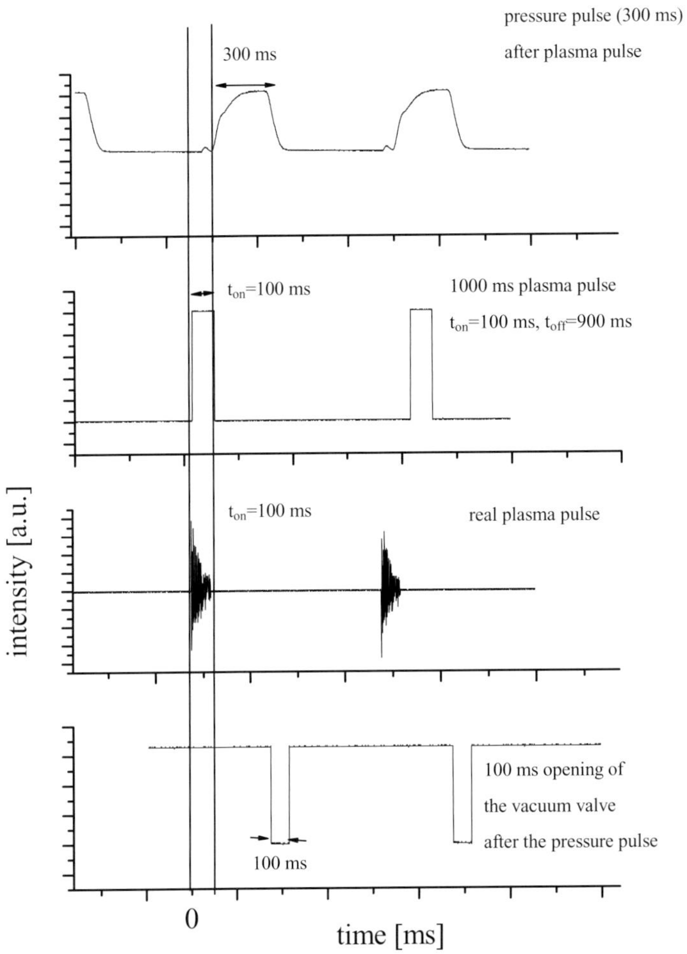

Figure 12.6 Principle of plasma and pressure pulse synchronization and measured response of the plasma system.

diene plasma polymers were deposited with much higher deposition rates than that of ethylene.

Using the additional pressure-pulses in the plasma-off period the deposition rate was increased by a factor of roughly 10^4. It must be emphasized that the difference between pulsed and pressure-pulsed plasma consists only of the much greater monomer supply in the plasma-off period. The plasma pulse itself is the same. The higher deposition rate of pressure-pulsed plasma can only be explained by advanced chemical polymerization during the plasma-less period.

Another important effect of the pressure-pulsed plasma is that the pressure pulse enhances deeper penetration of layer or functional groups, forming plasma

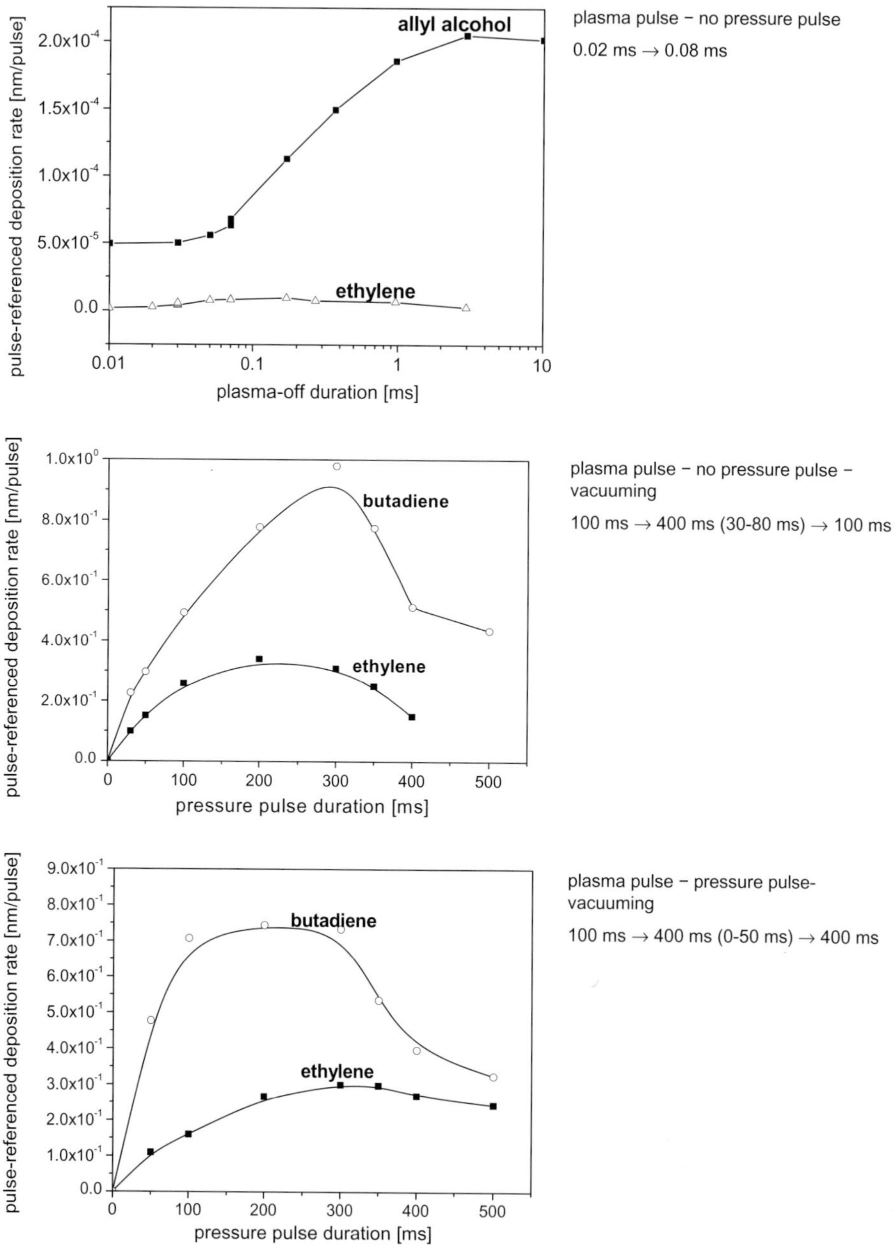

Figure 12.7 Pulse-referenced deposition rates of ethylene, butadiene, and allyl alcohol (10 W).

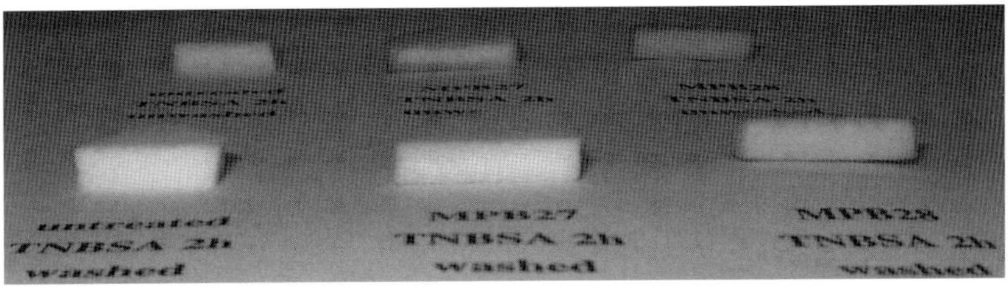

Figure 12.8 Left to right: untreated, NH_3 plasma-treated (continuous-wave), and pressure pulsed NH_3 plasma-modified porous polypropylene membrane material. The NH_2 groups formed were colorized by reaction with 2,4,6-trinitrobenzenesulfonic acid.

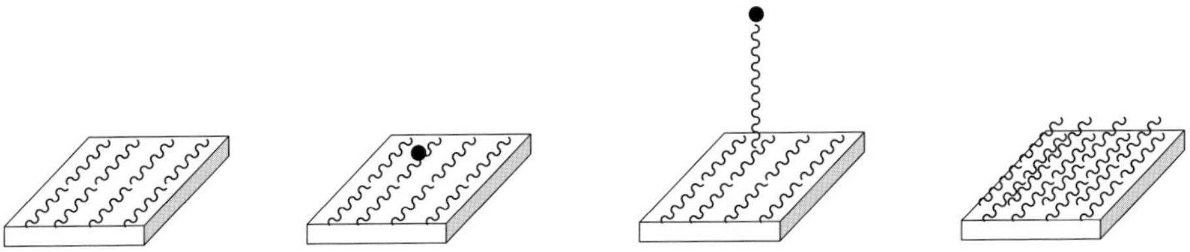

Figure 12.9 Single-point chain growth compared with growth in plane (layer by layer).

species in porous polymer or inorganic systems, such as pores in membranes, capillaries in paper, capillaries in blood purification devices, porous catalyst targets, and so on (cf. References [47–49]).

An instructive example is ammonia plasma treatment and the depth of occurrence of amino groups within the bulk of porous polypropylene membrane materials indicated by homogeneous yellowing over the complete cross-section of the membrane (Figure 12.8).

A partial overlap of plasma and pressure pulse was also tested, but did not produce significant changes in deposition rate.

Interestingly, the very low deposition rate of all pulsed-plasma polymerization processes allows chain growth only at one single-point during one pulse. No layer-by-layer growth is possible at each pulse. Homogeneous layer-by-layer growth needs much higher deposition rates. The pressure-pulsed plasma polymerization and its high deposition rates allow such layer-by-layer growth (Figure 12.9).

12.7 Differences between Radical and Pulsed-Plasma Polymerization

The previous section reported on dramatic differences between continuous-wave plasma polymerization and classic radical chain growth polymerization. Radical polymerization is initiated by production of radicals and the presence of reactive monomers in the liquid or gas phase at high pressure. Radicals are often produced by thermal decomposition of labile initiator molecules, such as peroxides and azo compounds, or by redox systems, or hydrogen peroxide–Fe(II) (Fenton's reagent). These radicals are the starting point of the addition of monomer molecules, forming from monomer addition to monomer addition a new radical. This continuous propagation of chains starting from one mother radical leads to high molar masses. It ends with inhibition or termination reactions to stable and often linear products with masses in the range 10^6–10^7 g mol^{-1}. Terminations are disproportionation, chain transfer, and radical recombination as described before.

The polymer-producing reactions in low-pressure plasma operate in the range 10^{-4}–10^{-5} bar in contrast to chemical polymerizations in the gas-phase, which are performed at 10^0–10^2 bar. Thus, a difference in pressure and therefore in particle density of about 10^6 orders of magnitude is characteristic. It is no wonder that classic processes at high pressure have enough monomers to maintain a continuous chain-growing reaction with high kinetic chain lengths. This is not possible under low-pressure conditions with the plasma-off intermediates. It can be also assumed that it is also not possible in any adsorption layer because of the fixed surface position determined by adsorptive interactions or hydrogen bonds among adsorbed molecules, which means that the distance between adsorbed molecules is often far from the necessary (near) distances for chain growth reactions. The application of pressure pulses may improve the chances of short kinetic chain propagation polymerizations. However, the frustrating results of ethylene polymerization (absence of more than two original units in the polymer) make it improbable. What are the reasons for this?

It has been noted that the high deposition rates seen with pressure-pulsed plasma are only explainable by significant contributions from chemical chain growing polymerization. Nevertheless, the structure of these layers is far from that of chemically produced polymers. The reactions in low-pressure plasma are due to inelastic collisions of electrons with atoms or molecules and the plasma particle bombardment of the substrate surface. Special conditions are present by charging of deposited layers, producing self-bias and ion acceleration. Arguably, this bombardment during the plasma pulses is responsible for structural damage of originally intact structures. High significance is also assigned to the energetic plasma UV radiation. Vacuum plasma UV irradiation of the growing polymer layer may also contribute to its damage.

Thus, the plasma state produces high energy species and a very high radiative energy flux per molecule in the plasma. Therefore, all kinds of chemical reactions can be performed in the plasma state. The result is a strong irregular and undefined structure and composition of polymers produced in the plasma compared to classic polymers. As early as 1971 Westwood pointed out that there is a large difference, by a factor of 10^6, in energy doses, when expressed in terms of *G*-value, between plasma-chemical polymerization and high-energy irradiation initiated polymerization [50, 51]. In a glow discharge much energy is consumed, that is, ca. 10^7–10^{11} J kg^{-1}, in comparison to ca. 2.6×10^6 J kg^{-1} using γ-irradiation for (radical) polymerization of styrene. More exact details and more general differences and similarities between radiation- and plasma-induced polymerization were pointed out by Stiller and Friedrich [52].

12.8 Surface Structure and Composition of Pulsed-Plasma Polymers

The dominance of a chemically formed defined structure and composition in plasma polymers was expected using the pulsed-plasma technique. Polystyrene produced by this technique (pp-PS) should be composed of a mixture of regular and irregular sequences, with regular structures dominating significantly in comparison to cw-plasma-produced polystyrene (cw-PS). Indeed, regular classic structures were identified in pp-PS.

NEXAFS (near-edge X-ray absorption fine structure) and XPS (X-ray photoelectron spectroscopy) spectra of pp-PS show good agreement with those of classic PS (Figures 12.10 and 12.11).

The C1s → π^*_{ring}-resonance characteristic for the aromatic ring in the NEXAFS spectrum is slightly decreased in intensity but shows the same general structure as the reference PS (Figure 12.10).

Both methods are surface-sensitive techniques and have a sampling depth of about 3–5 nm. The thin surface skin pp-PS examined shows a structure similar to that of the reference PS. The fine structure of shake-up satellites and valence bands is slightly noisier but agrees well with spectra of reference polystyrene. The small deviation of the pp-PS spectra may be influenced by some crosslinking.

XPS-valence band spectra and also the C1s peaks of pulsed-plasma PS and reference PS show excellent agreement (Figure 12.11).

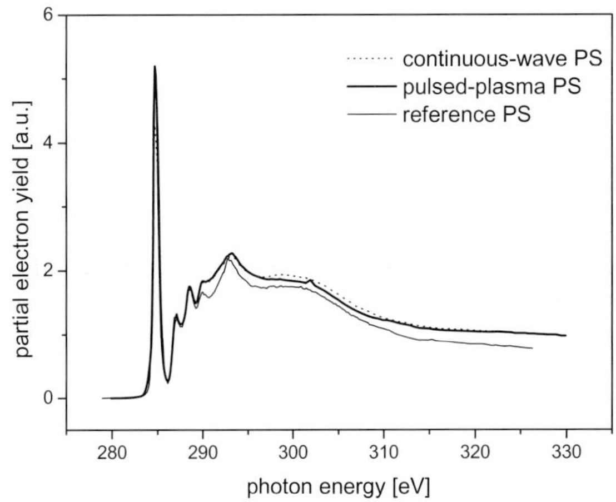

Figure 12.10 NEXAFS-C_K edge spectra of pulsed-plasma PS (pp-PS) and reference PS.

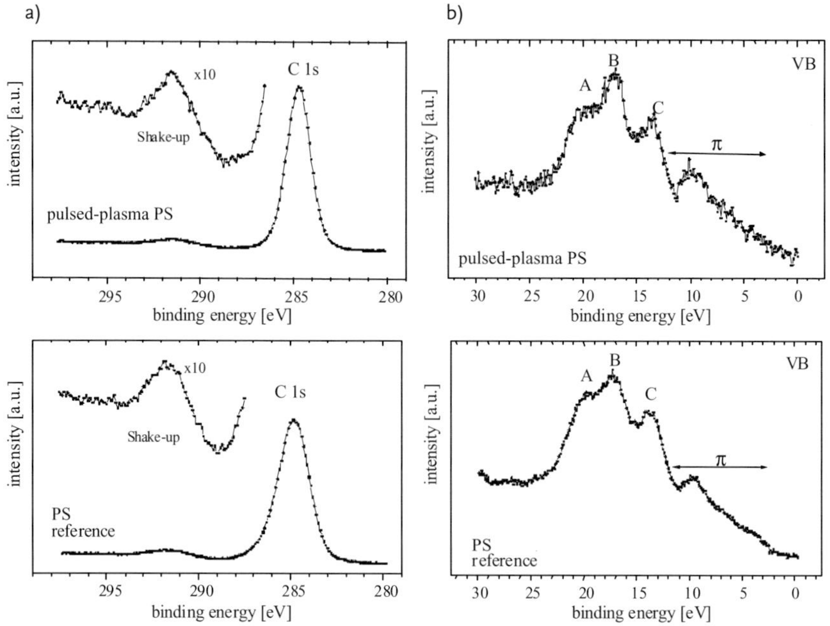

Figure 12.11 Comparison of (a) C1s- and shake-up satellites and (b) valence band (VB) spectra of pp-PS and commercial reference PS (A, B=C- atom orbitals, C=C-, molecular orbitals from σ and π bonds).

In contrast, pulsed-plasma produced polyethylene (pp-PE) and 1,3-polybutadiene (pp-PB) show significant differences in their NEXAFS spectra in comparison to their classic analogues (Figure 12.12).

Polymer formation using acetylene gas is only explainable by assumption of a fragmentation and recombination mechanism and, thus, the structure must be generally irregular. All three materials show intense C1s $\rightarrow \pi^*_{ring,C=C,C\equiv C}$ NEXAFS-resonances characteristic for the existence of high concentrations of C–C double and triple bonds in the resulting plasma polymer. Pulsed-plasma polymerized 1,3-butadiene shows a strong shake-up satellite in the C1s XPS signal. The intensity of this peak was comparable to that of pulsed-plasma polymerized PS (pp-PS). Pulsed-plasma polymerized polyethylene was similar to reference polypropylene and did not show the characteristic features of polyethylene as demonstrated by valence band spectra (Figure 12.13).

However, the survey scan and C1s peak of pp-PE did not show differences to commercial PE (Figure 12.14).

Pulsed-plasma polymerized polyacetylene showed different behavior to butadiene and ethylene plasma polymers in NEXAFS and XPS spectra.

ToF-SSIMS (time-of-flight static secondary ion mass spectrometry) using Ga-ion bombardment (static secondary ion mass spectrometry) of pp-PS and reference

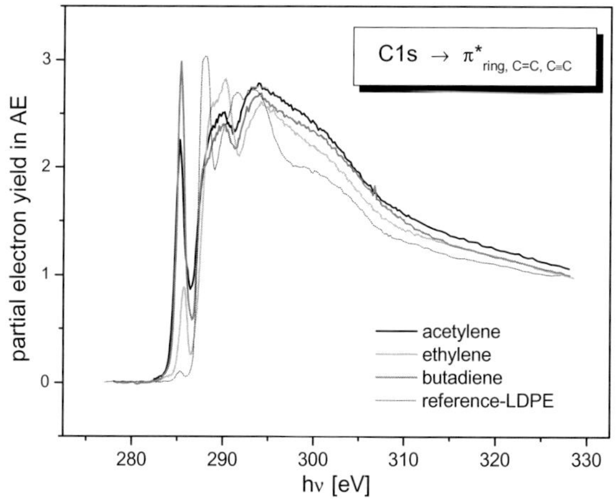

Figure 12.12 NEXAFS C1s → π* transitions of pulsed-plasma polymerized acetylene, ethylene, and 1,3-butadiene in correlation with a low-density polyethylene (LDPE) reference.

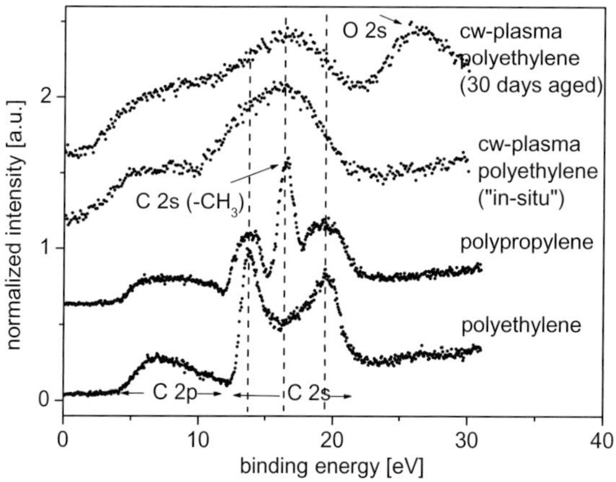

Figure 12.13 Valence band spectra of plasma-produced polyethylene and after its exposure to air.

PS ($5500\,g\,mol^{-1}$) shows some similarities in their patterns. Characteristic fragments of the styrene repetition unit and of the aromatic ring were found. However, any characteristic molar mass distributions were absent. The difference between the periodical intensity maxima was 13 amu (Da), indicating strong fragmentation of rings ($5 \times CH$) and backbone ($1 \times CH$). However, the reference PS exhibits similar behavior in the negative ion spectrum as well as in that of positive ions. In reference PS the most intense peak was located at 77 amu (C_6H_5) and in pp-PS

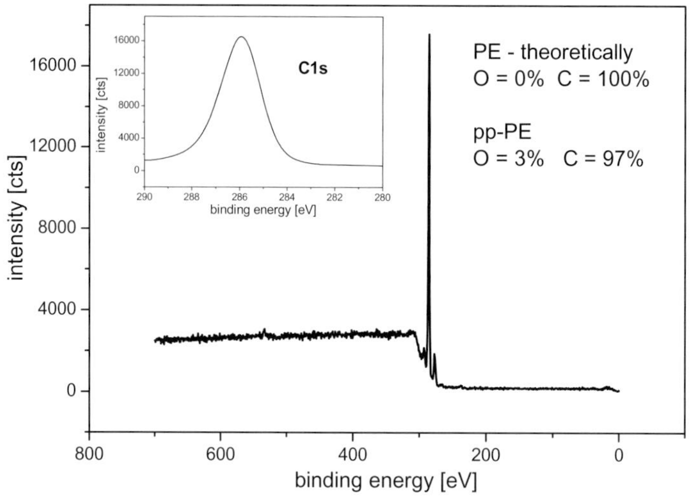

Figure 12.14 Survey scan and C1s peak of pulsed-plasma polymerized ethylene (pp-PE).

at 105 amu (C_8H_9). Nevertheless, most of signals in both spectra are nearly identical; however, the intensity distribution differs.

The same behavior was observed in the IR spectra considering the aromatic CH and the aliphatic CH_x stretching vibration around 3000 cm^{-1}, as shown before. In particular, the intensity of all absorptions in the range 2840–2970 cm^{-1}, referenced to the band at 701 cm^{-1}, is lowered by ca. 30–50% compared to the reference PS (Figure 12.15).

The loss in aliphatic CH_2 groups in relation to aromatic CH groups is eye-catching. Moreover, a strong absorption at 2962 cm^{-1} due to the $\nu_{as}CH_3$ vibration of methyl groups appears in the plasma polymer. The formation of CH_3 groups during plasma polymerization was always discussed in connection with the "quasi-hydrogen" plasma, indicating bond scissions in the backbone, hydrogen abstraction and possibly branching.

Pulsed-plasma polymerized layers of acetylene, ethylene, and 1,3-butadiene present strong deviations from their classic analogues in IR-ATR (ATR, attenuated total reflectance) spectra of about 2500 nm sampling depth, representing, therefore, more bulk properties. Non-classic polymerization processes are evident by the appearance of exotic groups in classic polymers such as 3300 cm^{-1} ($\nu\equiv CH$) and 2150 ($\nu C\equiv C$), at 1900 cm^{-1} (cumulenes) as well as 3060–3160, 1600, 1500, and 720 cm^{-1} (aromatics) (Figure 12.16).

The weak hint of the first appearance of ρCH_2 ($n \geq 4$) in pp-PE was discussed before and shows slightly more chemically defined structures in plasma polymers when using the pulsed-plasma mode.

Side-reactions with C=O-containing groups are present but become less important when the pressure was increased, as demonstrated by the lowering in inten-

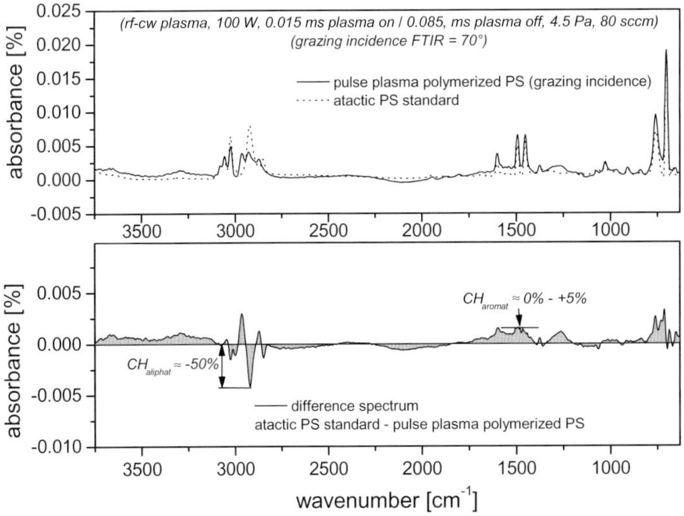

Figure 12.15 Comparison of IR spectra of pulsed-plasma PS (cw-rf plasma, 50 W; 0.015 ms plasma on/0.085 ms plasma off; 4.5 Pa; 80 sccm) and commercial PS standard (150 000 g mol^{-1}) using the grazing-incidence-reflectance FTIR ($\alpha = 70°$) normalized to γCH = 701 cm^{-1}.

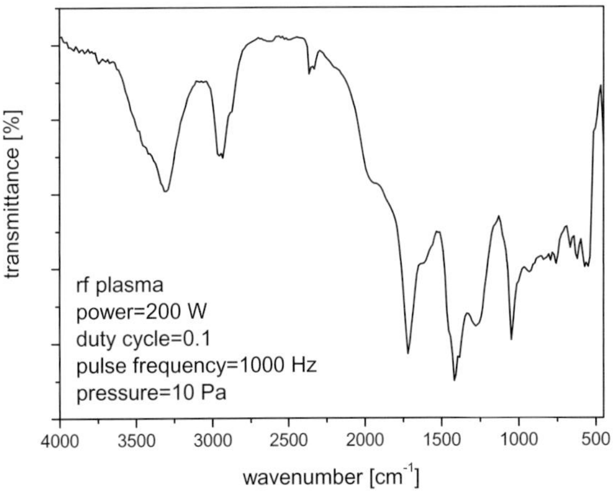

Figure 12.16 FTIR-ATR spectrum of pulsed-plasma polymerized acetylene.

sity of NEXAFS O1s $\rightarrow \pi^*$ resonance (Figure 12.17). This was explained by a lower energy input per monomer molecule (Yasuda factor) [35].

Olefinic C=C double bonds also play an important role in pp-PE as shown in Figure 12.17 (C1s $\rightarrow \pi^*$ resonance) and Figure 12.18. Figure 12.18 shows the increase in intensity of the C1s $\rightarrow \pi^*$ resonance with both duty cycle and wattage.

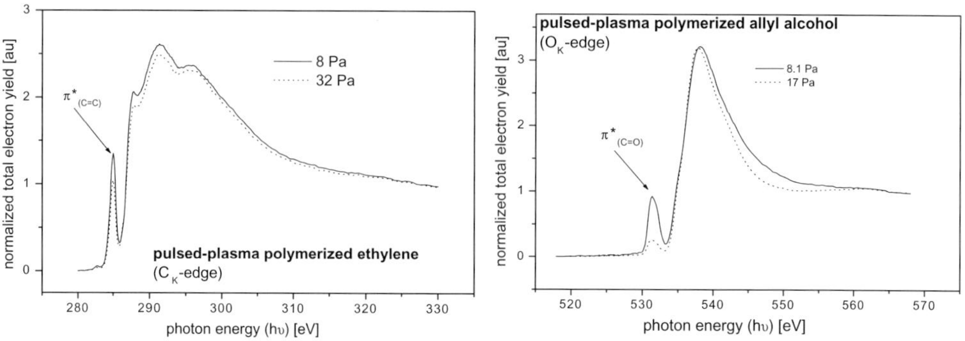

Figure 12.17 NEXAFS C_K edge spectra of ethylene (a) and O_K edge spectra of allyl alcohol (b) for different pressures.

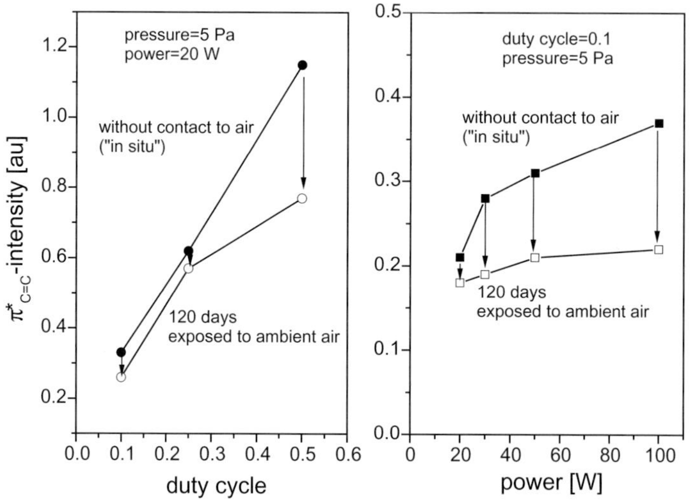

Figure 12.18 Intensities of NEXAFS C1s → π* resonances of C=C bonds in pp-PE before and after 120 h storage.

The concentration of C=C double bonds increases strongly with growing duty cycle and power. During storage of pp-PE about 10–50% of C=C double bonds disappear. This can be explained by a slow oxidation of C=C double bonds during long-time exposure to air. Thus, post-plasma oxidation of polymers is caused by oxidation at C-radical sites and addition onto double bonds as well as their epoxidation or ozonization.

Molar masses of the soluble fraction of plasma polymers were measured using size exclusion chromatography (SEC; or gel-permeation chromatography, GPC).

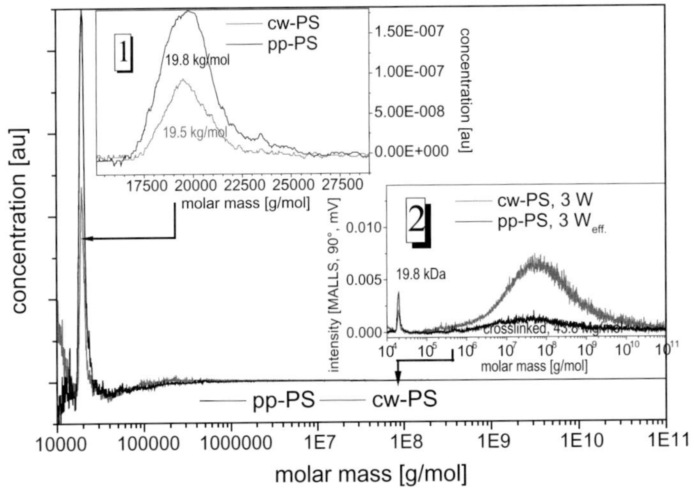

Figure 12.19 Thermal field flow fractionation (ThFFF) of pp-PS (power = 30 W, duty cycle = 0.1, pulse frequency = 1000 Hz) and cw-PS (3 W) after removing gel particles by filtration.

Ultrahigh molar masses and gel particles were determined using field flow fractionation (asymmetric field flow fractionation = AFFFF or F^4 and thermal field flow fractionation = ThFFF). SEC works near the exclusion limit because of gel particles, which must be removed by filtration. The soluble fraction of pp-PS also contained monomers, dimers, trimers, and so on up to octamers. The maximum molar masses were measured in the range 20–30 kg mol^{-1} (polymerization degree 200–300). ThFFF detected masses in the range of greater than 5×10^7 g mol^{-1} using a multi-angle laser light scattering detector (MALLS) or evaporative light scattering detector (ELSD) (Figure 12.19).

The ThFFF elugram of the unfiltered pp-PS showed also fractions in the range 20–60 kg mol^{-1} (Figure 12.20). All elugrams show the detector signal as the *y*-axis and not the concentration and, thus, high molar masses are overrepresented. Therefore, in Figure 12.19 the absolute concentration (number of species) of high molar mass species is low; nevertheless, high mass is accumulated in such a condensed (crosslinked) particle. Therefore, 99.9% of all detected molecules in Figures 12.19 and 12.20 are accumulated in the peak at about 20 kg mol^{-1} and the majority of mass is found in the Mg mol^{-1} region.

The asymmetric field flow fractionation showed for 3 W cw-PS (continuous wave produced polystyrene) and 30 W pp-PS (duty cycle = 0.1) also a main peak at 20 kg mol^{-1}. Using the cw-mode species with higher molar mass are also found (Figure 12.21).

Summarizing all the results of surface-sensitive and bulk analysis dramatic differences between all sort of plasma polymers (cw or pp) and the reference (standard) polymers are obvious for the bulk. This fact is important because

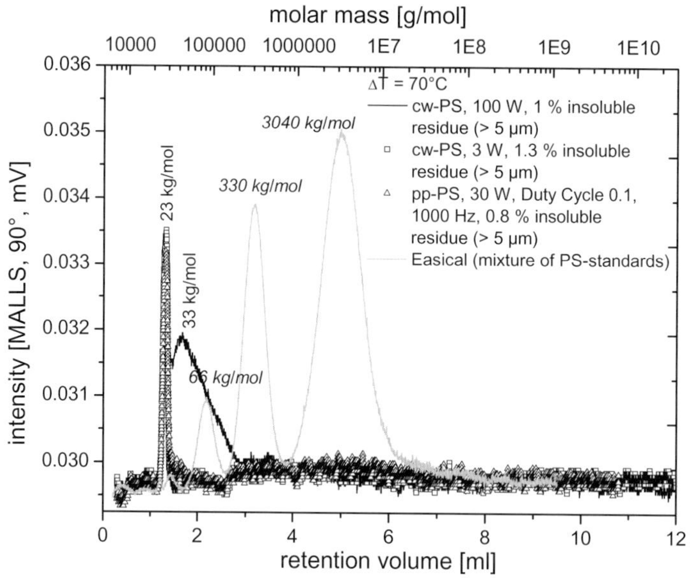

Figure 12.20 ThFFF-elugram of cw- and pp-PS without filtration.

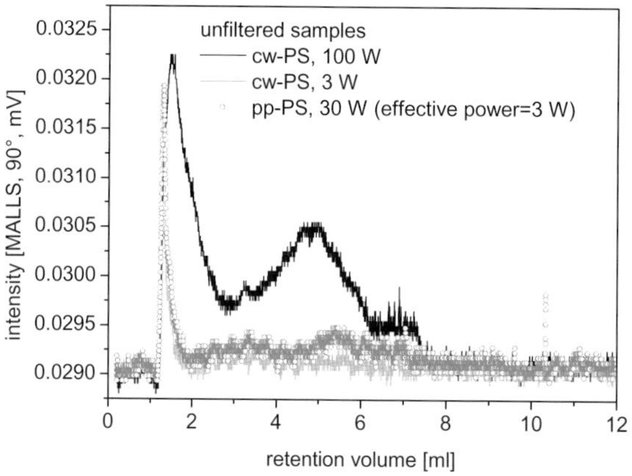

Figure 12.21 Asymmetric field flow fractionation (AFFFF, known also as cross-flow FFF) of pp- and cw-PS (power = 30 W; duty cycle = 0.1; pulse frequency = 1000 Hz) referenced to a mixture of PS standards.

surface-sensitive methods–XPS, NEXAFS, IR-ATR, ToF-SIMS (time-of-flight secondary ion mass spectrometry)–have predicted a few, sometimes many, similarities to commercial polymers. However, SEC and FFF as well as thermogravimetric (TG) investigations confirm the complete irregularity in structure, composition, and properties compared to reference polymers. There are only small differences in the thermal degradation behavior between cw- and pp-PS but significant deviations compared to commercial PS (Figure 12.22).

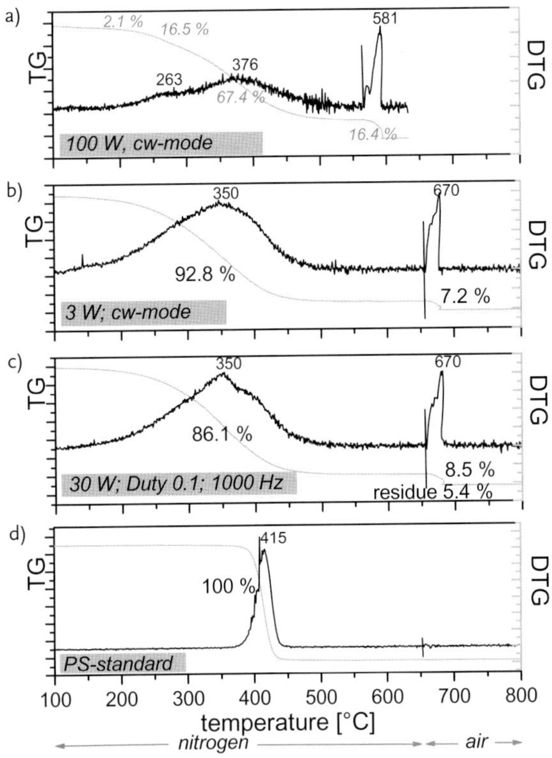

Figure 12.22 TG and DTG (differential thermogravimetry) plots of cw- (a, b), pp- (c), and reference (d) PS.

The weight loss of cw-PS begins early in a nitrogen atmosphere (263 °C) and extends over a long temperature range. It can be assumed and derived from FFF results that the broad molar mass distribution is responsible for this thermal behavior. The non-degradable carbon is maximal in the cw-PS sample produced with 100 W (16.4%). This carbon can only be removed by oxidation (air) at 581 °C (maximum). This behavior is characteristic of crosslinked polymers containing tertiary and quaternary carbon atoms. Low-wattage and pulsed-plasma produced plasma polymers have non-degradable carbon residues in the range of 7–8%, which is characteristic for lower crosslinking and branching. Thus, cw-PS (100 W) was more or less insoluble in tetrahydrofuran but pp-PS (30 W) and cw-PS (3 W) were predominantly soluble in THF.

UV/visible spectra of cw- and pp-PS as well as reference-PS (PS standard, $M_N = 226\,\text{kg mol}^{-1}$) are plotted in Figure 12.23. The spectra of low-wattage cw- and pp-PS are similar. A few similarities to the reference material are obvious. The fingerprint structure of these materials is diminished in the spectrum of the high-wattage cw-plasma.

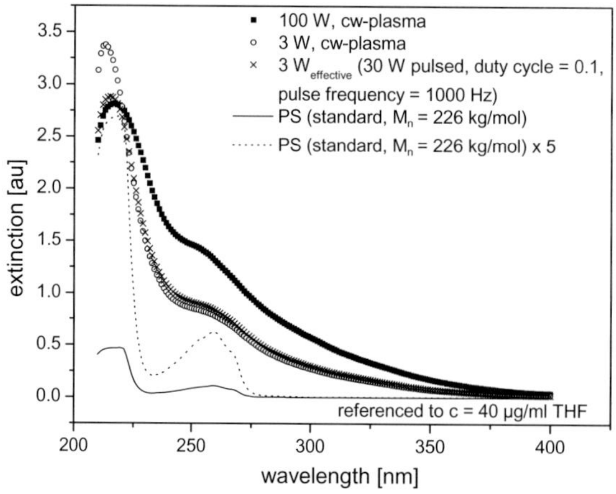

Figure 12.23 UV spectra of plasma-produced PS.

Notably, all plasma polymers show much increased absorption between $\lambda = 200$ and 350 nm in comparison to commercial reference-PS. This is due to the increased concentration of unsaturations in plasma polymers. It can be speculated that these double bonds are also localized in the polymer backbone. Removal of the phenyl group requires the scission of only one C–C bond. The alternative is cracking of the ring and formation of hexatriene intermediates, which may undergo crosslinking.

Dielectric relaxation spectroscopy (DRS) detects all kinds of dipoles in the plasma polymer (functional groups, radicals, defects). The dipole absorbs energy from the applied electromagnetic field. Polar groups and segments begin to move, oscillate, or rotate. Therefore, DRS is an excellent indicator of the regularity of polymer structure and the occurrence of defects [53]. The relaxation behavior of polymer layers (300 nm) deposited within a capacitor can be measured in terms of the dependence on frequency of the applied electrical field or temperature (at fixed frequency).

Plotting dielectric loss (ε'') versus temperature at constant frequency for a pulsed-plasma produced acrylic acid–styrene (1 : 1) copolymer, and heating this sample twice revealed a strong difference between the two thermal cycles (heating to 475 K and cooling down) (Figure 12.24). The increase of log ε'' at temperatures greater than 425 K is due to the movement of charge carriers. At first glance, the cooling down and second heating are similar. The might be due to crosslinking by radical recombination and/or peroxide formation of radicals by reaction with oxygen from air.

Such post-plasma oxidation might be auto-oxidation, forming O-functional groups, which significantly contribute to high ε''-values in plasma polymers.

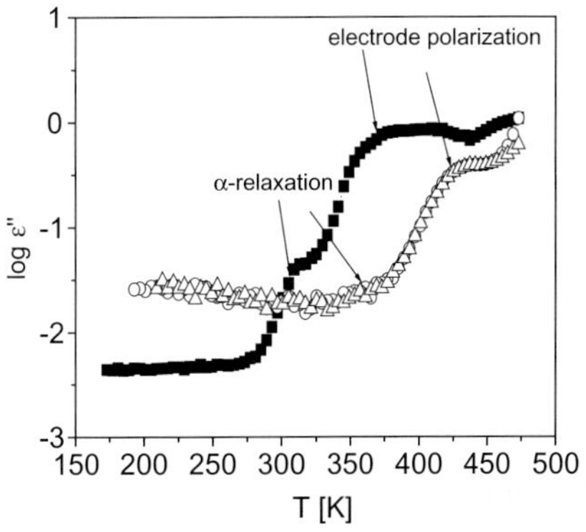

Figure 12.24 Dielectric loss (ε'') versus temperature at a frequency of 1 kHz for pulsed-plasma produced from acrylic acid and styrene (1 : 1 molar ratio) on (■) first heating, (○) cooling, and (Δ) second heating.

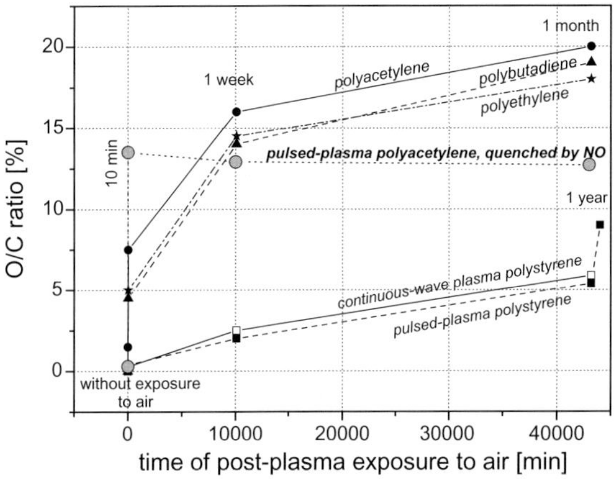

Figure 12.25 Post-plasma oxygen incorporation of plasma polymers on exposure to ambient air as well as after NO gassing.

12.9 Plasma-Polymer Aging and Elimination of Radicals in Plasma Polymers

All plasma polymers, in particular those produced from ethylene, acetylene, or styrene, undergo post-plasma oxygen introduction when exposed to ambient air (Figure 12.25) [18, 36, 54].

To prevent extensive formation of radicals during plasma polymerization "soft" plasma conditions (low wattage, pulsed-plasma mode) can be applied. Nevertheless, high concentrations of radicals are produced. Two post-plasma processes can quench the radical sites and their unforeseeable oxidation reactions. The first variant is gassing with NO (Figure 12.25) [1].

Evidently, the gaseous radical scavenger NO can completely stop the post-plasma oxidation of the plasma polymer and can hold the oxygen percentage constant over long periods of exposure to air. The reaction is: $R^{\bullet} + {}^{\bullet}NO \rightarrow R\text{-}NO$, followed by further oxidation to $R\text{-}NO_2$ or $R\text{-}O\text{-}NO_2$ [55]. The thus produced NO_x groups are hydrophilic.

The second variant, quenching with bromine, seems to be more efficient and produces a more hydrophobic surface, which is important for hydrophobic fluorine-containing plasma polymers. The reaction is in sum: $2R^{\bullet} + Br_2 \rightarrow 2R\text{-}Br$. Spontaneous bromination of aliphatic structures is not possible (S_N1 or S_N2 reaction, endothermic reaction) but olefin double bonds are brominated (A_E) and also bromination by electrophilic substitution at aromatic rings (S_E) and at side chains by radical mechanism (S_{Ar}) may occur to a distinct extent.

The aging of plasma-produced polyethylene is clearly seen by changes of the C1s peak when exposed to ambient air for many days (Figure 12.26).

Oxygen introduction is obvious by asymmetric strong broadening of the C1s peak at the side of higher binding energies. Special attention may be paid to an accentuated shoulder at 289 eV, assigned to carboxylic groups (COOH).

The aging of plasma-polymerized ethylene is also obvious in the valence band spectrum (Figure 12.13). Also in the valence band spectra, extensive oxygen occupation is evident after aging of pulsed-plasma polyethylene for 30 days. It was mentioned earlier that the valence band spectrum of pp-PE is similar to that of reference PP.

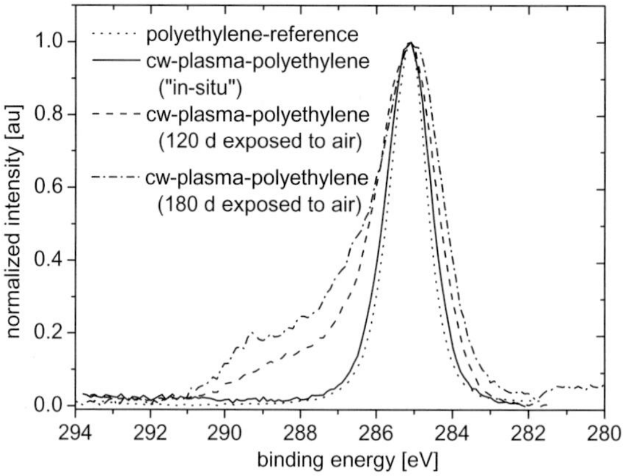

Figure 12.26 Changes in C1s peak of plasma-polymerized ethylene during exposure to air.

12.10
Functional Groups Carrying Plasma-Polymer Layers

Plasma polymerized organic thin films have proven to be of great use in the area of specifically engineered surfaces. These films have received considerable attention because of their potential applications in several technologies such as biocompatibility or adhesion promotion [56–60]. In the plasma polymerization process, gaseous or volatile compound are fragmented, excited, and/or ionized in glow discharge plasma, and then re-assembled on a surface. Past studies on plasma polymerization focused on continuous-wave discharge plasmas. However, in this high-energy mode a great fraction of functional-group bearing units are fragmented and many of the functional groups are destroyed. Therefore, in this process a very small percentage of the monomer functional group could be retained in the resulting plasma film. To retain higher degrees of functionality and better quality films the pulsed discharge technique was the method of choice. Details were presented in the previous section. An additional advantage of plasma polymerization is that it can be performed on almost any surface. Another route is the deposition of classic polymers carrying functional groups, such as poly(acrylic acid), poly(vinyl amine), or poly(vinyl alcohol), as intact ultrathin layer by electrospray (ESI) or with insignificant degradation in the additional presence of a plasma (APCI – atmospheric-pressure chemical ionization).

Using the pulsed-plasma mode with low duty cycle and introducing only low wattage into the plasma the retention of functional groups should be maximal and, thus, only one sort, or one dominant sort, of functional group should be formed. In such a case the stoichiometry of the monomer determines the maximal concentration of functional groups at, ideally, 100% retention. Vinyl monomers used for layer formation show a maximal concentration of functional groups of 50 groups per 100 C and for allyl monomers this is 33 groups per 100 C atoms.

Such plasma polymers can be deposited as about 50–300 nm thick layers onto all kinds of solids such as polymers, metals, inorganics, powders, or fibers. Figure 12.27 depicts schematically several types of monosort functional groups produced by plasma polymer coating using easily polymerizable monomers [61–67].

Functional group carrying plasma polymer layers have been produced from acrylic acid, allyl alcohol, allylamine, glycidyl methacrylate, acrylonitrile, acryl amide, allyl bromide, allyl iodide, vinylsulfonic acid, crotonaldehyde, acid chlorides, and so on [66–69].

12.10.1
Allyl Alcohol

The OH groups at surfaces can be easily produced in relatively high selectivity (90% retention of OH groups referenced to the monomer OH/100 C ratio) using allyl alcohol (AAl) or hydroxyethyl methacrylate (HEMA) [2, 70]. Allyl alcohol is difficult to polymerize chemically, as discussed in detail before [29]. However, using the pulsed low-pressure radiofrequency plasma and low wattages (3–30 W)

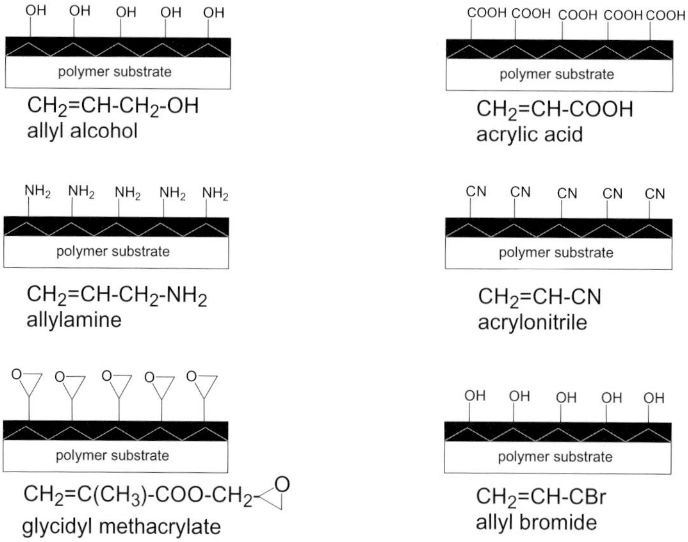

Figure 12.27 Schematic structure of functional group carrying plasma polymers.

allyl alcohol can be polymerized easily with a high retention degree of OH groups (pp-PAAl) [45]. The yield in OH groups amounts to about 30 OH/100 C as compared to a theoretically possible 33 OH/100 C, that is, the selectivity is >90%. A survey scan and the C1s signal are as expected (Figure 12.28). Only two weak additional components were found in the C1s peak.

The C–O singly bonded at a binding energy of about 286.3 eV decreased in intensity if the plasma dose was increased. This subpeak, produced by C-OH, might also be contributed to by traces of C-O-C, epoxy, and C-O-OH groups. The percentage of this subpeak compared to the whole C1s signal was lowered at high wattages in favor of increasing subpeaks at higher binding energies (Figure 12.29).

At about 287.5 eV undesired aldehyde and/or ketones were also present, often designated misleadingly as "carbonyls." Other "carbonyls" are also clearly identifiable at 289 eV, that is, carboxylic acid groups and esters.

Compared to the theoretical O/C ratio (0.33) for an ideal poly(allyl alcohol) polymer, higher O/C values (0.38) were often observed for plasma-polymerized films prepared under "mild" deposition conditions [39]. During plasma polymerization, so it was argued, two simultaneous processes, A and B, are present. Process A is the "polymerization" of the excited monomer molecules that leads to the retention of C-OH groups on the film and process B is the fragmentation of monomer molecules and the subsequent recombination of these fragments. The excess of oxygen observed in the films can be related to process B.

The IR-ATR spectrum confirms the existence of O-H and C-O stretching vibrations of high intensity (Figure 12.30). Low to moderate intensity is ascribed to the

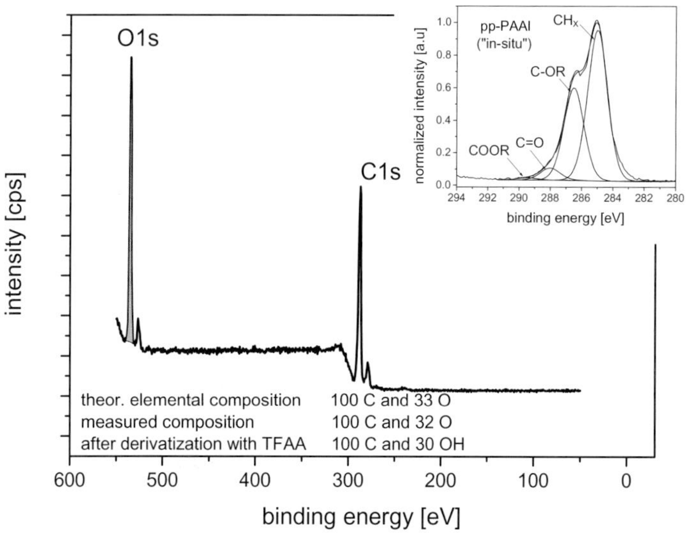

Figure 12.28 XPS survey scan and C1s peak of pulsed-plasma polymerized poly(allyl alcohol) (pp-PAAl) (low dose).

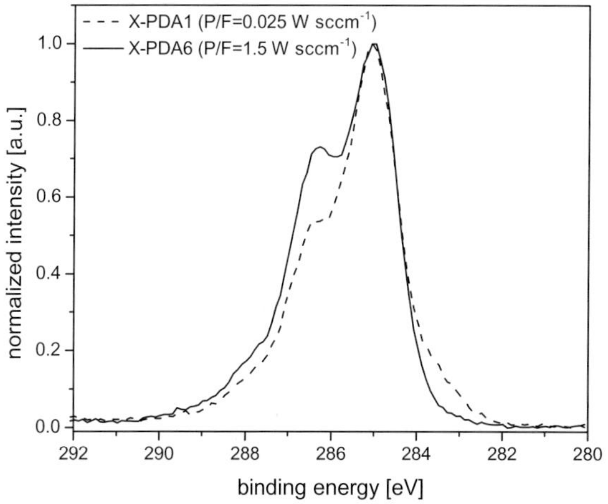

Figure 12.29 C1s signals of pp-PAAl using low and high energy doses.

(undesired) C=O stretching mode at $1714\,cm^{-1}$, thus confirming the XPS data on the newly formed higher oxidized species also for layers of pp-PAAl that are 2.5 μm thick, which is the sampling depth of the used ATR technique.

Considering plasma deposited allyl alcohol films, it is found, using ToF-SIMS, that the summarized intensity of all positively charged aromatic ions Σ_{Arom} increases

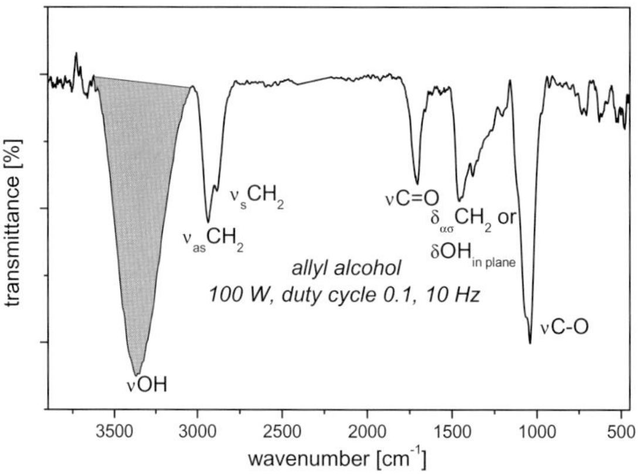

Figure 12.30 FTIR-ATR transmittance spectrum of pulsed-plasma polymerized allyl alcohol.

with the energy dose (Yasuda factor) [38]. This does not indicate the existence of aromatic structures within pulsed-plasma polymerized allyl alcohol (pp-PAAl) but, rather, at least the existence of unsaturations (C=C double bonds) [1, 65, 71]. An increase in branching and/or crosslinking of the plasma deposited poly(allyl alcohol) films with increasing plasma dose (Yasuda factor) was derived from the ratio of the total yield of C_6–C_8 hydrocarbon secondary ion clusters to the total yield of C_2–C_8 clusters $[\Sigma(C_6–C_8)/\Sigma(C_2–C_8)]$ [72]. Although the $[\Sigma(C_6–C_8)/\Sigma(C_2–C_8)]$ ratio does not discriminate between branching and crosslinking its value was found to increase with either of these characteristics.

SSIMS could contribute to a discussion on the retention of hydroxyl groups in plasma polymerized allyl alcohol films. Ameen *et al.* discussed this problem by comparing the positive ToF-SSIMS spectra of poly(vinyl alcohol) (PVA) and poly(ethylene oxide) (PEO) [65]. Oran *et al.* found that there is only a low conversion rate from the monomer's hydroxyl groups into other oxygen functionalities like ether groups in the plasma [38]. It seems that the positive secondary ions at *m/z* 31, 57, 101, 115, and 173 can be used as key fragments to prove the retention of hydroxyl groups in the plasma deposited allyl alcohol films [38]. The maximum yields of these secondary ions were found at low energy dose (P/F ratio of 0.05 W sccm^{-1}) and the lowest yields at high dose (P/F = 2.5 W sccm^{-1}). The oxygen content of the films was evaluated using the $(O^- + OH^-)/CH^-$ intensity ratio. This ratio was found to correlate with the atomic O/C ratio measured by XPS (Figure 12.31) [38, 39].

Mix *et al.* had derivatized all remaining OH groups within the plasma polymer by consumption with trifluoroacetic anhydride (TFAA):

$$\blacksquare\text{-OH} + \text{O=C(OCF}_3)_2 \rightarrow \blacksquare\text{-OOC-CF}_3 + \text{CF}_3\text{-COOH}$$

and had found a decrease of OH-group concentration from 31 OH/100 C at 30 W to 23 OH/100 C at 300 W (Figure 12.32) [46].

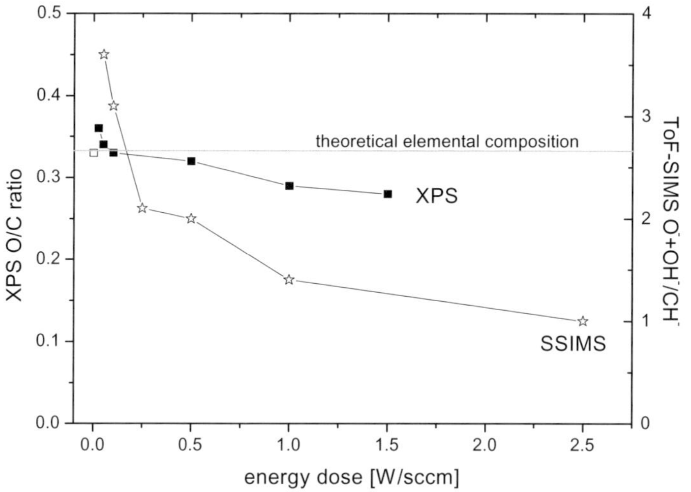

Figure 12.31 O/C ratio in pp-PAAl, as measured by XPS and ToF-SIMS, versus energy dose.

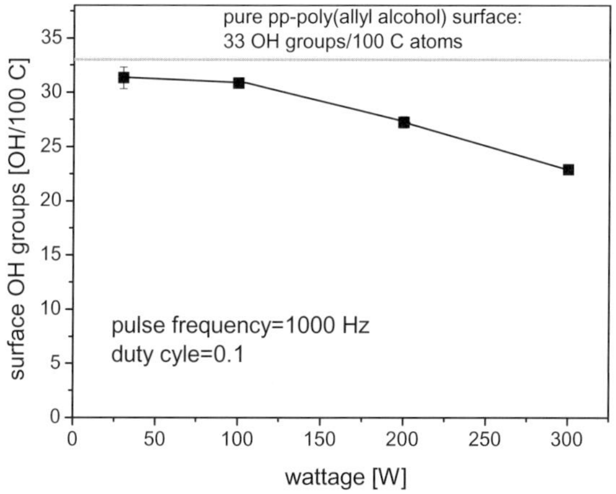

Figure 12.32 Dependence of TFAA-labeled OH groups of pulsed-plasma polymerized allyl alcohol on wattage.

The concentration of surviving OH groups is plotted in Figure 12.33 against the duty cycle of pulsed-plasma polymerization at a high energy dose (300 W) [46].

Figure 12.34 presents the C K-edge NEXAFS spectra of reference poly(vinyl alcohol) [73], reference poly(vinyl methyl ketone) [73] and a representative plasma polymerized allyl alcohol sample prepared for NEXAFS analysis.

The principal features seen in the displayed spectrum are representative. An important feature in the spectrum of the plasma polymerized sample is the

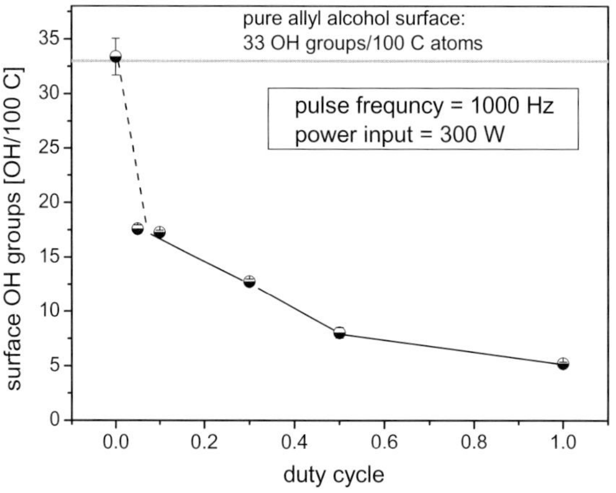

Figure 12.33 Dependence of TFAA-labeled OH groups of pulsed-plasma polymerized allyl alcohol on duty cycle.

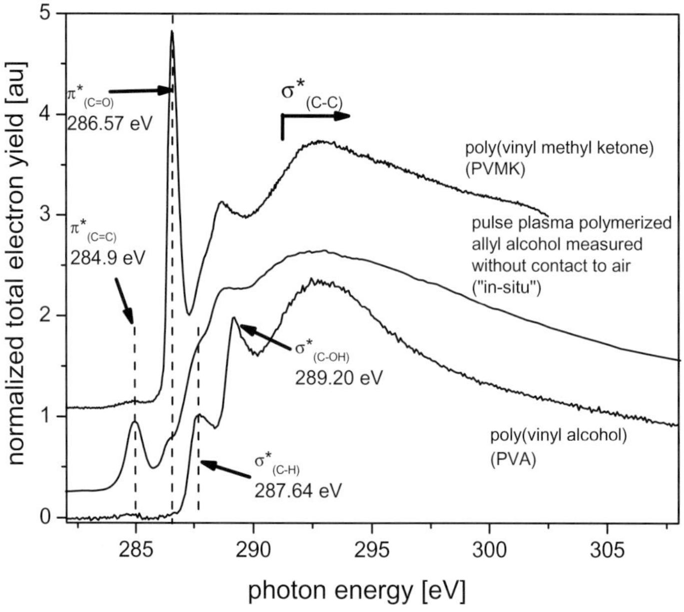

Figure 12.34 NEXAFS-C_K-edge spectra of poly(vinyl methyl ketone) (PVMK) reference, pp-PAAl (mild plasma condition = 1.0 W $sccm^{-1}$), and (poly(vinyl alcohol) (PVA) reference.

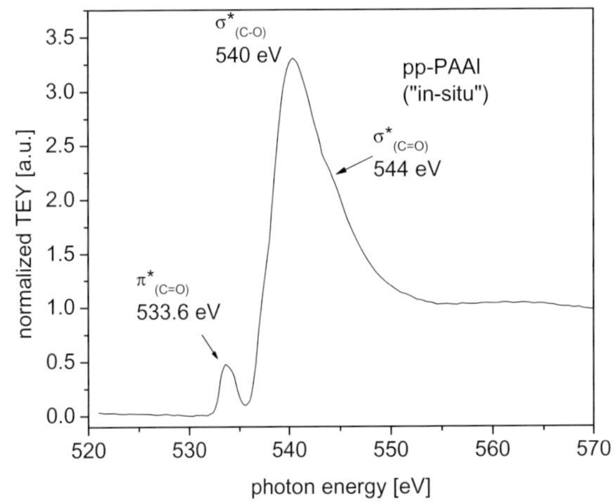

Figure 12.35 NEXAFS O_K-edge spectrum of pp-PAAl (mild plasma condition = 1.0 W $sccm^{-1}$).

presence of a C1s → $\pi^*_{(C=C)}$ feature [photon energy ($h\nu$) ≈ 284.9 eV] characteristic for the presence of C=C bonds. Comparison of the area under this feature also provides a specific semi-quantitative measure of the number of C=C bonds in samples prepared under different plasma conditions. A comparison with the reference spectra highlights some other features in the C K-edge NEXAFS spectrum of pp-PAAl: a shoulder at ~286.7 eV represents a C1s → $\pi^*_{(C=O)}$ resonance, another shoulder at ~287.6 eV represents a new C-H* resonance, a C1s → $\sigma^*_{(C\text{-}O)}$ at ~289 eV and C1s → $\sigma^*_{(C-C)}$ resonances above 292 eV were observed. The O K-edge NEXAFS spectra of the plasma-polymerized samples were also measured (Figure 12.35) to obtain information on the presence of C=O species. The spectral feature at photon energy 533.6 eV represents the O1s → $\pi^*_{(C=O)}$ resonance and can be used to obtain specific semi-quantitative information on the number of C=O bonds in the plasma polymerized samples. This spectrum also shows a $\sigma^*_{(C\text{-}O)}$ resonance at 540 eV and a shoulder representing the $\sigma^*_{(C=O)}$ resonance at 544 eV.

The IR spectra confirm the loss of OH groups with increasing wattage during the pulsed-plasma polymerization and vice versa the increase of the carbonyl feature at 1714 cm^{-1} with increasing of wattage (Figure 12.36).

A similar tendency is obvious if assigning the cw-plasma "hard" plasma conditions and the low-duty-cycle pulsed-plasma "mild" plasma conditions (Figure 12.37). The lower the duty cycle (duty cycle = 0.05) the higher the OH stretching mode, the higher the duty cycle the lower the OH concentration. This behavior corresponds to the XPS results of TFAA-labeling of OH groups.

Both C–O and C–OH related groups decrease in concentration with increasing power input, while the concentration of C=O-containing groups (groups of higher oxidation state) increase, as FTIR evaluation confirms (Figure 12.38).

Figures 12.39 and 12.40 present the dielectric loss versus temperature at 1 kHz for varied duty cycle and pressure, respectively. The dielectric loss for pp-PAAl is

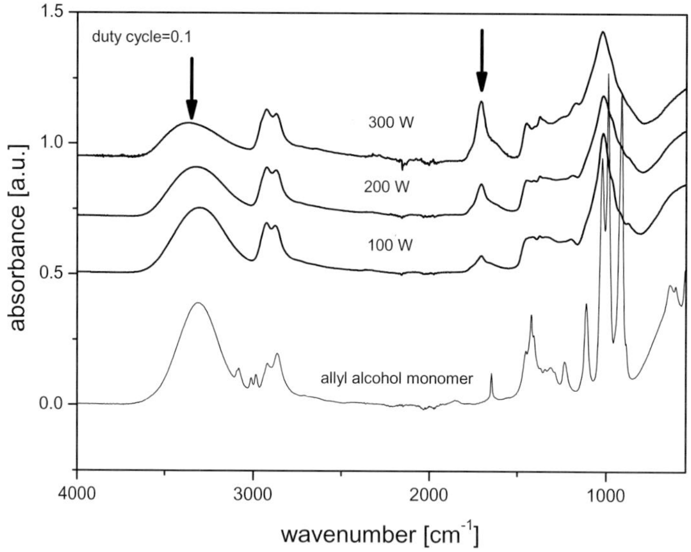

Figure 12.36 Variation of FTIR-ATR spectra of pp-PAAl with wattage.

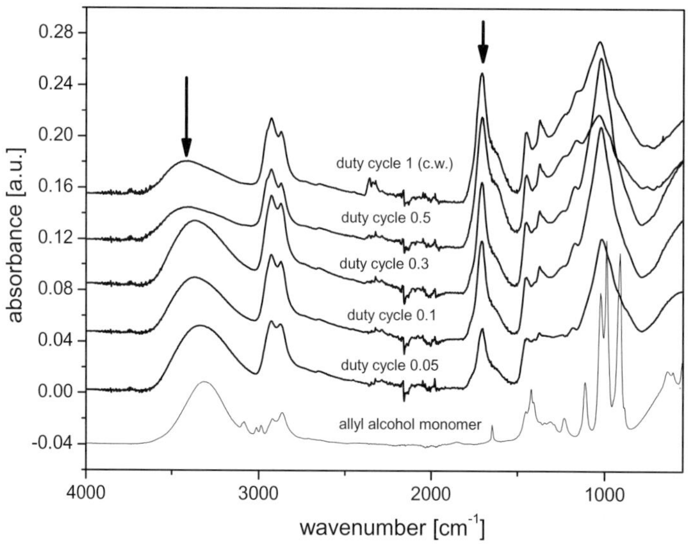

Figure 12.37 Variation of FTIR-ATR spectra of pp-PAAl on duty cycle at constant wattage (300 W).

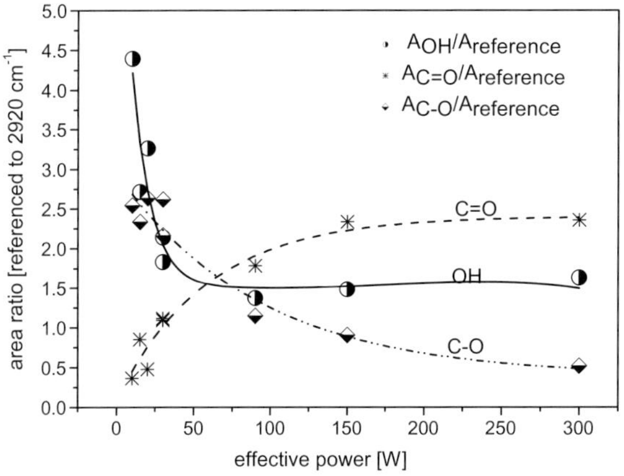

Figure 12.38 Dependence of IR bands characteristic for OH, C=O, and C–O groups on effective power (referenced to plasma on).

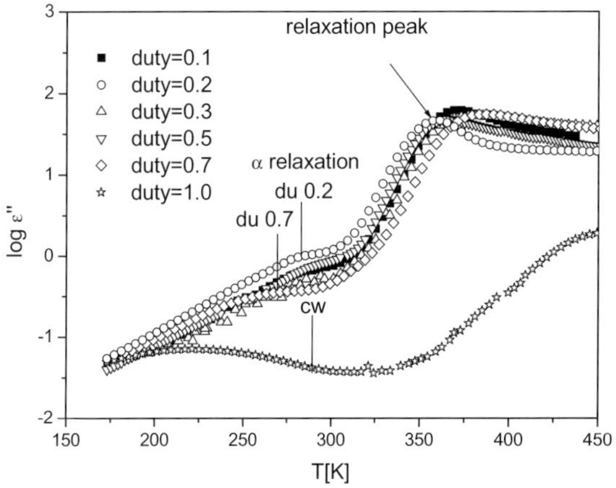

Figure 12.39 Dielectric loss ε″ of pp-PAAl produced using different duty cycles (power = 100 W, pressure = 10 Pa, monomer flow rate = 20 sccm, and pulse frequency = 1 kHz).

much higher than for cw plasma, while low pressures show slightly lower values. The high dielectric loss of pulsed-plasma polymers (duty cycle < 1) is unexpected because it was assumed that supernumerous radicals were saturated by reactions in the plasma-off period. It must be considered that chain propagation did not annihilate radicals but that such chain growing may contribute to bury and trap radicals.

The assigned α-relaxation is related to the glass transition of the system and is called a dynamic glass transition. The dynamic glass transition is defined as the

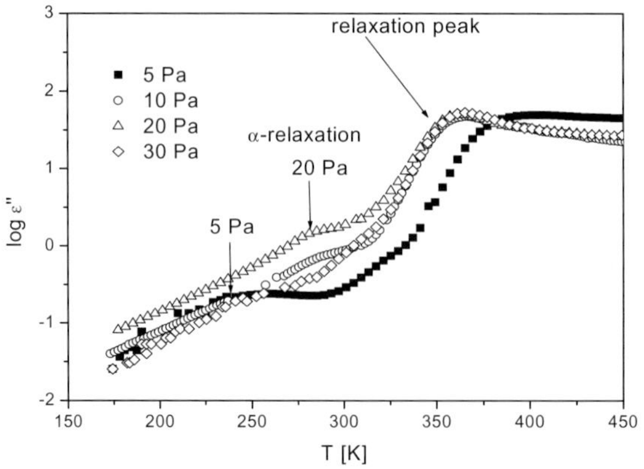

Figure 12.40 Dielectric loss ε″ of pp-PAAl produce using different plasma pressures (power = 100 W, duty cycle = 0.5, monomer flow rate = 20 sccm, and pulse frequency = 1 kHz).

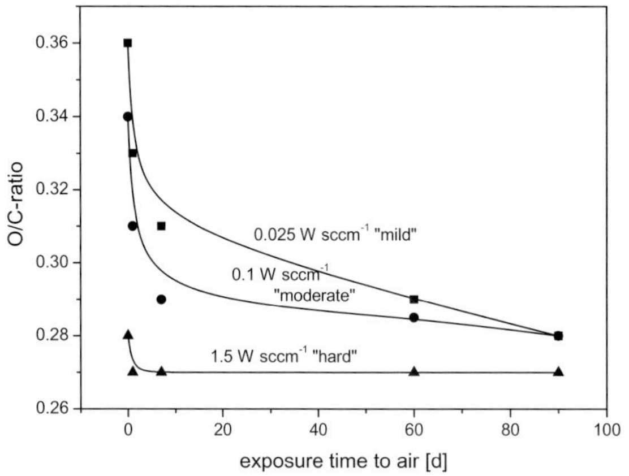

Figure 12.41 Aging (change in O/C ratio) of pp-PAAl films on exposure to ambient air.

temperature of maximal loss at the selected frequency. The glass transition is a cooperative phenomenon and for polymers the α-process corresponds to micro-Brownian motion of the segments that form the chain [74].

To investigate aging, freshly prepared films were examined without and with exposure to air in detail. In the case of plasma deposited allyl alcohol films the aging in air is probably based on two competing processes, that is, oxygen uptake (Figure 12.41) and diffusion of low molecular weight oxidized material (LMWOM) together with re-orientations of oxygen functional groups (Figure 12.42). For samples prepared under "mild" conditions, the re-orientation and the diffusion of LMWOM is more effective than the oxygen uptake process because the samples are less crosslinked and the amount of radicals is comparably low ($R^{\bullet}$ + $^{\bullet}O{-}O^{\bullet}$

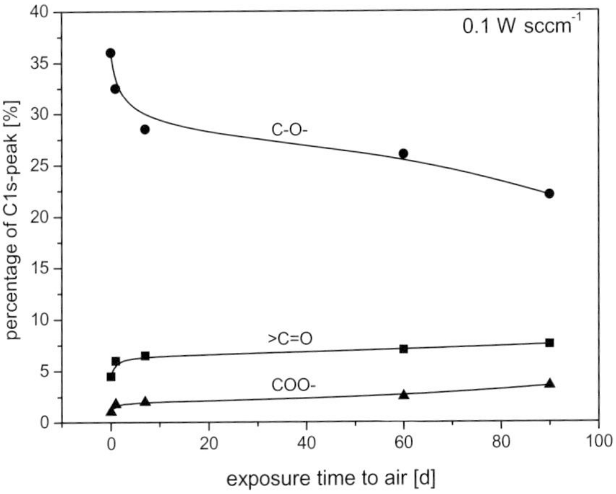

Figure 12.42 Changes in C1s peak fitting during aging of pp-PAAl.

→ R–O–O$^\cdot$). For samples prepared under "harder" plasma conditions, crosslinking and the amount of surface radicals are increased and the oxygen uptake becomes the dominant process.

Among the different O-species present in pulsed-plasma polymerized allyl alcohol the C–O (dominated by OH groups) species diminish slightly from the surface during exposure to ambient air (Figure 12.42).

12.10.2 Allylamine

Plasma polymer layers containing primary amino groups were deposited from allylamine (CH_2=CH–CH_2–NH_2) or diaminocyclohexane [$C_6H_{10}(NH_2)_2$] [33, 61, 75–83]. The C1s peak and the stoichiometry of the pulsed-plasma produced poly(allylamine) (pp-PAAm) do not show significant deviations from the expected elemental composition and structure (Figure 12.43).

It is characteristic for pp-PAAm that the oxygen content increases strongly within a few hours of exposure to air (Figure 12.46) [33, 84, 85].

Derivatization of the primary amino groups with pentafluorobenzaldehyde (PFBA) or trifluoromethylbenzaldehyde (TFMBA) provides information on the real concentration of amino groups at the surface of the pp-PAAm layer (R-NH_2 + OHC-aryl-F_5 → R-N=C-aryl-F_5-PFBA and R-NH_2 + OHC-aryl-CF_3 → R-N=C-aryl-CF_3-TFMBA) [32, 86–89]. Both derivatization reactions were not complete and showed ca. 80% consumption of NH_2 bond on referencing the results to standards with known NH_2 concentration [24]. Post-plasma oxidations at α-C- and N-atoms may hinder completion of the derivatization reaction. The results of these derivatizations are scattered strongly, from 10 to maximal 18 NH_2/100 C [31, 90, 91].

Reactions with amino groups during the polymerization process were widely discussed (formation of secondary amines, imines, amides, nitriles, and isonitriles)

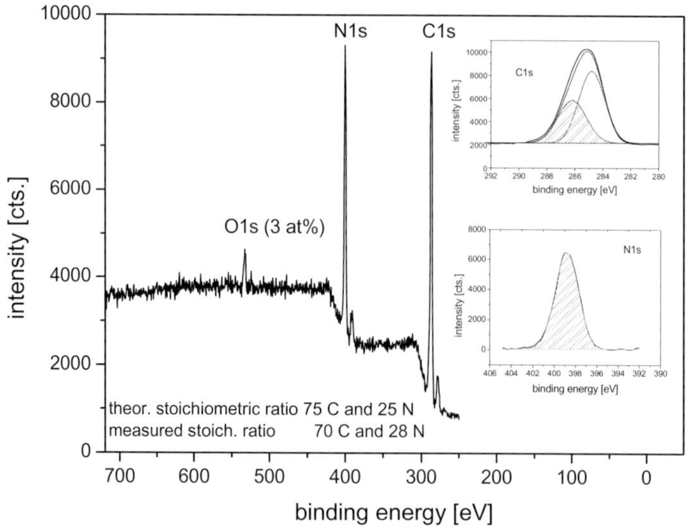

Figure 12.43 XPS-survey scan and C1s peak of pulsed-plasma produced poly(allylamine) (pp-PAAm).

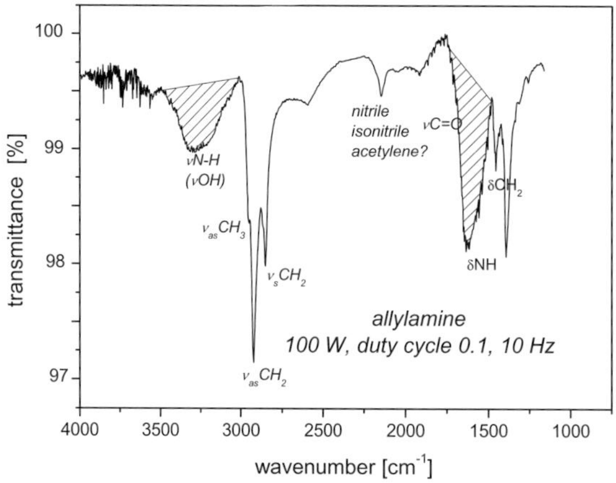

Figure 12.44 Infrared spectrum of pp-PAAm.

[24, 92, 93]. The amino groups can be protected by reversible reaction with special organic groups [94]. Characteristic absorptions found between 2100 and 2160 cm^{-1} were assigned to nitriles [$\nu_{C\equiv N} = 2240\,cm^{-1}$ in poly(acrylonitrile)], isocyanates ($\nu_{N=C=O} = 2240\,cm^{-1}$, might produced by post-plasma oxidation) [95], or preferably to un-symmetric substituted acetylenes ($\nu_{C\equiv C} = 2100–2160\,cm^{-1}$) [96] or isonitriles ($\nu_{N\equiv C} \approx 2100–2200\,cm^{-1}$) [97].

Figure 12.44 shows a typical IR-transmittance spectrum of pp-PAAm with the characteristic band at 2170 cm^{-1}.

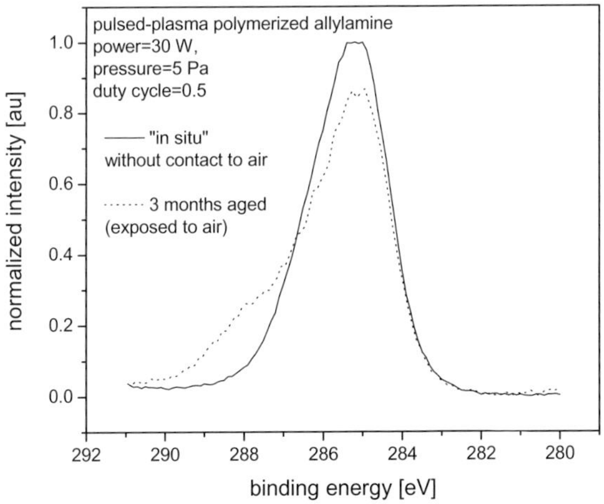

Figure 12.45 XPS C1s peaks of pp-PAAm referenced to the same peak area.

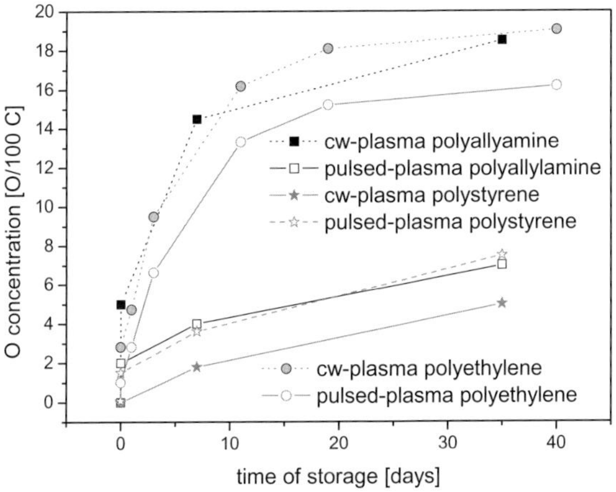

Figure 12.46 Aging behavior of pp-PAAm and other plasma polymers during exposure to ambient air.

Post-plasma oxidation of pp-PAAm can be clearly evidenced by changes in the respective C1s signals from samples without contact to air and samples exposed for three months to ambient air (Figure 12.45).

Comparison of peaks before and after aging shows a significant decrease in intensity for the hydrocarbon component, which might be represented also by an unsaturated polymer backbone at 285.0 eV. Moreover, the C–N/C–O component at ca. 286 eV is also decreased in intensity at the expense of C=O and COOH groups at higher binding energies.

Figure 12.46 depicts the post-plasma oxygen incorporation of pp-PAAm compared to that of other plasma polymers such as pp-PS and pp-PE as well as to those produced using the cw-plasma mode.

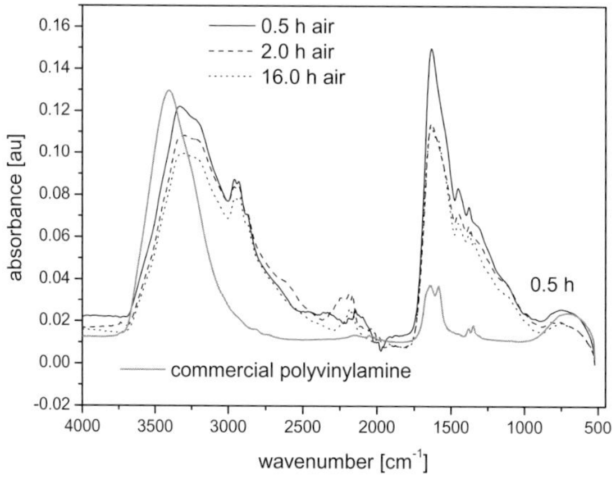

Figure 12.47 FTIR spectra of pp-PAAm (duty cycle = 0.1) in comparison to poly(vinylamine) reference material.

The IR-spectra also indicate the growth of C=O-related groups during exposure of pulsed-plasma produced poly(allylamine) to ambient air (Figure 12.47).

12.10.3
Acrylic Acid

Plasma polymer layers containing carboxylic groups can be produced using acrylic acid or other unsaturated acids or anhydrides (maleic anhydride) [69, 98–102].

As expected from a chemical point of view the polymer deposition rate of acrylic acid is high, indicating significant contribution of chemical chain growth. Therefore, Yasuda found a higher deposition rate using the pulsed-plasma mode than for the cw-mode [103]. The XPS-spectrum and the C1s and O1s signals correspond to those given for commercial poly(acrylic acid) (PAA) (Figure 12.48) [104].

The O/C-elemental ratio was slightly changed. The atomic oxygen percent changed from a theoretical 40 to 36 for pulsed-plasma polymerization. The expected carboxylic group concentration of 33 COOH/100 C was lowered to 22–28 COOH/100 C, as measured by derivatization. After hydroxyl groups, the retention degree was greatest for COOH at about 85%; the retention for OH groups >90% and for amino groups 50–65% [30]. Derivatization was performed by using trifluoroethanol in the presence of a carbodiimide for water elimination, and the fluorine concentration was measured by XPS (R-COOH + HO-CH_2-CF_3 $\rightarrow$ R-CO-O-CH_2-CF_3) [105].

The yield in carboxylic groups depends strongly on the duty cycle (Figure 12.49).

The lower the duty cycle is the higher the COOH retention, evidencing participation of a chemical chain growing reaction. At high duty cycle or under cw-plasma mode the COOH retention is minimal and near 10–12 COOH/100 C, that is, under such "hard" conditions only 33% of the monomer-introduced COOH groups have survived the plasma polymerization process.

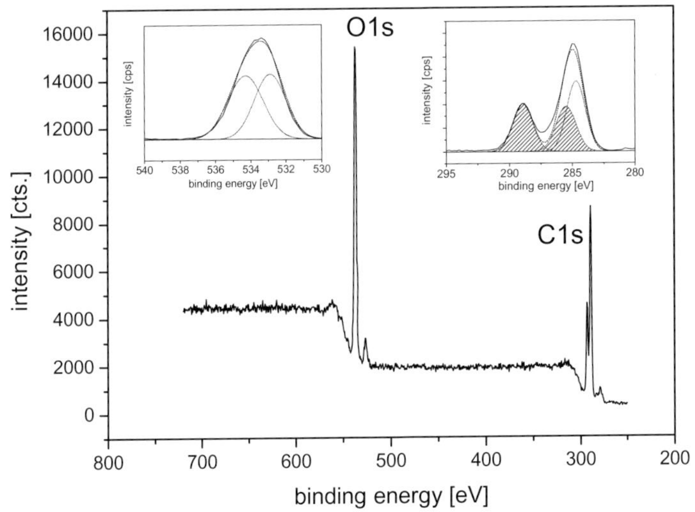

Figure 12.48 XPS survey scan and C1s peak of pulsed-plasma polymerized poly(acrylic acid) (pp-PAA).

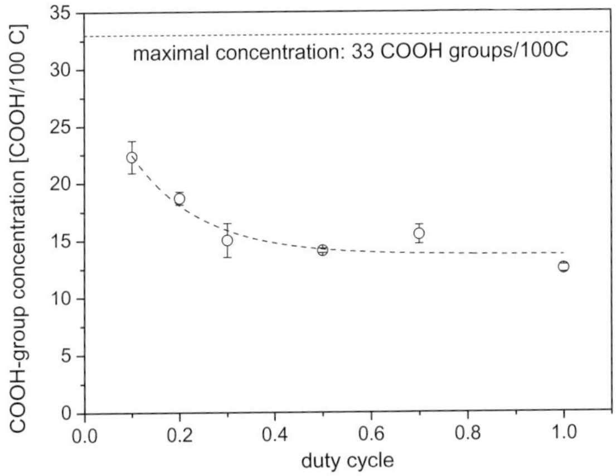

Figure 12.49 Concentration of carboxylic groups versus duty cycle. The dashed line is a guide for the eye. The error bars result from three different measurements with the same sample. The dotted line indicates the maximum possible concentration of carboxylic groups.

The IR spectrum differs from that of commercial poly(acrylic acid) (Figure 12.50). The ν_{OH}-stretching vibration shows after referencing to the $\nu_{C=O}$-band only 65% intensity in comparison to the commercial PAA, that is, the COOH groups (ν_{OH} stretching) did not survive the plasma process completely and a fraction of them was converted into carbonyl species other than carboxylic groups, such as esters, aldehydes, or ketones, present in a small subpeak at 287.8 eV in the C1s signal.

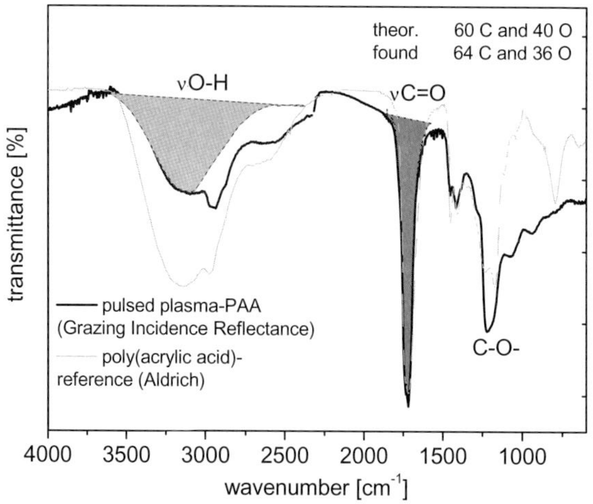

Figure 12.50 Infrared spectrum (grazing incidence reflectance, 50 nm) of pp-PAA (duty cycle = 0.1, power = 30 W) and of PAA reference.

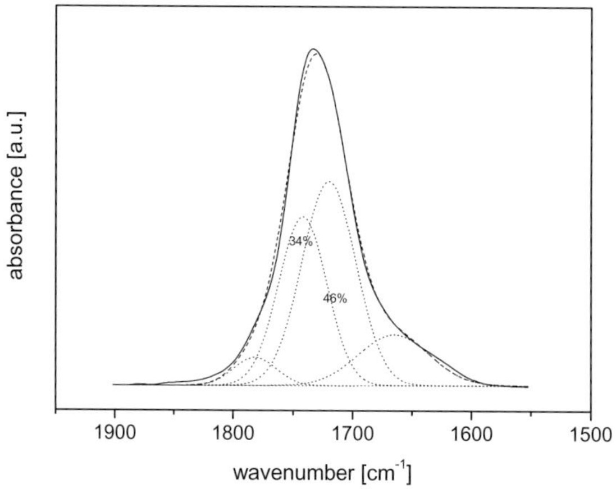

Figure 12.51 FTIR spectrum of pp-PAA (duty cycle = 0.2, fit of four components of the C=O stretching vibration).

The carbonyl absorption at ca. 1710 cm^{-1} was investigated in more detail (Figure 12.51).

The peak fitting of $\nu_{C=O}$ was performed using four (speculative) components, anhydride, ester, acid, and double bonds. Using the normal positions of these bands [95, 96], which is a raw approximation, then a considerable fraction of $\nu_{C=O}$ has its origin in ester groups. This contribution was located at 1742 cm^{-1} and may be produced by self-condensation of acrylic acid (under plasma conditions): CH_2=CH-COOH → ~CH_2-CH_2-CO-O~ [106, 107]. This may be the reason for peak shifting from 1703 to 1735 cm^{-1} seen between commercial poly(acrylic acid) refer-

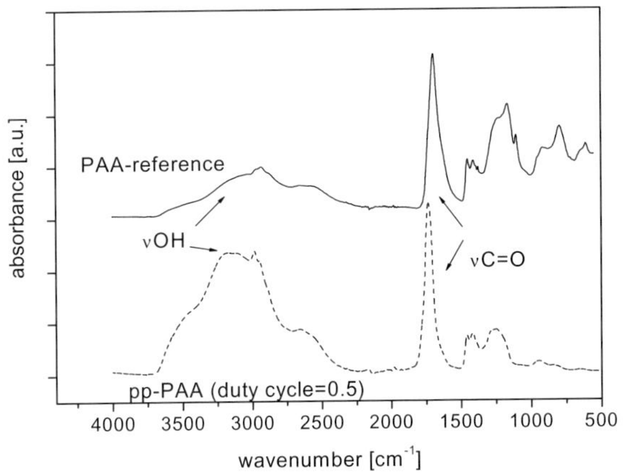

Figure 12.52 FTIR absorbance spectrum of pulsed-plasma polymerized poly(acrylic acid) (pp-PAA) deposited with a duty cycle of 0.5 compared with the spectrum of commercial PAA (reference).

ence material and pulsed-plasma polymerized PAA (cf. Figures 12.51 and 12.52). In the range 1727–1745 cm^{-1} ester groups are found, such as from poly(ethylene terephthalate) or poly(methyl methacrylate). The $\nu_{C=O}$ of acrylic acid is positioned at 1712 cm^{-1} [95].

Figure 12.53 plots the differential scanning calorimetry (DSC) results of the dependence on duty cycle. From high-molar mass reference PAA to low-molar mass reference PAA the glass transition temperature decreases with growing duty cycle. The difference in T_g to that of high-molar mass reference material amounts to between 50 and 80 °C. The inset of Figure 12.53 compares the DSC curves from the second heating run for high-molar mass reference material and pulsed-plasma produced poly(acrylic acid) deposited with duty cycles of 0.1 and 0.7. For each material, a step-like change in the heat flow is observed, which indicates the glass transition. The glass transition temperature (T_g) is estimated from the inflection point of the heat flow and is plotted versus the duty cycle in Figure 12.53.

For all pulsed-plasma deposited poly(acrylic acid) samples the T_g values are much lower than those for commercial PAA. It may be concluded that during the plasma polymerization a highly branched structure with many dangling bonds is produced, which may act as internal plasticizer. Another interpretation is that of the formation of low-molar mass products and macrocycles, which also plasticize the pp-PAA. With increasing duty cycle the number of these dangling ends increases, or more low-molar mass fragments were produced, lowering the glass transition temperature. Extrapolation of this dependence to duty cycle = 0 meets the value of T_g of PAA 1800.

Discussing the dielectric behavior of pulsed-plasma polymerized acrylic acid in more detail, a good impression on differences to reference PAA can be obtained from Figure 12.54.

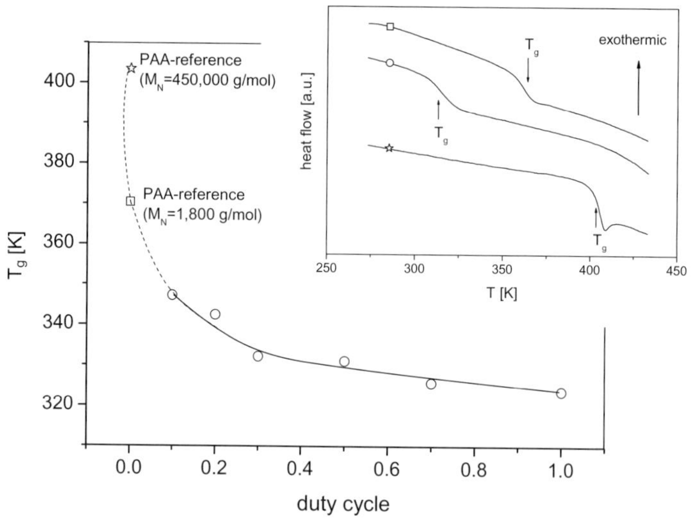

Figure 12.53 Glass transition temperature (T_g) versus duty cycle for conventional and pp-poly(acrylic acid). The solid line is a guide for the eye. The dashed line is the extrapolation of the dependence obtained for pp-poly(acrylic acid) to reference PAA. Inset: heat flow versus temperature for reference and pp-poly(acrylic acid) for the second heating run at a rate of $10\,K\,min^{-1}$: (○) pp-PAA duty cycle = 0.1; (□) pp-PAA duty cycle = 0.7; and (open star) PAA-reference ($450\,000\,g\,mol^{-1}$).

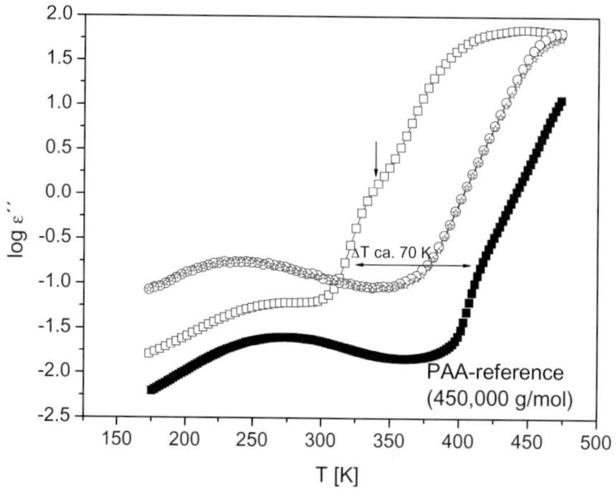

Figure 12.54 Dielectric loss versus temperature at a fixed frequency of 1 kHz of a pulsed-plasma deposited sample (duty cycle = 0.5): (○) first heating, (□) first cooling, and (open stars) second heating and comparison to PPA-reference material (■) ($450\,000\,g\,mol^{-1}$; first heating).

This plot presents the dielectric loss as a function of temperature for a fixed frequency (isochronal plot) and for different thermal histories for pp-PAA produced with a duty cycle of 0.5. The first run is a heating run of the as-prepared sample from low to high temperatures. The second run is the subsequent cooling from high to low temperatures. A pronounced hysteresis is observed between those two runs. The third heating run is more or less similar to the cooling cycle. For all runs at low temperatures the β-relaxation is observed as a well-defined peak in the spectra. For the first heating run a shoulder is observed at higher temperatures than the β-relaxation. At the first glance it seems similar to the behavior observed for conventional polymerized poly(acrylic acid). Closer inspection shows that this shoulder, probably due to the glass transition, is shifted to lower temperatures for the plasma polymerized poly(acrylic acid). This is in agreement with the results obtained by DSC (cf. Figure 12.53).

12.10.4 Acrylonitrile

Polymer layers containing nitrile groups have been deposited using acrylonitrile as monomer [24]. Plasma polymerized acrylonitrile shows a few abnormalities caused by side- and post-plasma reactions [66]. Using the pulsed plasma and low wattage such side-reactions were slightly repressed [108]. The XPS survey scan for cw and for pulsed-plasma produced poly(acrylonitrile) (pp-PAN) shows undesired oxygen introduction (Figure 12.55).

In Figure 12.56 pulsed and cw-plasma-produced PAN are compared with a commercial PAN reference material.

The nitrile absorption at $2245\,cm^{-1}$ for reference PAN is changed due to side-reactions and, thus, signals for unsaturated and N-containing undesired groups

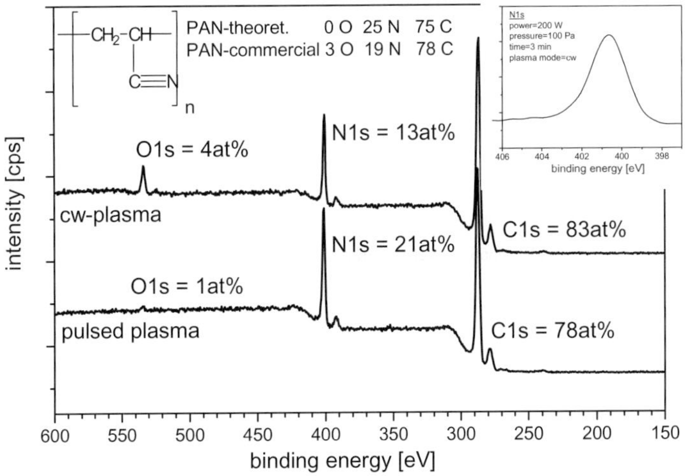

Figure 12.55 XPS-survey scan for pulsed and cw-plasma polymerized poly(acrylonitrile) (PAN).

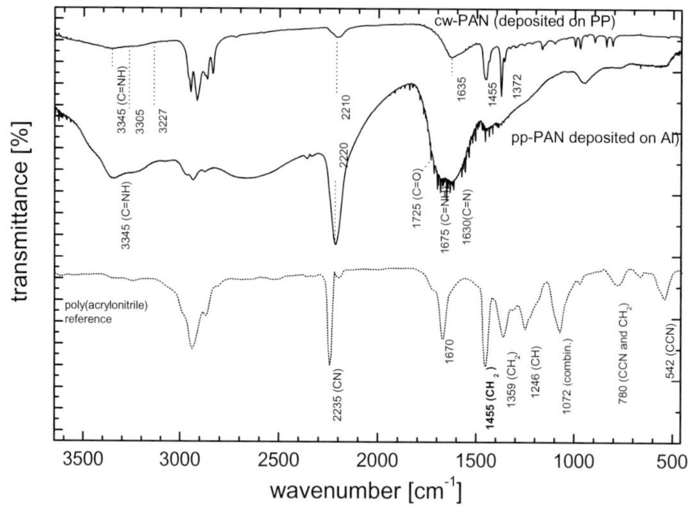

Figure 12.56 FTIR spectra of cw and pulsed-plasma polymerized acrylonitrile compared to that of a PAN reference.

appear (Figure 12.57). Besides nitrile, there may also occur isonitriles, isocyanates, mono- and disubstituted acetylenes, cumulenes (2025 cm^{-1}), and ketenes.

The situation is plotted also in Figure 12.58. As well as the two bands at 2245 and 2195–2000 cm^{-1}characteristic absorptions are also in the range 1600–1660 cm^{-1}.

These absorptions may be due to imines (R-C=NH); however, corresponding moderate absorptions in the range 3300–3400 cm^{-1} are not clearly identified. On the other hand, substituted imines (azomethine, Schiff's base) may be present with higher probability in this range, because there is no additional absorption in the 3000–3500 cm^{-1} region. Thus, it is probable that the C≡N triple bond has reacted to form azomethines or singly bonded C–N species (secondary, tertiary amines). The absence of a strong $\nu_{C=O}$ makes the presence of amide groups improbable.

Table 12.2 summarizes the maximal retention of functional groups for COOH, OH, and NH_2 [30].

The loss in oxygen for pp-PAAc indicates decarboxylation/decarbonylation of the COOH group (cf. Norrish I reaction) [109].

12.11
Vacuum Ultraviolet (VUV) Induced Polymerization

Vacuum ultraviolet (VUV) photo-polymerization by filtering of short-wavelength plasma UV radiation has also been used for depositing thin polymer layers. Low-pressure glow discharge plasmas emit intense resonance (fluorescence) radiation

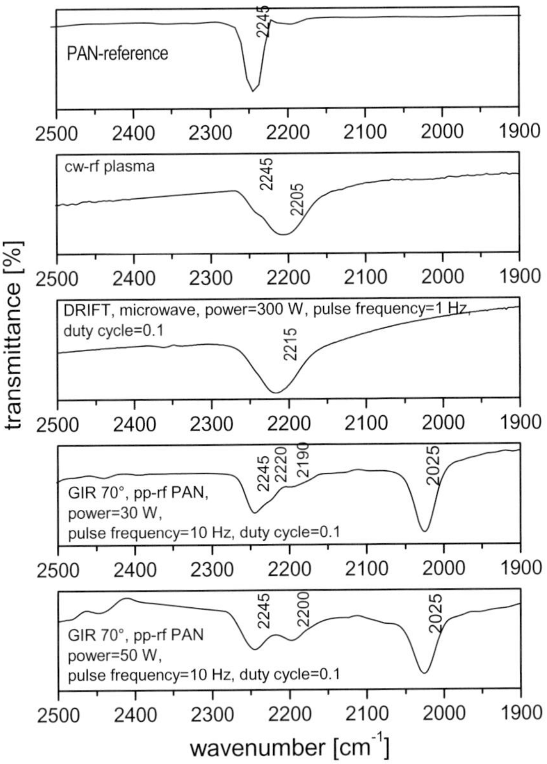

Figure 12.57 Variation in νC≡N for different types of plasma polymerized poly(acrylonitrile)s (pp-PANs) in comparison to reference poly(acrylonitrile) (PAN).

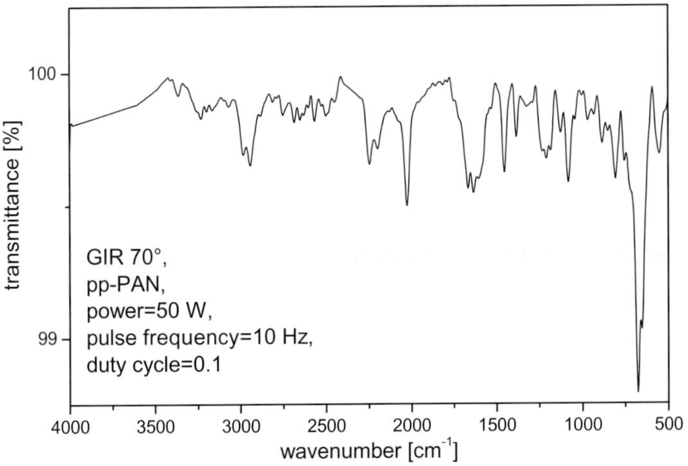

Figure 12.58 Grazing incidence reflectance spectrum of pp-PAN.

Table 12.2 Maximal yield of functional groups for pulsed-plasma polymerized monomers.

Monomer	Changes in elemental composition (orig. 100)	Percentage of group from C1s peak (%)	Retention yield measured by derivatization		
			%	Per 100 C (measured)	Per 100 C (theoretical)
Acrylic acid	90 (O)	80	85	≈28	33
Allyl alcohol	100 (O)	75	95	≈31	33
Allylamine	100 (N)	100	≈50	≈16	33

besides continuum radiation (recombination radiation and possibly Bremsstrahlung) [110]. These line are localized in the range 58–110 nm for He, 74–100 nm for Ne, 105–155 nm for Ar, and 148–200 for Kr. Important (high energetic) resonance lines are positioned at 105 and 107 nm (Ar) and 92 and 93 nm (Ar^+), 73 and 74 nm (Ne), as well as 46.1 and 46.2 nm (Ne^+), and 58 (He) and 30 nm (He^+) [111]. Thus, extreme UV (EUV) is included (10–120 nm), which is now gaining importance for photoelectron spectroscopy, solar imaging, and lithography. UV radiation is also produced by excimer laser or lamps [112, 113]. Here, fluorine (F_2) 157 nm, argon fluoride (ArF) 193 nm, krypton fluoride (KrF) 248 nm, xenon chloride (XeCl) 308 nm, or xenon fluoride (XeF) 351 nm are used.

Such VUV radiation was held responsible for polymer crosslinking [114]. Using Kr and Xe resonance irradiation plasma polymer layers with a high percentage of primary amino (70–80 NH_2/100 C) groups were produced by "copolymerization" of ethylene with ammonia with low energy consumption [115]. Other aspects of plasma polymerization by VUV are described in References [116–120].

Radiation in the VUV region covers energies from 6.2 eV (200 nm) to 12.4 eV (100 nm) and 124 eV (10 nm, only found for multiply charged ions). Thus, vinyl (double) bonds were easily activated ($\pi \rightarrow \pi^*$, C=C → C–C, ≈1.0–2.5 eV), covalent C–C and C–H bonds were easily broken ($\sigma \rightarrow \sigma^*$ transitions, ≈3.8 eV) and also $C\text{–}C_{arom}$ (5.8 eV) and can recombine as polymer similar to that of plasma polymer.

12.12
Plasma-Initiated Copolymerization

12.12.1
Reasons for Copolymerization

Plasma polymerization of volatile organic compounds is a technically available method used to produce thin, pinhole-free, and uniform films. Such films have become increasingly important with the miniaturization of technical and electronic devices. While in the past interest was focused preferentially on inert films exhibiting protective properties towards environmental influences as well as coatings with anti-scratch properties or barrier properties for separation of gases and liquids, now the need to produce thin films with monotype functional groups at the surface has come to the fore. An increasing number of such monofunctionalized plasma polymers have been produced in view of the interaction with biological materials such as peptides, proteins, living cells, or blood [121–123].

In this context the plasma polymerization of functional groups carrying monomers and their homo- as well as copolymerization with "neutral" inserted and alternately bonded co-monomer units gains in importance. Several groups have investigated the experimental conditions of the plasma polymerization of functionalized monomers with the aim of attaining the highest possible retention degree of the pristine functionality of the used monomers [24]. Ameen *et al.* have reported a maximal retained 21 OH groups per 100 C atoms using allyl alcohol deposits [63]. Using a pulsed plasma, Friedrich *et al.* could increase the retention of OH groups to 31 OH/100 C, which is near the theoretical limit (33 OH/100 C) [24], characterizing the surface composition by XPS and applying the derivatization of OH groups with TFAA [86].

The retention of functionality during plasma polymerization is also manifest in that the deposited plasma polymer with the highest possible yield is usually associated with very gentle plasma parameters ("soft" plasma), characterized by a low Yasuda factor of the plasma used [124]. This coefficient can be also expressed in terms of energy in eV per monomer molecule. It influences the properties of the resulting polymer layer and the adhesion to the substrate. For "soft" plasmas such as used here this coefficient should amount to <10 eV per monomer [125].

By employing plasma copolymerization in combination with other experimental conditions such as pulsed regime, pulse frequency, low power, and low duty cycle an efficient instrument to generate and tune surfaces with a defined concentration of monofunctional groups was achieved [58, 65, 87, 126]. While the plasma homopolymerization of allyl alcohol has been very extensively investigated [63, 87, 127] data for allyl alcohol copolymerizations have been published only by two groups. Short *et al.* investigated the copolymerization with dienes [126] and Friedrich *et al.* produced and characterized allyl alcohol copolymers with ethylene, acetylene, and butadiene as co-monomer [87] as well as with styrene [46, 128]. Moreover, copolymers with acrylic acid were also in the focus [109]. From a chemical viewpoint, acrylic acid, butadiene, styrene, and with some limitations also ethylene were proper co-monomers. It was argued that acetylene and all allyl co-monomers can only react by fragmentation–recombination copolymerization. It is important to distinguish between the predominantly "plasma-initiated chemical copolymerization" and the non-chemical "plasma copolymerization," which can be characterized as a simple mixing of chemically inactive co-monomers or gases reacting only by fragmentation of these precursors followed by random recombination to irregular formed structures.

As emphasized earlier [30], the ambition was to find plasma parameters that allow maximization of the chemically produced fraction of the resulting copolymer. Thus, it was expected that more chemically defined structures would be present using this pulsed-plasma deposition. A precondition for the dominance of chemical copolymerization is the existence of polymerizable groups with sufficient reactivity (vinyl, acryl, diene, etc.). The composition of a chemically produced copolymer is adjusted by the "copolymerization parameters" [106]. Here, the copolymerization parameters of allyl alcohol, acrylic acid, ethylene, butadiene, acetylene, or styrene are very different, thus it should be difficult to find plasma conditions and a co-monomer precursor ratio where genuine copolymers are formed. Because of the higher reactivity styrene homopolymers should be preferentially produced. Using the chemically dominated pulsed plasma and realizing the "plasma-initiated chemical copolymerization" the chemical rules of copolymerization become of significance. Now, similar to the classic polymerization it is recommended that the so-called "copolymerization parameter" be considered. Generally, the copolymers were formed from functional group carrying and "chain-extending" co-monomers. By varying the ratio of co-monomers, the number of functional groups of copolymers can be adjusted (Figure 12.59).

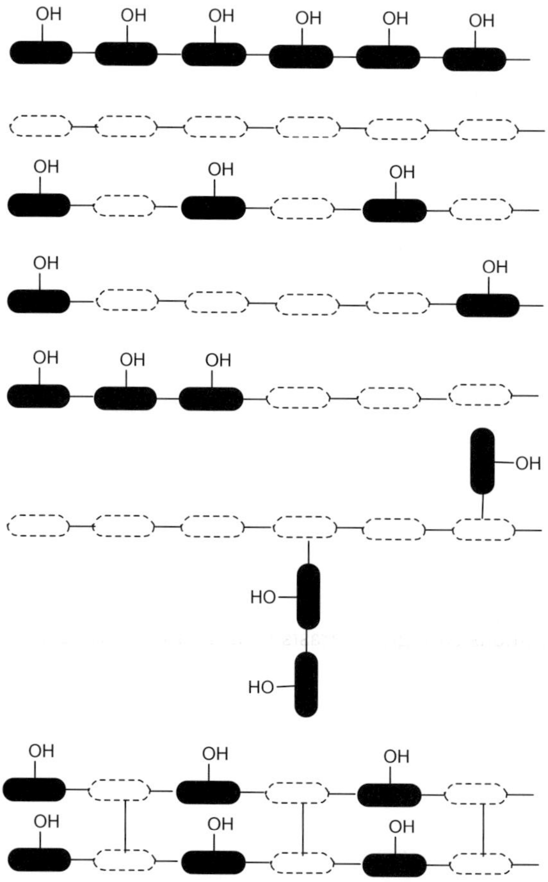

Figure 12.59 Scheme of possible copolymer structures.

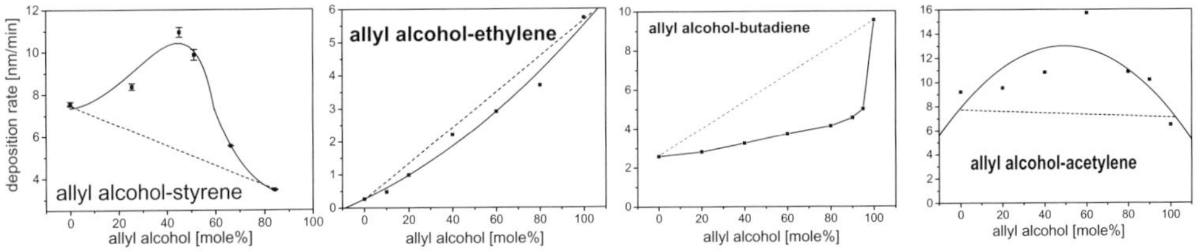

Figure 12.60 Dependence of copolymer deposition rates on the types of co-monomers and their ratio in the co-monomer mixture.

The copolymers were characterized by XPS after derivatization, FTIR, NMR, ThFFF, SEC, and dielectric relaxation spectroscopy (DRS) as known for homopolymers. Of course, the SEC results are produced using the soluble fractions of the deposited plasma polymers.

12.12.2 Copolymer Kinetics

The dominance of the chemical mechanism of copolymerization was suggested by the characteristic dependence of deposition rates on the types of co-monomers and their ratio in the precursor mixture (Figure 12.60). The nonlinear dependence and the special shape of the curves favor the chemical copolymerization. Thus, a good agreement was found between the Fineman–Ross approach and the measured kinetics as shown in Figure 12.60 [106].

In nearly all cases deviations from linearity of deposition rate on co-monomer ratio in the precursor mixture are observed. The deposition rates were measured using quartz microbalances and appear as a complex function, as known from copolymer chemistry. Therefore, the usual copolymer kinetics were adapted such as Fineman–Ross and Alfred Price's Q-e scheme [106]. This coincidence confirmed the influence of chemical processes during pulsed-plasma initiated copolymerization.

Interestingly, allyl alcohol–ethylene and butadiene copolymers show the highest deposition rate at high fractions of allyl alcohol in the precursor mixtures. Acetylene and styrene present the opposite behavior. The same was observed for allylamine copolymers. This special behavior of allyl monomers under low-pressure pulsed-plasma conditions strongly contrasts with their chemical polymerization and copolymerization tendency [106].

12.12.3 Allyl Alcohol Copolymers with Ethylene, Butadiene, and Acetylene

In the same manner the concentration of retained functional groups in the copolymer layer was nonlinearly dependent on the composition of the precursor mixture and also on the types of co-monomers, as shown for allyl alcohol copolymerized with ethylene, butadiene, and styrene (Figure 12.61).

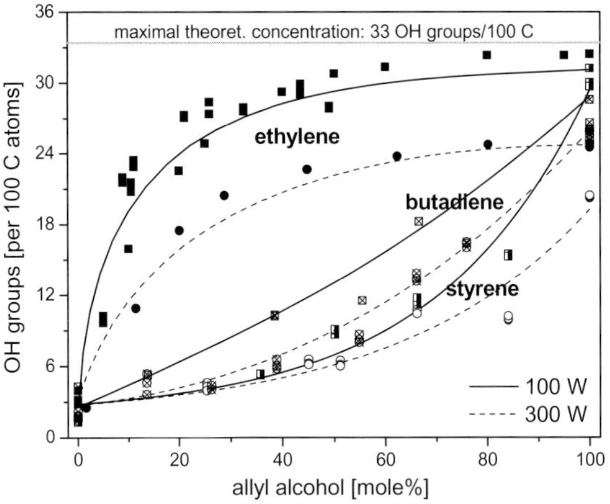

Figure 12.61 Copolymerization of allyl alcohol with ethylene, butadiene, and styrene and measurement the resulting OH group concentrations after derivatization with trifluoroacetic anhydride (TFFA) (cw-rf, 25 Pa).

The dependences of OH group retention in the copolymer on allyl alcohol concentration in the precursor mixture with ethylene, butadiene, or styrene did not show the same characteristic behavior as the deposition rate shown in Figure 12.60, confirming special behavior

In such a way it was actually possible to adjust the number of functional groups, by consideration of the nonlinear behavior as depicted in Figure 12.61, by mixing the corresponding co-monomer ratio. However, in the case of a strongly different copolymerization tendency, expressed by a strong difference in copolymerization parameters, homopolymerization of the most reactive co-monomer dominates over a broad range of precursor mixtures. Thus, it is difficult to adjust a given concentration of functional groups.

If the copolymerization parameters of both co-monomers are similar, as in the case of allyl alcohol and ethylene, the copolymerization proceeds easily. The formed copolymer shows a linear combination of both the ethylene and allyl alcohol C1s signals (Figure 12.62).

NEXAFS spectra of copolymers with different compositions confirm the undesired post-plasma formation of carbonyl groups (Figure 12.63).

As can seen from the NEXAFS O_K-edge spectra, the concentration of undesired C=O features increases with growing percentage of ethylene in the co-monomer precursor mixture. Thus, it can be assumed that only a minority of OH groups bound in the allyl alcohol is transformed into C=O groups; however, the majority is incorporated in the ethylene sequences by post-plasma oxidation at exposure to air.

The IR-specific OH absorbances depend on the composition of the co-monomer precursor mixture in similar manner as the XPS results (Figure 12.64).

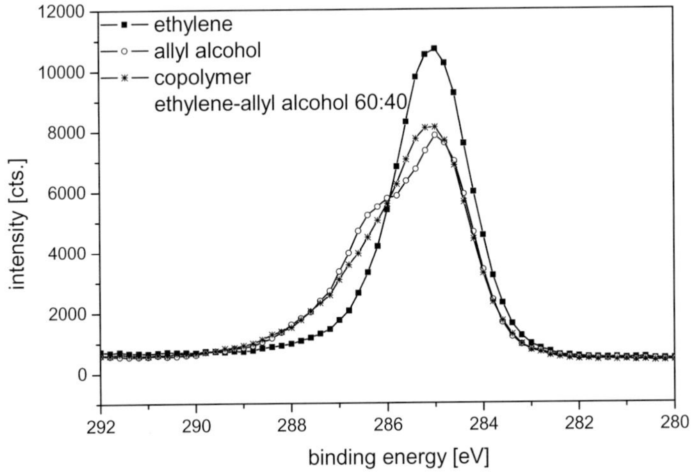

Figure 12.62 XPS C1s signals of ethylene, allyl alcohol, and allyl alcohol–ethylene copolymer (rf, 1000 Hz, 0.1 duty cycle, 100 W, 26 Pa).

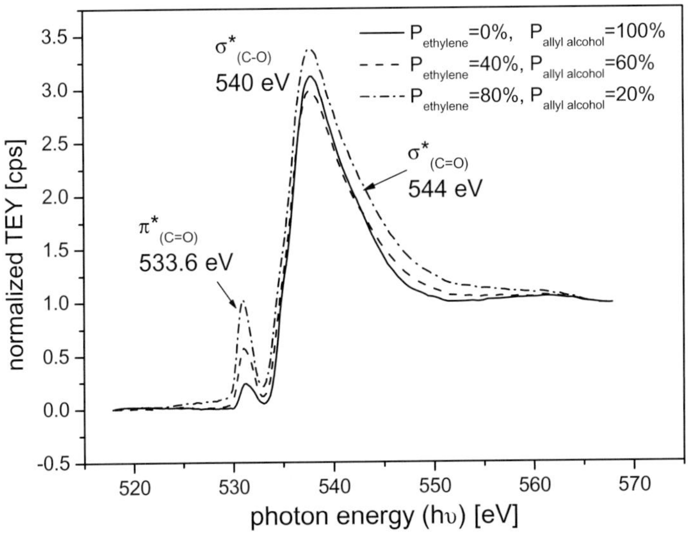

Figure 12.63 O_K K-edge and $\pi^*_{C=O}$-resonance of pulsed-plasma polymerized allyl alcohol–ethylene copolymers of different composition (rf, pulse frequency = 1000 Hz, duty cycle = 0.1, power = 100 W, and pressure = 6 Pa).

Allyl alcohol copolymers with ethylene or butadiene also show a nearly identical surface energy (polar contribution) dependence on OH-group concentration, thus showing the equivalency of both types of copolymers and the comparable presence of OH groups at the surface (Figure 12.65).

The stability of copolymers was measured using DRS. The samples were successively heated and ε* was measured isothermally versus frequency. Figure 12.66

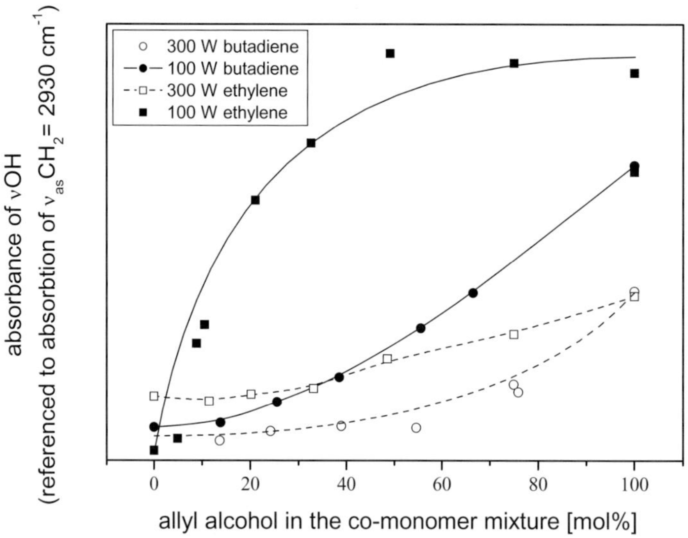

Figure 12.64 Variation of IR-measured OH absorbances with the composition of the co-monomer precursor mixture of allyl alcohol and either ethylene or butadiene.

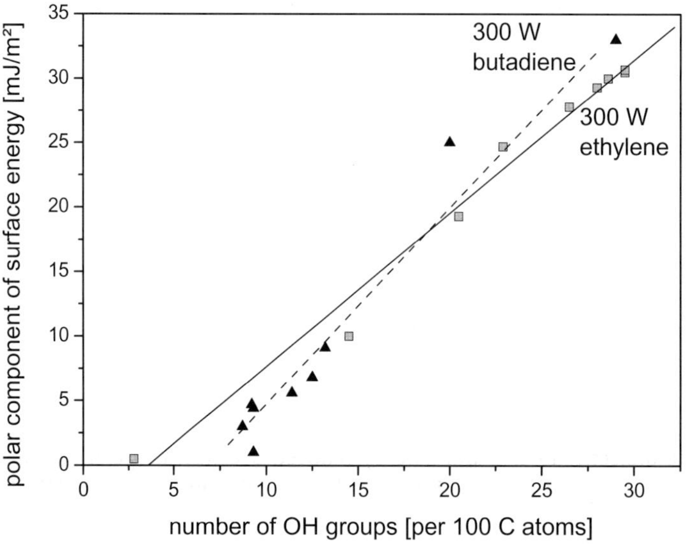

Figure 12.65 Correlation between the polar component of surface energy and the concentration of OH groups in different allyl alcohol copolymers copolymerized with either ethylene or butadiene.

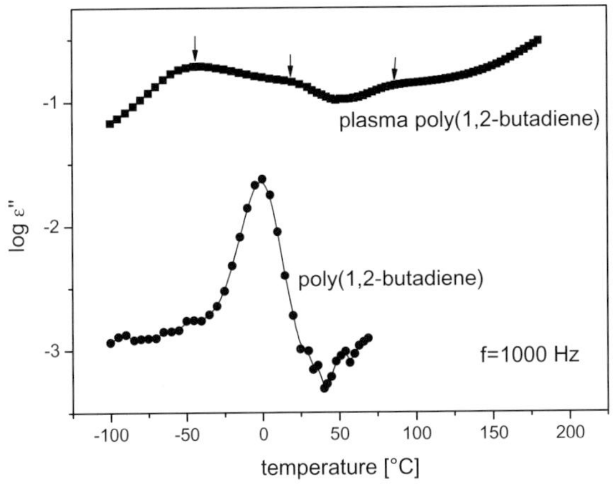

Figure 12.66 Dielectric loss ε″ versus temperature at a frequency of 1000 Hz for a pulsed-plasma homopolymerized polybutadiene (■) and an anionic polymerized poly(1,2-butadiene) (●) (arrows indicate peaks).

indicates that the microstructure of the product obtained by the pulsed-plasma polymerization of butadiene is quite different from that obtained by conventional polymerization techniques. Probably, a highly crosslinked structure is obtained because the plasma process can activate both double bonds present in 1,3-butadiene. Additionally, 1,2- or 1,4- trans/cis-sequences can be established. Secondly, the dielectric loss of the pulsed-plasma polymer is at least one order of magnitude higher than that of poly(1,2-butadiene). Thus, it has to be concluded that the plasma process produces several unsaturated radicals, which can react with oxygen on exposure to air. The consequence is the formation of various oxygen-containing polar groups that are not present in a conventionally synthesized polybutadiene.

This line of argument leads to the hypothesis that the plasma synthesized products are not thermally stable because the unsaturated radicals can recombine at high temperatures. This is demonstrated by the data in Figure 12.67, where the first heating run is compared with a second one for the pulsed-plasma polymerized copolymer with 20/80 (1,3-butadiene/allyl alcohol) in the formulation. In the second run the dielectric loss is dramatically reduced because the polar radicals are converted into less polar groups. Moreover, the relaxation process around 75 °C seems to disappear in the second run, indicating also a change in structure. Therefore, only the low-temperature process is further analyzed and discussed in terms of dependence on the formulation. With increasing temperature the position of this low temperature peak shifts to higher frequencies and increases in intensity (Figure 12.68).

The temperature dependence of the relaxation rate can be described by the Arrhenius law (Figure 12.69).

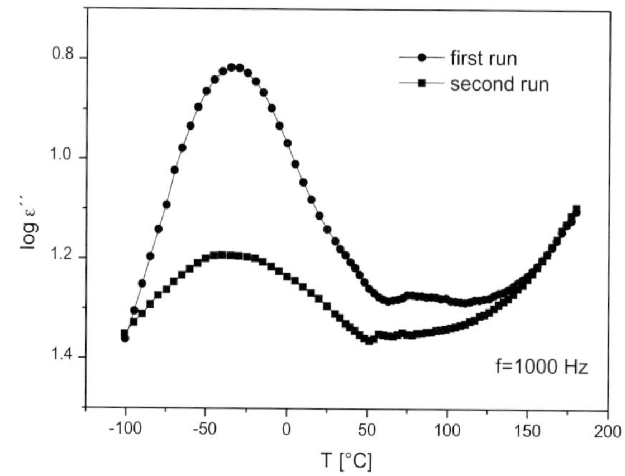

Figure 12.67 Dielectric loss ε″ versus temperature at a frequency of 1000 Hz for a pulsed-plasma copolymer with 20 : 80 (1,3-butadiene : allyl alcohol) in the formulation: (●) first run and (■) second run.

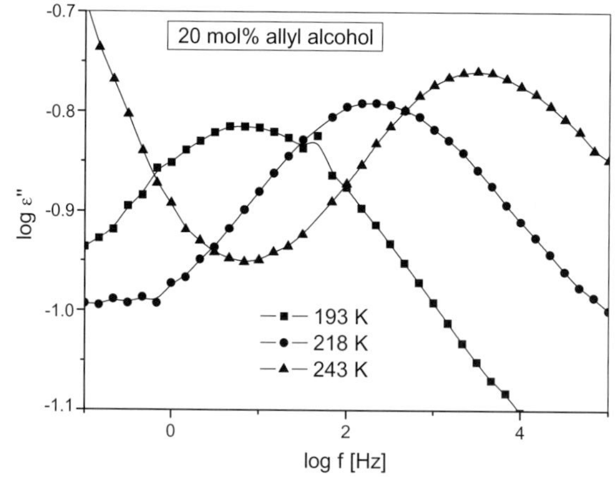

Figure 12.68 Dielectric loss ε″ versus frequency at the indicated temperatures for a pulsed-plasma copolymer with 80 : 20 (1,3-butadiene : allyl alcohol) in the formulation.

For the plasma-polymerized polybutadiene an activation energy (E_A) of 40 kJ mol^{-1} is obtained, which compares well with that for the β-relaxation of conventionally polymerized 1,2- and 1,4-polybutadiene (41 and 37 kJ mol^{-1}, respectively [129]). This indicates localized motions of groups like the β-relaxation in conventionally synthesized polymers [130]. This is also consistent with the increase of Δε with temperature (see inset Figure 12.69).

The E_A increases systematically with the content of allyl alcohol (Figure 12.70). This effect points to additional interaction caused by the allyl alcohol groups,

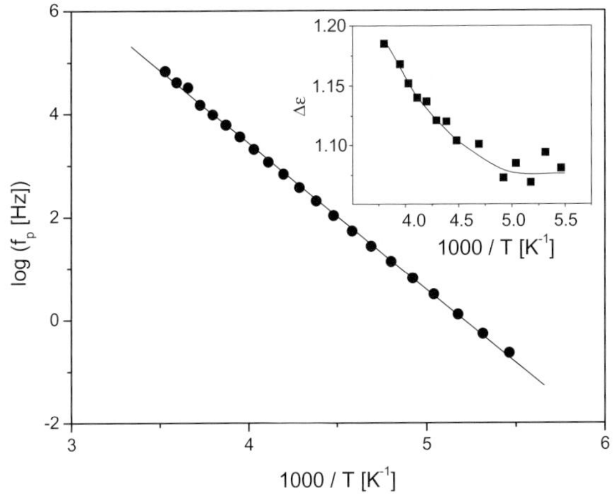

Figure 12.69 Relaxation rates (f_p) versus inverse temperature for a pulsed-plasma copolymer with a 10 : 90 (1,3-butadiene : allyl alcohol) formulation. The line is a fit of the Arrhenius equation to the data. Inset: temperature dependence of the dielectric strength ($\Delta\varepsilon$) for the same polymer.

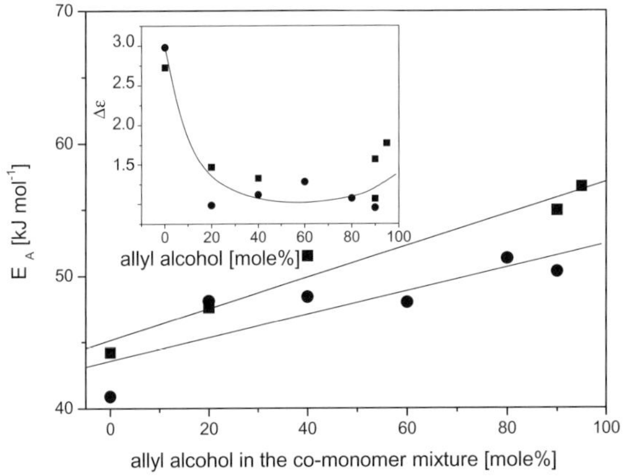

Figure 12.70 Dependence of activation energy (E_A) on the composition of the formulation: (■) first and (●) second run. The line was obtained by a linear regression of the data. Inset: composition dependence of the dielectric strength at $T = -55\,°C$: (■) first and (●) second run. The line is a guide for the eye.

which hinders the molecular fluctuations. Moreover, this systematic variation also indicates that the two monomers are incorporated according to their concentrations in the formulation.

The dependence of the dielectric strength ($\Delta\varepsilon$) on the concentration of allyl alcohol in the mixture is plotted in the inset of Figure 12.70. The value of $\Delta\varepsilon$ is

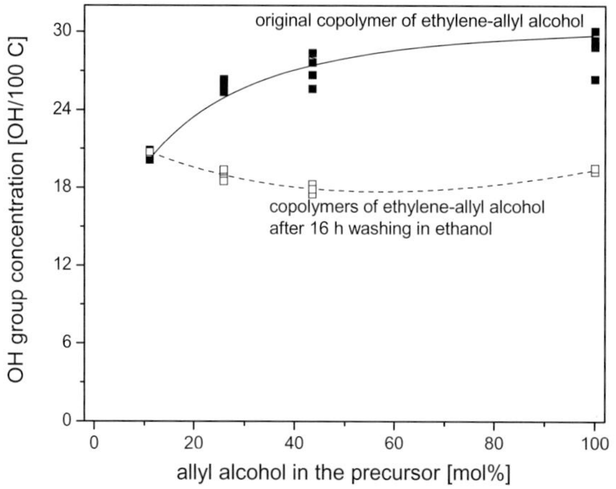

Figure 12.71 Extraction of differently composed allyl alcohol–ethylene copolymers in ethanol for 16 h and the resultant loss in OH group concentration.

proportional to the mean squared dipole moment (μ^2) and to the mobility of dipoles. Pure plasma-polymerized polybutadiene shows a high $\Delta\varepsilon$. A minimum $\Delta\varepsilon$ was found around the equimolar composition of the two monomers. At the first glance this is unexpected because the dipole moment of allyl alcohol should be higher than that of butadiene. Therefore, it is concluded that the increasing interaction of hydroxyl groups restricts the molecular mobility. For higher percentages of allyl alcohol an increase of $\Delta\varepsilon$ was measured, which is interpreted as an increase of the dipole moment (μ) at high concentrations of OH groups. This interpretation is in agreement with the dependence of the activation energy on the concentration of allyl alcohol. Finally, it should be stressed again that both dielectric quantities vary systematically with the composition. Because dielectric spectroscopy is a "bulk method" this result points to a homogeneous composition of the layers.

The permanence of functional groups and the solubility of the plasma produced copolymer layers in the polar solvent ethanol have been tested. Preliminary investigations showed that the copolymers were not completely soluble. Complete solubility is a measure of the absence of crosslinks. Copolymers of allyl alcohol and ethylene showed a significantly higher resistance towards complete dissolution for all mixture ratios of co-monomers (Figure 12.71). Only the copolymer with 90% ethylene in the precursor mixture was nearly resistant towards intense extracting with ethanol.

12.12.4
Allyl Alcohol Copolymers with Styrene

Allyl alcohol–styrene (2 : 1) copolymer presents a different dependence of quantities of OH groups on the duty cycle than does pure allyl alcohol (Figure 12.72). At

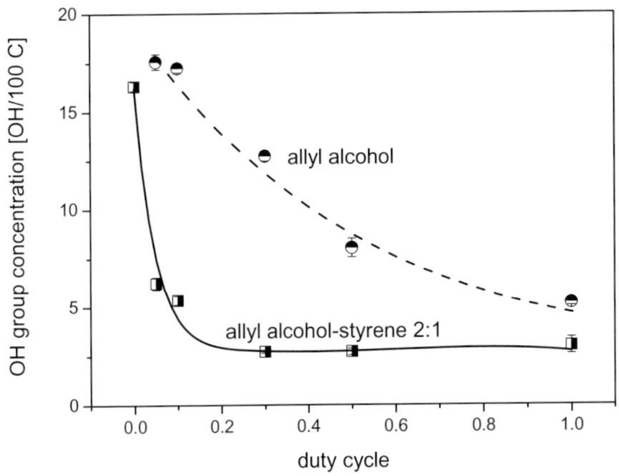

Figure 12.72 Variation of the concentration of hydroxy groups with duty cycle in plasma-polymerized allyl alcohol–styrene (2:1) and pure allyl alcohol under the same plasma conditions (1000 Hz, 300 W), as measured by XPS after derivatization with TFAA.

duty cycles >0.3 a low but constant OH concentration of only 3 OH groups/100 C atoms was found in comparison to the 14 OH/100 C theoretically calculated. At lower duty cycles (<0.3) higher OH concentrations were measured. Consequently, only a small part of allyl alcohol was incorporated into the deposited layer, which consisted predominantly of polystyrene. This behavior apparently was due to preferred excitation of the vinyl bond in styrene. The necessary energy to start a radical polymerization reaction is about 1 eV [45]. In comparison to this the allyl double bond requires significantly higher activation energy. Thus, Kurosawa *et al.* found, using a quartz microbalance that the styrene deposition at 100 W and 100 Pa (cw plasma) occurred eight times faster than the allyl alcohol deposition [127]. Using the pulsed-plasma mode and also a microbalance Friedrich *et al.* reported that the styrene deposition rate was tenfold greater than that of allyl alcohol [131].

After removing the polymer layers from glass substrate, FTIR spectra were recorded using the ATR mode. Soft polymers were formed at low wattages and very brittle polymers were obtained at higher power input.

Figure 12.73 shows the FTIR spectra of plasma polymerized allyl alcohol–styrene copolymer (2:1) obtained with different duty cycles. In contrast to pure allyl alcohol, for plasma polymers deposited under the same experimental conditions the OH band grows slightly with increasing duty cycle.

The respective area ratio A_{OH}/A_{CH2} increases linearly with the duty cycle (Figure 12.74).

This IR result is in contrast to that obtained by XPS after derivatization (Figure 12.72). The only plausible explanation is that the composition of the XPS measured surface layer (5–7 nm) differs from that of the "bulk" measured by ATR. The thickness of the ATR-measured polymer layer (scratched off and mixed) was ca. 1000 nm. A possible explanation for this experimental finding is the formation of

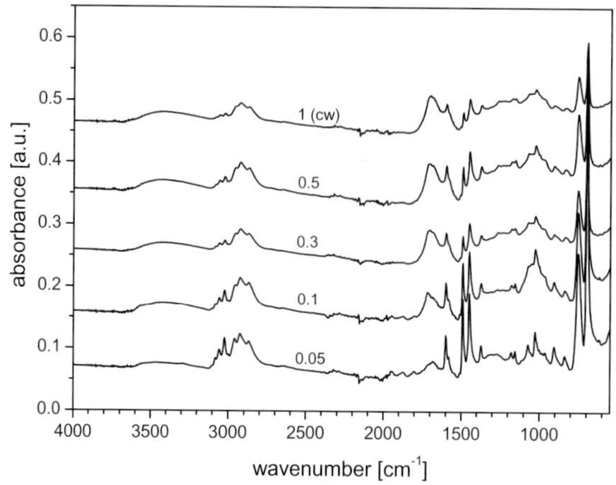

Figure 12.73 FTIR spectra of plasma-polymerized allyl alcohol–styrene (2:1) copolymers deposited at 300 W but with different duty cycles.

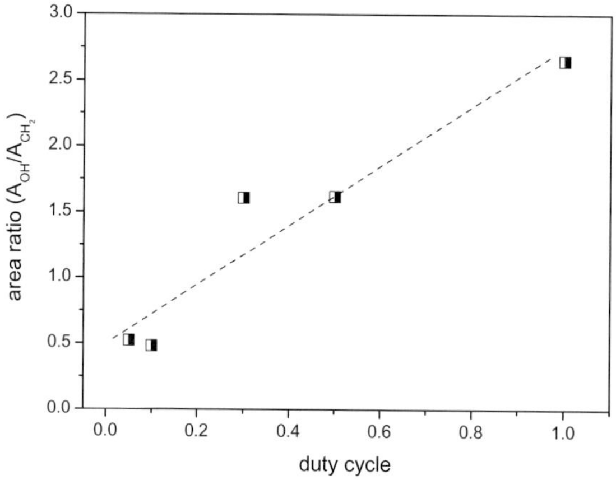

Figure 12.74 Area ratio A_{OH}/A_{CH2} of plasma-polymerized allyl alcohol–styrene copolymers (molar ratio of monomers in the precursor 2:1) versus the applied duty cycle at 300 W rf peak power.

a styrene-rich skin caused by the post-plasma polymerization of styrene after turning-off the plasma.

In other words, in the plasma-off time styrene molecules react much faster than allyl alcohol monomers with the remaining radicals at the surface, thus forming a thin polystyrene skin on the surface. The minimum energy needed to induce such skin formation lies in the region of 90 W for the system considered and duty cycles less than 0.3.

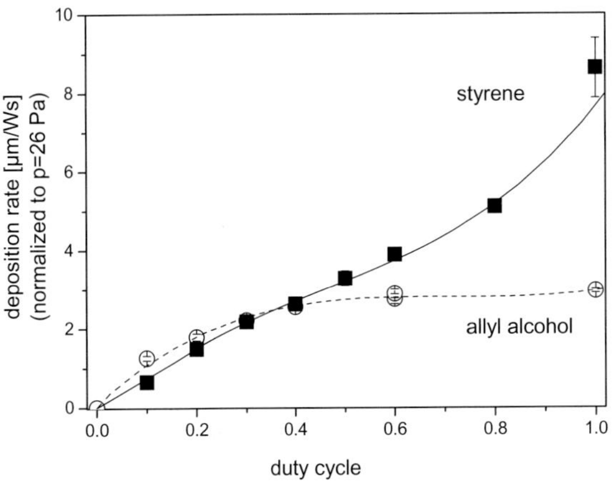

Figure 12.75 Dependence of deposition rates of (pure) allyl alcohol and styrene, normalized to 26 Pa and 1 W power input, on duty cycle.

To prove the theory of the formation of a styrene-rich skin deposited in the plasma-off time some deposition experiments were carried out, the results of which are shown in Figure 12.75.

The deposition rates of styrene and allyl alcohol (normalized to a power input of 1 W and a pressure of 26 Pa) versus duty cycle show different behavior. Up to duty cycles of 0.3 the allyl alcohol deposition rate is slightly higher than that of styrene. However, at higher duty cycles (>0.3) the styrene deposition rate dominates strongly.

CHN analyses of allyl alcohol–styrene copolymers confirm, as for all plasma polymers, a hydrogen loss during plasma polymerization of about 20% (and additionally insignificant N-introduction) (Figure 12.76).

Thermal field flow fractionation (ThFFF) was applied to measure the existence of high- or ultrahigh molecular weight fractions in plasma polymers. Compared to size-exclusion chromatography (SEC) the advantage of this method is the absence of any obstruction or blocking of chromatographic columns because of the use of a flow channel with much greater dimensions. However, also, this method needs (partial) solubility; however, gel particles do not disturb the process. Notably, the applied detection (MALLS – multi-angle laser light scattering) measures all scattering centers independently of the molar mass, for example, very often ultrahigh molecular weight molecules, which are very rare, possess a great number of scattering centers within a single molecule and, consequently, this fraction is overrepresented in the fractograms. The copolymer spectra show characteristic molar mass distributions (Figure 12.77).

The pure styrene polymer product possesses a well-sized molar-mass distribution near 1 MDa (1000 kg mol^{-1}). The allyl alcohol–styrene (2 : 1) copolymer has a similar mass distribution as the homopolymer of polystyrene. Changing to a 6 : 1

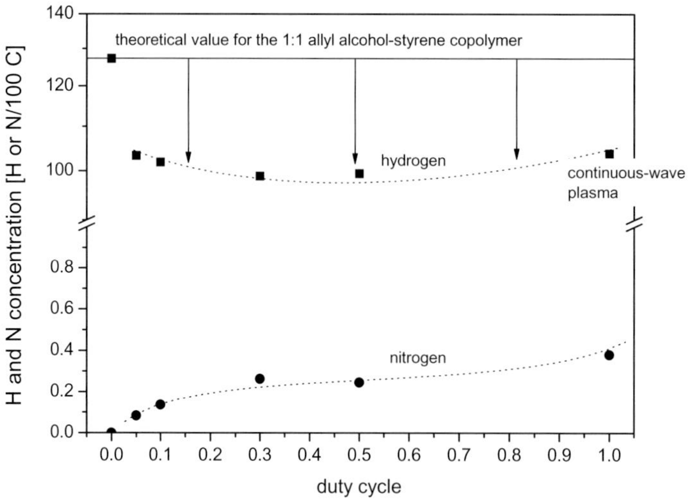

Figure 12.76 CHN analyses for allyl alcohol–styrene (1 : 1) copolymers produced using different duty cycles.

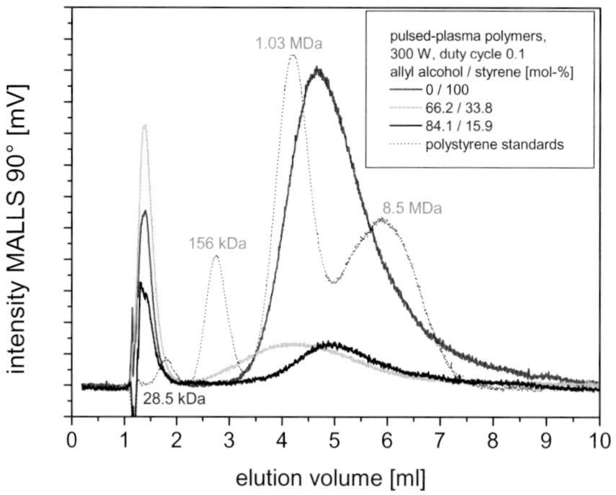

Figure 12.77 ThFFF fractograms of plasma-polymerized poly(allyl alcohol)–polystyrene copolymers, measured in tetrahydrofuran (THF), superposed with a three-component polystyrene reference material (THF, rf, 300 W, 26 Pa, 1000 Hz, and 0.1 duty cycle).

allyl alcohol–styrene copolymer the distribution maximum shifts slightly to higher values in the low MDa region as seen in Figure 12.77. Thus, it can be concluded that the copolymers consist predominantly of polystyrene. Only the 6 : 1 copolymer should include more allyl alcohol sequences. Using 300 W, high molar mass copolymers were produced (>10^6 g mol^{-1}). Using 100 W, the co-monomers of allyl alcohol and styrene were predominantly soluble. Figure 12.78 depicts the respective elugrams.

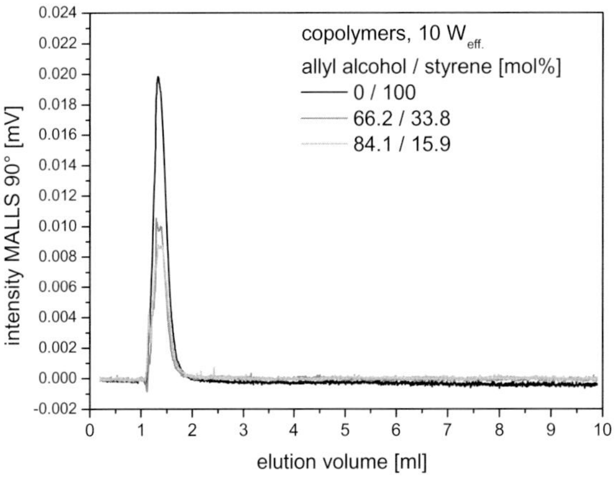

Figure 12.78 FFF fractograms of different poly(allyl alcohol)–polystyrene copolymers (THF, rf, 100 W, 26 Pa, 1000 Hz, and 0.1 duty cycle).

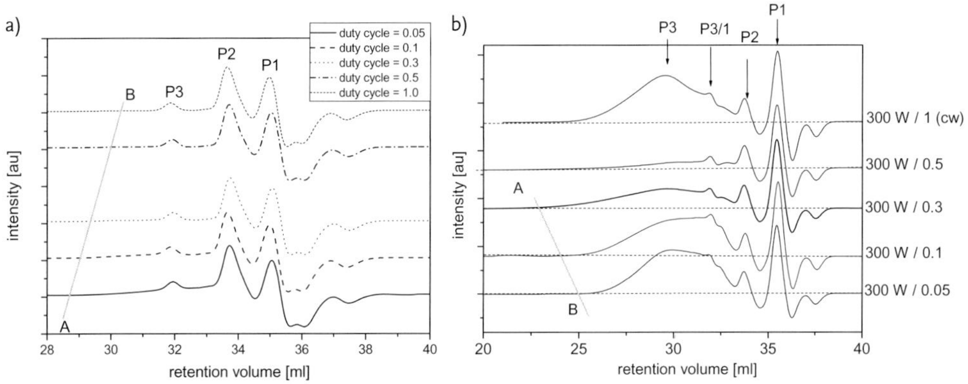

Figure 12.79 SEC chromatograms of plasma poly(allyl alcohol) (a) and polystyrene (b) homopolymers, measured in THF using different duty cycles.

To obtain more information on the molar masses and their distribution in plasma polymers SEC [also known as gel-permeation chromatography (GPC)] was used. To find a reference point commercial polystyrene standards were also measured under the same experimental conditions. In the case of allyl alcohol polymers and copolymers a calibration is difficult (because of lack of suitable calibration standards) and, therefore, interpretation of the chromatograms is complicated. Moreover, association effects caused by hydrogen bonding of the OH groups of the polymers may complicate the chromatographic process further. Additionally, aggregates and gel particles must be completely removed from the polymer solution to avoid blocking of the chromatographic columns. Consequently, only the well-soluble fraction of more or less linear polymers in the low or moderate molar mass region was measured. The allyl alcohol homopolymers show characteristic chromatograms with several peaks (Figure 12.79).

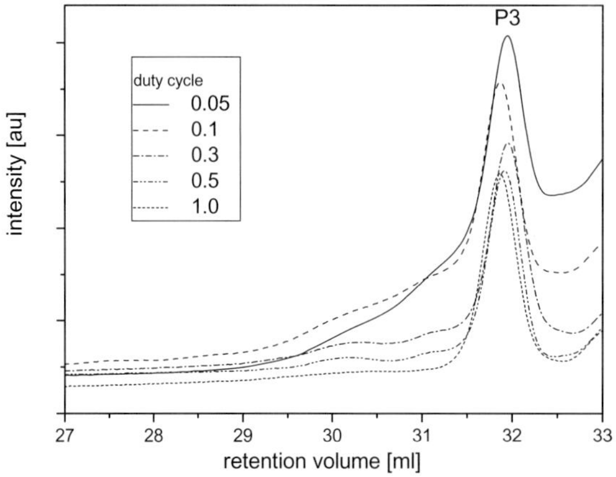

Figure 12.80 P3 region of SEC chromatograms of plasma-polymerized allyl alcohol homopolymers, measured in THF.

The peak maxima of pp-PAAl are labeled as P1 to P3. Peaks P1 and P2 are caused by the solvent and measuring system, respectively. Peak P3 results from the deposited polymer. The line A–B denotes the beginning of peak P3 for samples polymerized at different duty cycles. As expected, low duty cycles produce more soluble polymer fractions and with increasing duty cycle the soluble fractions are reduced. The molecular weight distributions were broadened with lowering the duty cycle (soluble fraction) as shown by the line A–B in Figure 12.79 and in a magnified view plotted in Figure 12.80.

However, the application of cw plasma (duty cycle = 1) also produces soluble polymers. On the other hand, the expected stronger fragmentation of the monomer molecule and therefore a more randomly crosslinked structure decreased the soluble fraction. The occurrence of bimodal molar mass distributions was characteristic for all duty cycles. Thus, the general mechanism of poly(allyl alcohol) formation should be the same at all duty cycles (measured only for the soluble fraction).

The plasma polymerized polystyrene samples were also dissolved in THF for 24 h to investigate the soluble part of the plasma polymers. The results are shown in Figure 12.79 on applying calibration with polystyrene standards.

The P3 peak of pulsed-plasma polymerized styrene shows a broad distribution of molar mass. The peak is particularly broadened at duty cycles from 0.05 to 0.3 as indicated by the A–B line. The polymer P3 peak at duty cycle 0.5 is minor. The styrene polymerized in the cw mode shows a completely different molar mass distribution, comparable with that species polymerized at duty cycle 0.05. Thus, the interpretation is that the polymer formation process becomes more chemically dominated at smaller duty cycles.

When allyl alcohol and styrene in the molar ratio 2 : 1 were copolymerized (Figure 12.81) characteristic features of both homopolymers were found in the

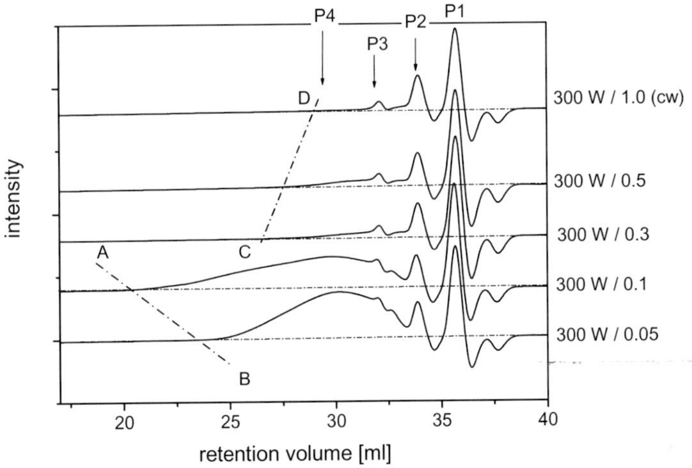

Figure 12.81 SEC chromatograms of plasma copolymers of allyl alcohol–styrene deposited at 300 W rf peak power and different duty cycles (monomer ratio in the precursor 2 : 1).

chromatograms. Here, background information is needed; the application of low duty cycles also produces polymers with a relatively large fraction of homopolymer/copolymer soluble in tetrahydrofuran (THF). On increasing the duty cycle, beginning at 0.3, only small fractions of soluble copolymers were found. In addition, in the continuous wave mode (cw, duty cycle = 1.0), plasma deposited copolymers exhibited only a very small fraction of soluble copolymers. Therefore, the base of SEC analyses changes strongly with duty cycle. Moreover, these molar mass distributions (duty cycles 0.3–1.0) are quite similar to that of the allyl alcohol homopolymer (cf. Figure 12.79). The molar mass distributions of the copolymers produced with the lowest duty cycles (0.1 and 0.05) correspond strongly to that of the polystyrene homopolymer. (cf. Figure 12.79). Also in this case, considering also the IR results, lower duty cycles indicate a slightly different copolymerization mechanism, which may again be attributed to the more dominate route of pure chemical formation rather than the fragmentation–polyrecombination mechanism [24].

The exact molar mass distribution of P3 of plasma produced poly(allyl alcohol) and polystyrene homopolymers was assigned after using polystyrene standards. The peak P3, which represents the molar mass distribution of the (soluble) plasma polymer, was estimated to be about 700–4000 $g\,mol^{-1}$ for poly(allyl alcohol) homopolymers, with a maximum at 790 $g\,mol^{-1}$ (Figure 12.82). The exact molar mass distribution of pulsed-plasma produced polystyrene homopolymers was assigned after measuring the polystyrene standards as seen in Figure 12.82. Here, peak P 3 represents the highest molar mass of about 2700 $g\,mol^{-1}$. The peak P3 contains a shoulder named P 3/1, which may be attributed to an octamer of styrene at about 820 $g\,mol^{-1}$.

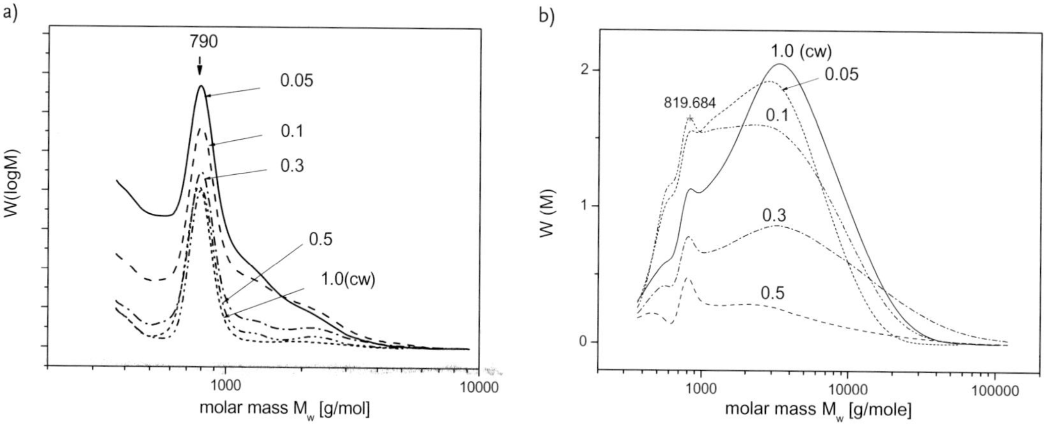

Figure 12.82 Differential molar mass distribution of plasma-polymerized (at different duty cycles) poly(allyl alcohol) (a) and polystyrene (b) homopolymers, dissolved in THF.

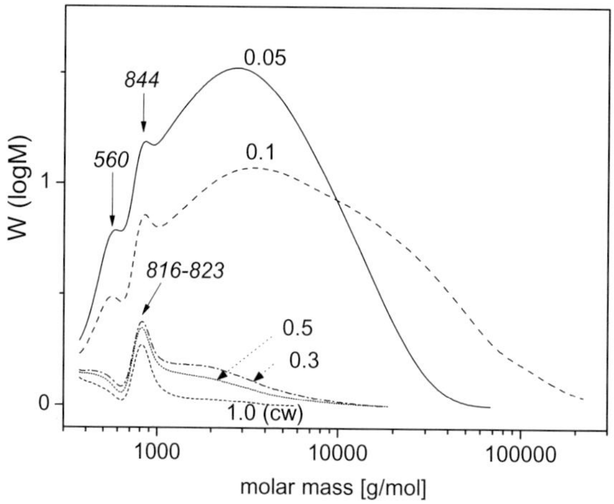

Figure 12.83 Differential molar mass distribution of plasma-polymerized (at different duty cycles) allyl alcohol–styrene (2 : 1) copolymers, dissolved in THF.

Figure 12.83 demonstrates the molar mass distribution of allyl alcohol–styrene copolymers (2 : 1). As noticed before, two groups of different behavior could be distinguished. The copolymers produced with lowest duty cycles exhibit a very high yield of the highest molar masses in the region of 600–50 000 g mol^{-1} (duty cycle 0.05) or 600–110 000 g mol^{-1} (duty cycle 0.1) with a maximum at about 3000 g mol^{-1}. Additional discrete shoulders at 560 and 844 g mol^{-1} are also observed. In the group that was produced at high duty cycles (>0.3), or in the continuous wave mode, molar masses in the range 700–8000 g mol^{-1} with a maximum at 820 g mol^{-1} were found (Figure 12.83).

These results do not give concrete hints as to the actual mechanism of copolymer formation or branching and so on. Nevertheless, obviously, at lower duty cycles the polystyrene sequences strongly dominate and at higher duty cycles those of poly(allyl alcohol) dominate, as already shown by FTIR spectroscopy. Thus, it is concluded that respective OH group functionalization of allyl alcohol–styrene copolymers switches very sensitively from homopolymerization of styrene (or allyl alcohol) to copolymerization in dependence on the co-monomer ratio in the precursor mixture as well as in dependence on duty cycle, wattage, and so on. Therefore, a well defined and reproducible deposition of copolymer layers is very difficult and cannot unambiguously be predicted theoretically.

12.12.5
Acrylic Acid

Copolymers of acrylic acid and styrene show characteristic deposition rates in terms of dependence on co-monomer ratio in the precursor mixture (Figure 12.84). Generally, acrylic acid has a much higher deposition rate than styrene under the plasma conditions used [109].

The yield in carboxylic groups, measured by XPS after esterification with trifluoroethanol ($R\text{-}COOH + HO\text{-}CH_2\text{-}CF_3 \rightarrow R\text{-}COO\text{-}CH_2\text{-}CF_3$), corresponds to the deposition rate. At a low percentage of acrylic acid in the precursor mixture the concentration of COOH groups remains low (Figure 12.85). However, the deposition rate of acrylic acid–styrene is strongly increased for low acryl acid percentages in contrast to pure styrene.

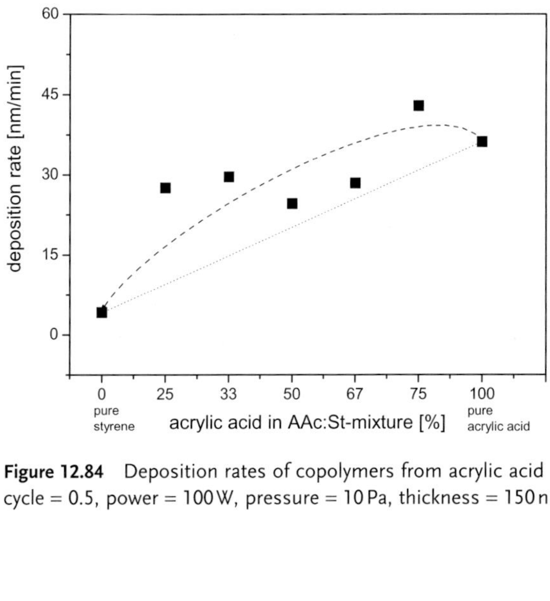

Figure 12.84 Deposition rates of copolymers from acrylic acid (AAc) and styrene (St) (duty cycle = 0.5, power = 100 W, pressure = 10 Pa, thickness = 150 nm).

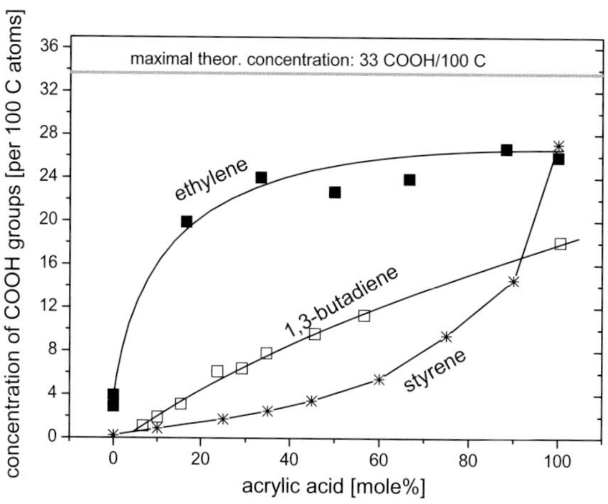

Figure 12.85 Yield of COOH groups for different kinds of acrylic acid copolymers and different copolymer compositions produced in the pulsed-plasma mode.

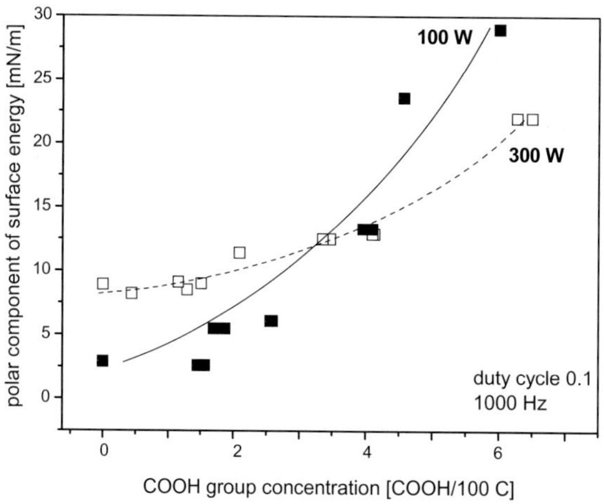

Figure 12.86 Dependence of the polar contribution to surface energy on the concentration of COOH groups deposited by pulsed-plasma initiated copolymerization of acrylic acid and butadiene in different co-monomer ratios (precursor mixture).

Acrylic acid–butadiene copolymers show an exponential increase of surface energy with the concentration of inserted COOH groups (Figure 12.86), which is far from the expected linear correlation between polar component and COOH concentration. A superproportional concentration of CH_2–CH sequences at the surface of poly(acrylic acid) is assumed, for example, a portion of COOH groups points away from the surface towards the bulk. Thus, especially at low

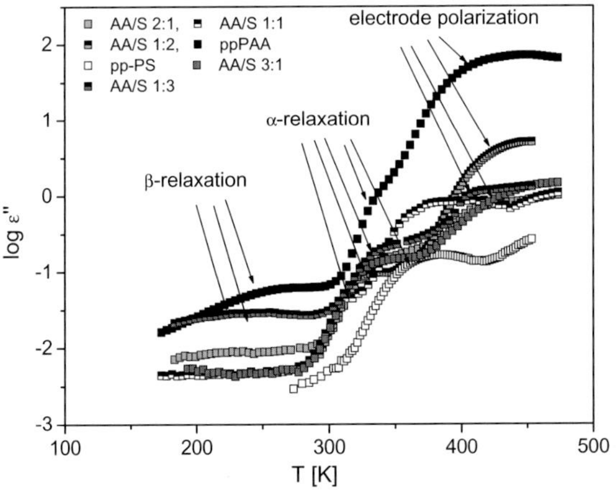

Figure 12.87 Dielectric loss versus temperature at a fixed frequency of 1 kHz of pulsed-plasma deposited samples of acrylic acid–styrene (AA/S) copolymers with different ratios of co-monomer.

concentrations of COOH, the polar component becomes smaller than expected (cf. Figure 12.86).

The dielectric loss versus temperature at a fixed frequency (1 kHz) and varied ratio of the co-monomer is plotted in Figure 12.87.

Again, at low temperatures the β-relaxation is observed in pure commercial and plasma-polymerized PAA; however, it becomes reduced on increasing the styrene percentage. At higher temperatures, in the range of the glass transition, measured by DSC, an ill-defined shoulder is visible that corresponds to the dynamic glass transition (α-relaxation) in polymers and copolymers on heating up.

12.12.6 Copolymers with Allylamine

Figure 12.88 shows the dependence on precursor composition of the deposition rate of allylamine–ethylene (acetylene, butadiene) mixtures for 100 and 300 W or 10 and 30 W effective power if using a duty cycle of 0.1.

The deposition rate versus precursor composition shows characteristic behavior for each "co-monomer."

The influence of plasma conditions on the structure of the allylamine plasma polymer has been demonstrated by using IR spectroscopy (Figure 12.89).

The most eye-catching anomaly of all IR spectra of plasma polymerized allylamine is again the appearance of the CN band at about $2200\,cm^{-1}$. Moreover, the NH stretching region around $3400\,cm^{-1}$ is superposed by OH stretching as well as in the region of $1700\,cm^{-1}$ by C=O features. The existence of several O-containing groups in the plasma polymer was evidenced before.

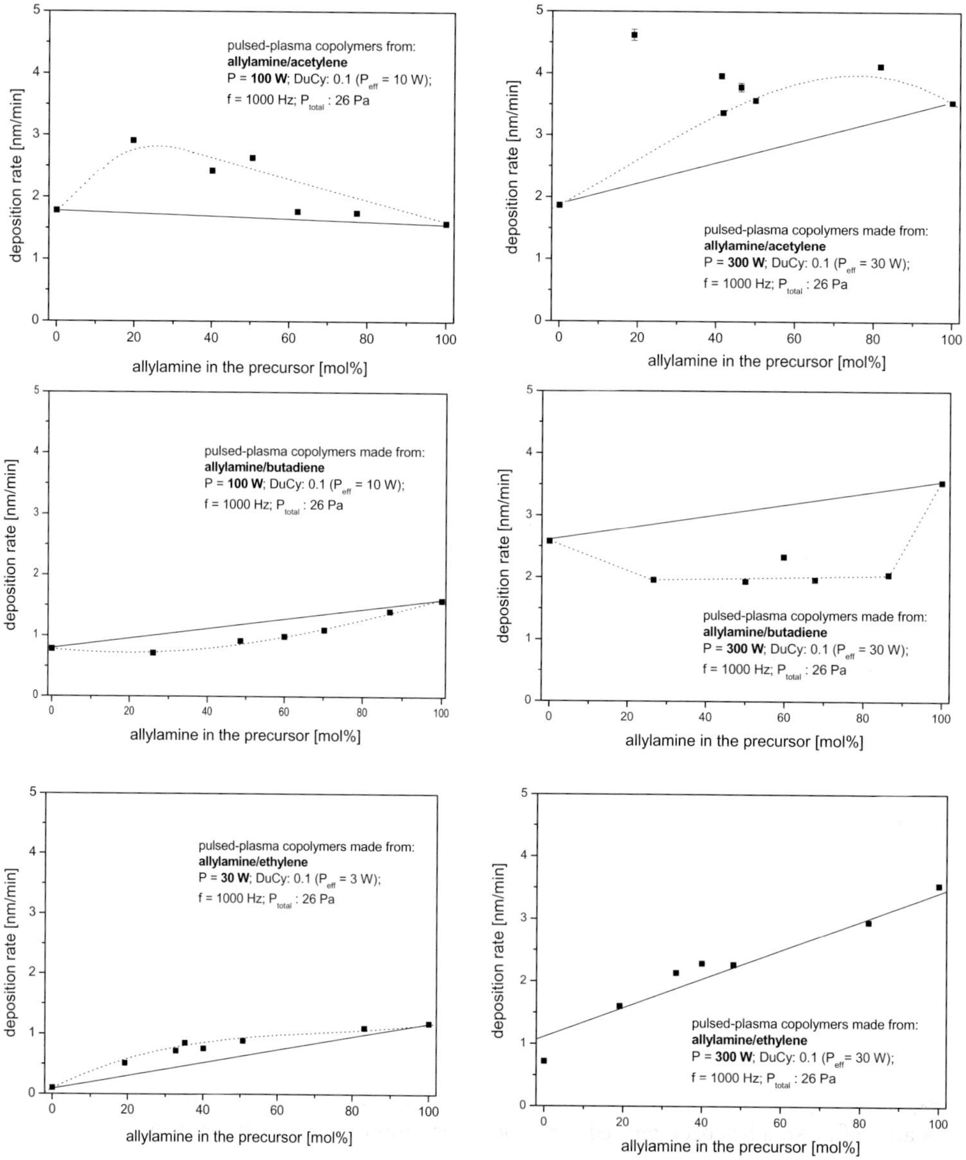

Figure 12.88 Deposition rates of co-monomer mixtures of allylamine versus precursor composition.

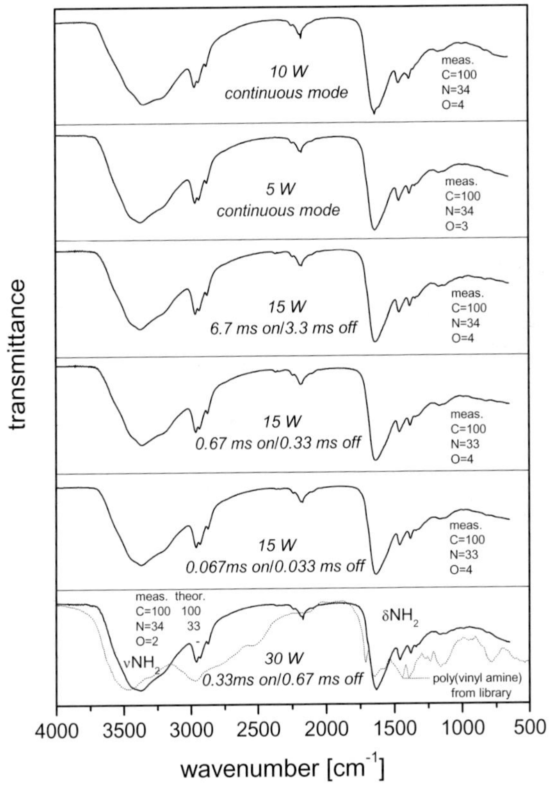

Figure 12.89 Influence of different plasma conditions on the IR spectrum of allylamine plasma polymer compared to a commercial poly(vinyl amine).

12.13
Graft Polymerization

The activation of inert polymer surfaces (polyolefins) and making them reactive by exposure to plasma to start chain-growth from a C-radical site without further assistance of plasma has been termed "graft polymerization" and sometimes also "copolymerization." Such graft polymerization can also be initiated by irradiation, sensitized photochemical activation, and vacuum-UV- or ozone-pretreatment [132–134]. There are different possibilities for grafting monomers, molecules with functional groups, or polymers onto substrates such as polymers or inorganics and starting from there new chain propagation (Figure 12.90).

Several possibilities of grafting have been discussed and tested.

The grafting of vinyl- or acrylic monomers onto C-radical sites at the backbone of polymer substrate molecules and initiation of a chain-growth polymerization as side-chain was often dreamt of. Graft copolymers are formed [135]. The chain

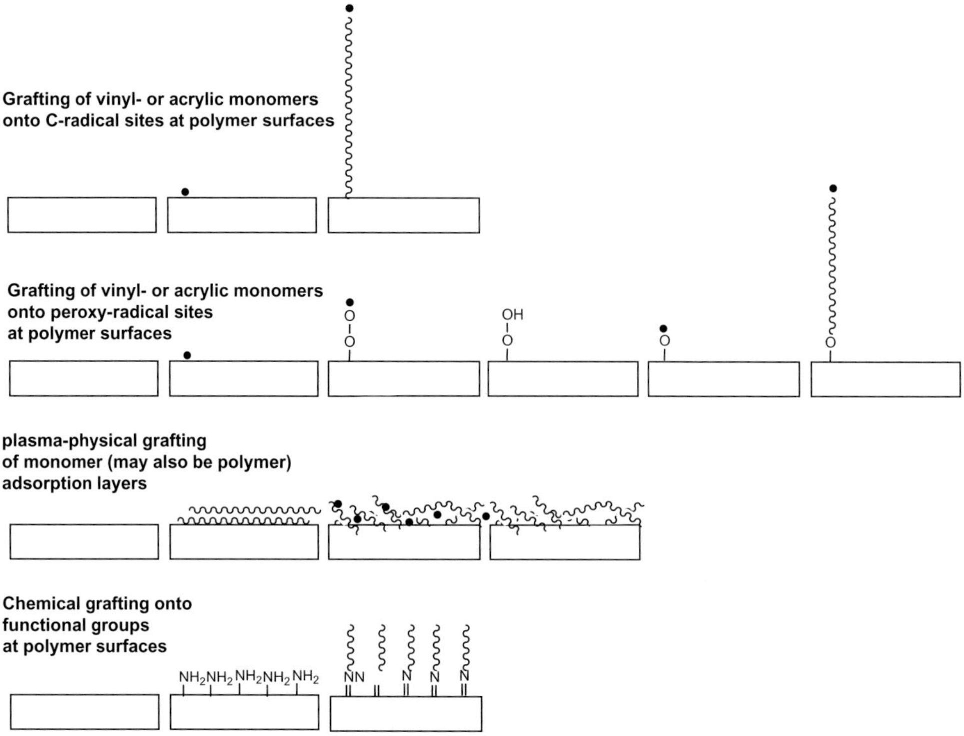

Figure 12.90 Variants of plasma-chemical grafting onto polymer surfaces.

growth is terminated by radical recombination, chain transfer, or disproportionation [136, 137]. By following the same principle and using additionally functional groups at the surface of the polymer substrate for adhesion promotion well-adhering layers with special functionalities were produced [138]. Very often the formation of peroxy functions awaited exposure of the plasma-exposed surface to the ambient air. The plasma produced C-radical sites react with biradical molecular oxygen by recombination and formation of proxy radicals:

$$\mathrm{CH_2} + \mathit{plasma} \rightarrow \mathrm{CH^\bullet} + 0.5\mathrm{H_2};\ \mathrm{CH^\bullet} + {}^\bullet\mathrm{O\text{-}O^\bullet} \rightarrow \mathrm{CH\text{-}O\text{-}O^\bullet};\ \mathrm{CH\text{-}O\text{-}O^\bullet} + \mathrm{CH_2}$$
$$\rightarrow \mathrm{CH\text{-}O\text{-}OH} + \mathrm{CH^\bullet};\ \mathrm{CH\text{-}O\text{-}OH} + h\nu \rightarrow \mathrm{CH\text{-}O^\bullet} + {}^\bullet\mathrm{OH};\ \mathrm{CH\text{-}O^\bullet}$$
$$+\, n\mathrm{CH_2{=}CH\text{-}R} \rightarrow \mathrm{CH\text{-}O\text{-}CH_2\text{-}CHR)}_n{}^\bullet$$

Peroxy radicals can react with vinyl or acryl monomers before or after their decomposition by irradiation to alloy radicals, which then start the graft polymerization to linear side-chains as shown later [139]. Impregnation with monomers (or polymers) or adsorbing monomers and their exposure to plasma for grafting of adsorption layer onto polymer substrate was another variant of grafting [140]. The polymerization of a condensed monomer layer on a (cooled) substrate was introduced as early as 1972 by Hollahan [141]. This must also be remembered in the CASING (crosslinking by activated species of inert gases) technique, which was also proven to bond impregnate polymer layers onto polymer substrates (see Section 6.10).

Chemical grafting as a defined reaction following ordinary chemical reaction pathways is another and often probed variant by reaction between (monosort) functional groups and those of molecules, oligomers, or polymers. A broad variety of new molecular architectures at polymer substrate surfaces can be realized [2, 30, 91, 142]. Proved and tested reactions of functional groups were OH with silanes, isocyanates, alkyl halides and acids, NH_2 with alkyl halides, isocyanates and aldehydes, COOH with alcohols, epoxy groups, amines, and so on.

Radical-based graft reactions have many disadvantages. The generation of radicals by plasma exposure of polymer surfaces is not tunable, predictable, and reproducible. Therefore, no clear dependence of radical yield on plasma parameters exists. Traces of oxygen, halides, and so on inhibit strongly the radical yield. Primary alkyl-chain radicals are very short-lived (half-life of ca. 10^{-10}s [143]) and therefore not suited for grafting of vinyl or acrylic monomers. Moreover, the main source of radicals is the plasma vacuum-UV irradiation, which produces radicals in surface-near layers or in the bulk. Radicals at the topmost surface react with monomer molecules. This route is most often closed for radicals below the surface. These trapped radicals bring about some degradation effects during their life-time. First peroxy radicals are produced by reaction of free radicals with molecular oxygen from air. Peroxy radicals are produced as shown before in an uncontrolled manner. The decay of hydroperoxides or peroxide starts the graft reaction as also denoted before [144]: $RH + h\nu \rightarrow R^{\bullet} + 0.5H_2$, then $R^{\bullet} + {}^{\bullet}O\text{-}O^{\bullet} \rightarrow R\text{-}O\text{-}O^{\bullet}$ followed by $R\text{-}O\text{-}O^{\bullet} + R'H \rightarrow R\text{-}O\text{-}OH + R^{\bullet}$ and $R\text{-}O\text{-}OH \rightarrow R^{\bullet} + {}^{\bullet}O\text{-}OH$ or $R\text{-}O\text{-}OH \rightarrow R\text{-}O^{\bullet} + {}^{\bullet}OH$ or $R\text{-}O\text{-}O\text{-}R' \rightarrow R\text{-}O^{\bullet} + R'\text{-}O^{\bullet}$ or $R(O)O\text{-}OH \rightarrow R(O)^{\bullet} + {}^{\bullet}O\text{-}OH$ or $R(O)O\text{-}OH \rightarrow R(O)O^{\bullet} + {}^{\bullet}OH$. This type of initiation is also not defined and leads to linkages of different stability such as: $R\text{-}O^{\bullet} + CH_2{=}CH\text{-}X \rightarrow R\text{-}O\text{-}CH_2\text{-}C(X)^{\bullet}$ or $R\text{-}O\text{-}O^{\bullet} + CH_2{=}CH\text{-}X \rightarrow R\text{-}O\text{-}O\text{-}CH_2\text{-}C(X)^{\bullet}$ or $R^{\bullet} + CH_2{=}CH\text{-}X \rightarrow R\text{-}CH_2\text{-}C(X)^{\bullet}$ followed by chain growth. It is assumed that the density of alkoxy radicals is low but the chain length of grafted monomers is high [135]. Biomedical application and membrane modification were in the foreground but also adhesion promotion uses such peroxide intermediates [134, 143–152].

The *in situ* grafting without any intermediate contact to air of vinyl- and acrylic monomers on plasma-produced radicals was already reported by Bamford in 1960 [136, 153]. Later, Bradley exposed textiles to plasma and after this activation he filled the reactor with acrylic acid for grafting onto. The thus treated textiles showed improved properties [154]. This type of grafting was repeatedly tested but without permanent industrial production [155–157]. Such grafted polymers were tested as membranes [158, 159] or for impregnation of wool and synthetic fibers and textiles [160] and Kevlar fibers [161]. The disadvantage of this technique is that the radical can be transferred to the monomer and then a homopolymerization starts and the formed polymer chain has no anchoring to the substrate: $substrate^{\bullet} + M \rightarrow substrate + M^{\bullet}$ and $M^{\bullet} + M \rightarrow P^{\bullet}$. Without radical transfer the growing chain would be anchored to the substrate: substrate$^•$ + M $\rightarrow$ substrate-M$^•$ and substrate-M$^•$ + M $\rightarrow$ substrate-P$^•$.

In contrast, the method of plasma-introduced monosort functional groups at polymer surfaces to produce highly dense and covalently bonded groups, which can be consumed, was often tested as a reproducible chemical process [162]. Thus, grafting with much higher density and permanence towards solvents, aging, mechanical stress, and so on is possible [2, 32, 67]. Moreover, the graft reaction does not depend on radicals; in contrast to radical production and reaction it is well controllable. Additionally, the grafted molecules do not possess vinyl or acrylic double bonds, only reactive functional groups. This pure chemical grafting follows chemical rules and leads to the stable and permanent anchoring of molecules via covalent bonds.

12.14
Grafting onto Functional Groups

For high densities of grafts on polymer surfaces a high-density population of reactive functional groups of one sort is necessary. The following plasma processes allow such monosort functionalization (Scheme 12.1):

- Oxidation of polyolefin surface by exposure to the O_2-plasma, followed by reduction of all carbonyl-containing functional groups to OH groups using B_2H_6, $LiAlH_4$, $NaBH_4$, Vitride® (sodium bis[2-methoxyethoxy]aluminum hydride), and so on and simultaneous hydroboration of C=C double bonds using diborane and hydrogen peroxide [163, 164];
- bromination of polyolefin surface using bromine, bromoform, or allyl bromide plasmas (J. Friedrich and I. Novis, unpublished results, 1989) [67, 163–166, 167];

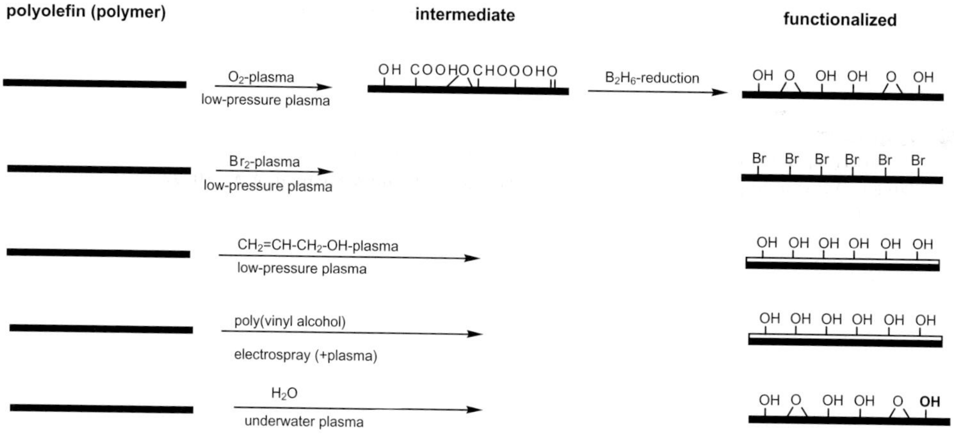

Scheme 12.1 Examples of the production of monosort functional groups at polymer surfaces.

- using underwater plasma to produce preferably OH groups [168];
- coating with plasma polymers or copolymers possessing monosort and a high density of functional groups [169, 170];
- coating with polymers using the electrospray ionization technique [168].

These monosort functional groups can be consumed chemically by easy and often tested reactions as mentioned before with mono- or bifunctional molecules, oligomers, and polymers (PEG = $10\,000\,g\,mol^{-1}$) [30, 168].

For covalent coupling onto OH-, COOH-, and NH_2-groups different reactions are well suited [91, 171]:

$$\blacksquare\text{-OH} + (\text{RO})_3\text{Si-R}'\text{-X} \rightarrow \blacksquare\text{-O-Si(RO)}_2\text{-R}'\text{-X} \quad \text{silanization}$$

$$\blacksquare\text{-OH} + \text{OCN-R-NCO} \rightarrow \blacksquare\text{-O-CO-NH-RNCO} \quad \text{urethane bond}$$

$$\blacksquare\text{-NH}_2 + \text{OHC-R-CHO} \rightarrow \blacksquare\text{-N=CH-R-CHO} \quad \text{Schiff's base (azomethine)}$$

$$\blacksquare\text{-NH}_2 + \text{OCN-R-CNO} \rightarrow \blacksquare\text{-NH-CO-NH-R-CNO} \quad \text{urea bond}$$

$$\blacksquare\text{-COOH} + \text{HO-R} \rightarrow \blacksquare\text{-COOR} \quad \text{ester bond}$$

$$\blacksquare\text{-COOH} + \text{H}_2\text{N-R-NH}_2 \rightarrow \blacksquare\text{-CO-NH}_2\text{-R-NH}_2 \quad \text{amide bond}$$

$$\blacksquare\text{-Br} + \text{H}_2\text{N-R-NH}_2 \rightarrow \blacksquare\text{-NH-R-NH}_2 \quad \text{nucleophilic substitution}$$

$$\blacksquare\text{-N}_3 + \text{R}_1\text{-C}\equiv\text{C-R}_2 \rightarrow \blacksquare\text{-substituted triazine} \quad \text{click-chemistry.}$$

References

1 Friedrich, J., Mix, R., Kühn, G., Retzko, I., Schönhals, A., and Unger, W. (2003) *Composite Interface*, **10**, 173–223.
2 Yasuda, H. and Hsu, T. (1977) *J. Polym. Sci., Polym. Chem. Ed.*, **15**, 81.
3 Yasuda, H. (1976) *J. Macromol. Sci. Chem., A*, **10**, 383.
4 Friedrich, J., Retzko, I., Kühn, G., Unger, W., and Lippitz, A. (2001) *Metallized Plastics VII* (ed. K.L. Mittal), VSP, Utrecht, pp. 117–142.
5 Savage, C.R. and Timmons, R.B. (1991) *Polym. Mater. Sci. Eng.*, **64**, 95.
6 Savage, C.R., Timmons, R.B., and Lin, J.W. (1991) *Chem. Mater.*, **3**, 575.
7 (a) Schüler, H. and Reinebeck, L. (1951) *Z. Naturforsch.*, **6a**, 271; (b) Schüler, H. and Reinebeck, L. (1952) *Z. Naturforsch.*, **7a**, 285; (c) Schüler, H. and Reinebeck, L. (1954) *Z. Naturforsch.*, **9a**, 350.
8 Schüler, H. and Stockburger, M. (1959) *Z. Naturforsch.*, **14a**, 981.
9 Schüler, H., Prchal, K., and Kloppenburg, E., (1960) *Z. Naturforsch.*, **15a**, 308.
10 König, H. and Hellwig, G. (1951) *Z. Physik*, **129**, 491.
11 Hafer, R. and Mohamed, A.A. (1957) *Acta Phys. Austr.*, **11**, 193.
12 Yasuda, H. and Lamaze, C.E. (1973) *J. Appl. Polym. Sci.*, **17**, 1519.
13 Yasuda, H., Marsh, H.C., Bumgarner, M.O., and Morosoff, N. (1975) *J. Appl. Polym. Sci.*, **19**, 2845.
14 Yasuda, H. and Hirotsu, T. (1977) *J. Polym. Sci., Polym. Chem. Ed.*, **15**, 2749.
15 Yasuda, H. and Hirotsu, T. (1978) *J. Appl. Polym. Sci.*, **22**, 1195.

16 Gross, T., Lippitz, A., Unger, W.E.S., Friedrich, J.F., and Wöll, C. (1994) *Polymer*, **35**, 559.

17 Koprinarov, I., Lippitz, A., Friedrich, J.F., Unger, W.E.S., and Wöll, C. (1998) *Polymer*, **39**, 3001.

18 Retzko, I., Friedrich, J.F., Lippitz, A., and Unger, W.E.S. (2001) *J. Electron. Spectrosc. Relat. Phenom.*, **121**, 111.

19 (a) Weidner, S., Kühn, G., Decker, R., Roessner, D., and Friedrich, J. (1998) *J. Polym. Sci.*, **36**, 1639; (b) Meisel, J. and Tiller, H.-J. (1972) *Z. Chem.* **7**, 275.

20 Yasuda, H. and Hsu, T. (1976) *J. Appl. Polym. Sci.*, **20**, 1769.

21 Nakajima, K., Bell, A.T., Shen, M., and Millard, M.M. (1979) *J. Appl. Polym. Sci.*, **23**, 2627.

22 Vincant, J.W., Shen, M., and Bell, A.T. (1978) *ACS Polym. Prepr.*, **19**, 453.

23 Friedrich, J., Kühn, G., Mix, R., Fritz, A., and Schönhals, A. (2003) *J. Adhesion Sci. Technol.*, **17**, 1591.

24 Friedrich, J., Kühn, G., Mix, R., Retzko, I., Gerstung, V., Weidner, S., Schulze, R.-D., and Unger, W. (2003) *Polyimides and Other High Temperature Polymers: Synthesis, Characterization and Applications* (ed. K.L. Mittal), VSP, Utrecht, pp. 359–388.

25 Friedrich, J.F., Kühn, G., and Mix, R. (2005) *Plasma Proc. Polym.*, **1**, 3.

26 Krüger, S., Schulze, R.-D., Brademann-Jock, K., and Friedrich, J. (2006) *Surf. Coat. Technol.*, **201**, 543–552.

27 Haupt, M., Barz, J., and Oehr, C. (2008) *Plasma Proc. Polym.*, **5**, 33.

28 Retzko, I., Friedrich, J., Lippitz, A., and Unger, W.E.S. (2001) *J. Electron. Spectrosc. Relat. Phenom.*, **121**, 111–129.

29 Elias, H.-G. (1996) *Macromolecules*, Wiley-VCH Verlag GmbH, Weinheim.

30 Friedrich, J., Kühn, G., and Mix, R. (2006) *Prog. Colloid Polymer Sci.*, **132**, 62–71.

31 Friedrich, J., Wettmarshausen, S., and Hennecke, M. (2009) *Surf. Coat. Technol.*, **203**, 3647–3655.

32 Friedrich, J., Kühn, G., Mix, R., and Unger, W. (2004) *Polym. Proc. Plasmas*, **1**, 28–50.

33 Friedrich, J., Mix, R., and Kühn, G. (2003) *Surf. Coat. Technol.*, **174–175**, 811–815.

34 Oran, U., Swaraj, S., Friedrich, J., and Unger, W.E.S. (2004) *Polym. Proc. Plasmas*, **1**, 123–139.

35 Oran, U., Swaraj, S., Friedrich, J., and Unger, W.E.S. (2004) *Polym. Proc. Plasmas*, **1**, 141–150.

36 Swaraj, S., Oran, U., Lippitz, A., Schulze, R.-D., Friedrich, J., and Unger, W.E.S. (2004) *Polym. Proc. Plasmas*, **1**, 134–140.

37 Oran, U., Swaraj, S., Friedrich, J.F., and Unger, W.E.S. (2006) *Appl. Surf. Sci.*, **252**, 6588–6590.

38 Oran, U., Swaraj, S., Friedrich, J.F., and Unger, W.E.S. (2005) *Plasma Proc. Polym.*, **2**, 563–571.

39 Swaraj, S., Oran, U., Friedrich, J.F., Lippitz, A., and Unger, W.E.S. (2005) *Plasma Proc. Polym.*, **2**, 572–580.

40 Mix, R., Friedrich, J.F., and Kühn, G. (2005) in *Plasma Polymers and Related Materials* (eds M. Mutlu, G. Dinescu, R. Förch, J.M. Martin-Martinez, and J. Vyskocil), Hacettepe University Press, Ankara, pp. 107–114.

41 Oran, U., Swaraj, S., Friedrich, J.F., and Unger, W.E.S. (2005) *Surf. Coat. Technol.*, **200**, 463–467.

42 Kobayashi, H. (1973) *J. Appl. Polym. Sci.*, **17**, 885.

43 Jesch, K.F., Bloor, J.E., and Kronick, P.L. (1966) *J. Polym. Sci. A-1*, **4**, 1487.

44 Kronick, P.L., Jesch, K.F., and Bloor, J.E. (1969) *J. Polym. Sci. A-1*, **7**, 767.

45 Friedrich, J., Mix, R., and Kühn, G. (2005) in *Plasma Processing and Polymers* (eds R. d'Agostino, P. Favia, C. Oehr, and M.R. Wertheimer), Wiley-VCH Verlag GmbH, Weinheim, pp. 3–21.

46 Mix, R., Gerstung, V., Falkenhagen, J., and Friedrich, J. (2007) *J. Adhesion Sci. Technol.*, **21**, 487–507.

47 Friedrich, J., Throl, U., Gähde, J., and Schierhorn, E. (1982) *Acta Polym.*, **33**, 405–410.

48 Throl, U., Gähde, J., Friedrich, J., and Schierhorn, E. (1982) *Acta Polym.*, **33**, 561–566.

49 Throl, U., Gähde, J., and Friedrich, J. (1982) *Acta Polym.*, **33**, 667–673.

50 Westwood, A.R. (1971) *Eur. Polym. J.*, **7**, 363.
51 Friedrich, J., Kühn, G., and Gähde, J. (1979) *Acta Polym.*, **30**, 470–477.
52 Stiller, W. and Friedrich, J. (1981) *Z. Chem.*, **21**, 91–118.
53 Kremers, F. and Schönhals, A. (2007) *Broadband Dielectric Spectroscopy*, Springer, Berlin.
54 Swaraj, S., Oran, U., Lippitz, A., Schulze, R.-D., Friedrich, J.F., and Unger, W.E.S. (2005) *Plasma Proc. Polym.*, **2**, 310.
55 Rabek, J.F. (1996) *Polymer Photodegradation*, Chapman & Hall, New York.
56 Partridge, A., Harris, P., Hirotsu, T., and Kurosawa, S. (2000) *Plasmas Polym.*, **3**, 45.
57 Lopez, G.P., Ratner, B.D., Rapoza, R.J., and Horbett, T.A. (1993) *Macromolecules*, **26**, 3247.
58 Johnston, E.E. and Ratner, B.D. (1996) *J. Electron. Spectrosc. Relat. Phenom.*, **81**, 303.
59 Friedrich, J.F., Retzko, I., Kühn, G., Unger, W.E.S., and Lippitz, A. (2001) *Surf. Coat. Technol.*, **142–144**, 460.
60 Denes, F.S. and Manolache, S. (2004) *Progr. Polym. Sci.*, **29**, 815.
61 Müller, B.M. and Oehr, C. (1999) *Surf. Coat. Technol.*, **116–119**, 802.
62 (a) Choukourov, A., Biederman, H., Slavinska, D., Trchova, M., and Holländer, A. (2003) *Surf. Coat. Technol.*, **174–175**, 863; (b) Wickson, B.M. and Brush, J.L. (1999) *Colloids Surf.* **156**, 201.
63 Ameen, A.P., Short, R.D., and Ward, R.J. (1994) *Polymer*, **35**, 4382.
64 Alexander, M.R. and Duc, T.M. (1998) *J. Mater. Chem.*, **8**, 937.
65 Rinsch, C.L., Chen, X., Panchalingam, V., Eberhart, R.C., Wang, J.-H., and Timmons, R.B. (1996) *Langmuir*, **12**, 2995.
66 Lefohn, A.E., Mackie, N.M., and Fisher, E.R. (1998) *Plasma Polym.*, **3**, 197.
67 Teng, M.-Y., Lee, K.-R., Liaw, D.-J., Lin, Y.S., and Lai, J.-Y. (2000) *Eur. Polym. J.*, **36**, 663; (b) Wettmarshausen, S., Mittmann, H.-U., Kühn, G., Hidde, G., and Friedrich, J.F. (2007) *Plasma Proc. Polym.* **4**, 832–839.
68 Denaro, A.R., Owens, P.A., and Crawshaw, A. (1970) *Eur. Polym. J.*, **6**, 487.
69 Calderon, J.C. and Timmons, R.B. (1998) *Macromolecules*, **31**, 3216.
70 Lai, C.Y. and Chao, Y.C. (1990) *J. Appl. Polym. Sci.*, **39**, 2293.
71 O'Toole, L., Beck, A.J., and Short, R.D. (1996) *Macromolecules*, **29**, 5172.
72 Lianos, L., Quet, C., and Duc, T.M. (1994) *Surf. Interface Anal.*, **21**, 14.
73 Dhez, O., Ade, H., and Urquhart, S.G. (2003) *J. Electron. Spectrosc. Relat. Phenom.*, **128**, 85.
74 Adam, G.,and Gibbs, J.H. (1965) *J. Chem. Phys.*, **43**, 139.
75 Gombotz, W.R. and Hoffman, A.S. (1988) *J. Appl. Polym. Sci. Appl. Polym. Symp.*, **42**, 285.
76 Mix, R., Kühn, G., and Friedrich, J. (2005) in *Adhesion Aspects of Thin Films*, vol. 2 (ed. K.L. Mittal), VSP, pp. 123–144.
77 Meyer-Plath, A. (2004) *Vakuum Forschung Praxis*, **16**, 118.
78 Prucker, O., Stöhr, T., and Rühe, J. (2006) *Vakuum Forschung Praxis*, **18**, 25.
79 Choukourov, A., Biederman, H., Slavinska, D., Trchova, M., and Holländer, A. (2003) *Surf. Coat. Technol.*, **174–175**, 863.
80 Burns, N.L., Holmberg, K., and Brink, C. (1996) *J. Colloid Interface Sci.*, **178**, 116.
81 Barbarossa, V., Contari, S., and Zanobi, A. (1992) *J. Appl. Polym. Sci.*, **44**, 1951.
82 Oran, U., Swaraj, S., Lippitz, A., and Unger, W.E.S. (2006) *Plasma Proc. Polym.*, **3**, 288.
83 Li, Z.-F., and Netravali, A.N. (1992) *J. Appl. Polym. Sci.*, **44**, 319.
84 Calderon, J.G., Harsch, A., Gross, G.W., and Timmons, R.B. (1998) *J. Biomed. Mat. Res.*, **42**, 597–603.
85 Rostami, H., Iskandarani, B., and Kamel, I. (1992) *Polym. Comp.*, **13**, 207–212.
86 Everhart, D.S. and Reilley, C.N. (1981) *Anal. Chem.*, **53**, 665.
87 Friedrich, J.F., Mix, R., and Kühn, G. (2005) *Surf. Coat. Technol.*, **200**, 565–568.
88 Friedrich, J., Mix, R., Schulze, R.-D., and Kühn, G. (2005) in *Adhesion* (ed. W.

Possart), Wiley-VCH Verlag GmbH, Weinheim, pp. 265–288.
89 Friedrich, J. (2010) Plasma-polymerization, in *Vakuum-Plasma-Technologien Beschichtung und Modifizierung von Werkstoffoberflächen Section 2.5* (eds G. Blasek and G. Bräuer), Eugen-G.-Leuze-Verlag, Saulgau, pp. 325.
90 Swaraj, S., Oran, U., Lippitz, A., Friedrich, J.F., and Unger, W.E.S. (2005) *Surf. Coat. Technol.*, **200**, 494–497.
91 Resch-Genger, U., Hoffmann, K., Mix, R., and Friedrich, J.F. (2007) *Langmuir*, **23**, 8411–8416.
92 Krishanamurthy, V., Kamel, I.L., and Wie, Y. (1989) *J. Polym. Sci., Part A: Polym. Chem.*, **27**, 1211.
93 Bell, A.T., Wydeven, T., and Shen, M. (1975) *J. Appl. Polym. Sci.*, **19**, 1911.
94 Fanghänel, A. and Schwetlick, K. (2002) *Organikum*, Wiley-VCH Verlag GmbH, Weinheim.
95 del Fanti, N.A. (2008) *Infrared Spectroscopy of Polymers*, Thermo Fisher Scientific, Madison, WI.
96 Doerffel, K. *et al.* (1973) *Strukturaufklärung-Spektroskopie Und Röntgenbeugung*, VEB Deutscher Verlag für Grundstoffindustrie, Leipzig.
97 Colthup, N.B. (1950) *J. Opt. Soc. Am.*, **40**, 397.
98 Bradley, A. and Czuha, M., Jr. (1975) *Anal. Chem.*, **47**, 1838–1840.
99 Oehr, C., Müller, M., Elkin, B., Hegemann, D., and Vohrer, U. (1999) *Surf. Coat. Technol.*, **116–119**, 25.
100 Hsieh, Y.-L., Pugh, C., and Ellison, M.S. (1984) *J. Appl. Polym. Sci.*, **29**, 3547.
101 Cho, D.L., Gölander, C.-G., and Johansson, K. (1990) *J. Appl. Polym. Sci.*, **41**, 1373.
102 Sciaratta, V., Vohrer, U., Hegemann, D., Müller, M., and Oehr, C. (2003) *Surf. Coat. Technol.*, **174–175**, 805.
103 Yasuda, H., Hirotsu, T., and Polym, J. (1978) *Sci. Polym. Chem. Ed.*, **16**, 743.
104 Beamson, G. and Briggs, D. (1992) *High Resolution XPS of Organic Polymers*, John Wiley & Sons, Ltd, Chichester.
105 Alexander, M.R. and Duc, T.M. (1999) *Polymer*, **40**, 5479–5488.
106 Elias, H.-G. (1990) *Macromolecules*, Hüthig&Wepf, Basle, p. 221.
107 Akiyama, Y., Fujita, S., Senboku, H., Rayner, C.M., Brough, S.A., and Arai, M. (2008) *J. Supercrit. Fluids*, **46**, 197.
108 Munro, H.S., Grünwald, H., and Polym, J. (1985) *Sci. Polym. Chem. Ed.*, **23**, 479.
109 Fahmy, A., Mix, R., Schönhals, A., and Friedrich, J.F. (2011) *Plasma Proc. Polym.*, **8**, 147–159.
110 Brockhaus (1968) *Atom-Struktur Der Materie*, VEB Bibliographisches Institut, Leipzig.
111 Clark, D.T. and Dilks, A. (1978) *J. Polym. Sci.*, **16**, 911.
112 Basov, N.G., Danilychev, V.A., Popov, Y., and Khodkevich, D.D. (1970) *J. Exp. Theor. Phys. Lett.*, **12**, 329.
113 Eliasson, B. and Kogelschatz, U. (1988) *Appl. Phys. B*, **46**, 299.
114 Hudis, M. (1972) *J. Appl. Polym. Sci.*, **16**, 2397.
115 Girard-Lauriault, P.-L., Truica-Marasescu, F., Petit, A., Wang, H.T., Desjardins, P., Antoniou, J., Mwale, F., and Wertheimer, M.R. (2009) *Macromol. Biosci.*, **9**, 911–921.
116 Truica-Marasescu, F., Pham, S., and Wertheimer, M.R. (2007) *Nucl. Instrum. Methods Phys. Res., Sect. B*, **265**, 31.
117 Truica-Marasescu, F. and Wertheimer, M.R. (2008) *Macromol. Chem. Phys.*, **209**, 1043–1049.
118 Ruiz-Bucio, J.-C., Girard-Lauriault, P.-L., Truica-Marasescu, F., and Wertheimer, M.R. (2010) *Radiat. Phys. Chem.*, **79**, 310–314.
119 Ruiz, J.C., St-Georges-Robillard, A., Thérésy, C., Lerouge, S., and Wertheimer, M.R. (2010) *Plasma Proc. Polym.*, **7**, 737–753.
120 Skurat, V. (2003) *Nucl. Instrum. Methods Phys. Res., Sect. B*, **208**, 27–34.
121 France, R.M., Short, R.D., Duval, E., and Jones, F.R. (1998) *Chem. Mater.*, **10**, 1176.
122 Curtis, A., Forrester, J., McInnes, C., and Lawrie, F. (1983) *J. Cell. Biol.*, **97**, 1500.
123 Griesser, H.J., Chatelier, R.C., Gengenbach, T.R., Johnson, G., and Steele, J.G. (1994) *J. Biomater. Sci. Polym. Ed.*, **5**, 531.
124 Yasuda, H. (1985) *Plasma Polymerization*, Academic Press, Orlando.

125 Friedrich, J., Gähde, J., Frommelt, H., and Wittrich, H. (1976) *Faserforsch. Textiltechn./Z. Polymerenforsch.*, **27**, 517.

126 Beck, A.J., Jones, F.R., and Short, R.D. (1996) *Polymer*, **37**, 5537.

127 Kurosawa, S., Hirokawa, T., Kashima, K., Aizawa, H., Han, D.S., Yoshimi, Y., Okada, Y., Yase, K., Miyake, J., Yoshimoto, M., and Hilborn, J. (2000) *Thin Solid Films*, **374**, 262.

128 Mix, R., Falkenhagen, J., Schulze, R.-D., Gerstung, V., and Friedrich, J.F. (2009) *Polymer Surface Modification: Relevance to Adhesion*, vol. V (ed. K.L. Mittal), Brill, Leiden, pp. 317–340.

129 Hofmann, A., Alegria, A., Colmenero, J., Willner, L., Buscaglia, E., and Hadjichristidis, N. (1996) *Macromolecules*, **29**, 129.

130 Schönhals, A. (2002) in *Broadband Dielectric Spectroscopy* (eds F. Kremer and A. Schönhals), Springer Verlag, Berlin, p. 225.

131 Friedrich, J., Retzko, I., Kühn, G., Unger, W., and Lippitz, A. (2001) in *Metallized Plastics 7: Fundamental and Applied Aspects* (ed. K.L. Mittal), VSP, Utrecht, pp. 117–142.

132 Suzuki, M., Kishida, A., Iwata, H., and Ikada, Y. (1986) *Macromolecules*, **19**, 1804.

133 Kang, E.T., Neoh, K.G., and Tan, K.L. (1992) *Macromolecules*, **25**, 6842.

134 Uyama, Y., Kato, K., and Ikada, Y. (1998) *Adv. Polym. Sci.*, **137**, 1–37.

135 Geckeler, K.E., Gebhardt, R., and Grünwald, H. (1997) *Naturwissenschaften*, **84**, 150–151.

136 Bamford, C.H., Jenkins, A.D., and Ward, J.C. (1960) *Nature*, **186**, 712.

137 MacCallum, J.R. and Rankin, C.T. (1971) *J. Polym. Sci. B*, **9**, 751–752.

138 König, U., Nitschke, M., Menning, A., Eberth, G., Pilz, M., Arnhold, C., Simon, F., Adam, G., and Werner, C. (2002) *Colloids Surf. B*, **24**, 63.

139 Zhang, M.C., Kang, E.T., Neoh, K.G., and Tan, K.L. (2001) *Colloids Surf. A*, **176**, 139.

140 Kou, R.-Q., Xu, Z.-K., Deng, H.-T., Liu, Z.-M., Seta, P., and Xu, Y. (2003) *Langmuir*, **19**, 6869.

141 Hollahan, J.R. (1972) *Makromol. Chem.*, **154**, 303.

142 Kühn, G., Ghode, A., Weidner, S., Retzko, I., Unger, W.E.S., and Friedrich, J.F. (2000) in *Polymer Surface Modification: Relevance to Adhesion*, vol. 2 (ed. K.L. Mittal), VSP, Utrecht, pp. 45–64.

143 (a) Alfassi, Z.B. (1988) *Chemical Kinetics of Small Organic Radicals*, CRC Press: Boca Raton, FL, Bd. 1–4; (b) Gupta, B., Hilborn, J.G., Bisson, I., and Frey, P. (2001) *J. Appl. Polym. Sci.* **81**, 2993.

144 Hirotsu, T. (1987) *Ind. Eng. Chem. Res.*, **26**, 1287.

145 Inagaki, N., Tasaka, S., and Goto, Y. (1997) *J. Appl. Polym. Sci.*, **66**, 77.

146 Tan, K.L., Woon, L.L., Wong, H.K., Kang, E.T., and Neoh, K.G. (1993) *Macromolecules*, **26**, 2832.

147 Ulbricht, M. and Belfort, G. (1995) *J. Appl. Polym. Sci.*, **56**, 325.

148 Hirotsu, T. (1996) *Pure Appl. Chem.*, **A33**, 1663.

149 Ulbricht, M. and Belfort, G. (1996) *J. Membrane Sci.*, **111**, 193.

150 Ulbricht, M. and Belfort G. (1995) *Surface. J. Appl. Polym. Sci.*, **56**, 325–343.

151 Vasilets, V.N., Hermel, G., König, U., Werner, C., Müller, M., Simon, F., Grundke, K., Ikada, Y., and Jacobasch, H.-J. (1997) *Biomaterials*, **18**, 1139.

152 Zou, X.P., Kang, E.T., and Neoh, K.G. (2002) *Surf. Coat. Technol.*, **149**, 119.

153 Bamford, C.H. and Ward, J.C. (1961) *Polymer*, **2**, 277.

154 Bradley, A. and Fales, J.D. (1971) *Chem. Technol.*, 232–237.

155 Wertheimer, M.R. and Schreiber, H.P. (1981) *J. Appl. Polym. Sci.*, **26**, 2087.

156 Siminescu, C.I., Denes, F., Macoveanu, M.M., and Negulescu, I. (1984) *Makromol. Chem., Suppl.*, **8**, 17.

157 Meichsner, J. and Poll, H.-U. (1981) *Acta Polym.*, **32**, 203–208.

158 Ihm, C.-D. and Ihm, S.-K. (1995) *J. Membr. Sci.*, **98**, 89.

159 Wavhal, D.S. and Fisher, E.R. (2002) *J. Membr. Sci.*, **209**, 255.

160 Millard, M.M. and Pavlath, A. (1972) *Textile Res. J.*, **42**, 460.

161 Yamada, K., Haraguchi, T., and Kajiyama, T. (1996) *J. Appl. Polym. Sci.*, **60**, 1847.

162 Friedrich, J., Wettmarshausen, S., Hidde, G., and Hennecke, M.

(2009) *Surf. Coat. Technol.*, **203**, 3647–3655.

163 Nuzzo, R.G. and Smolinsky, G. (1984) *Macromolecules*, **17**, 1013.

164 Kühn, G., Weidner, S., Decker, R., Ghode, A., and Friedrich, J. (1999) *Surf. Coat. Technol.*, **116–119**, 796–801.

165 Kühn, G., Retzko, I., Lippitz, A., Unger, W., and Friedrich, J. (2001) *Surf. Coat. Technol.*, **142–144**, 494–500.

166 Friedrich, J. (1991) in *Polymer-Solid Interfaces* (eds J.J. Pireaux, P. Bertrand, and J.L. Bredas), Institute of Physics Publishing, Bristol, pp. 443–454.

167 Kiss, É., Samu, J., Tóth, A., and Bertóti, I. (1966) *Langmuir*, **12**, 1651–1657.

168 Friedrich, J.F., Mix, R., Schulze, R.-D., Meyer-Plath, A., Joshi, R., and Wettmarshausen, S. (2008) *Plasma Proc. Polym.*, **5**, 407–423.

169 Friedrich, J., Loeschcke, I., and Gähde, J. (1986) *Acta Polym.*, **37**, 687–695.

170 Friedrich, J.F., Retzko, I., Kühn, G., Unger, W.E.S., and Lippitz, A. (2001) *Surf. Coat. Technol.*, **142–144**, 460–467.

171 Hoffmann, K., Resch-Genger, U., Mix, R., and Friedrich, J., (2006) *J. Fluoresc.*, 441.

Index

The Plasma Chemistry of Polymer Surfaces: Advanced Techniques for Surface Design, First Edition.
Jörg Friedrich.

q

r

s